Oil and Gas: Crises and Controversies 1961 – 2000

Studies and Commentaries by
Peter R. Odell
Professor Emeritus of International Energy Studies
Erasmus University, Rotterdam

Volume 2: Europe's Entanglement

Multi-Science Publishing Company Ltd.
Brentwood, England.

ISBN 0 906522 18 8

to Ann van der Kaag, the late Elly van Reijn,
Eibellien Visser and Carol Klüg
whose efforts in preparing these manuscripts
over the years went far beyond the call of duty

Contents

Section III – The Evolution of Europe's Gas Markets

Section IV – On Oil and Gas Politics, Policies and Structures

List of Figures

List of Tables

Foreword

In Volume 1 of this book – on Global Issues – regional implications were sometimes examined, but without any special focus: except, of course, for considerations relating to the suppliers of oil and gas to international markets, most notably with respect to the continuing significance of the Middle East.

The choice of Europe as the region for exclusive attention in this second Volume of my papers and commentaries reflects, first, my almost full-time residence and employment over the 40 years since 1961 in the United Kingdom and the Netherlands; and my familiarity with the continent as a whole through a continuing observation and evaluation of energy-important events and, more specifically, through my participation in numerous conferences and seminars on energy issues across Europe and by liaison and cooperation with colleagues concerned with energy studies in virtually every other country of the continent.

For the first 30 years from 1961 Europe was divided by an "iron curtain" separating western Europe from the countries to the east lying within the sphere of influence – and control – of the Soviet Union. Under these political conditions not only were contacts with the latter group of countries severely circumscribed, but those nations' developments so reflected the requirements of Soviet-style central planning that their energy sectors came to bear little resemblance to those which were emerging in the rest of the continent. Pan-European studies thus became difficult and inappropriate. Hence the studies and commentaries in this volume dating from before 1990 are almost exclusively concerned with the western part of the continent.

Since 1990 the political and economic barriers have been broken down. Nevertheless, integration by the east European countries into the rest of Europe has necessarily proceeded relatively slowly (except for East Germany) so that the "weight" of the commentaries and analyses on the oil and gas sectors of the economies of that part of Europe still remains modest. Moreover, the boundary of Europe as far as the geographical scope of this book is concerned has continued to exclude the territories of the former Soviet Union. Indeed, these enter the analyses only as an energy-supplying region external to Europe: in a like manner – and in competition with – the oil and gas supplying regions of North Africa and the Middle East. Turkey is also specifically excluded, in spite of its political and military links with Europe, in that its role vis a vis energy is essentially as a transit country, not only for oil and gas from the Gulf countries, but also from the Caspian basin. This role has placed Turkey in a very special sort of contemporary energy relationship with Europe; and one, moreover, that seems likely to become steadily more important over the next two decades.

Historically, the content of this volume is determined by the time-span of the author's contribution to European energy studies, viz. from the early 1960s to date. Given the particular "personalised" start-point – which does not happen to coincide with a break-point in the extended evolutionary process of the European energy economy – it seems appropriate that this Foreword should briefly set out the relevant pre-1961 antecedents.

a. Pre-1945 Components

These date as far back as the late 18th century, since when there was a slow evolution of Europe's traditional coal-based energy system. Dependence on indigenously-mined coal remained of the essence throughout the period during which people and industry became concentrated on Europe's coalfields[1]. Even the late 19th century introduction of electricity made only slow headway in industrial and residential applications[2]. It was, indeed, only in a few locations, with advantageous geographical conditions for the exploitation of low-cost hydro-power, that the pre-1920s use of electricity became the basis for the location of some energy intensive industries and was also cheap enough to secure its widespread use in the commercial and residential sectors.

Elsewhere, most of non-urban Europe remained beyond the local electricity supply networks through the 1930s and even into the post-World War II period. Meanwhile, coal continued to dominate the supply of primary energy – such that in 1937 oil still supplied only 7% of total energy used in Europe; and natural gas, under 0.5%. Countries and regions without coal (or hydro-electricity) remained energy-poor.

b. The post-1945 Period

In western Europe, post-World War II recovery and economic rehabilitation was predicated on the basis of an "energy-mix" input as before 1939, but with an anticipated eventual addition of nuclear power[3]. Significantly, the first two European integration treaties the E.C.S.C. and the European Treaties, respectively,[4] pre-dated the Treaty of Rome. They were specifically concerned with coal (plus steel) and with atomic power. In 1952, coal still provided 90 and 95%, respectively, of the UK's and Western Germany's primary energy – and all but 13% of Europe's total energy demand.

Very considerable efforts – and investments – were made between 1945 and 1955 to resuscitate and stimulate indigenous coal production. Eventually, the industry just about recovered to its pre-1939 size. Nevertheless, increasing depletion costs, resulting from the need to exploit geologically more difficult coal resources, and rapidly rising real labour costs in a labour-intensive industry, in the context of full employment, quickly led thereafter to the dethroning of coal in what was becoming a rapidly expanding Western European energy market[5]. Oil rapidly substituted its use in most sectors of the economy.

Concurrently, the other main pre-1939 thrust in the energy sector, viz. the growth of electricity, was resumed. Networks of supply were expanded and intensified so that virtually the whole of the region's population – except in the remotest areas – secured access to electricity. Its range and intensity of use was steadily expanded in the commercial, industrial and even the agricultural sectors. By the early 1960s, after three-quarters of a century of development, electricity finally became the norm across the whole of western Europe, not only for lighting, but also for power and even heating applications[6].

Meanwhile, in eastern Europe, the adoption of Soviet-style central planning, requiring the maximum possible exploitation of indigenous energy resources and the adaptation of forced industrialisation, also brought rapid, but contrasting, changes in this region's energy situation[7]. On the supply side, this involved the expansion of national coal, brown coal and lignite production and of what little indigenous oil there was to exploit (most notably in Romania). In marked contrast with the post-1955 decline in indigenous energy production in western Europe, east European energy output rose to almost 200 million tons oil equivalent by 1960; to 135% of that of 1937. Imports, mainly of Soviet coal and oil, did also increase, but they were complementary to national energy production, rather than at the expense of indigenous resources' exploitation, as was the case for Western Europe[8].

The contents of this volume present a selection of the post-1961 contributions of the author to the analysis of Europe's evolving energy sector. These are ordered in four Sections, each dealing with a particular theme. Section I (pp.1 to 132) is concerned with the concurrent explosion of energy demand and the diversification of supply. First and foremost, oil effected its take-over of western Europe's supply of energy and thus became a new commanding height in the continent's political economy. With its powerful corporate structure and its access to very low cost external supplies, the oil industry was able to successfully offer energy at significantly lower energy costs to consumers – so subordinating the severe social consequences of the demise of coal and the political risks of dependence on imported oil to purely economic considerations[9].

Falling real oil prices – and, later, those of gas too – not only overwhelmed coal as a competitor, but also constrained the ability of indigenous nuclear power to achieve the expansion that had been anticipated. The belated recognition of the dangers of a 65+% imported oil dependence came too late to 'save' Europe from the adverse consequences of the oil supply and price shocks of the early 1970s. The crisis prospects that oil-dependence thus generated for Europe's continued development were, however, ultimately avoided: albeit only by the fortuitous success by the oil industry in discovering and developing unexpected – and well-nigh unbelievable – volumes of indigenous resources of north-west Europe. In the historical context of preceding discoveries of a large number of small oilfields in the countries located around the North Sea that had proved to be of little significance (except for their contribution to Germany's oil needs in World War Two), the discovery of the Slochteren gasfield in the northern part of the Netherlands in 1959 was initially little more than casually presented in public; it is worthy of note that within Shell, the company responsible for the discovery, it was talked about "with bated breath", such were the "in-house" unconfirmed reports of the size of its potentially producible reserves. Chapter II-1, concerned essentially with the phenomenon, was published in 1969 – ten years on from the Slochteren discovery which, by this time, had been renamed the Groningen gasfield – appropriately so, given that the reservoir was now known to underlie the whole of the north-east part of the Dutch province of that name. The chapter presents the exploitation of this mega-giant gasfield as the most significant development in the post-1945 history of the European energy sector. Subsequent chapters offer periodic reviews of its importance – not only because of its designation as the largest gasfield in the world outside the Soviet Union, but also because it was the geological indicator for the existence of a world-rated hydrocarbons province stretching northwards for almost 1000 kilometres under the North Sea[10]. Chapters II-2 to II-7 thus comprise a time-series of studies and commentaries on the oil and gas prospects of this massive European hydrocarbons province over the remaining decades of the century: and they also speculate on the implications therefrom for the prospects for the European energy sector for up to 60 years in the future!

These show that my interpretations of the future of Europe's hydrocarbons were sometimes overly optimistic especially *vis a vis* the ability and/or willingness of national and European Community energy policy makers to take decisions which would have maximised the contributions of this new world-standard hydrocarbons province to the continent's oil and gas supplies. In due course, however, subsequent events demonstrate that such optimism was, indeed, largely justified. This is shown in the last two chapters in Section II (II-8 and 9) which offer retrospectives on the first 30+ years of the exploitation of north-west European oil and gas; together with views on the province's continuing future prospects in the early 21st century.

Section III (pp. 307 to 476) gives attention specifically to the evolution of natural gas as Europe's third major energy source: and to its marketing in the context of oligopolistic, monopsonistic and other governmental/corporate constraints on its use. It also characterises the 15 year long period from the mid-1970s when there were widely held "beliefs" that gas was *either* a "scarce resource" – contrary to the evidence of its ready availability from both indigenous and external sources; *or* an indigenous energy source which was too "noble" to use in applications as mundane as steam-raising facilities for power stations and energy intensive industry.

Happily, after the necessarily gloomy presentations in Chapter III-3–5, the text can continue from Chapter III-6 with the subsequent re-appraisals of the prospects for natural gas. These show that my earlier predictions of natural gas as Europe's most important future energy source became very generally recognised in industry circles, even if not yet fully so amongst national and Euro-governmental energy policy makers. Their full "conversion" is now, however, simply a matter of time!

Section IV (pp. 477–624) is concerned with a melange of

European energy sector politics, policies and structures. Chapter IV-1 and 2 recall the "crisis" outlook for Europe's politico-economic prospects arising from foreign oil's domination of our energy supply and stress the need at the time to implement measures to enhance efficiency in energy use and the effective development of indigenous oil and gas resources. The attitude of the oil industry to the creation of an integrated and freely trading Europe is examined in Chapter IV-3. This shows how the international oil companies – almost exclusively responsible for Europe's oil supply, refining, distribution and marketing – had, in effect, already created European-wide integrated operational systems which were far ahead of the inter-governmental agreements for economic integration. Aspects of British policies towards oil and of Dutch attitudes to the exploitation of its low-cost gas are then presented in Chapter IV-4 to 6. They highlight the institutional constraints on indigenous oil and gas exploitation.

Finally, in the concluding chapters of the Volume the implications of some very important political developments – including the demise of the Soviet Union and the loss of its control over the countries of Eastern Europe – are considered; as is the rise of the now dominant ideology of the need for energy markets' liberalisation.

By the end of the 20th century a combination of demand, supply and structural elements in the oil and gas dominated energy economy of Europe have, I argue, led to a level of energy security which it would have been impossible to contemplate in the dark days of the Cold War and of dependence on oil imported from OPEC member countries, especially in the Middle East. Europe's own growing oil and gas resources and the still increasing levels of production from them, combined with ready access to sea-borne oil from various world regions and to some 70% of the world's resources of natural gas through emerging pipeline systems, provide sound foundations for the continuation of a European oil and gas-based economy for well into the second quarter of the 21st century – and possibly well beyond that. In this respect Europe is in a more favourable position than either of its rival groupings of industrialised nations in the Americas and the Western Pacific Rim.

As in Volume 1 of this book, the selection of 36 contributions in this volume have been made from a much larger number of studies and commentaries which I published over the past 40 years. Should a greater range and/or depth of description and analyses be required, then this can be achieved through the references to my other publications and to those of others. These have been given on a chapter-by-chapter basis. Except where severe editing has been required to limit the length of individual chapters, the contributions presented in this Volume are much as they were on original publication. All my opinions and interpretations have been maintained so as to show how prospects appeared to be at the time of writing – rather than indulging in the re-interpretation of events and their significance with the benefit of hindsight. Forecasts, in particular, remain as they were made, so indicating failures – such as the much too optimistic view which I took over the speed with which the countries concerned would allow their gas and oil resources to be developed – as well as successes, in terms, for example, of long-term future of gas in Europe. Original graphs, diagrams, maps, tables and bibliographies have also been generally re-presented – except when duplication needed to be avoided. In these cases references have been given to previous presentations in the volume of Figures and Tables.

Finally, I would wish to acknowledge my debts to the many academic colleagues and research students and the equally large number of governments' officials, oil and gas companies' employees and to consultants, directly or indirectly concerned with European energy issues. Their contributions to my understanding and knowledge of European oil and gas matters have been very significant. We didn't always see 'eye to eye' on issues of importance, but our discussions were invariably of utility. A number of companies and institutions concerned with oil and gas in Europe generously contributed towards the costs involved in undertaking the compilation of studies for this volume. I gratefully acknowledge their help, but all have requested anonymity. None of them placed any limitations on what I might include in – or exclude from – the contents.

London, 13 May 2002

References

1 Pounds, N.J.G. and Parker, N.W., *Coal and Steel in Western Europe*, Faber and Faber Ltd., London, 1957

2 Schumacher, D. et al, *Energy; Crisis or Opportunity?* MacMillan Publishers Ltd., Basingstoke, 1985

3 Jensen, W.G., *Energy in Europe, 1945–1980*, G.T. Foulis and Co Ltd., London, 1967

4 Jensen, W.G., *ibid*; and Lucas, N.J.D. *Energy and the European Communities*, Europa Publications, London, 1977

5 Jensen, W.G., *ibid*; and the Organisation for European Economic Cooperation, *Towards a New Energy Pattern for Europe*, Paris, 1960

6 Jensen, W.G., *ibid*.

7 Dienes, L. and Shabad, T., *The Soviet Energy System: Resources, Uses and Policies*, Halsted Press, New York, 1979; Park, D., *Oil and Gas in COMECON Countries*, Kogan Page, London, 1979; and Voloshin, V.I., "Electric Power in the Comecon European Countries", *Energy Policy*, Vol. 18, No 8, 1990, pp. 740–6

8 Park, D., op cit; Hoffman, G., *The European Energy Challenge; East and West*, Duke University Press, Durham, N.C., 1985

9 Adelman, M.A., "Security of Eastern Hemisphere Fuel Supply", M.I.T. Department of Economics Working Paper, No. 6, December, 1967: reprinted as Chapter 22 in his book, *The Economics of Petroleum Supply*, The M.I.T. Press, Cambridge, Mass, 1993, pp 469–482

10 Glennie, K.W., Brooks, J. and Brooks, J.R.V., "Hydrocarbon Exploration and Geological History of North West Europe" in Brooks, J. and Glennie, K.W. (Eds.), *Petroleum Geology of North West Europe*, Graham and Trotman, London, 1987, Vol 1., Chapter 1, pp. 1–10.

Section I

Oil and Gas take over the European Energy Economy

Chapter I – 1

Oil's Onslaught on Europe

A. Demand, Imports and Refining*

1. Introduction

Western Europe is the world's second largest oil consuming region with a total demand in 1960 of about 166 million tons. Five of the world's ten largest consuming countries (excluding the United States and the Soviet Union) are in this area. These are the United Kingdom, with a consumption of almost 40 million tons; Western Germany and France, each using almost 30 million tons of oil; Italy with a demand of 20 million tons and Sweden which consumed over 11 million. Since 1950 consumption in Western Europe has been increasing very rapidly, with an average annual rate of increase of almost 14 per cent. The incremental demand each year has, indeed, for the last few years exceeded the additional amounts of oil sold in the United States. Even more remarkedly the additional 23 million tons of oil sold in Western Europe in 1960 was only about 5 million tons less than the additional quantities sold in the whole of the remainder of the non-communist world.

This growing significance of the Western European oil market has, of course, had an important impact on those producing areas which are particularly well suited to supply it.

North Africa and the Middle East are in the most favourable locations in this respect, so that the rapid expansion and the buoyancy of their Western European outlets have been the main factors that led to

*** Edited selections from the author's book, *An Economic Geography of Oil*, G. Bell and Sons Ltd., London, 1965, pp64–65, 86–92 and 111–121.**

continued investment designed to secure increases in these regions' productive capacity throughout this period. The Soviet Union is in a similarly favourable position. This will be much enhanced when the massive forty-inch pipeline from the Volga-Urals fields through to Eastern Europe and points on the Baltic and the East European/West European border is completed in 1963. The Soviet Union plans to put almost 10 million tons of oil a year for Western Europe through this line which will significantly reduce the costs of transporting Soviet oil (it will replace a longer haul by pipeline, barge or rail and tanker from the Black Sea ports) and hence further improve its competitive position in the market.

2. Oil outcompetes Coal

Petroleum products in Western Europe have, over the past decade, made increasingly significant inroads into the essentially coal-based economy of the region. Indeed, coal has been unable to compete for the expanding energy markets. At first, this was because of difficulties in increasing the supply of coal and, more recently, because oil prices moved favourably relative to the price of coal in the latter's traditional markets. These factors, together with the expanding non-competitive uses of oil, particularly in the automotive field, have led to the rapid rate of increase in oil consumption noted above. In Western Europe as a whole the share of oil in the total energy consumption has increased from 13 per cent in 1950 to 30 per cent by 1960. The changes in the main countries over this period are shown in Table I-1.1.

The general success of oil in expanding its share of the competitive markets in countries as dissimilar as those listed in the Table, with respect to both their indigenous supplies of energy and in the kinds of economic policies their governments have followed, has arisen from two main sets of factors. In the early years after the war, and even up to 1955, the demand for energy exceeded the supplies available. In particular, the coal industry was unable to expand its output sufficiently quickly to match demand. The growth of world oil supplies in this period was, therefore, very welcome and encouragement was given to the use of oil rather than of coal. For example, in the United Kingdom the Government encouraged both the building of power stations designed to burn oil and the conversion of exiting coal-fired stations to oil-firing. In this way oil was intended to replace the use of some 8 million tons of coal per annum. Other European countries which had formerly relied on imported coal now found that supplies were unobtainable, except at high cost from the United States, and they also proceeded with converting major coal using industries to oil.

TABLE I-1.1
Oil's contribution to total energy consumption in major Western European countries 1950–60
(in millions of metric tons of coal equivalent)

Country	1950 Energy Consumption	1950 Percentage of Oil	1955 Energy Consumption	1955 Percentage of Oil	1960 Energy Consumption	1960 Percentage of Oil
United Kingdom	222.9	9.5	254.5	13.6	259.0	24.1
West Germany	123.0	3.3	169.2	6.7	202.9	19.6
France	78.6	16.8	93.7	23.0	109.4	30.0
Italy	20.5	24.7	34.7	41.7	58.6	56.1
Sweden	15.0	34.5	19.9	57.7	26.1	71.4

Source: U.N. Statistical Papers – Series J. Note that hydro-electricity is included in terms of the coal equivalent of the electricity actually produced, viz.:1000 kWh=0.125 tons of coal.

Since 1955 the post-war shortage of fuels has disappeared as economic recession brought a check to rising energy demand. In this context industrialists and others put energy costs under close scrutiny, in an effort to reduce the costs of production so that exports might be made more competitive. Thus, increased efforts were made to find the cheapest energy supplies and to use energy more efficiently. In this situation the general and continuing tendency of coal prices to rise relative to those of oil led to the development of a marked preference for oil use over much of Western Europe. This preference has, moreover, been further strengthened by the actual decline in oil prices under the impact of increased supplies both from new areas (for example, the Soviet Union and North Africa) and from new suppliers (such as United States companies with Venezuelan and other supplies surplus to their western hemisphere requirements) which had not formerly sought outlets in the European market.

Petroleum fuels have thus gradually become competitive over a widening range of uses even in areas of Europe in close proximity to the continent's coalfields. It was estimated in April 1961, that only 30 per cent of Britain's annual production of almost 200 million tons of coal was competitive with the ex-refinery price of fuel oil. In such a situation, the conversion of power stations (in which use oil has only modestly greater efficiency in use than coal and one in which the question of cleanliness and convenience does not arise) from coal-to-oil-firing in all areas, except those in juxtaposition to the low-cost coal mines of the East Midlands obviously became an attractive development, if the generating authority were to take only commercial factors into account. The government's imposition, of a 2.2d per gallon tax on fuel oil (equivalent to a tax of over £2 per ton – about 25 per cent, on the ex-refinery price) significantly altered the relationship in favour of coal so that over 95 per cent of coal production now became competitive. In view, however, of the inevitability of rising costs in the coal industry,[1] and of the likelihood that the country will eventually secure cheaper supplies of oil as new companies move in and thus upset the oligopoly of refining and marketing operations in this country, it would seem that this will prove no more than a temporary setback for oil in the process of securing additional outlets.

In other parts of Europe the situation has already moved even more strongly in favour of oil, because of the generally higher cost of coal production, the lower prices for oil resulting from a greater degree of competition amongst oil suppliers and the existence of refinery capacity independent of the major international oil companies. As a result of these factors, oil prices have fallen well below the prices posted by the major companies. In Italy discounts of up to 62 cents a barrel (representing a cut of about 30 per cent) off the posted price are reported as having been obtained on crude oil imported by independent refining companies. In Germany there have been discounts of 54 cents a barrel and discounts of between 30 and 40 per cent have also been reported from Belgium. In the face of such competition the European coal industry has found great difficulty in maintaining its markets, let alone in securing a share of the incremental demand for energy.

In Germany, the imposition of a fuel oil tax in 1960 designed to protect the coal industry was largely ineffective as, in general, the tax was absorbed by the suppliers and the prices to consumers were not raised. Thus, coal production in West Germany fell away from 153 million tons in 1956 to 142 million tons in 1961, when the industry was only working at 75 per cent of capacity.★ The situation has not yet been stabilized and a report[2] to the German government in 1962 suggested that production

★ Domestic coal production has been affected not only by competition from oil, but also by imports of cheaper coal mainly from United States.

should be cut back to about 125 million tons by 1975.

Likewise, in Belgium, the high cost mines have been unable to compete with oil and there have been many closures and an overall attempt at the rationalization of the industry. Production declined from a post-war peak of almost 30 million tons in 1957 to 21 million tons in 1961.

A similar situation has arisen even in Austria, where oil prices are higher as a result of the transport costs on imported products, mainly obtained from refineries on the Italian Adriatic coast and in the Ruhr. Consumption of oil products rose at the expense of coal, in spite of the imposition of a fuel oil tax in 1960. As a result coal output declined from 7 million tons in 1957 to 5.7 million tons in 1961. In spite of this significant fall in production, stocks of coal in 1961 rose to a level equal to about one-third of the annual output and the coal industry accumulated deficits of about £1 million per annum.

3. Policy Implications

Overall, after rising to a post-war peak of about 600 million tons in 1957, European coal consumption fell away quite rapidly. By 1960, given the increasing competition from oil, there appeared possibilities of the collapse of significant sectors of the coal industry, with accompanying social problems, and the likelihood of greatly increased reliance on imported oil despite the political and supply risks entailed. The implications of this have perhaps been given the closest attention in West Germany, where the failure of fiscal measures to stabilize the situation has already been noted. The German government commissioned a full-scale enquiry into the problems of future demand for and supply of energy. This enquiry was commissioned following a survey made by an American oil consultant, Mr. W. Levy, for the German Confederation of Industries[3]. Although essentially an 'oilman', Levy argued in favour of protection of coal against fuel oil on the grounds that coal is at present uneconomic because fuel oil prices are abnormally low, thereby exerting a short-term competitive pressure which cannot be maintained in the future. He argued that this short-term pressure is sufficient to cause unnecessary and unwelcome long-term disruptions in coal-producing capacity and went on to recommend (to the consternation of the oil companies) that in order to prevent this development, actual energy prices should be geared to a more realistic long-term trend; and that since fuel prices had deviated from this, the immediate remedy should be to increase prices by additional taxation. Additionally, he suggested that the electricity generation market should be reserved to the coal industry in order to produce a growth sector for coal and to increase the security of electricity supply.

The report was attacked by the oil industry on the grounds that it would increase the price of energy and thus make national output less competitive with that from countries, such as Japan and Italy, using the cheapest fuels available. The industry also pointed out that the recommendations would remove the incentive to rationalization in the coal industry, which because of rising costs would not, it claimed, be a viable competitor with fuel oil in the foreseeable future. Nevertheless, even though the report may not have considered all the implications of its recommendations, it did indicate the kind of action that governments in coal-producing countries would have to take if they wish to maintain the output of coal at levels sufficiently high to minimize social problems and to avoid great reliance on imported energy.

In the United Kingdom, the decision to convert all dual-fired power stations using oil back to coal (in spite of the protests of the electricity authorities), the refusal to accept supplies of Soviet oil which would weaken the price structure and intensify competition and the imposition of a fuel oil tax were all designed to offer a measure of protection to indigenous coal production. The major oil companies, in part because they appreciated the difficulties facing the coal industry and recognized the need for security of energy supplies, in part because they wished to avoid even harsher government action against them at a later and perhaps more critical stage and, in part, because they also wanted to avoid an unduly rapid growth of demand for fuel oil, which gives a lower rate of return than other products and which could lead to imbalance in their refinery operations, were not too unhappy to see the power market reserved for coal. However, they seem likely to be inhibited from co-operating in this way by the growing involvement of the small oil companies in supplying oil to European markets. In general terms, the latter aim at maximizing their sales in order to secure some return on the capital they have invested in producing and/or refining facilities. They are thus unlikely voluntarily to withdraw from the business or even prepared to moderate their objectives. Thus, as in the United States, where the oil industry was at first asked to curb its imports of oil and then forced to do so by legislation, it seems likely that voluntary restraint in Europe will also be unworkable and that specific government action – or inter-governmental action through the European Economic Community – will be necessary, if the coal industry is to gain a respite from the competition from oil.

Throughout much of Europe, therefore, the prospects for

government directed or influenced energy policies, which will help to determine the rate at which the demand for oil will increase, seem fairly certain. As a result, the rapid rise in oil consumption in the 1950s (an average of 12.7 per cent per annum, with an even faster rise in the last three years of the decade), is likely to be contained to some degree in the 1960s. From this kind of approach, however, one would expect the Netherlands, the Scandinavian countries and Italy to be excluded, as they are dependent on imported fuels to a very large degree. In these countries the main consideration would appear to be continued efforts to minimize the unit cost of imported energy. Their markets will thus represent increasingly important outlets for the supplies of cheap crude and products that will continue to be available.

The division between the indigenous energy producers, on the one hand, and the energy importing countries, on the other, may be a stumbling block in the moves towards economic integration in Europe. The coal-producing members of the E.E.C. might reasonably anticipate openings for their surplus production in other parts of the Community – certainly in preference to coal imported from outside the area and possibly in preference to imported oil. Italy, and any other countries in a similar position which decide to join the Common Market, can be expected to resist such pressure and insist on having access to the cheapest supplies of energy that are available.*

Thus, politico-economic decisions that will have to be taken both by individual European governments and by European organizations will obviously affect the future strength of growth in oil demand in the region. The overall effect of such decisions, coupled with the less rapid rate of economic growth in the future than in the immediate past, and the generally more efficient use of fuel seems, on balance, likely to reverse the process in which Western Europe has steadily increased its share of oil consumption in the world outside North America and the Soviet bloc. From 1950 to 1960 Western Europe's oil consumption rose from 36 to 53 per cent of the global total. In the next decade its share may well fall back to under 50 per cent, as demand in the developing countries increases more quickly than in Europe.

4. Downstream Developments in European Oil

Even before the Second World War it was possible to discern a move of refining towards the growing oil consuming areas of Europe. New capacity was established in Italy, Germany, Belgium and the

* Common Market energy policy was still being negotiated at the time of writing

Netherlands. At this time the sales of individual companies were developing to a volume at which local refining by these companies was becoming a more attractive proposition. Shell-Mex and B.P. Ltd. in the United Kingdom, for example, sold about 5 million tons of oil products in 1938 and in Germany, Shell's sales approached 1.5 million tons.

This early development of new trends in the pattern of refining was naturally curbed by wartime conditions. During the war it became necessary to make use of whatever facilities existed to provide for the essential needs of the armed forces and of industry and commerce. The major consideration in the supply and refining of petroleum was the need of the war effort. Thus, United States capacity, together with that of the Venezuelan/Caribbean export complex, was used – and expanded as necessary – to meet the demands of the Atlantic and western hemisphere areas.

In the period since 1945, two main locational trends can be distinguished in the post-war refining industry outside the United States. The earlier, and the more significant, of these was the reinstatement of the trend in Europe, viz. renewed growth of refining capacity in the major consuming areas of Europe.

By 1939 the capacity of the European refining industry was only 17 million tons. In marked contrast the industry's capacity by 1962 was about 250 million tons, mostly working at a very high load factor, with under-utilized capacity restricted mainly to Italy and Western Germany. During this period, Western Europe's share of total refinery capacity outside North America and the Soviet bloc increased from 20 to over 40 per cent. Figures I-1.1 and 2 show the pattern of refinery development in Western Europe in the period since 1950. Three significant trends can be seen; first, the increasing numbers of refineries; secondly, their increasing size; and third, their increasingly widespread distribution. One outstanding result of this development has been the increase in the percentage of total demand for petroleum products met from local refineries from only 25 per cent in 1939 to more than 85 per cent in 1962, in spite of a six-fold increase in total consumption. Several factors have combined to produce this result.

First, the post-war trend in the Western European consumption of petroleum products has led to an increase in the proportion of fuel oil in the total supplies required. The pattern of supplies required began to match more closely the most economical pattern of refining the commoner crudes. Thus, European refineries became much more likely to be able to find local outlets for all their products so that prior need to back-haul the fuel oil not required in Europe has thus been eliminated.

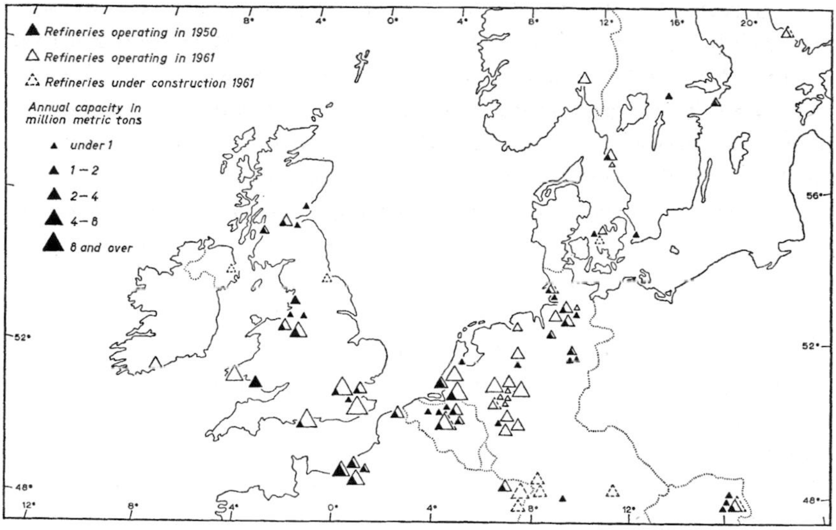

Figure I-1.1: The Pattern of Refining in Western Europe – North

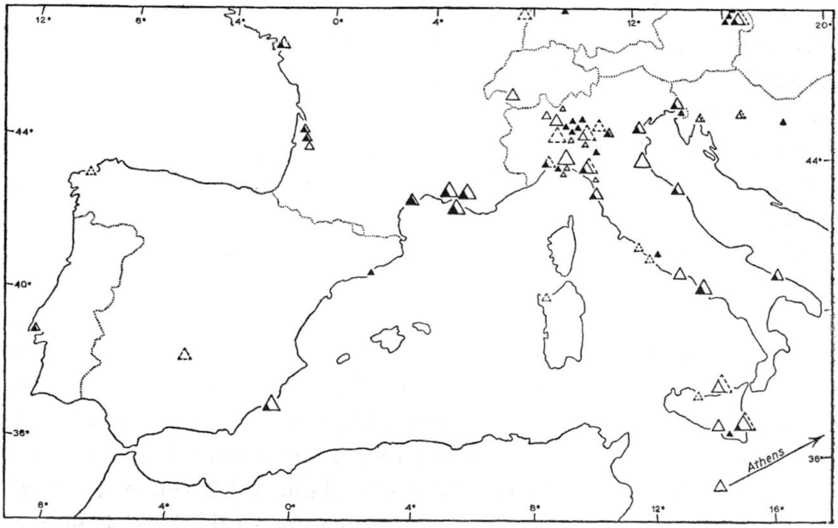

Figure I-1.2: The Pattern of Refining in Western Europe – South

Second, the increased demand for all products in all European countries has produced a situation in which much larger refineries can be built. Thus, refineries having a capacity of some 2–4 million tons a year, so producing quite significant economies of scale,[4] became feasible in most countries – even in small countries such as Norway, Denmark and Ireland, where the overall demand for petroleum products in each case is of the order of 2 million tons. Although this total demand may be divided among several companies, arrangements have often been made either for the construction of a jointly owned and operated refinery to supply all the country's needs, or for the smaller distributors to draw their supplies from the refinery built by the market leader. Alternatively, because of the relatively short distances between countries in Europe and the general feasibility of shipping by sea, some companies have been able to build a refinery of this order of magnitude to serve the collective needs of its marketing affiliates in neighbouring countries.

Although it is estimated that it is still somewhat cheaper to refine crude in even larger refineries, with a capacity of up to 7 million tons per annum which can, in general, only be built in producing, rather than in consuming areas, the additional savings on such refineries can be more than offset by the higher cost of shipping products than crude. Thus, a marketing company with annual sales of 2 million tons seems likely to achieve a net saving by having its refinery on the spot and importing its crude requirements, rather than by buying and transporting its products from a larger and somewhat lower cost refinery in a producing area.

This trend towards more economic refining in Europe has also been assisted by technical developments in the refinery industry which permit a much greater degree of flexibility in refinery operations. The out-turn of products from a given crude can be varied and, hence, the output of the market oriented refinery can, within quite wide limits, be adjusted as necessary to match the changing demands of the market. This flexibility eliminates the need to find alternative export outlets for surplus products, or to import other products to make up the local deficiencies, as was the case when a refinery's out-turn of products was much more rigidly determined by the simple separation of the hydrocarbon components of a crude on the basis of the differences in their physical properties. The importance of this development is seen in the way in which the Western European refineries have gradually adjusted their output to match a more rapidly increasing demand for fuel oil and diesel oil than for the lighter products. Table I-1.2 shows how the demand for main products has changed in some of the largest Western European countries between 1950 and 1960.

TABLE I-1.2
The changing demand for main products in major Western European countries 1950–60
(Each product as a percentage of total demand)

Country	Gasoline		Kerosene		Gas/Diesel Oil		Fuel Oil	
	1950	1960	1950	1960	1950	1960	1950	1960
United Kingdom	35.6	19.4	9.3	4.0	17.9	15.5	21.2	44.3
West Germany	29.8	20.0	7.2	n.a.	30.2	41.6	12.0	23.8
Italy	15.6	13.7	4.3	1.0	17.7	14.8	54.0	59.0
France	27.1	23.3	1.2	1.6	23.9	35.7	36.5	24.4
Sweden	16.8	12.8	5.0	2.0	30.1	33.6	41.5	44.5

Source: Oil Industry Trade Journals

Government policies in most Western European countries have also sought to attract refinery capacity. Such policies have involved the use of both the 'carrot' and the 'stick'. The parlous state of most of the countries' economies at the end of the war necessitated strict limitations on the amount of foreign currency that could be spent on importing oil. In particular, the shortage of dollars meant that a continuation of the practice of importing a very large percentage of Europe's needs for oil products from the United States and from other 'dollar' refineries in the western hemisphere could not be countenanced, and pressure was therefore exerted by governments for the expansion of refining in Europe. It has been estimated, for example, that from 1949 to 1955 a foreign exchange saving of the order of £160 million was made as a result of the gradual change from product to crude oil imports into the O.E.E.C. countries of Western Europe.[5] Therefore, by making dollars available for importing American refining equipment, by permitting crude oil to enter duty free, while charging significant duties on petroleum products, together with the political pressure that governments are generally in a position to bring to bear on oil companies, the development of local refining capacity was given an important fillip.

Finally, both governments and oil companies have become increasingly concerned with the need to ensure the continuity of oil supplies in the face of various political uncertainties in the producing areas of the world. Before 1939, a shortage of oil would, in general, have been no more than an inconvenience for most West European countries. A similar shortage today would cause severe dislocations in the economy as a result of the growth in importance of road transport, the use of oil in rail transport and the great development of diesel oil and fuel oil as industrial and commercial fuels.

5. Strategic Issues

It is, of course, partly the fear for the dislocation of overseas supplies of oil that provides the impetus to national energy policies aimed at reducing the dependence of a country on imported fuels. An additional way of reducing such dependence is to locate oil refineries at home. With refineries located in Europe the only necessity, in the event of the disruption of supplies from an oil producing area, would be to find an alternative source of crude. That would be difficult but certainly not impossible, given the success of oil exploration and development in providing a large reserve of shut-in capacity which can be quickly utilized. With refineries located at the sources of the crude, dislocation involves not only finding an alternative crude supply, but also finding somewhere else to refine it. This could in fact be a physical impossibility given a large dislocation, as refinery capacity is not, in general, developed greatly in excess of estimated needs.

The nationalization in 1951 of the assets of the Anglo-Iranian oil company and the subsequent closure of the world's largest refinery at Abadan (processing Anglo-Iranian's output and providing some 25 million tons of products to the United Kingdom (about 20 per cent of its petroleum requirements), effectively emphasized the dangers of dependence on overseas supplies of refined products and stimulated a wave of refinery development in Western Europe. The wisdom of the policy could be seen by the existence of adequate refinery capacity in Western Europe in 1956, whereby the problem of supplying the area with the required quantities of oil at the time of the Suez crisis was greatly simplified. Had the refinery facilities Europe needed still been located largely in the Arab countries, some at least of them might have been closed down, together with the Suez Canal and the pipelines to the Mediterranean. Even if they had remained in operation, the much more complicated revised supply pattern necessitated by the closure of the Canal might have proved too difficult to organize quickly and effectively. The validity of this is indicated by a comment by P.H. Frankel who suggests that even the re-organization of crude supply routes proved almost too complex an operation in the face of the many interests that were involved. In an article which analysed the problems of oil supplies attendant upon the Suez crisis he observed:

> *'The repercussions of the crisis were less dramatic than they were at one time expected to be. This was due to the warm winter, the fact that demand was also for some other reason below estimate and to the skill with which all parties concerned handled their problems... However, most people who are aware of the facts involved know how narrowly we escaped being faced with a very ugly situation.*[6]

Concern expressed by the governments of consuming countries over the location of too much refinery capacity in the producing areas has been shared by the oil companies, anxious for the security of their investments in parts of the world which have become much less stable with the rise of nationalism. This is particularly the case for the Middle East, where the pre-war influence and control by Great Britain and France has declined to such a degree that investments made by oil companies can no longer be guaranteed. The failure to take any counter action at Abadan in 1951 and the debacle of Suez in 1956 illustrate this point. Recurrent political crises, often backed by threats against the oil 'imperialists' have necessarily reduced the confidence of the oil companies, which, with enormous capital assets tied up in producing and associated activities, have naturally been hesitant to commit even more of their funds to the same areas for the development of refinery capacity. It became preferable for them to spread the risk by investing in refineries located in Europe, where the possibilities of expropriation are minimal and where the refineries are available to utilize crude imported from whatever parts of the world are not affected by a political crisis.

B. Oil: the New Commanding Height*

1. Introduction

In 1965, the oil industry in Britain remains, in the words of the major oil companies' monthly journal, *Petroleum Press Service*, "essentially private and this in spite of the fact that the government as long ago as

* Extract from an edited version of the author's *Fabian Society* Pamphlet with the same title (No. 258, London, 1965). Limitations on the length of the chapter in this volume have necessitated the exclusion of the detailed changes suggested for the organization of the UK's oil industry at the critical moment in the mid-1960s when Britain was poised to become a major oil and gas producing country.

1914 acquired a stake in Anglo-Persian – now the British Petroleum Company"[7]. Apart from the Ministry of Power's recent allocation of exploration and development licenses to various companies in the British sector of the North Sea, whereby the country will secure some royalty and other payments if oil or gas is discovered and produced, the oil industry in Britain is largely untouched by statutory and publicly-known forms of direction or control over its activities.

Some companies would argue, however, that there is, in fact, an *informal* control over their activities, arising from "the close relationships" which they have with the Ministry of Power and other Departments of State. Such informal arrangements are, however, unsatisfactory as the public cannot judge how effectively its interests are defined or defended. The informal arrangements have not, for example, prevented our paying between £30 and £40 million per annum more for our crude oil imports than is justified by the world oil market situation. Though there may be agreements between the Treasury and some of the companies concerned designed to mitigate the effect of this on our balance of payments, this is too important a matter of public interest to relegate to "informal" arrangements. The facts of the situation should be revealed and publicly argued and determined before decisions are taken on them.

Although the Ministry of Power's functions are defined as including the responsibility for ensuring a co-ordinated fuel industry including oil, successive governments have chosen not to have the Ministry fulfil this function and thus the oil industry remains free to determine its own policy towards imports, investment decisions and price levels. The absence of government intervention has produced a situation in which Britain has become a major importing nation of refined products, as refinery construction has lagged behind the growth in demand, with consequent adverse effects on the balance of trade. In 1964 almost 25 per cent of the oil products used were imported at a cost of £167 million. A domestic refining industry capable of meeting our total needs for products would have given a £40 million saving on these import costs. British consumers also pay higher prices for their oil products than in most other European countries.

This freedom for oil in Britain is in marked contrast to the situation which exists in every other major oil importing and consuming nation in the non-communist world. In France, the oil industry, consisting of a range of privately owned as well as state owned companies, is firmly and effectively controlled by a Director General of Energy. In Japan, the Ministry of Trade and Industry (MITI) controls the main activities of the oil companies, having the right to fix price levels and to control investment. This year, for example, the investment proposals by the Japanese oil industry for the expansion of refineries and storage and marketing facilities were cut by no less then 28.7 per cent in order to match what MITI considered to be the oil expansion needs of the country. In addition, MITI also exercises quantitative controls over imports of crude oil and products by the companies refining and marketing oil in the Japanese market. In the US and Italy private oil companies operate under systems of public control and surveillance while even in West Germany the government has recently decided to restrict the freedom of the international oil companies operating in the country and now plans to place controls on the development of refineries and pipe lines.

The contrasting position in this country was summed up by a speech, made by Mr. F.J. Erroll, the Minister of Power in the last Conservative Government, at the 1964 annual dinner of the Institute of Petroleum;

> *"We welcome oil companies to this country from anywhere in the free world... We are delighted to see them investing their money here and we believe that the competition they bring will benefit our consumers.* All we ask *of the newcomers is that as soon as their business fully justifies it they* should become full members of the club *and take a real investment stake in our community."*[8]

Not much less could have been required of the oil companies operating in the United Kingdom. The government's policy did not even ask – let alone require – them to fit in with any national considerations and did not even try to persuade them to co-ordinate their activities, either with each other, or collectively with the other energy supply industries, such that national requirements for fuel and power could be met in the most expeditious manner. As fundamental difficulties for Britain's coal industry arose from this absence of a policy towards oil, the Conservatives reacted by a series of short-term expedients, such as virtually banning imports of Soviet oil and American coal and by imposing a 2d per gallon tax on fuel and diesel oils, equivalent to a purchase tax of up to 33 per cent on the ex-refinery price. These measures merely had the effect of postponing the need for decisions on the oil industry's role in the country's energy economy. A continuation of such a *laissez faire* attitude towards oil cannot be justified by any government dedicated to the idea of socio-economic planning for the greater good of the whole community.

Regretfully, at the other end of the political spectrum the Labour movement cannot be accused of a pre-occupation with oil matters. On the contrary, it has shown remarkably little concern for the industry which appears never to have been included on any list for state ownership and never to have been the subject even of investigations for policy recommendations during the Party's long period in opposition between 1951 and 1964. The present Labour Government cannot, therefore, be said to have come to power with any preconceived ideas as to what it should do about oil, apart from a generally unexpressed feeling of antipathy towards the industry because it seems to epitomise "monopoly capitalism" and possibly "neo-colonialism" as a result of its great influence on the economics and politics of independent, but relatively weak, nations.

For the important section of the Labour movement, however, representing the coal mining areas of the country, antipathy towards oil has had a rather more immediate cause. It views oil as public enemy number one, the insidious influence of which causes markets for coal to contract, so leading to pit closures with the loss of jobs and other hardships that accompany such traumatic experiences for the relatively closed mining communities. In that these areas invariably return Labour members of the House of Commons, the impact of their justifiable fear of oil has perhaps led to Labour appearing as a pro-coal and, as a necessary corollary, an anti-oil party.

A second point to bear in mind is that previous Labour Governments were in office when oil was of limited importance in the nation's economy, so that the companies were able to stand aloof from the party political battles on economic policy. Indeed, the UK's use of oil as a general purpose source of energy has developed only with the post-war Conservative Governments. From 1945 to 1951 oil meant little more than petrol; a product to which there was no alternative. Apart from this, whatever other oil we could afford to import was welcomed as a means of eking out our limited domestic availability of energy in the immediate post-years. In 1951, when the Labour Government left office, oil provided less than 10 per cent of the country's total energy consumption and barely 4 per cent of the sectors of the energy market in which it was in competition with coal.

Since then, however, oil has developed as an industrial, commercial, and domestic fuel and as a source material for both chemicals and town gas. It thus now impinges on the economic life of the country at a multitude of points ranging from basic industries such as iron and steel, to the whole range of manufacturing industry, to transport, both rail and road, and to the domestic sector where free standing kerosene heaters and oil fired central heating systems have enjoyed a phenomenal rate of growth. Thus, Britain has become a two-fuel rather than a one-fuel economy, with oil's share of total energy consumption already standing at 34 per cent. It seems certain that by 1987 it will account for over 40 per cent of the country's total energy requirement and, in the absence of any major discoveries of natural gas in the North Sea, it will provide over 50 per cent of the country's energy by the middle of the 1970s.

2. Politics and Economics

Oil has thus become one of the major "commanding heights" of the economy with an influence on all sectors of the country's life. British energy policy, even under a Labour Government, can no longer be based on the assumption that domestic coal both can and must provide the bulk of our energy requirements and that oil is only important in meeting demands which coal cannot meet for technical reasons or to tide the country over temporary energy shortages. Policy considerations for the future have to take into account the fact that oil can do almost anything that coal can do and that, in most cases, it can do it a great deal more cheaply.

Oil in Labour's Energy Policy

After one year in office, the Labour Government seems to have accepted this proposition. In spite of several parliamentary statements by the Minister of Power, to the effect that coal will form the base load of our fuel supplies for as far ahead as we can see, the government's actions suggest that it accepts that only an attenuated coal industry will be a viable proposition. Thus, the Minister in presenting his proposals for the capital reconstruction of the coal industry to the House of Commons on 1 July 1965 said:

> *"the coal market has contracted... and discussions in the Energy Advisory Council suggest some further contraction may be unavoidable in the years ahead"*[9]

The degree of immediate contraction foreseen is spelled out in the National Plan which estimates a 170 million ton market for coal in 1970 compared with over 190 million tons in 1965. As the White Paper on *Fuel Policy*[10] clearly indicates, this decline in coal consumption is related entirely to competition from oil.

Having accepted that there will be a more rapid expansion in the market for oil in Britain than there would with the continued protection and/or subsidisation of the coal industry, the Labour Government now appears to stand without an adequate policy for this new commanding height of the economy. The Energy Advisory Council, appointed in January 1965 "to consider and advise the Minister of Power about the energy situation and outlook and the plans and policies of the fuel and power industries in relation to national objectives for economic growth", appears not to have been instructed to investigate and make recommendations on the operations and organisational structure of the oil industry in Britain. Moreover, on 6 April 1965 the Minister, in a written reply to a question as to how many official advisers and review bodies had been appointed to his department since October 1964, answered "none, except for the Energy Advisory Council"[11]. The Ministry of Power must thus surely stand alone among the Departments of State concerned with economic questions not to have sought advice as to how long-term policies inherited from successive Conservative Governments could be radically adjusted to meet the needs of a progressive Britain. Yet in few other sectors of economic life is radical adjustment as vital as it is towards the oil industry.

It is thus not surprising that the long awaited White Paper on Fuel Policy, designed to provide the framework for the Labour Governments' promised co-ordinated energy policy, could, as far as its recommendations on the oil industry are concerned, have been equally well presented by a Conservative Minister of Power. It concludes:

> *"A continual rapid growth in oil consumption is likely for many years to come. This growth carries with it risks to security of supply and increasing direct costs to the balance of payments, but any undue restrictions on the use of oil would hinder technological advance and increase costs. The Government believes that it is in the national interest to accept a rapid growth in the use of oil, but will take measures to mitigate the security and balance of payments disadvantages by ensuring adequate stocks in this country and by encouraging the oil companies to diversify their sources of supply and to develop the United Kingdom refining industry. It is also taking steps to ensure that exploration for oil and natural gas deposits in the North Sea is vigorously pursued, for few developments could be of more benefit to the national economy than the discovery of major reserves with British jurisdiction"*[12]

This contains nothing to which any previous Minister of Power would have objected. It confirms that this increasingly important "commanding height" of the economy has not yet been subjected to any effective analysis as to how it would fit it into an economy under Labour direction.

Shell and BP

The failure to initiate studies of the oil industry may reflect a continuation of the attitude within government circles that with Shell and BP in control of market the state can reasonably expect adequate co-operation through "informal arrangements", given the influence that the state thinks it has over both these companies: in the case of BP by the Government's ownership of 51 per cent of the company's shares and, in the case of Shell, by virtue of the company's position as part of the "establishment' so that it can be viewed almost as a branch of the Civil Service. One should note in passing that even at the time when Shell and BP were the essence of the British oil market, the expectations of co-operation were pitched much too high. After all, Shell is mainly non-British in its ownership (the British share-holding in the Royal Dutch Shell Group is less than 40 per cent) and it has legitimate commercial interests which are not necessarily compatible with the United Kingdom's national interests. On the other hand, BP vows its unfettered belief in the efficacy of free enterprise and appears to conduct its operations as though the 51 per cent state holding does not exist.

In May 1965 Harold Wilson revealed in the House of Commons that BP made no written reports to the Government on its decisions. "Reports from the Government appointed directors on the board of BP" he said, "are quite frequent, but they are mainly of an oral character... and mainly informal.[13] The Government, in effect, relies entirely on informal verbal exchanges for its information on BP's policy decisions. This situation suggests that the Government has, in fact, no control over the company's activities – an even more surprising and disturbing position than that suggested by the usual definition of the relationship between the company and the Government, viz."under a gentleman's agreement, the state does not interfere in the *day to day* management of the concern"[14]

The ability or otherwise of the Government to influence Shell and BP is, however, rapidly becoming a less significant issue as these companies are of declining relative importance in the British oil industry. Their share of retail petrol sales fell from 51.4 per cent in 1953 to 45 per cent in 1964[15], while, in other markets, where competition from new companies has been much more intense, the decline in their market share has probably been much more pronounced. Almost 60 per cent of the British oil industry is now foreign controlled and the situation is moving even more rapidly in favour of the non-British companies, with, for example, about two thirds of new refining capacity in the country planned or projected by foreign oil concerns.

Foreign Owned Oil Companies

In the past ten years Shell and BP have faced effective competition from the other major international oil companies, such as Standard Oil of New Jersey, Texaco, Socony-Mobil, and, as a result, these companies have secured a gradually increasing share of the British market. Although these US companies have associated companies registered in this country to look after their United Kingdom operations, the parent companies own no allegiance to the United Kingdom, such as that of Shell and BP, and are therefore free to determine their actions by reference only to their international corporate interests and those of the United States. An example of international corporate decisions adversely affecting Britain was Socony-Mobil's decision in 1965 to meet its requirement for additional refining capacity in Western Europe by building a new four million ton refinery at Amsterdam, rather than by expanding the still relatively small refinery at Coryton on the Thames owned by its subsidiary, Mobil Oil. As a result of this, Britain lost both possible export opportunities for oil products and the benefits of enhanced economic activity necessarily associated with the construction and operation of a larger refinery.

Likewise, Standard Oil of New Jersey's subsidiary in this country, the Esso Petroleum Company Limited, has no say in its crude oil purchasing policies. It is allocated supplies by its international parent and is charged for its crude at the price which is formally "posted", rather than at the price at which Esso Petroleum might buy its oil were it free to obtain crude oil independently of this international tie. This is demonstrated in the contrast between the cost of supplies of Libyan crude imported into Britain, imported almost exclusively by Esso, at an average f.o.b. price of $2.21 per barrel, and that imported by Italy at an average price of $1.86. As the UK's imported 68.8 million barrels of Libyan crude oil in 1964, this 35 cents per barrel premium paid by the UK produced an additional foreign exchange cost of the order of £8 million. Similar arrangements also apply to the British subsidiaries of the other international major oil companies, and collectively lead to a significant burden on Britain's balance of payment. Excluding the impact of high crude oil prices paid by Shell and BP refineries, on the grounds, which are not entirely justified, that this does not involve a foreign exchange loss as these are British companies, the overall burden to the UK's trade balance can be calculated to be about £20 million per year. Moreover, the high crude oil import prices which the UK's subsidiaries of the international companies have to pay make it difficult for them to show any profits on their operations here given a gradually falling price level for most oil products sold in the country. Their local taxable income is thus severely restricted – so that their contribution to the national exchequer is very modest.

The adverse impact on the British economy of these long-standing foreign companies in the British market is clear. Yet it would have been relatively easy to correct, as the companies are susceptible to pressure, as shown by action taken in many other countries of the world. They could have been persuaded to amend their crude oil import policies in order to ameliorate the adverse impact on Britain's economy. In the past three years or so the situation has, however, been made even more complex by the establishment of operations in Britain by more than a dozen new foreign oil companies – anxious to take advantage of the rapidly expanding market hitherto controlled by the oligopoly of the major companies. They not only compete for business in the retail gasoline market, but also compete in the wholesale trade for the supply of large quantities of gas and diesel oil to commercial and industrial users. It is in these markets, where prices can be negotiated directly with customers who provide their own tankage, that the newcomers have most likely had their greatest successes, but quantitative data on this is not available.

3. A National Oil Agency

The structure of the UK oil industry which has been described in the paper indicates a need for a fundamental change whereby the new "commanding heights" of the economy can be brought under national control. It perhaps suggests, as far as one particular industry is concerned, the validity of the plea for "a more rational and objective basis for decisions in industry" made by Jeremy Bray, MP in *The New Economy*.[16] He argues that with the code of raw capitalism dead, risks in the economy need be spread more widely, "thus blurring the distinction between public and private enterprise". A recommended policy for oil would not be concerned with the question of private v. public enterprise *per se* in that both state and private companies have positive and effective roles to play in the co-ordinated and scientific development of this particular sector of the economy.

In the first place, much more investigation is required into the viability of possible developments, particularly as many of them will have repercussions on other aspects of our national economy. For example, a serious study is required of the likely overall effect on Britain's balance of payments which could arise from changes both at home and abroad to the mechanisms for pricing crude oil imports into this country by British, American and other foreign oil companies. Such reactions could include decisions by other consuming countries also to try to force down their import prices for crude oil such that the profits of the British based oil companies, Shell and BP, would be so squeezed that their foreign exchange earnings would be reduced with a possible adverse effect on the UK's balance of payments.

There are, however, now very few countries which have continued to pay the high crude oil prices which successive British governments have condoned. One exception is the United States, but there the Treasury is currently examining the situation. The argument that we, as a nation, ought to pay higher imported prices because we may be considered to be an oil producing country, by virtue of the overseas foreign exchange earning activities of Shell and BP, is no stronger than the argument that we should be prepared to protect the domestic coal industry because this saves foreign exchange which would otherwise be used to import alternative fuels. In both cases energy consumers suffer, and our exporters are placed at a competitive disadvantage compared with their foreign rivals. This resultant loss of exports seem likely to more than offset the foreign exchange advantage arising from the protection of British energy supplying organisations such as BP and the National Coal Board.

Secondly, study is required to show how necessary and desirable changes in oil policy could be effectively implemented. Such studies could, we suggest, be best conducted by a newly formed and expert *National Oil Agency* (NOA), the appointment of which should be the first and urgent task of a government which accepts the need for radical change in oil policy. Once established for this purpose it would then have the continuing responsibility for operating the system accepted by the Government and for ensuring its success.

That work would represent only the immediate task of the NOA. In the medium term its work would be supplemented by the need to evaluate the development of North Sea oil and gas over the medium term. Even a very rapid successful search is unlikely to lead to more than a very modest contribution from the North Sea to the country's energy economy in the rest of the decade, with the only possible practical problem being the choice of the most appropriate way of incorporating a limited amount of off shore UK crude production into purchases by British refineries. This is not very likely before 1970.

In the meantime, however, the willingness of some 23 groups of oil, and other, companies to spend at least £100 million, and probably twice as much, under the terms of the Continental Shelf Act, can have only a beneficial effect on the economy in general. The successful results of their searches, and the search of other companies which may be attracted to new areas which the Government has recently announced as being available for exploration and development licences, will need only to be taken into consideration at a somewhat later date when the immediate problems of our oil economy have been tackled.

Finally, in the long term, the U.K. National Oil Agency seems certain to find itself operating in a world in which governments everywhere in the non-Communist nations are gradually assuming a greater degree of control over oil matters. Even today, in oil producing country after oil producing country, one sees the establishment and the expansion of state companies designed to take over at least part of the operations and part of the work of oil expansion and development, including refining and marketing, previously conducted entirely by the private international oil companies. Agreements reached between private companies and governments to permit the former to search for oil in the latter's territories, increasingly stipulate that the government shall be involved in the operations once oil has been discovered. This involves not only the traditional share in the profits of the producing company, but also participation in the refining, transport and marketing operations of that company in its world-wide operations.

In the light of this inevitable development, our own National Oil Agency may, within the next decade, find itself involved in negotiations not only with companies which aim to operate in the British market, but also with the state oil authorities of the oil producing countries. In such a situation it would be essential for Britain, as a major consumer of oil products, to be represented in all negotiations by an *expert* and *official* oil agency, rather than by the employees of private companies whose main concern is quite naturally and logically aimed at securing short term advantages for their own company, rather than in protecting the national interest.

Over the long term, therefore, one would envisage a range of functions for the British National Oil Agency. It will be responsible for the control and direction of British oil import policy; for the control and direction of refining and distribution within the country; for ordering the systematic exploitation of whatever oil wealth may be found in our off-shore areas; and for representing Britain's interests in negotiations with other nations of the world with an interest in oil. Its potential importance to the British economy is thus self-evident.

References

1 D.L. Munby, 'Investing in Coal', *Oxford Economic Papers* (New Series), Vol II, No. 3, October, 1959, p.250

2 The Baade-Friendensburg report on Germany's supply of energy. Summarised in *Petroleum Press Service*, June 1962, p. 203

3 W.J. Levy 'Report on Coal-Oil Competition in Germany 1961'. Summarized in *Petroleum Press Service*, October 1961, p.363

4 P.H. Frankel and W.L. Newton, "The Location of Refineries", *Institute of Petroleum Review*, Vol 15, No 175, July 1961

5 A. Melamid, "Geographical Distribution of Petroleum Refining Capacities", *Economic Geography*, Vol. 31, No. 2, 1955, p.168

6 P.H. Frankel, "Oil Supplies in the Suez Crisis" *Journal of Industrial Economics*, Vol IV, No.2, February 1959, p. 273

7 *Petroleum Press Service*, Vol.42 , No.8 , August 1965, p.297

8 *Petroleum Times* Vol.68, 6 March, 1964, p. 107

9 *Hansard*, House of Commons, 1964–5, Vol. 715, col. 838

10 Department of Fuel and Power, *Fuel Policy* White Paper, (cmd. 2798), HMSO London 1965

11 *Hansard*, House of Commons, 1964–5, Vol. 716, col. 41

12 Department of Fuel and Power, *op cit*, para. 66

13 *Hansard*, House of Commons. 1964–5, Vol. 710, cols. 111–3

14 *Petroleum Press Service*, *op. cit.*, p. 297 (author's italics)

15 Monopolies Commission, *Report on the Supply of Petrol to Retailers*, HMSO, London p.23

16 Jeremy Bray, M.P. *The New Economy*, Fabian Tract, London, 1964

Chapter I – 2

Oil Policies in Western Europe*

Coal's Inability to Cope

Before the war the economy of Western Europe was, in the main, based upon the use of coal as its primary source of energy, with very little of the diversification into oil and gas that occurred in the U.S.A. Between 1939 and 1945, however, the European coal industry was very badly hit by wartime dislocation and destruction and much of the productive capacity in countries such as West Germany, Belgium and France was out of action because of difficulties either in the mining areas themselves or in associated transport facilities. Even in places which were not directly affected by land fighting or intense bombing – for example, most of Britain's coal-mining areas – the coal-mining industries were, nevertheless, affected by the running-down of facilities, the lack of capital investment and the difficulties in obtaining labour. Thus, in 1946 coal output in Western Europe totalled only 340 million tons (with Britain contributing nearly 60 per cent) and the prospects for a rapid growth of production were anything but bright. Most of the industrialized countries were thus brought face to face with a probable serious deficiency in the total energy supply required for post-war construction.

*** Originally published as Chapter 5 in the first edition of *Oil and World Power*, Penguin Books, Harmondsworth, 1970, Chapter 5**

Imported Oil Products come to the Rescue

In such a situation the possibility of obtaining oil supplies from overseas appeared to offer a sure and almost immediate solution to the impending energy crisis, and increasing quantities of petroleum products started to move to Western Europe from the U.S.A., the Caribbean and the Middle East. In this early post-war period, oil imports had to be largely in the form of immediately usable products, as most of the limited amount of refining equipment that had been built in Europe before the war had, along with the coal mines and the railways, suffered physical damage and could thus only be brought back into production as rebuilding took place. Moreover, rebuilding was, in any case, only going to solve part of the problem. The pre-war refining capacity of Western Europe had been small, as the economic advantage then lay in refining oil at source and in shipping the products to their markets in Europe. Not even relatively undamaged Britain could offer refining facilities to meet Europe's needs and it too had to import products from the U.S.A. and elsewhere.

Oil Refineries Encouraged

This dependence of Europe on imported petroleum products – paid for in part by U.S. aid – did not persist for more than a few years. First, the larger countries and then the smaller ones began to investigate the possibilities of curtailing the impact of growing oil imports on their balance of payments, as most of the oil available came either from dollar areas or through American companies and hence produced a serious drain on Europe's very limited dollar resources. The initial investigations concentrated on ways of encouraging the development of large oil refineries, so that the import of oil products could be substituted by the import of crude oil and its processing in Europe. This would ensure a reduction in the unit foreign exchange costs of Europe's oil requirements, even allowing for the dollar cost of much of the equipment required in the new oil refineries. Thus governments used both the carrot and the stick to persuade the oil-supplying companies to go along with them in the implementation of this policy. The companies were, for example, given loans at less than the then current rates of interest for financing the construction of refineries and they were guaranteed the availability of foreign exchange to import the necessary equipment. But, at the same time, they were told that once the refineries had been established preference would be given through tariff or quota arrangements to crude-oil imports over product imports, with the result that companies which failed to build refineries could well find themselves forced or priced out of the markets.

The companies, however, needed little encouragement to follow this official policy. Development in transport and refining technology, coupled with the rapid growth in European demands for oil, persuaded them that the establishment of refineries in Western Europe offered a lower-cost and more profitable way of meeting these demands than the traditional importation of oil products. The growth in the size of crude-oil tankers brought down the unit cost of crude-oil transportation very considerably, while the size of demand in the industrialized parts of Western Europe now meant that larger refineries could be constructed, thus ensuring the achievement of significant economies of scale in the new projects. And as oil moved in to take up markets in those areas of Europe which had formerly depended almost exclusively upon coal, so the rise in demand for fuel oil outstripped that of all other products. Indeed, it increased sufficiently to ensure the local use of all the fuel oil that the new refineries could make. In contrast, the pre-war demand for fuel oil had been so small (it had been unable to compete with indigenous coal) that even the small refineries had been obliged to re-export some of this product. This had meant higher transport costs on the operations overall, so making them more expensive than the alternative of refining the production near the sources of the crude oil.

Thus, from the points of view both of West European governments and of the large international oil companies there were strong economic motives for the construction of large oil refineries in the years immediately after the war. By 1950, the first of the post-war constructions – for example, at Fawley in the U.K., Pernis in the Netherlands and Marseilles in France – were already successfully operating, mainly on crude oil imported from the rapidly expanding fields of the Middle East. The economic arguments for refinery construction in Europe were soon to be reinforced by strategic and political ones. Before the war, the major oil-producing areas had been securely under British and French control, but now the Middle East was entering on a period of extreme nationalism in which the influence of Britain and France declined very markedly. The consequent political instability both of the region, in general, and of individual oil-producing countries, in particular, persuaded both the importing nations and the large international companies against dependence on refineries in such places. It became clear that it would be better to limit investment in the oil-exporting areas to that required in the oil-producing operations, and to spread the immediate risk of interruptions to supply by building refineries in countries where the degree of political stability and the guarantees against expropriation seemed greater than in the Middle East. Various events soon emphatically demonstrated the political and strategic considerations involved. These included the closure of the pipeline from Iraq to Haifa and the consequent closure of the refinery there, as a result of the formation of the state of Israel in 1948 and the Arab nations' refusal to supply crude oil. Shortly afterwards, in 1951, the Anglo-Iranian oil company was nationalized. As a result the country's oil-producing facilities and the world's largest refinery, at Abadan, were closed down for a period of more than three years.

As oil became more important in the economies of the Western European countries, the degree of risk they were prepared to accept over oil supplies became smaller. This was accompanied by an escalating interest in the rapid development of local refineries. Thus, by the mid-1950s Western Europe was, by and large, self-sufficient in refinery capacity, and the continent's main estuaries and other favourable locations were dotted with refineries (see Fig.II-2.1). The pattern of European trade in oil had now changed from one in which the main imports were oil products to one in which the crude oil requirements for the continent's new refineries were imported. By this time the annual saving of foreign exchange achieved by the development of a European refining industry was of the order of £50 million.

Oil Prices under Scrutiny

Western Europe, however, was not satisfied merely with the successful establishment of a refining industry. There was also serious concern over the international companies' controlled system of oil pricing, for this placed Europe under an unnecessarily high burden of foreign exchange costs for its growing requirements. In the post-war period there still persisted an element of the international oil cartel established by the major oil companies in the early 1930s as a result of the world depression. This had created a supply system which kept prices higher than they need have been in certain markets. It had also restricted the development of low-cost sources of oil because of the effect that the system had achieved in equalizing the delivered price of oil to Western Europe no matter what its source.

The system worked in roughly the following way. Posted prices for crude oil and oil products were established at U.S. Gulf ports and freight rates were established for the transportation of oil from these ports to the ports of Western Europe. Lower-cost oil from the growing producing nations of Venezuela and the Middle East was only made available by the international companies at the posted U.S. Gulf prices on a grade-for-grade basis. The delivered price of this oil sold to Western Europe was then, moreover, calculated by the companies concerned as though it was shipped from the Gulf of Mexico, rather than at a price reflecting the cost of transport from the country of production. Thus, oil in Western Europe was not only tied to prices which reflected the relatively high cost of the regulated production of oil within the U.S.A., no matter what the costs of production were in the Middle East or in Venezuela whence most of Europe's supplies originated, but they also reflected higher than actually incurred transport charges.

Western Europe thus paid, on average, about $4 for each barrel of crude oil imported and even higher prices for imports of oil products. As might well be imagined, the governments of Western Europe were not at all happy with this situation. In 1950, as a result of pressure by these governments and by the U.S. authorities responsible for the Marshall Plan, by means of which some of Europe's oil imports were being financed, the first break in this system occurred when the companies agreed to post separate prices for oil produced in Venezuela and in the Middle East. Though these posted prices were at first not much lower than those of the equivalent grades at the Gulf ports of the U.S.A., they quickly became so as oil price levels in the U.S.A. rose in the post-war inflationary situation. The eventual effect of this was, of course, a more favourable situation for Western Europe as imports were increasingly derived from non-U.S. suppliers. In addition, separate freight rates reflecting the use of larger, lower-cost tankers were established for direct movements from the Caribbean and from the Middle East. As a result of these important changes the average delivered price of oil per barrel in Western Europe came down to under $3.50 in the early 1950s.

Even this lower price was still well above the supply price of oil – that is, the lowest price at which companies concerned in oil production would need to sell this oil in order to achieve an adequate return on their investments, including due allowance for the degree of risk involved in the enterprise. The existence of such a favourable general price level made the oil industry a highly profitable one and thus encouraged the entry of new producers in many parts of the world. The potential producers looked particularly to marketing opportunities in Western Europe, where, with rapidly expanding economies and the increasing substitution of oil for coal, the consumption of oil was increasing at a rate of some 15 per cent per annum. New crude supplying companies – both American and European – started to ship oil to Western Europe, with the result that the posted-price system and the freight-rate system, which together had fixed

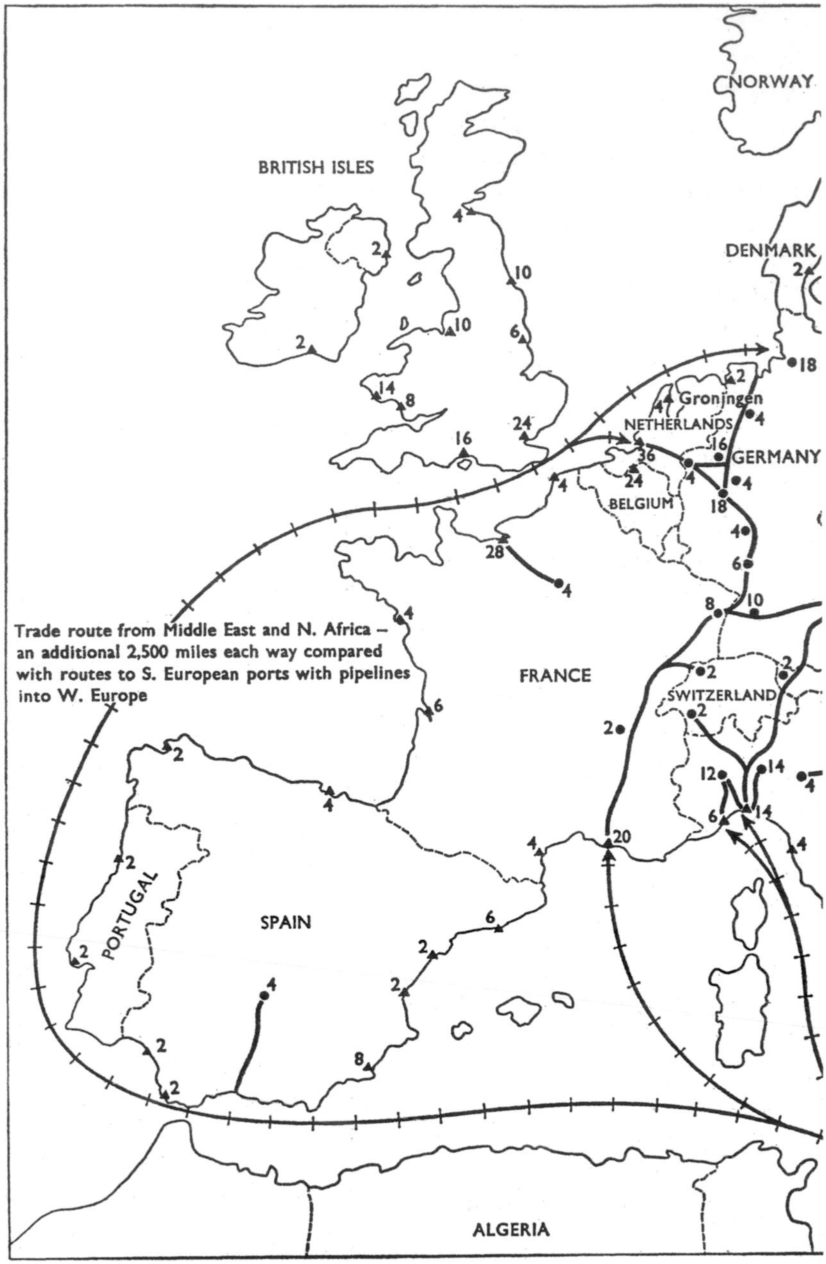

Fig I-2.1 The Oil Industry in Western Europe in 1969

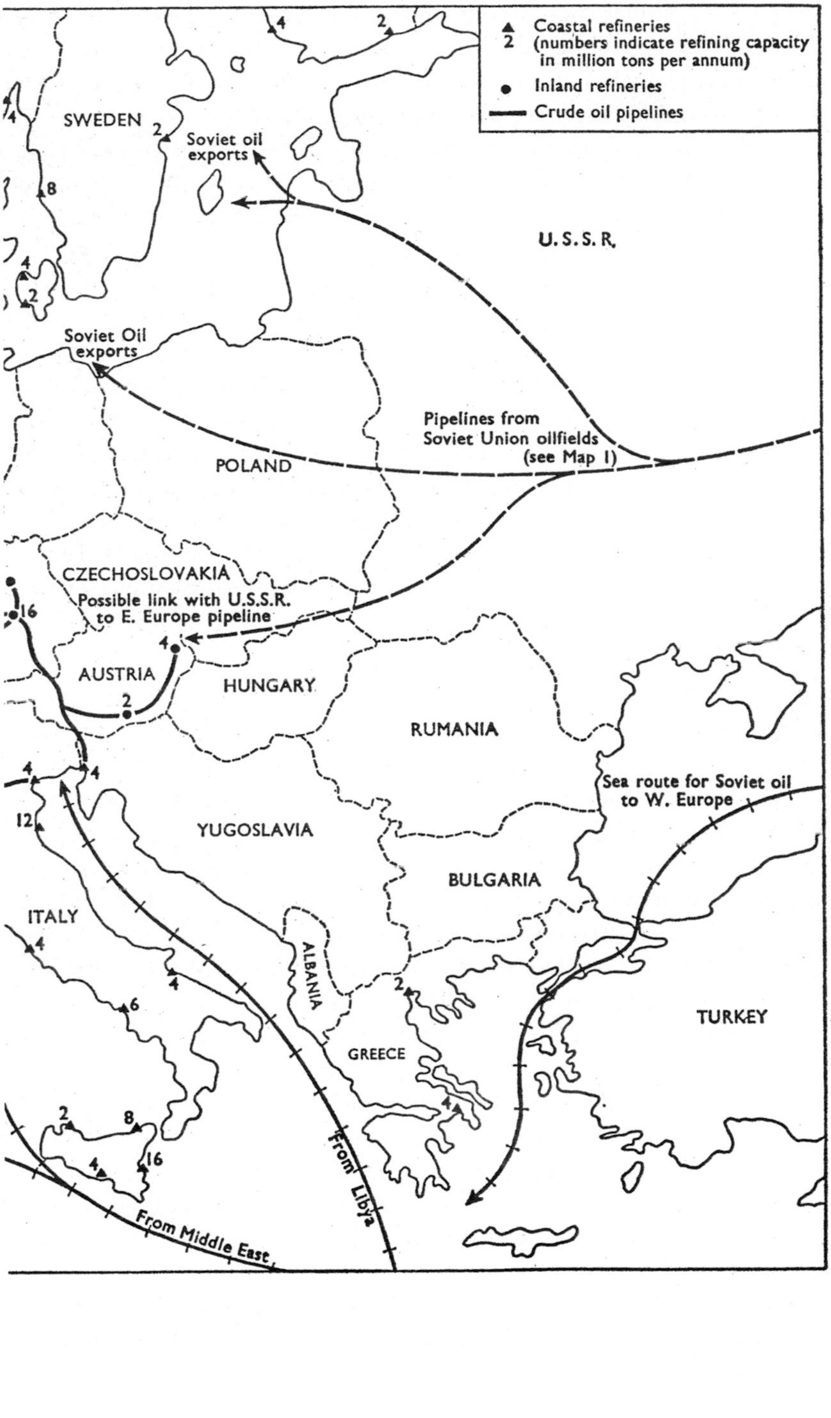
▲ Coastal refineries
2 (numbers indicate refining capacity in million tons per annum)
● Inland refineries
Crude oil pipelines
SWEDEN
Soviet oil exports
U.S.S.R.
Soviet Oil exports
Pipelines from Soviet Union oilfields (see Map 1)
POLAND
CZECHOSLOVAKIA
Possible link with U.S.S.R. to E. Europe pipeline
AUSTRIA
HUNGARY
RUMANIA
Sea route for Soviet oil to W. Europe
YUGOSLAVIA
BULGARIA
ITALY
ALBANIA
GREECE
TURKEY
From Libya
From Middle East

an 'appropriate' price for the delivery of any grade of oil to any port in Europe, came under such pressure that they eventually broke down.

Marketing companies and refineries in Europe not tied to the major international oil companies which 'organized' the systems started to shop around for their supplies and increasingly found them at prices well below those quoted officially on the basis of posted prices and agreed freight rates. At first, the activities of such companies usually affected only the fringes of national markets, but as their business started to boom the international oil companies themselves were forced to respond in order to keep their own local companies competitive and able to make profits on their local operations. They were thus obliged to lower the prices at which they transferred oil to their affiliates in European countries, with the result that practically the whole of the market and not just the fringe business started to enjoy lower prices – which naturally led to even greater growth in oil consumption. Indeed, since the late 1950s oil use in Europe has gone ahead at a more rapid rate than ever before, steadily forcing coal into a position where it makes a steadily-decreasing contribution to Western European energy. In 1966 oil replaced coal as the most important source of Western Europe's energy.

Country Differences

Western European countries thus achieved both of their immediate post-war objectives, viz. of increasing the size of their own refining industries and of securing their oil imports at prices more closely related to the cost of production in the areas from which most for Europe's oil originates – the Middle East and, more recently, North Africa. There have, of course, been country-to-country variations in the implementation of these policy objectives. Italy, Scandinavia and Western Germany pushed the policies to their ultimate conclusion, whereas Britain and France tended to remain relatively isolated from this main stream of development as a result of their own particular interests in wider aspects of the international oil industry.

France gave preference to oil from the franc zone, notably Algeria, and maintained prices at a level which made this possible. The government exercised strict control over the oil industry throughout the post-war period (as, indeed, it was already doing before 1939), in order to ensure that national policies were respected by the companies concerned.

Britain, meanwhile, the headquarters of two of the international oil companies, tended to view itself as having a vested interest in the maintenance of prices on an international level and this, coupled with the dominant 90 per cent share of the international companies in the

British domestic market, ensured, at least until 1965, that the country's import prices remained well above the levels applying throughout most of the rest of Europe. In 1965 the average import price per barrel of oil into Britain remained some 25 per cent above the average import price into the rest of Western Europe – to the detriment of the country's balance of payments and a continuing factor in explaining Britain's economic difficulties. However, competition eventually came even to Britain as new companies moved in to take advantage of its highly profitable oil market. The established companies were thus obliged to reduce their prices in order to meet the competition. By the late 1960s the oil market within Britain was thus probably not much less competitive than that in most other European countries, although Britain had still not enjoyed the balance of payments advantage of reduced import prices of crude oil to the same degree as Germany, Italy and Scandinavia.

The Consequential Decline of Coal

But there was another aspect of the energy situation in Western Europe that affected attitudes to oil, viz. the impact of oil on the continent's indigenous energy supplies. This, of course, was limited to those countries where coal had been important. As we have already shown, the difficulties over indigenous coal in the early post-war period led to oil being welcomed enthusiastically as a means of overcoming the energy shortage. But the coal industries of Britain, West Germany and Belgium and, on a smaller scale, those of France and the Netherlands steadily overcame these difficulties. Re-capitalization of the mines enabled mechanization and later automation to be introduced. Thus production and labour productivity increased and the continent's reduced demand for coal could easily be met. By this time, however – in the later 1950s – the demand for coal was already declining steeply as a result of its substitution by oil in many end uses. By 1958 the situation of post-war energy shortage – in which every ton of coal that could be won from the ground was more or less assured of a market at a price which was government controlled in order to prevent it rising too high – had changed.

Thereafter, the European coal industry had to face up to the prospect of decline occasioned by its inability to compete with oil in a wide variety of uses – in domestic home-heating, in electric-power generation and in basic industries such as iron and steel, etc. Coal from the high-cost mines of the continent proved impossible to sell and thus every major coal-producing nation had to have a programme of mine closures, usually accompanied by measures designed to prevent social distress in the mining areas. But the programme of closures in almost every case moved too slowly to keep the relationship between supply and demand in equilibrium, so that other measures were required in order to prevent the situation getting out of hand. Thus, the coal-producing nations were obliged to give protection to their coal industries, either through subsidies on coal production or by the imposition of taxes on oil products competing with coal in end uses other than transport.

For example, in Britain a policy of rationalization on the part of the N.C.B., whereby many mines were closed down and production reduced from its post-war peak of 227 million tons in 1957 to only 155 million tons in 1969, was insufficient to produce equilibrium between the oil and coal industries. In the early 1960s the government was thus obliged to put a 2.0d. per gallon tax on all oil used for heating purposes. This tax – since raised by stages to 2.42d. – may seem to be insignificant compared with the contemporary tax of 3s.9d. on each gallon of petrol used for motoring, but even so it represented a purchase tax on the ex-refinery price of fuel oil of up to 50 per cent. According to estimates made by the Ministry of Power the tax maintained the output of coal at a level some 18 million tons per annum higher than it would otherwise have been. Such taxation measures – common to all the major coal-producing nations of Western Europe – were sometimes accompanied by other measure designed to force "captive customers', such as government departments and other state-owned entities, to burn coal, even when oil was cheaper. But all this did no more than ameliorate the adverse situation for coal, whose economic viability has been seriously and probably irreversibly undermined by the growth of oil.

Even now – in 1969 – the process of substitution of coal by oil has certainly not come to an end. Independent evaluations made in the United Kingdom indicate that, on the basis of stable oil prices coupled with continued upward pressure on the costs of coal production, no more than 80 to 100 million tons of Britain's coal output is likely to be competitive with oil unless protection is afforded to it. This situation was paralleled throughout the continent with the prospect of the virtual close-down of the French and the Dutch coal industries; and reductions of over two-thirds in Belgium's coal industry and by 60 per cent of the West German industry. Oil provided over 55 per cent of Western Europe's total consumption of energy by the end of the 1960s. Western Europe can, therefore, no longer be termed as having a coal-based economy, but rather as having an economy based on the availability of two sources of fuel and power, with the competitive position running strongly in favour of yet more oil use throughout almost the whole of the continent.

A Changed Geography of Energy Supply

Over the post-war period the growing use of oil in an increasing number and variety of end-uses, technical developments in refining and inland transportation and changes in the political structure of Western Europe, such as the establishment of the European Common Market, have all combined to produce a changing geography of the oil industry within the continent. Before the war the pattern was a simple one. Oil products were distributed in relatively small quantities by road and rail to the centres of inland demand from the coastal import terminals or from the small refineries at ports such as Rotterdam, Hamburg and Marseilles. The immediate post-war development confirmed this general pattern, but by then the oil was, of course, moving mainly from the expanded refining centres which were being established at convenient points for crude-oil imports around the coasts of Europe (see Fig I-2.1). In some cases, the quantities involved grew sufficiently quickly to justify the construction of products pipelines to take the oil from the refineries to the major internal cities of the continent. Important examples of this development were the oil-products lines laid from refineries at the mouth of the Seine to Paris and, in the U.K., the lines from refineries on the Thames and the Mersey to distribution centres in the Midlands.

At a later stage, however, it became clear that this method of supplying oil products was not the cheapest way of reaching some inland centres, with their growing requirements of fuel oil which could not be pipelined economically over long distances. Thus possibilities were investigated of establishing inland refining centres which would be supplied with crude oil through pipelines running from import terminals on the coasts. First, refineries were built in the Ruhr, fed by crude-oil pipelines from the Rotterdam area and northern Germany. More recently, in the mid-1960s, refining centres were established in Bavaria, eastern France and Switzerland, dependent upon crude-oil supplies coming in by pipeline from southern European ports. Marseilles and Genoa have been particularly important in this respect. The use of crude-oil import terminals to serve the refineries in the centre of Western Europe does, of course, markedly reduce the length of ocean transportation for this crude oil – it had only to be brought the relatively shorter distance from the crude oil supply points in the Persian Gulf/Eastern Mediterranean or from North Africa to the southern Mediterranean ports of Europe. Tankers are thus saved a journey of some 4,500 miles around the coasts of Western Europe to the ports of northern France, the Netherlands and Germany (see Fig II-2.1). With the continued development in pipeline technology it could well be that, within a few years, even northern Europe (for example, the northern parts of Germany and Scandinavia) are also fed by pipelines coming up from the Mediterranean, rather than by tankers bringing the crude oil in to nearby deep-water ports. Though the growth of these new patterns has depended largely upon the developing technology of pipelining and refining, some of the change has undoubtedly arisen from the movements towards economic and political integration in Western Europe whereby the barriers against the movement of goods between the nations of the Common Market have been largely eliminated. In this situation, geopolitical problems associated with frontier crossings which might have arisen with the international movement of oil within Western Europe have been greatly reduced.

Oil and European Integration

But this inter-relationship between European integration and the oil industry, though important in contributing to lower costs for the industry and thus to the possibility of lower prices to consumers – which is the sort of favourable effect that customs unions are 'supposed' to have – has not provided the main topic for discussion between European 'oilmen' and the politicians and administrators of the Common Market. When the latter was formed oil was still relatively unimportant. Its predecessor, the European Coal and Steel Community, while being made responsible for the development of a European coal policy, was given no such authority over the oil industry, which at that time was probably considered to be of too little importance to justify such governmental interest. But since the oil industry's rise to importance in the mid-1950s, many complicated and long-lasting discussions have taken place on the establishment of an overall energy policy, including oil, for the member countries. However, because of the basic disagreements which have arisen between the interests of those members of the European Economic Community with no indigenous energy supply industries to protect and those which have, energy policy has remained one of the unresolved and difficult sectors on the pathway towards European economic integration. There can be little doubt that an agreed energy policy for the whole of the area will ultimately emerge and that when it does the political changes generated will eventually have repercussions on attitudes towards oil within the community. The trend currently is in the direction of more effective and more comprehensive government control. The French experience in this respect, dating back some thirty years, perhaps gives the best idea of what could happen.

Control of the Oil Industry

It seems likely, therefore, that a common energy policy for the E.E.C. will involve some degree of protection for the remaining domestic energy industries and that the oil industry will continue to be run by individual companies – both public and private – but along lines which are either laid down or subject to supervision by European civil servants. Such control – which is reflected in similar advancing measures of government control among the non-member status of the Common Market – should not even be too upsetting to U.S. firms which are used to effective and comprehensive government intervention at home. It will, moreover, produce a situation within which they can be certain of securing an adequate return on their investments, and thus take away much of the normal commercial risk to which they would otherwise be subjected. This will apply particularly in the case of 'Community' companies – that is, companies whose parent company 'belongs' to a member nation, as opposed to foreign-owned companies – for they are likely to be accorded preferential treatment on both political and strategic grounds.

The development of comprehensive energy policies in the Common Market and in other countries of Western Europe, has, nevertheless, generally been opposed by the oil companies, which have seen them as a means of restricting their growth and freedom of action., But in the 1970s such policies could well become a means whereby the interests of the greatly expanded oil industry – with an established position to defend – are given some protection against a more recent development in the energy economy of Western Europe, which could otherwise threaten not only their growth prospects, but also their future levels of profitability.

The Threat from Indigenous Natural Gas

This development is the recently discovered availability of large quantities of natural gas within Western Europe itself (see Fig II-2.2). Gas has been available as a primary energy source throughout most of the post-war period in northern Italy and south-western France, but the quantities involved have been relatively small and capable only of providing a limited amount of the energy required in limited areas of the two countries in a period in which energy demand was moving ahead very rapidly. Thus, the oil companies hardly felt the effect of the competition. The gas field of Groningen in the northern-most part of the Netherlands is, however, a very different matter. This field was not

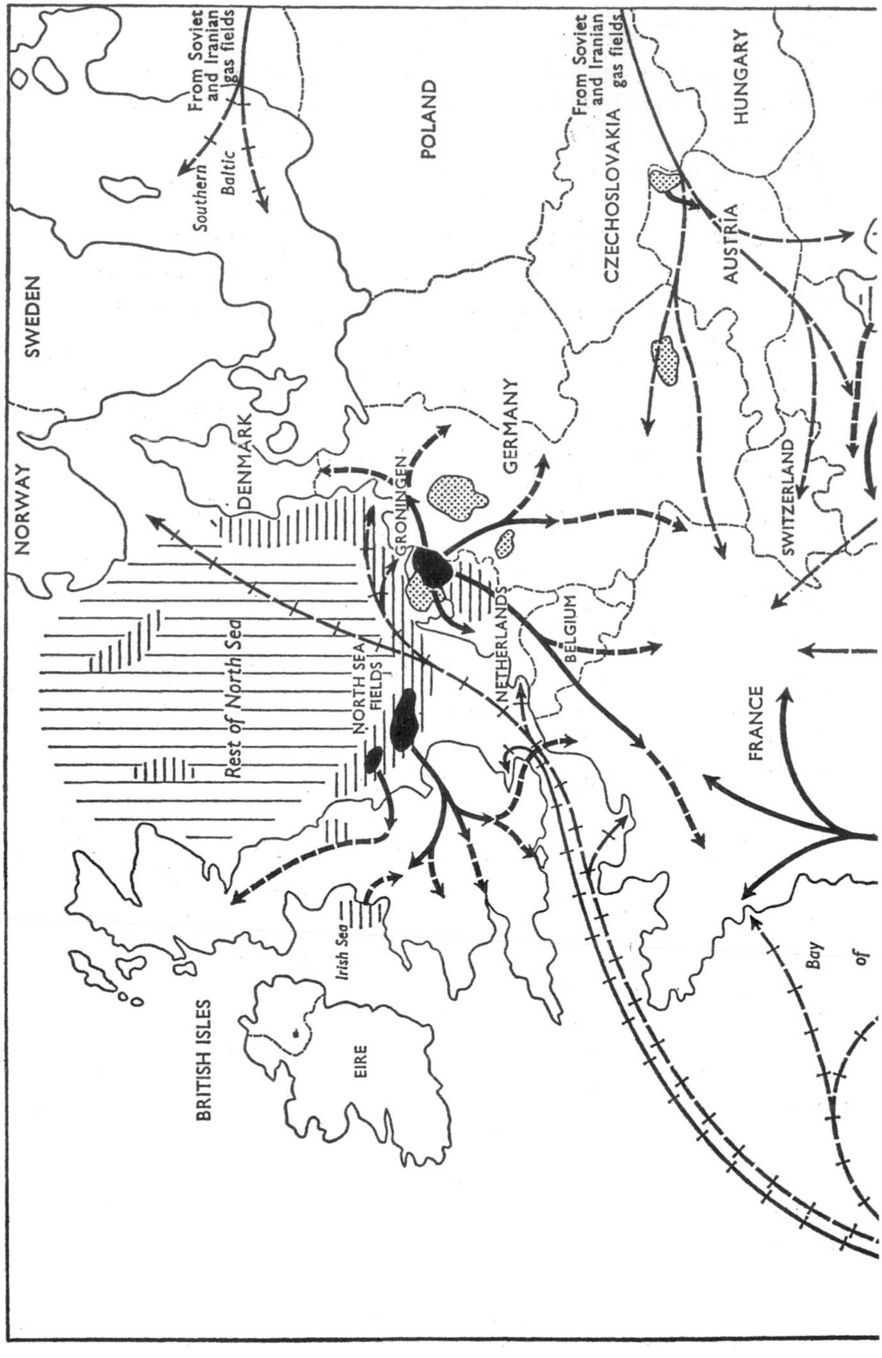
From Soviet and Iranian gas fields
Southern Baltic
POLAND
SWEDEN
NORWAY
DENMARK
GRONINGEN
GERMANY
From Soviet and Iranian gas fields
CZECHOSLOVAKIA
HUNGARY
AUSTRIA
SWITZERLAND
Rest of North Sea
NORTH SEA FIELDS
NETHERLANDS
BELGIUM
FRANCE
Irish Sea
BRITISH ISLES
EIRE
Bay of

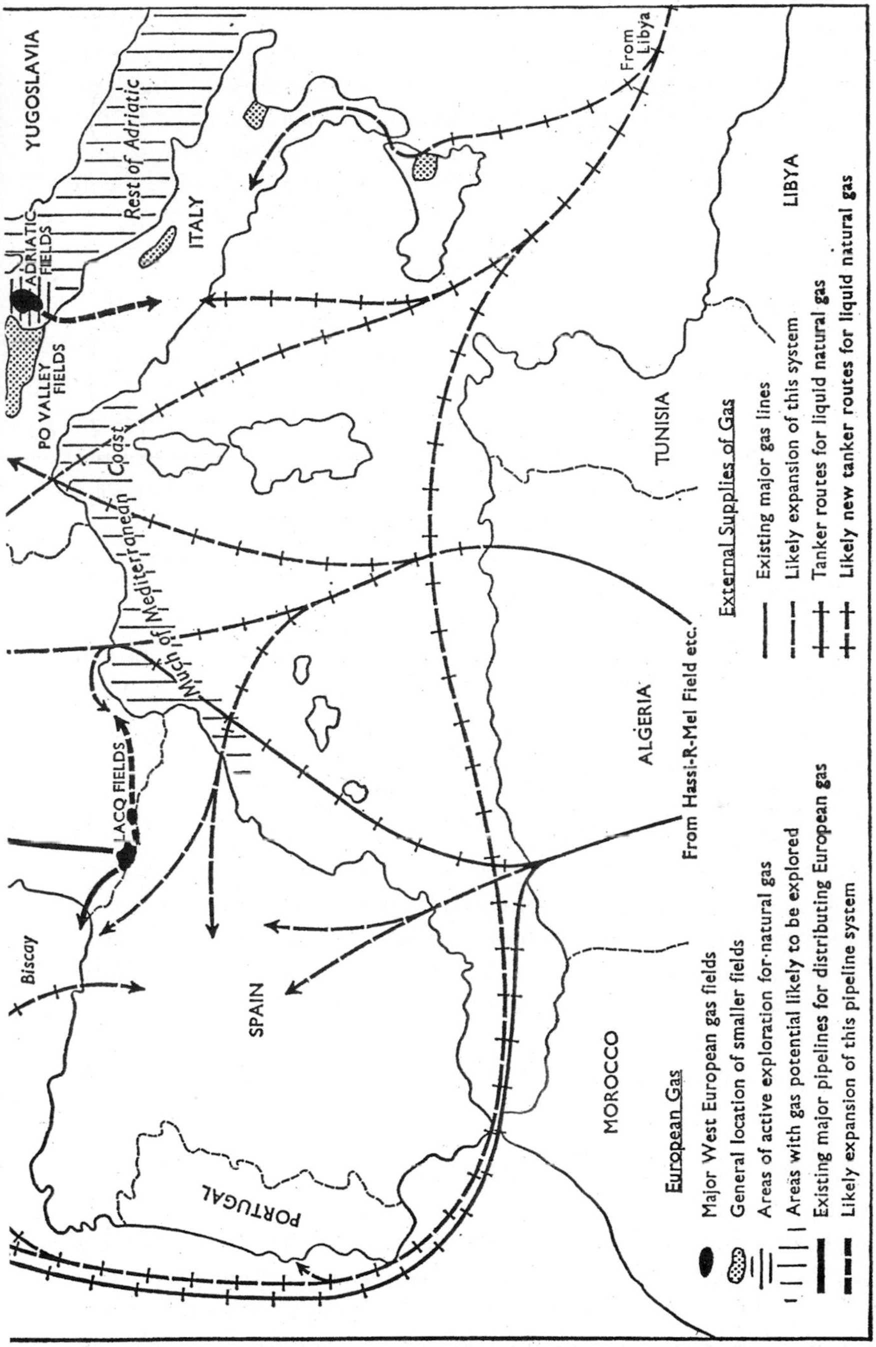

Fig II-2.2: Natural Gas in Western Europe

discovered until 1959 and was only slowly evaluated. It is now known to be the world's largest exploitable gas field outside the Soviet Union, exceeding in size even the largest fields in Texas. It has already been shown capable of producing up to 100 billion cubic metres of natural gas daily (equivalent of over 90 million tons of oil per annum). As the premium or high-price markets for this gas in the Netherlands and nearby parts of France, Belgium and West Germany are satisfied, it will increasingly be made available for use in industrial processes and by other consumers who turn to it in preference even to the cheapest fuel oil.

While the most important source of western European gas was confined to the Netherlands, however, the situation never threatened to get out of hand so far as competition with oil was concerned, for the Dutch field was very much in the hands of two of Europe's major oil companies, Shell and Esso. With the cooperation and agreement of the Dutch government, also a part-owner of the resources and also anxious to maximize its short-term gains from the gas, it was priced at such a level for export sales that it would not become quickly attractive to major oil consumers. The joint state/oil company enterprise – NAM – responsible for marketing the gas in foreign countries – France, Belgium and Western Germany – has thus ensured that its price at the Dutch border was high enough to limit its ability to compete against fuel oil, once transport and distribution costs to foreign consumers had been added on.

Competition for Groningen Gas

This initial prospect for the orderly marketing of gas is now undergoing change. This arises in the first place from the development of the North Sea's gas resources. When it became evident, as a result of the discovery of the Groningen field, that the whole of the North Sea Basin was a potentially rich gas-bearing basin, the nations of Europe with shorelines on to the North Sea got together to discuss the problem of ownership of these resources. They quickly reached an agreement on the division of the North Sea into a set of nationally controlled areas. Exploration for the resources lying beneath the bed of the sea is now being conducted in areas belonging to Britain, the Netherlands, West Germany, Denmark and Norway.

By 1969 the search under the British section of the North Sea had already been successful enough to determine a potential productive capacity equal to more than half of that of which the Dutch were already capable. By 1975 there seems little doubt that Britain will be producing the energy equivalent of some 40 million tons of oil a year in the form of North Sea gas. In the case of Britain the quantity of gas available, the fact that it is being sold by a national entity entirely outside the control of the oil companies and the relatively low price at which it is being sold will certainly mean that gas will be used to replace some of the fuel oil used by consumers who turned to it in preference to coal in an earlier part of the post-war period. This will have the effect of slowing down very significantly the growth of the oil industry in Britain over the next ten to fifteen years, as gas moves up to take a position of providing as much as one-quarter of the total primary energy requirements. Repetition of the Dutch and British successes in the search for natural gas in some of the other continental shelf areas could well produce enough gas in north-western Europe to provide one-fifth or more of the total energy demand within the whole of this densely peopled and industrialized part of the continent.

Gas Imports to Europe

For more easterly and southerly parts of Western Europe there are yet other sources from which this third primary source of energy can be made available. This involves importing international gas on a very large scale both from the known gas fields of Northern Africa – via a pipeline or liquefied natural gas tankers from the massive fields of Algeria and Libya across to southern Italy and possibly southern Spain – and also via large-diameter pipelines running into Austria, Switzerland and West Germany and Italy, too, from the virtually limitless gas fields of the U.S.S.R. (see Fig II-2.2). The development of pipeline technology has made this very long-distance transmission of gas into the markets of Western Europe economically feasible so the gas will eventually become available in the consuming countries at prices which enable it to compete with oil products in a wide variety of end-uses.

Europe's Three-fuel Economy

Thus, Europe's energy economy of the future may well become based on three fuels, in much the same way as have the fuel economies of the U.S.A. and the U.S.S.R. This must be partly at the expense of the existing markets for oil – markets, that is, which oil secured from Europe's other indigenous resource, coal, in the earlier post-war period. Such a development would be welcome in almost every country of Western Europe, for many governments have, from time to time, expressed great concern over their countries' undue reliance on imported energy from politically unstable parts of the world.

While oil, however, remained as the only alternative to coal, and an alternative which offered energy to industry, commerce and domestic consumers at prices well below those which could be provided by domestic coal, there was little that Europe could do, or could afford to do, to reduce this dependence, except by ensuring the expansion of the refining industry at home, and the purchase of oil supplies from as many different overseas sources as possible. Natural gas introduces a new variable, for not only is it indigenous, and hence welcome as a means of reducing the continent's high degree of dependence on overseas oil-producing countries, but it is also cheap to produce – even in comparison with the much-reduced oil prices achieved in Western Europe by the mid-1960s. Western European governments will thus want to ensure that its development goes ahead as rapidly as possible and for this reason will undoubtedly continue to accept that the exploration and development work should be undertaken by companies with long experience and the requisite expertise.

Such companies are, in the main, the oil companies which we have met previously as producers of oil and gas in other parts of the world. But in other parts of the world, at least until very recently, these companies have been able to secure the ownership of the energy resources produced from their operations in return for a payment of royalties and other taxes. They have then been in a position to charge whatever prices for their products the markets would bear. Western European countries, on the other hand, have a long history of national control over basic sectors of their economies and thus seem likely to insist on the oil companies operating only as contractors to produce the gas. The gas will then be purchased from the companies at an appropriate supply price designed to give the latter an adequate return on their investment – no more, no less – and then made available through public utilities to consumers at prices which do no more than reflect the supply price of the product, plus the costs of transport and of distribution. In such circumstances, there seems little doubt that natural gas will greatly enlarge its markets, partly at the expense of imported oil, and so provide the main new element in the 1970s and the 1980s within Europe's energy economy.

The Limitation on Oil's Role

When the potentialities of natural gas are coupled with the increasingly likely large-scale development of nuclear power stations (a phase already under way in Britain, which will complete at least one large – 1,200 MW or more – new atomic power station each year in the 1970s), whose construction will, at least in part, replace oil-fired thermal power stations, Western Europe can be seen to be on the brink of containing the increase of oil's contribution to its energy economy. And it seems not impossible that the markets in absolute terms for some products will decline. This is particularly so in the case of fuel oil – against which coal will still be protected especially in electricity generation – which will lose potential outlets to nuclear-power developments and for which natural gas will be substituted in many areas and end-uses. This had, indeed, already started to happen in the Netherlands by 1968 as natural gas became the preferred fuel in most end-uses. However, the total oil requirement is likely to continue to rise slowly overall because of its use in sectors – particularly motor transport – where there are as yet no satisfactory alternatives to oil-fired internal-combustion and diesel engines. This likelihood of a changing pattern of product demand will necessitate changes in refining programmes in order to reduce the fuel-oil output, hitherto maximized in response to earlier market demands. Refining may thus become even more closely tied in with petrochemicals (whose feed-stocks and fuel requirements can absorb the refinery output for which other markets do not exist) and thus re-establish the preference for refineries in coastal locations, where petrochemical developments are usually more appropriate.

The changes will have serious repercussions for the oil industry itself, but for Western Europe its post-war oil crises – arising out of political difficulties in the supply areas of the world – should quite soon be a spectre of the past. In this respect there are also, of course, the recently improved prospects of diversifying supply sources by importing Australian, Alaskan and northern Canadian, as well as African oil. After a couple of decades of great concern over oil by European planners and governments, the continent can now afford to adopt a somewhat more relaxed attitude. Coupled with the lessening of fears over the security of supplies, there can also be a relaxation of the European-wide surveillance of price levels. Given the pressure of competition both between oil supply sources and between oil products and other energy sources, oil prices are more likely to take adequate care of themselves – that is, as seen from the point of view of European governments concerned with oil import costs, and of European consumers concerned with the cost of oil at factory gate or delivered into a domestic storage tank.

Europe's 'battle for oil' is now all over bar the shouting. Within the framework of constraints introduced by overall energy policies (national and European Community) and of intensive competition for markets, the companies concerned in the trade are going to have to be content with small profit margins on the highest possible volume sales they can achieve. And with such small margins – which, moreover, are likely to be constantly subject to pressures arising from increasing competition from other energy sources – the companies concerned will need to devote increasing attention to the problems of alternative continental-scale supply and distribution patterns. A wrong choice in this respect could easily eliminate the profit element entirely. The changes in the technology and the economics of refining and transportation and the changing geographical and product requirement patterns of oil demand further complicate the situation. So much so, in fact, that there seem to be striking differences of opinion between companies, which, though making use of the same information, arrive at different conclusions. Thus, for example, one company (Gulf Oil) plans to serve its European refineries indirectly through a massive terminal in Southern Ireland at which oil will be received in 300,000-ton tankers and dispatched to various European countries in 50,000–100,000-tonners. Another company (C.F.P.) sees the trans-European (south to north) pipeline system, with refineries strung out along it in consuming centres, as the primary pattern of distribution at which to aim. As a result of such contrasting plans, the geography of the European oil industry seems likely to become even more complete in the period that lies ahead.

Finally, the structure of the industry poses another set of problems. In particular, its low degree of 'Europeanization', in face of the still-increasing private U.S. companies' dominance of the European oil scene, creates a situation which may not be acceptable to a large part of European political opinion in a gradually strengthening economic framework. In the 1970s this could lead to some efforts to 'Europeanize' the industry in a way not dissimilar from the 'nationalization' of the older-established energy industries in an earlier period of Europe's political history.

Bibliography

Apart from publications on oil by national governments, the European Common Market, the Council of Europe, the Organization for Economic Cooperation and Development and the U.N. Economic Commission for Europe also publish regular studies on oil in Europe. An early E.C.E. document, *The Price of Oil in Western Europe* (U.N., Geneva, 1955) was one of the most influential in affecting European policies towards the industry. Surprisingly, however, there is not yet a book which deals comprehensively with the rapid growth of the European oil industry in its economic and political environment, but

W.G. Jensen, *Energy in Europe, 1945–80* (Foulis, London, 1967), goes part way towards this. A European who had a marked impact on the oil industry was Enrico Mattei, who ran E.N.I. until his death – see P.H. Frankel's *Mattei Oil and Power Politics* (Faber & Faber, 1966). Europe's oil difficulties arising from political and military upheavals in its main supply area have been analysed by H. Lubell in *Middle East Oil Crises and Western Europe's Energy Supplies* (John Hopkins Press, 1963). *The Petroleum Times* (published fortnightly in London) is particularly concerned with European oil and gas developments.

Chapter I – 3

Europe's Oil Dependence*

Introduction

The dangers and difficulties for Western Europe inherent in its heavy dependence on imported oil for its energy requirements are well known. The winter of 1971–71 saw the beginning of a new series of danger periods for energy supplies, marked by a combination of upward pressures on oil prices occasioned by the much higher taxes on oil production in the oil exporting countries and the resolve of the oil companies to pass on these increases and more, to their customers. The latter, at the same time, are at some risk of their supplies being cut off should the countries and companies not get what they want. Future danger periods of this kind hardly seem likely to coincide again with the highly favourable offsetting circumstances such as those which prevented trouble over the last year, viz. an exceptionally mild winter throughout most of the continent which eliminated much of the normal increased demand for heating oils, the lack of as much economic growth as had previously been expected throughout Western Europe which meant that the demand for energy was well below that predicted, and, for the first time, the availability of large supplies of Groningen natural gas to many customers in the Netherlands, Belgium, West Germany and elsewhere at prices well below the going prices for other forms of energy so providing them with a substitute for oil. As a result, most of Western Europe emerged from the 1971–72 winter with oil prices well below the

*** Published in the *National Westminster Bank Quarterly Review*, London, August 1972, pp.6–21**

levels that the new higher international crude oil prices would have indicated as likely and with oil tanks throughout the continent full of products that customers had not needed to buy. Inadvertently, and largely at the expense of the oil companies, Western Europe must have very nearly achieved the 90 days' oil stocks that the OECD Oil Committee has recommended should be gradually built up to give some protection against emergencies.

In the likely absence of a similar combination of favourable circumstances in the future, Western Europe may reasonably expect oil supply and pricing problems in the next few seasons of peak energy demand. Pessimism on this score is further compounded by expressions of opinion on the possibility of the world's oil resources being incapable of sustaining even the short-term rate of growth in production, to meet all the demands made upon it by a developing world. This pessimism is excessive. The current anxieties in the United States about the 'energy gap' which have affected thinking on the subject in Western Europe, are really concerned with the sorts of energy-supply problems that we have been living with ever since 1945, when there was the first realization that Europe's indigenous energy resources were not going to be able to meet the energy needs of an expanding European economy.

Indigenous Energy Ample

Today, paradoxically, we stand on the threshold of a period when Europe will enjoy the energy advantage that the United States has always – until now – enjoyed, viz. a resource base of low-cost indigenous energy capable of sustaining all, or almost all, of the continent's needs. Thus, looking beyond the next few years, we can predict a future in which Europe's energy options exceed those to which we became used over the last 15 years and the limitations of which still pervade our thinking. In brief, we are in the process of moving from being a continent deficient in energy (or at least deficient in energy which can be sold at reasonable prices for, it should be remembered, indigenous coal has not been abandoned because it is no longer available, but simply because it is too expensive in relation to imported rivals), to one in which the bulk of our needs will be derived from oil and gas fields located within the geographical confined of Western Europe itself.

Table I-3.1 presents the 1970–80 energy supply outlook, first as it has been generally forecast and as it is still portrayed in existing official estimates; and secondly, as it could now develop, given the rapid exploitation of Western Europe's newly found oil and gas wealth. The result of the change would be a complete reversal by 1980 of the one to two relationship between indigenous and imported energy. Beyond that date there is the prospect that the role of indigenous energy will grow steadily stronger as, first, nuclear power finally achieves large-scale development throughout the continent, so increasing the share of primary electricity above the revised figure of 9 per cent shown for 1980. And, second, as the amounts of oil and gas produced within Western Europe rise at least fast enough to sustain the increasing demand for these dominant energy sources and thus limit imports of oil and gas to no more than the volumes needed in 1980. Even in 1980 these will be below the peak figures of imports required by Western Europe in the mid-1970s.

Assumptions Made

There is, of course, nothing axiomatic about these developments as they depend upon decisions, both by European governments and by oil and other companies, to exploit the North Sea hydro-carbons province at a rate rapid enough to enable the annual production figures indicated here to be attained. However, we shall here assume that both economic and political decisions, together with the capital resources of Western Europe, will ensure that such production levels are achieved – unless the 'conservationists' somehow persuade us that we should be better off if we left the oil and gas in the ground and continued to depend on resources brought into Europe from other parts of the world!

Here we are concerned with two other questions. First, whether there are yet sufficient indications of oil and gas resource bases adequate to sustain the production levels of which we are talking; and, second, if so, what the consequential impact of indigenous oil and gas development will be on overall supply-patterns of Western Europe's oil needs.

The first issue is that of the resources of natural gas which in cautious official thinking are still presented, as shown in Table I-3.1, as capable of meeting no more than 10 per cent of total Western European energy demand by 1980. The pessimism appears to be derived from earlier post-war experience with gas discoveries in Austria, Italy and France where relatively small and isolated occurrences of natural gas were "over-sold', with consequential later difficulties in maintaining the production levels to which the fields had been committed. In contrast, over-optimism has never been a fault in the evaluation of the resources and production potential of the massive Groningen field in the Netherlands whose total resources have been more or less continually up-rated over the last decade and whose currently declared reserves of over 2,300 milliard (10^9) cubic metres have so far been barely touched by

TABLE I-3.1
Western Europe – Changes in energy consumption patterns 1970–80 arising from greatly increased indigenous oil and gas production

(in millions of metric tons coal equivalent – mtce)

	1970		1975				1980			
	Actuals		Official estimates		Author's estimates		Official estimates		Author's estimates	
	mtce	%	mtce	%	mtce	%	mtce	%	mtce	%
Total	1,475	100	1,700	100	1,750	100	2,250	100	2,300	100
of which										
a. Indigenous										
Primary electricity	40	3	75	4	55	3	210	10	150	7
Coal, etc	385	26	315	19	250	14	205	9	200	9
Oil	25	2	40	2	100*	6	50	2	500†	22
Natural gas	95	6	150	9	355‡	20	215	10	500§	22
	545	37	580	34	760	43	680	31	1,350	60
b. Imported										
Coal etc.	45	3	55	3	50	3	80	3	50	2
Oil	880	60	1050	62	880	50	1450	64	775	33
Natural gas	5	–	15	1	60	4	40	2	125	5
	930	63	1120	66	990	57	1570	69	950	40

* *Estimated by totalling the existing levels of production from existing fields, projected production from Ekofisk and other fields in Norwegian waters, plus British North Sea output from Forties and Auk fields, small amounts from other North Sea Fields and new Spanish offshore fields.*

† *This equals 330 million tons of oil. By 1980 the fields mentioned above will be producing about 50% of this; the other 50% will be produced from other fields already discovered in British, Norwegian and Dutch waters or from fields which will be discovered in the next three years in the North Sea, Celtic Sea and Irish Sea etc. Good seismic indications of a large number of possible discoveries already exist.*

‡ *See P R ODELL,* Natural Gas Prospects in W. Europe for 1975, Petroleum Times, *October 8, 1971 for the build-up of production in different west European countries.*

§ *This modest increase over 1975 natural gas production levels (from 260 to 365 milliard cubic metres) will be achieved largely by the production of associated gas from the discovered oilfields and partly by increased output of non-associated gas in the Netherlands and British off-shore areas. See also Table I-3.2.*

exploitation. From a declared maximum annual production potential of only 40–50 milliard cubic metres a few years ago, the argument over Groningen's production now concerns the establishment of an optimum rate lying between 85 and 105 milliard m^3. This alone would meet at least 57 per cent and possibly 70 per cent of the total European use of gas officially estimated for 1975. But, elsewhere in Europe, already discovered and declared reserves of associated and non-associated gas amount to roughly the equivalent of another one and a half Groningens – with a production potential amounting to over another 150 milliard cubic metres as shown in Table I-3.2. In other words, the gas resources required to supply something like 20 per cent of Western Europe's total energy demand by 1975 have already been discovered and evaluated. Beyond this, a further expansion of the gas resource base requires very little by way of new discoveries of non-associated gas. As shown in Table I-3.2, only minor developments in this respect are expected by 1980. Much more important in building up the gas reserve base, so as to enable it to sustain a 40 per cent increase in annual production between 1975 and 1980, is the inevitable availability of associated gas from the oilfields which are to be developed over the period. Thus, the gas reserves figures for new areas are not shown in Table I-3.2 as they are irrelevant in determining the gas production potential for 1980. This will be largely a function of the degree to which the oilfields are exploited. Given current knowledge of the size of the oilfields in the middle sector of the North Sea, the expectation of a gas production rate of 80 milliard cubic metres by 1980 is a modest evaluation of the potential, though it does assume that the re-injection as well as the flaring of the associated gas will be kept to a minimum.

Natural Gas in 1980

The forecast of 40 milliard cubic metres from the northern part of the North Sea and the rest of the continental shelf is speculative, but, in light of the reported discoveries in these areas (as, for example by Shell/Esso in Block 211/29 between the Shetlands and Norway and by Marathon off the coast of the Republic of Ireland) and the degree of enthusiasm of companies anxious to secure concessions in these areas, the possibility of these levels of gas production not being achieved by 1980 is very low. Thus, for 1980 the total potential for gas production from known fields, plus very modest speculation about additional gas producing areas, places natural gas firmly in a position to sustain over 20 per cent of Western Europe's energy demand. Since gas is almost always the fuel preferred above all others, this will then lead to a diminution of

TABLE I-3.2
Western Europe: An estimate of its natural gas reserves and production potential in 1975 and 1980

Location	1975 Reserves in cubic metres x10⁹	Production potential in cubic metres x10⁹	Coal Equivalent metric tons x10⁶	1980 Reserves in cubic metres x10⁹	Production potential in cubic metres x10⁹	Coal Equivalent metric tons x10⁶
On Shore Netherlands	2500	120	135	2750	135	150
South North Sea – British Sector	1000	50	65	1250	60	75
Austria, France, Italy etc	600	35	45	600	35	45
On Shore West Germany	550	30	35	650	35	40
South North Sea – Dutch, German and Danish Sectors	600	30	40	1000	40	55
Middle North Sea – British and Norwegian Sectors	?*	30	40	?*	80	110
North North Sea – British and Norwegian Sectors	–	–	–	?*	25	35
Rest of Continental Shelf	–	–	–	?*	15	20
Totals	5750+	295	360	8500+	425	530

* *As these are mainly associated gas reserves, the production potential is a function of the rate of oil off-take, rather than of the volumes of gas in the fields. In 1975 we may assume the associated gas reserves to be at least 500 milliard cubic metres and in 1980 at least 2250 cubic metres.*

the quantities of other fuels demanded – including oil, the total demand for which in 1980 will, as can be derived from Table I-3.1, be reduced from officially expected levels of about 1,000 million tons to around 840 million tons.

Indigenous Oil

It is now further suggested that something like two-fifths of this oil demand will be produced from indigenous fields. That would mean an indigenous production in 1980 of some 330 million tons, compared with a present annual oil production in Western Europe of under 30 million tons. Little more than six months ago, such a suggestion would have been received with complete disbelief; today, merely with formidable scepticism! The earlier disbelief and the present scepticism come from failure to evaluate dynamically the patterns and attributes of the rapidly emerging hydro-carbon geology of the north-west European oil province. The southern end of this has, so far at least, produced only small (by world standards) additions to the oil resource base, but its middle and northern parts stand revealed, or at least estimated, on the basis of seismic and geological works, as a multi-field major oil province by world standards. Moreover, the fields already discovered are scattered over so large an area that between them lie many other unexplored locations with high probabilities for a succession of new discoveries. The ten fields that have already been proved appear – by themselves – to have reserves capable of sustaining an ultimate and long-term production of at least 150 million tons per annum; and plans for developing this potential are already well advanced. Beyond this, in the light of experience in other parts of the world concerning the behaviour of newly developing oil provinces, we should face a unique event in the history of oil exploitation if the continuing and intensifying searches in the North Sea failed to produce new oilfields in the next five years or so capable of at least doubling the production potential that has so far been revealed. Thus a figure of 300 million tons of oil production per year, additional to the present output in Western Europe, offers a reasonable hypothesis on which to work for establishing possible strategies for a decade hence.

To move from the realm of reasonable probabilities concerning the production potential of known oil reserves in the North Sea to speculation about the extension of the resource base as a result of exploration efforts elsewhere on Europe's continental shelf (including areas to the west of the United Kingdom where, in fact, initial seismic surveys are already under way), would serve to indicate an increased chance that, in the decade beyond 1980, Europe's increasing availability of indigenous oil will continue to shift the balance of the total supply of oil even more strongly in favour of the indigenous component. Thus, we can look forward to the imports of oil in 1980 as representing a declining residual element in the European supply of the commodity.

Neither the medium-term developments in our indigenous oil

supplies up to 1980, nor the less quantifiable developments after that will be in any sense automatic. The volumes produced depend, in the first place, upon the continued willingness and ability of the oil companies to pursue the search for and development of the resources in a more than ordinarily diligent way and, in the second place, upon the application of appropriate policies by both individual European oil producing countries and the enlarged European Economic Community.

Costs of Investment

Under a situation in which oil supply to Europe is at costs which are steadily rising – principally because of increased tax payments to the major oil producing and exporting countries – and which may well lead to an average delivered-to-refinery price of $4 per barrel by 1975, few companies seem likely to hesitate to commit as much investment as they possibly can to develop Europe's indigenous resources, even when the investment required is upwards of £3000 per daily barrel of producable capacity. Assuming at least four thousand days' production from this initial investment, even this seemingly high input of capital gives a per barrel cost of no more than $1.45, after allowing for interest charges on all the capital invested for the whole of the producing period. Allowing as much again, on average, for all other costs involved in keeping the daily barrel of oil flowing to the refinery, the average total cost per barrel of oil delivered to the refinery is $2.90, so giving a pre-tax gross profit of over $1 per barrel on the basis of the import price mentioned above. Given such a level of potential profitability, only the companies' over-riding commitments to governments in the traditional oil producing areas as part of their concession and other agreements will restrain them from exploiting European possibilities to the utmost.

Earlier commitments in respect of their operations elsewhere may, however, limit the amount of capital resources that companies can devote to the North Sea operations and it is to this possible financial limitation on the earliest possible maximization of indigenous production that European governmental and Economic Community policy makers should address their attention. Either they should ensure that sufficient additional financial resources are forthcoming to achieve maximum development rates, or they should stimulate indigenous production by insisting that it makes a certain minimum contribution to the total oil supply by, for example, encouraging inter-company cooperation. Equally positively, both governments and the Community should take action to stimulate the flow of private investment into the development of these significant indigenous resources. This will not be easy to achieve because Europe has lost its former familiarity with resource development, given its reliance on other parts of the world for supplies of energy and raw materials in recent decades.

Effects on Supply, Transport and Refining

If there is appropriate action by producing companies as well as by the authorities, the indicated 300 million tons per year level of oil production from indigenous sources could be achieved within a decade and, as shown in Table I-3.1, would greatly reduce the level of dependence on imported supplies. This is important enough in itself, but in this final part of this article, we shall attempt to indicate some of the important changes in the supply, transport and refining patterns of the European oil industry that would flow from such a development by the early 1980s.

In the first place, as seen in Table I-3.1, the 1980 use of oil in Western Europe appears likely to be about 15 per cent below the current official forecast of around 1,000 million tons for that year, although one must point out that American estimates of Europe's oil consumption in 1980 put the figure at significantly more, viz. 1,200 million tons[1].

In the author's view the high estimates – based simply on the extrapolation of oil use trends in Europe in the 1960s – have little realistic evaluation behind them, given the recent marked slow down in the rate of increase in energy demand (partly as a consequence of environmental consideration) and the tremendous increase in the availability and use of natural gas. However, since the much higher estimates still pervade both economic and political thinking about the options open to Europe over its future oil supply patterns, the alternatives presented here will appear to be that much more radically different and, thus, may hopefully stimulate a searching discussion about oil developments in Western Europe in the next critical ten years.

Our lower estimate of demand by 1980 will, in itself, cause a significant easing in the required expansion of oil transport and refining facilities. Assuming the average size of a new refinery or refinery expansion is six million tons of annual capacity, then two to three fewer such projects will be required each year from now until 1980. Assuming that all the oil would move from the Gulf via the Cape in the very largest possible tankers (of, say, 500,000 tons capacity) then, by 1980, at least

1 "Western Europe, which today consumes 12 million barrels of oil per day is expected to consume double that amount of oil by 1980." Statement by the US Under-Secretary of State, John N Irwin, to the OECD Council meeting on 26 May 1972.

thirty-five fewer such tankers would be required for the transport of crude oil. Significant though these changes are in their own right, they are, nevertheless, likely to be the least important elements which can be expected in the supply, transport and refining sectors of the European oil industry over the next decade.

Less Use of Large Tankers

The first additional component in a fundamental reappraisal of the oil supply pattern lies, of course, in the incorporation of the very large production potential expected from the new North Sea resources. As shown above, this could be up to 300 million tons per annum – a total which is, by coincidence, more or less the annual capacity of all the existing and planned refineries situated around the coastlines of the North Sea. Simply to equate the availability of crude oil from the North Sea oilfields with these existing and planned refinery outlets is, of course, to ignore the complications of individual company requirements and of quality considerations, to mention but two of the important constraints. Moreover, some of the oil, once it has been brought on-shore by pipeline to the nearest accessible landing points on the coasts of Scotland, Norway and Denmark, will then be taken by pipeline to refineries on the west coast of Britain or in the interior of West Germany. These points, however, are issues of detail within the framework of a generalized geographical hypothesis which can be stated as follows; viz, the production of North Sea oil will give a high degree of self-sufficiency to the whole of the North Sea basin as far as the demands of its oil refineries are concerned. Furthermore, one can hypothesize that this oil will be delivered to the refineries direct from the off-shore fields by means of pipelines or by short hauls of the crude oil in small tankers of up to no more than 70,000 tons. Some of the oil could, for various reasons, move a little further afield – to the Baltic or to Atlantic Europe, for example – but even in these cases the distances over which the oil would have to be moved are relatively so short (by the standards of traditional inter-continental oil supply movements) that it would not be possible to justify the use of tankers bigger then 100,000 tons for such deliveries, given the adverse relationship which would otherwise develop between loading/unloading times and actual transit times.

Thus, over one-third of Western Europe's current and planned refinery capacity could be served by pipelines direct from the off-shore fields or by tankers of less then 100,000 tons sailing either directly from floating loading terminals in the middle of the North Sea or from re-loading points at the on-shore terminals of the pipelines which will be constructed to bring the oil from the off-shore fields to the coast (see Fig I-3.1). Moreover, refinery expansion in this most important oil-consuming region of Western Europe will no longer require the proximity of very deep water berths of the kind needed for crude oil tankers of more than 300,000 tons.

Finally, in this context, as it is this part of Europe in which the future markets for oil will be most seriously affected by competition from greatly increased supplies of natural gas, the need for refinery expansion beyond that currently planned will be curtailed over the next decade. Thus, i[illegible] north-west Europe, future investment in refineries seems likely to b[illegible]ected towards the construction of new processing units at existing [illegible]ies in order to make these refineries more suitable for operation [illegible] lighter North Sea crudes. This will shift their output away [illegible]el oil which, of course, is more susceptible to competition from natural gas than most other oil products. In this respect, too, north-west Europe's oil industry will undergo fundamental changes in the next decade.

Transport of Oil from Outside Europe

The second element making for change in Europe's supply and refining pattern derives from changes in the transport of crude oil from outside Europe and operating in combination with the further expansion of an internal Western European oil transport system that has been developed over the last decade. This internal development has been the construction of south to north pipelines from points on the Mediterranean coasts of France and Italy, designed to carry crude oil originating from oilfields in Algeria and Libya and from terminals in Lebanon and Syria, handling oil brought by pipeline from Iraq and Saudi Arabia. These south European pipelines have already moved crude oil to newly-built refining centres in eastern France, Switzerland, south Germany and western Austria for a decade. Since 1967, as a result of the continued closure of the Suez Canal and the increased importance of Mediterranean crude, they have enjoyed a rate of expansion which has pushed their use beyond that originally expected. In 1971 their throughput exceeded 90 million tons.

The success of these lines to date – and the consequential growth of refining in inland areas of Western Europe which would otherwise have been served by oil products brought in from refineries on the Northern Sea coast of West Germany and the Netherlands – has created a probability that their capacity will be at least doubled by 1980. The stimulus for this comes partly from the continued growth in the demand

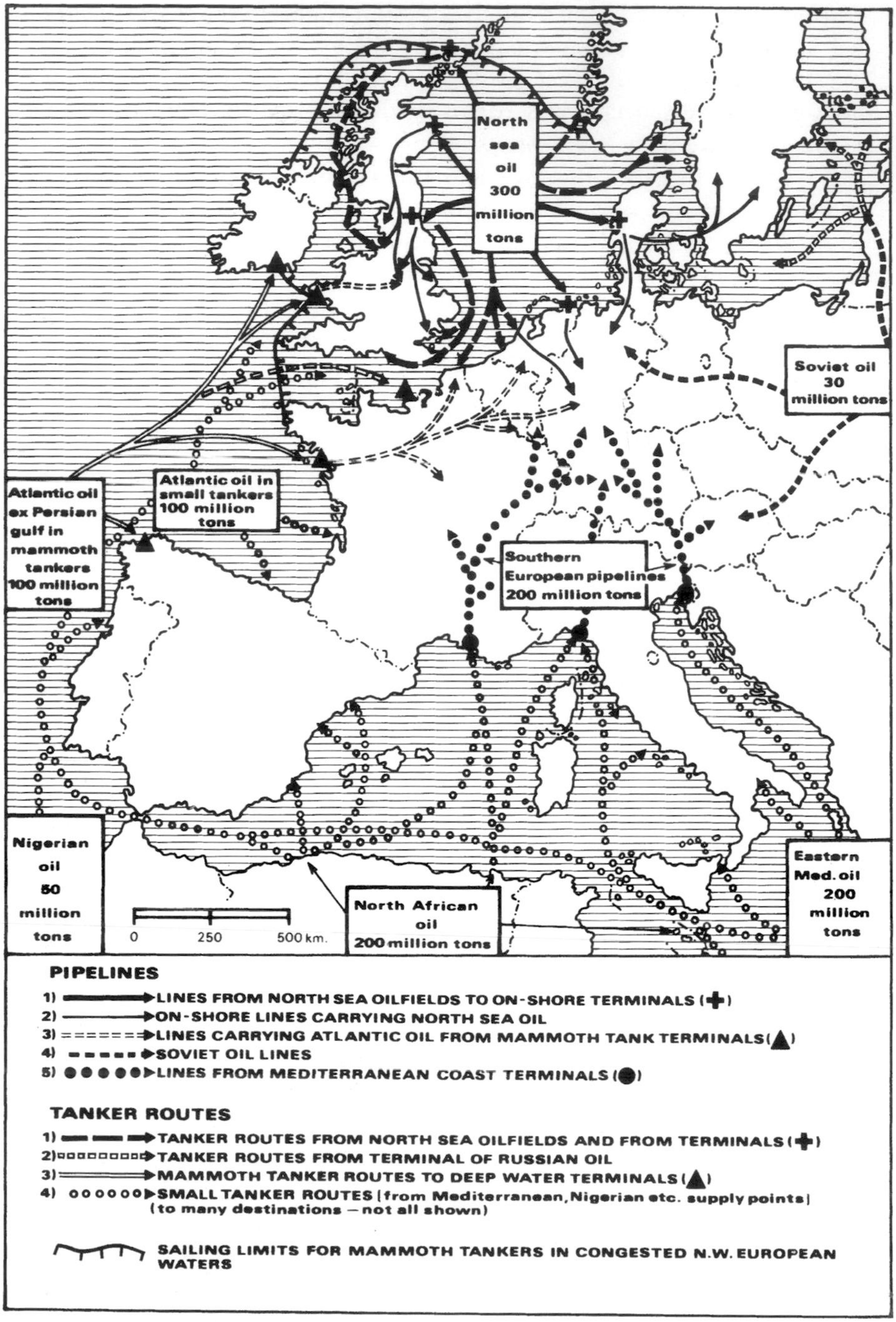

Figure I-3.1: A Forecast of Europe's Oil Transport Systems by the mid-1970s

for oil in the areas served by pipelines, in that competition there from natural gas is likely to be less significant than in consuming areas further north, owing to their more remote location from the main natural gas supply points. In greater part, however, the expansion possibilities arise because of expected developments in the Middle East and North Africa whereby hundreds of millions of tons more crude oil per annum can be made available at various points on the coastline of the east and south Mediterranean for loading into short-distance tankers destined for European Mediterranean coast refineries and the import terminals for the south European pipelines. Quite apart from a steadily increasing production from Libya, Algeria and, possibly, Tunisia and the continuing use of the trans-Arabian pipelines, major new quantities of oil will become available in 1973–74 on the Mediterranean coast of Egypt near Alexandria at the terminus of the trans-Egyptian pipeline from the Red Sea. From an initial 60 million tons per year, the capacity of this line is planned to increase within a couple of years or so to 120 million tons and, given sufficient demand in Western Europe for oil, to as much as 240 million tons per annum by the end of the decade. Sometime prior to that, moreover, the projected direct Iran-Turkey pipeline could also be completed and be capable of delivering at least 60 million tons of oil from Iran to the Mediterranean coast of southern Turkey from where, as in the case of Alexandria, tankers of up to 100,000 tons will ferry it to the off-loading points for the south European pipelines. In addition, the pipeline from Israel's port of Eilat on the Gulf of Aquaba to Haifa could be contributing another 50 million tons a year to the Mediterranean oil transport system – even without a peace settlement between Israel and the Arab nations.

In all, a total Mediterranean availability of at least 500 million tons of crude oil in 1980 is well within the bounds of existing and planned development. In reality, however, the controlling element on the speed of development of the European pipeline system seems more likely to be the size of the European demand for imported oil. Thus, the quantity available will certainly be more than enough to meet the possible 200 million ton annual capacity of the pipelines running north from the Mediterranean coasts of France, Italy and, possibly, Yugoslavia; plus the 200 million tons or so of oil which will then be required each year for refineries in Italy, Spain, southern France, Yugoslavia and Greece – assuming that such considerable imports are still necessary, following the recent discoveries of oil on the continental shelves of some of these south European countries. Meanwhile, the pipelines from the Mediterranean terminals will by then be transporting oil inland at lower real unit costs than at present (given the economies of scale in the expanded pipeline systems) and may, as a result, be able to serve refineries as far north as Frankfurt, Luxembourg and Nuremberg at which points this 'Mediterranean' oil will come into competition with crude oil and oil products moving south from the North Sea basin.

Recapitulation

To recapitulate, it is suggested that by 1980 North Sea basin oil together with oil flowing to Europe via short tanker hauls across the Mediterranean, of which about half will move into the interior of the continent by various south European pipelines, will provide the bulk of Western Europe's total demand for some 800 to 900 million tons of oil. Apart from a relatively very small flow of crude oil from the Soviet Union by pipeline to Austria and West Germany, the residual element in the overall supply and transport pattern will be the oil brought to Europe by tankers via the Western Approaches – which we will here term 'Atlantic oil' (see Fig I-3.1). It seems possible that by 1980 this will amount to no more than 200 million tons per annum as compared with the more than 300 million tons that currently come to Western European refineries via this route.

This suggested absolute decline in Atlantic oil must be compared with the general assumption to date that the flow of oil via Europe's Western Approaches would rise steadily in the future. This assumption, made on the basis of a simple trendline extrapolation of the increasing amount of crude oil which has been handled along this route ever since the end of the war, may thus be called into question, so raising doubts about many of the important medium-term and longer-term plans made by oil companies and their associated transport services. In the first place, it will very significantly reduce the currently expected mammoth-tanker fleet requirement for handling Europe's oil supplies. Thus, instead of the 400 to 500 million tons of crude oil that has hitherto been expected to come to Europe via the Western Approaches in tankers of between 250,000 and 500,000 tons, it may well be that tankers of this size will be appropriate for carrying no more than 100 million tons of crude oil per annum – that is, only about half the total amount of oil as forecast above for the Atlantic route by 1980. The other half, originating at Mediterranean supply points, and also in Nigeria and Venezuela, can be no more economically shipped to Western Europe in such large tankers than it can in smaller tankers as currently used.

Thus, smaller tankers already available appear more than able to cope with the demand on them for the next decade, given Western

Europe's reduced overall demand for oil and, most particularly, for oil brought in via the Atlantic. The 100 million tons of Atlantic oil shipped over only medium distances (of up to 3,000kms) can continue to move most economically in tankers no bigger than 100,000 tons and to existing terminal facilities. In general terms, one can think of this Atlantic oil moving without navigational restrictions to existing terminals and refineries at various points around the coast of the UK and the rest of north-west Europe. Thus oil will supplement supplies from the North Sea so as to provide the approximate mixes of crude oil required to optimize refinery operations in the light of changing market demands.

Policy for Terminals

Only in the case of the balance of 100 million tons or so of Atlantic oil which would be most economically moved by mammoth-tankers from distant sources of supply – namely the Persian Gulf via the Cape of Good Hope, Australia and perhaps even the Canadian Arctic – would the issue of very deep water unloading terminals in Western Europe need to be addressed. The capacity needed for the terminals capable of handling this traffic is, however, very different indeed from the existing plans for constructing such unloading terminals in the expectation that up to 500 million tons of oil per year will arrive in mammoth-tankers. These plans have resulted in a series of projects for large terminals at virtually all points around the coast of Western Europe where suitable deep water and on-land facilities are juxtaposed. Such locations stretch from north-west Spain, through France, south-west Ireland, Wales, Scotland and up the English Channel to the Maasvlakte, Foulness and Wilhelmshaven, located on the southern shores of the North Sea.

Navigational Problems

The hitherto expected demand for mammoth-tankers has occasioned studies of the navigational problems that they, or special shallow-draught giant tankers, would present along the congested cross-channel routes in the Irish Sea, the English Channel and the southern part of the North Sea. This, of course, has led to opposition from many communities concerned about possible crude oil pollution of coastlines through the enhanced risk of collisions and other accidents arising in such waters as the giant tankers, lacking navigational flexibility, have to manoeuvre amongst the increasing numbers of car and passenger ferries. In spite of this environmental concern, however, the mammoth-tanker terminals that have been projected appear to offer a potential capacity even in excess of the conventional estimate of a demand for 500 million tons per year by 1980. In the light of the fundamental changes in supply routes outlined in this paper, it seems that most of these projects would be better abandoned immediately.

European governments should be able to work together to determine appropriate navigational limits beyond which the mammoth-tankers should not be allowed to sail (see Fig I-3.1). Their concerns for the health of the marine life of their coastal waters and for traffic congestion in the narrow and shallow seas are no longer in conflict with the changing economic realities of European oil supply and transport. Thus, limitations on the movement in north-west European waters of tankers of, say, more than 100,000 tons will become generally acceptable. One benefit of the consequential decline in the demand for tankers will be a great reduction on the capital requirements of the mammoth-tanker development programmes of shipping companies and of the tanker subsidiaries of major international oil companies, so making more resources available for indigenous oil developments.

As shown in Fig I-3.1 the growth of the mammoth-tanker terminals by 1980 need be only quite limited. In fact, nothing more than some modest development of existing facilities at Bantry Bay and Milford Haven and new facilities in western France and in north-west Spain are needed. The present Bantry Bay terminal could be converted into a multi-company facility, and from it could originate, as now, an onward movement of smaller tankers into the shallower and congested European coastal waters. Supplies brought directly to the refineries in this way, together with the oil also moving in smaller tankers from Mediterranean points of origin, will ensure the right balance of crude oil supplies to a large number of Western European refineries which will become dependent for the greater part of their crude oil requirements on supplies from the North Sea. Milford Haven will continue to serve the needs of the refineries it currently serves, but its expansion will, in any case, be constrained by new refinery developments in Scotland and north-east England at places near to the landing points of North Sea oil. It could, however, provide part of the crude supply to an inland refinery complex in the Birmingham area where it would be blended with crude oil coming to the refineries by pipelines from North Sea sources.

In France, Le Havre, rather than a point somewhere on the west coast of the country, is best placed to serve as a break of bulk point for crude destined for refineries in the north and north-eastern parts of the country. It would also be an appropriate point of origin for a crude oil pipeline system taking oil to refineries as far away as Belgium and West

Germany where it could be blended with other crude oils moving in through Antwerp and Rotterdam in small tankers, or via pipelines from the North Sea oil terminals or from the Mediterranean. But agreed constraints on the movements of mammoth-tankers into the English Channel would push this terminal for the Low Countries to the north coast of France, with only relatively small consequential increases in total tanker and pipeline costs. Another terminal in north-west Spain, Portugal or even the Canary Islands seems unlikely to be necessary before 1980. Elsewhere in Western Europe the sooner the plans for the proposed mammoth-tanker terminals are folded away and put back in the appropriate pigeon-holes for possible dusting off and re-evaluation in the 1980s, when a new period of rapid change in energy supplies might be beginning, the better for the investors and governments concerned.

Conclusion

In the meantime, if the hypotheses advanced in this article are valid, much more appropriate uses for the scarce capital resources of the oil industry lie in stimulating North Sea and other European offshore exploration for additional oil and gas reserves, in speeding up the development of oil fields that are discovered and of their associated transport requirements, and in financing further pipelines from southern Europe to the interior of the continent. This would ensure that Western Europe's oil requirements would, as far as possible, be met from indigenous sources by the end of the decade.

Chapter I – 4

Europe and the Oil and Gas Industries in the 1970s*

Introduction

Fundamental changes in the last year or two in Western Europe's relationships with the international oil industry are of great political and economic significance and demand careful evaluation, particularly in light of the prospect of an enlarged Economic Community within eighteen months and the urgent need there will then be for an agreed energy policy. And whereas earlier energy policy discussions in the Community have centred on coal – Europe's traditional energy source – the renewed discussions will be largely concerned with oil and gas, which by the mid 1970s will account for over 75 per cent of the region's total energy use.

Several elements have combined to produce a new situation over oil and gas. First, the establishment of effective collaboration between the world's major oil exporting countries and the rapid success that has attended their collective action over the new tax and associated arrangements with the oil companies. Second, the success of the oil companies in winning the right from the United States Government to negotiate collectively with the oil producing countries and, as a by-product of such collective efforts, their ability to determine a strategy to eliminate the highly competitive marketing position in Western Europe which for the decade 1960–70 had led to very low oil product prices. And

* **Extracted from two articles published in successive issues of the *Petroleum Times*, viz. Vol.76, No. 1929, pp. 9–11 and No. 1930, pp. 100–105, January 1972.**

third, the proving of major reserves of hydrocarbons within Western Europe itself, most notably in the North Sea basins, but also in other off-shore waters elsewhere around the continent. The first two elements combine to produce significant short-term difficulties for Western Europe – in a very marked break with the recent past. The third element establishes, for both individual European governments and the enlarged EEC, somewhat longer term challenges to and opportunities for energy policy making.

The 1960s were tremendously turbulent for the world oil industry. This was so in respect of technical developments, particularly in the continuing moves to increasing scales of operation in refining and transportation, which significantly reduced the costs of delivering oil products to consumers around the world. For Western Europe, however, the reductions in consumer oil prices over this period were more a function of intense competition between a steadily increasing number of oil companies operating in the region. Companies had to accept much reduced profits per barrel of oil sold – a process which took the "fat" out of the comfortable situation previously enjoyed by the old-established seven major international oil companies under their traditional pricing and marketing policies. In this highly competitive situation it was only the ability of all the companies to sell a rapidly increasing number of barrels of oil each year that ensured they all survived. This they did because oil replaced hundreds of millions of tons of coal previously used in the hitherto mainly coal-dependent economies of Western Europe.

Collective Action

The question of what would happen to oil prices and to the profitability of oil companies once the process of oil/coal substitution inevitably came to an end with the near demise of the coal industries concerned, such that the oil companies had to fight each other even more strongly for a share of what would then be a less rapidly growing oil demand, looked, by the end of the 1960s, to be little more than a few years ahead. Some students of the industry – notably Professor Adelman of MIT – considered it likely to produce such pressures on the companies that they would try to force down the levels of royalties and taxes they paid to the governments of the oil producing countries. He argued that such efforts would be successful as the oil producing countries, too, had no option but to compete with each other to maintain their share of the more slowly increasing demand for oil – so as to keep up the flow of revenues to which they had become accustomed.

Perhaps the arguments of Professor Adelman and others helped to persuade the oil producing countries that the solution to this developing problem lay in collective action to maintain the high revenues per barrel of oil produced – even though this might mean that each of them produced fewer barrels. Or perhaps Venezuela, which had unilaterally sacrificed quantity for revenue-maintenance considerations since 1959, would finally persuade first Libya, and then the Middle East oil producers, that only their collective imposition of an agreed tax regime could prevent the erosion of the value of their oil revenues. Or perhaps it was simply the case of a tough, well-informed new Libyan government, which happened to control oil resources with such significant locational advantages over most other supplies – given the long continued closure of the Suez Canal – that started the whole ball rolling towards a radically new attitude on the part of the oil producers towards the oil consuming nations.

In any event, the member countries of the Organisation of Petroleum Exporting Countries (OPEC), after ten years of experiment and argument during which their successes made only minor differences to the nature of the international oil business, finally decided at their Caracas meeting in December 1970 to co-ordinate their policies and collectively to seek increased and guaranteed payments from the oil companies for each barrel of oil exported. This collective resolve enabled them to avoid the phenomenon which had always marked their individual action against the oil companies in the past – that of the "punishment" of the offending nation by the oil companies, with the punishment usually taking the from of a cutback in production levels and/or investment in the country concerned – with the consequential deficiency in total world supply made up by opening the producing valves a little wider in other "non-offending" countries.

In the short term collective OPEC action did pose a threat to Europe's essential energy supplies and, given the organization of the world oil industry in such a way that the interests of the consumers have been "taken care of" by the international oil companies, the required response to ensure the continuation of supplies had to come from the companies. This explains the willingness of the United States – backed by Britain and the Netherlands – to allow, and indeed even encourage, the oil companies to work together to meet this threat. And this would all have been very appropriate if the consumers' interests and those of the oil companies really were identical; but they were not, for the basic reasons that establish the difference between the interests of suppliers and those of consumers in any competitive commercial activities.

Thus, although the oil companies have been very ready to claim that their response to the oil producing countries' threats has been on behalf of oil consumers everywhere, and although governments – and other parties – in many consuming countries have "accepted" the oil companies as their representatives in the struggle against the oil producing nations, basic economic facts dictate otherwise.

Most important of all is the fact that the oil producing countries and the international oil companies have more interests in common than any other sets of interested parties in the whole oil world. This is something which the oil companies have appreciated for a long time, but about which they have been able to do nothing, given the earlier failure of the producing countries to work amicably together, and given constraints on collective action on their own part imposed by effective United States anti-trust legislation. The Caracas OPEC meeting eliminated the first barrier to collusion between producing countries and oil companies, whilst United States government action in lifting restraints on co-operation between companies, removed the second one almost immediately afterwards. The stage, by January 1971, was thus set for a joint "assault" on consumers' interests.

Public Relations

The higher posted prices "forced" on the oil companies by the producing countries, far from representing the threat to the viability of the oil companies, such as the action has generally been made out to be, actually guarantee higher profits to the companies, given their ability to relate oil product prices to the higher posted prices and given their success in keeping down the percentage of the gross profits in oil production payable to the governments. Thus, an increase in the posted price for a particular crude oil from $1.80 to $2.20 per barrel (the order of magnitude of the 1971 increases agreed in Teheran between the producing countries and the companies), accompanied by an increase in the government share of profits from 50 per cent to 55 per cent, gives the company concerned an increase of about 10 cents (or about 12½ per cent) in its own profit margin per barrel of oil sold – providing these sales are also at posted prices or at levels bearing the same relationship to them as hitherto. Further increases in posted prices – as already agreed for 1973, 1974 and 1975 – will further increase company profits on the assumptions made above.

The level to which posted prices were raised was thus a non-issue as far as the companies were concerned. To some degree, at least, the higher the better, providing they were able to do something about the following three critical matters:

(a) to keep down the producing governments' share of per barrel profits. By "giving way" on the levels of posted prices the companies were able to limit the increase in the government's share of profits to only 5 per cent – not only for the moment, but for the whole period up to 1975. This was a major success for the companies.

(b) to persuade governments and consumers around the world that each and every one cent increase in posted prices really meant a one cent increase in the oil companies' costs and that they had no alternative but to pass them on in full to oil consumers. This demanded a massive public relations campaign along lines which had been collectively agreed by the companies. This was duly undertaken and the objective rapidly achieved, such that there now appears to be an almost general acceptance of the claim that real and unavoidable costs in the oil industry have moved upwards to such an extent that price increases to consumers of up to 2p or 3p per gallon (= to 6 U.S. cents/American gallon) are fully justified. Such per gallon price increases are, however, at least five times, and even more than ten times, greater than the real cost increases involved though the posted price and tax changes. Such is the power of public relations.

(c) to ensure that the demand for oil products was kept sufficiently firm to allow the level of price increases mentioned above to be passed on to consumers. This involved the imposition of restraints on the quantity of oil supplied – a difficult task for the companies in that any overt joint action in this respect would quickly arouse not only the concern of the United States Justice Department, with its recollections of earlier battles with the oil companies over the oligopolistic trends, but also of the governments of certain consuming nations – particularly Germany and Japan – where, in spite of the public relations efforts by the oil companies, there are still suspicions that the oil companies are taking undue advantage of their new arrangements with the producing nations. However, less-than-overt action was possible as a result of the oil companies' opportunities legally to meet together collectively for their negotiations with the oil producing countries.

Of the range of price maintenance possibilities opened for discussion by the companies two particularly important ones appear to have been taken up. The first was a decision to strengthen collective controls over the volume of production from jointly-operated concessions. As these, in one way or another, cover most of the oil produced in the Persian Gulf, the output limitations thereby achieved are important, especially as the other two major exporting countries – Venezuela and Libya – can themselves be relied on to impose limits on increases in the rates of offtake. The second was a decision to limit the rate of expansion in European refinery capacity to a rate at which it roughly equated with oil company expectations as to the rate of increase in demand for oil products at the higher prices now aimed for. In as far as the fifteen or so largest oil companies control over 95 per cent of the refining capacity in Western Europe, the critical consuming area as far as internationally produced oil is concerned, and in so far as no other institutions are capable of doing anything about refining capacity in the short to medium term, this means that a "shortage" of oil products could theoretically be created and then sustained for some years. This gives the oil companies a continuing opportunity to relate their product prices to the posted prices for crude oil and thus able to achieve the higher level of profits they consider appropriate for their investment and other needs.

Diversifications

There can be little doubt of the success to date of the oil industry's strategy. The artificially contrived shortage of oil products has even been "explained" by what appear to be largely oil industry-inspired stories purporting to show that overall reserves of oil are now too small to meet the world's increasing oil needs. One needs to look no further than statistics published by British Petroleum* to see that the existing reserves/production ratio position – that is, the ratio showing the interrelationship between current proven reserves and current annual production – is in reality better than that of a decade ago in spite of the unprecedentedly high demand for oil throughout the 1960s. Last year alone, the 1970 Review shows, five times as much oil was discovered as was used. In other words any physical constraints on the availability of oil are not factors which influence contemporary decisions on production levels – and, moreover, they are not likely to be for the foreseeable future in economic terms.

* *BP Statistical Review of the World Oil Industry* – 1970

In respect of its profitability, the industry's claim that it needs higher profit levels than it enjoyed in the 1960s in order to find, transport, refine and market the oil that will be demanded in coming years is all very well and even made with some justification, but its arguments are somewhat undermined when one notes that oil company after oil company is busily reinvesting its profits not in oil, but in alternative ventures such as mining of potash, magnesium and bauxite, the manufacture of aluminium and plastic goods and the provision of shops, hotels, motels and other facilities in the rapidly expanding tertiary sector of Europe's economies.

In the long-term all these enterprises may be highly successful and profits from them might well be used to subsidise the production of oil. In the meantime, however, all the new activities will absorb increasing amounts of oil industry capital – all of which has to come out of the cash flow created from the oil activities of the companies', thus necessitating "higher-than-would-otherwise-be-necessary oil prices" for oil consumers for the 1970s. All these developments bode ill for Europe, as the world's major oil consuming region over the next few years. The situation can be summarised as one which has emerged out of collusion between oil producers – both countries and companies – against a continent whose governments have allowed essential energy supplies to be increasingly and mainly provided by a small number of oil companies, most of which are not only foreign-owned, but which also draw the overwhelming part of their supplies from a limited number of foreign countries. In most of these a rapidly emerging nationalism acts as a further restraint on the adequate and continuing off-take of an inherently plentiful – even if not bounteous – supply of what is essentially a low-cost, easily discovered and even more easily produced and transported commodity. Both accident and design have contributed to the occurrence of this basically unsatisfactory position for Western Europe in 1971.

Accident and Design

For example, the "accident" of continued Arab-Israeli conflict in the Middle East, with its consequential effects on the costs of moving oil from the Middle East to Europe, coupled with the "accident" of the overthrow of the old regime in Libya and its replacement by an aggressive, revolutionary government, have provided two of several important external events which have made it possible to hold Europe to ransom for its oil supplies.

On the other hand, it was "design" which led the oil companies to "kill" the coal industries of Western Europe and thus relegate the continued use of coal to a minor role in meeting the continent's total energy needs. By 1971 not even the stronger British coal industry – stronger, that is, than any other coal industry in Western Europe – is in any position to increase its share of the total energy market or even to maintain its present contribution in tonnage terms. The marketing opportunities for coal are now largely restricted, sometimes by means of appropriately selective pricing by the oil industry for sales in those sectors of the economy (notably the electric supply and the iron and steel industries) which are never likely to produce anything like enough profits to enable the coal industry to afford large new investments in new production facilities. Western Europe's willingness in the 1960s to allow its coal industries to fail has effectively shut-in the large remaining usable reserves of the commodity for the foreseeable future and has thus eliminated coal as a potential competitor against foreign controlled and supplied oil in the 1970s.

All these considerations spell economic difficulties for Europe. Moreover, in the short -term there is little effective counter action which can be taken to do more than modify to a small degree the effects of a basically adverse situation. The unit price of energy for most consumers can be expected to range from 50 per cent to 100 per cent above the levels which they enjoyed through most of the last decade. The break with the recent past is thus of fundamental proportions. Even the longer-term outlook for Europe's energy economy would, indeed, be grim were it not for certain other considerations which will now be discussed.

Appropriate Counteraction

Within a time-span of three years, consortia of large European oil consumers could upset the oligopolistic trends in the behaviour of the oil companies, as they try to restrain the expansion of refining capacity, by building jointly-owned or sponsored refineries. These could be designed to produce the range of products the oil-using participants require and could run on crude oil purchased directly from the state entities in the oil exporting countries. The latter are developing increased production capacity and are actively seeking markets. If the oil companies deliberately diversify away from oil, then bulk oil users in suitable locations, would certainly have an incentive to integrate into oil refining in the expectation that they would thereby secure their energy requirements at lower costs.

Already one consortium – consisting of a major airline, an international chemical company and a large electricity authority – is in negotiation with at least two oil exporting countries for the construction of a 10 million ton per annum purpose-built refinery, probably to be located in Ireland to take advantage of generous Irish government assistance for such ventures, and to secure the cost advantages to be gained from shipping the crude oil there in the very largest tankers. The plan is for the project itself to be financed in large part by the crude supplying country, which is anxious to participate in down-stream oil activities. But such a reaction to increasing oil prices by large consumers does not solve the problem for small oil users. They do not, of course, have the alternative of building their own refinery, but still have to look to existing suppliers with all that that implies in both political and economic terms.

It is in this context that Western Europe's large, recently discovered and still undervalued hydro-carbon resources must be viewed. It is, indeed, a remarkable commentary on the efficacy of certain vested interests, which have chosen to keep these indigenous resources under-declared, that have made oil-developments in Alaska better known than those in Western Europe itself. They have even given a greater weight to the prospects for oil from that part of the world as an alternative source for European consumers than to oil which will be produced within Europe itself.

Groningen

The "line" on how little to say about hydro-carbon developments in Europe was indeed set by the pretence for many years that the Groningen gasfield, discovered in 1959 in the extreme north of the Netherlands, was really little more than just another local occurrence which could not possibly have any general effect on the continent's energy situation. Even after it finally stood revealed almost a whole decade later as one of the world's largest gasfields with, moreover, near-perfect producing conditions, the strict control exercised over its development by Shell, Esso and the Dutch government have limited its exploitation. The gas was priced at levels high enough not only to create significant monopoly profits for its owners, but also to ensure that it did not compete effectively for oil markets. Thus, even as late as 1969 the strategy for its development envisaged a maximum annual production rate of only 40 Bcm – the energy equivalent that is of only about 32 million tons of oil per year – to be sold in markets where it would offer little or no competition to the bulk use of oil products. By United States standards of gasfield development, the field was capable of producing up to two-and-a-half times as much gas at costs low enough to make it fully competitive with lowest cost oil. Since 1969 a variety of events have, in fact, gradually created a situation in which the earlier planned marketing strategy has become unviable. There are now, belatedly, clear indications that a production of 100 Bcm will be achieved by the mid 1970s, when it will represent some 7½ per cent of Western Europe's total energy supply. Demand pressure from gas consumers in the Netherlands, Belgium and West Germany, the emergence of competition for gas markets from supplies of Russian gas and new market opportunities created for gas use by rising oil prices in Western Europe have all combined to bring Groningen fully "into action" as a major source of the continent's energy by the mid-1970s.

Other European Gas

Additionally, natural gas exploitation elsewhere in Europe (in the British, Dutch, Danish and Norwegian sectors of the North Sea, in North Germany and in the northern Adriatic, etc) have collectively, together with gas from Groningen, ensured a supply of indigenous gas which, could provide between 20 per cent and 25 per cent of the region's total anticipated demand for energy by 1975.

When the author first made this forecast over two years ago* it was discounted not only because of disbelief over the adequacy of the resources available to justify this level of production, but also because it required about a 50 per cent reduction in the price level fixed for gas to enable it to compete effectively with oil. By mid-1971, the rise in oil prices had already effectively created that price reduction as the price for gas, mainly committed by the producers on very long-term contracts, remained more or less stable. Moreover, the resources to sustain a 1975 production level of up to 250 Bcm have now been shown to exist. The ability of indigenous natural gas to achieve the level of penetration of the European energy market predicted above is no longer a question of "if" but of "when".

Can the Supply be Developed?

To achieve an accelerated rate of gas supply requires governments and other European institutions to ensure that enough additional indigenous investment is put into the exploitation of the resource. Elsewhere in the world where natural gas has been locally available in regions of high energy demand – notably the United States and the U.S.S.R. – gas has quickly become the preferred fuel. But in Europe the

* P.R. Odell *Natural Gas in Western Europe* de Erven F. Bohn NV, Haarlem, 1969

juxtaposition of gas producing and energy consuming areas makes the potential for the rapid development of gas use particularly high. Attitudes and policies towards production possibilities should reflect this.

For example, in the case of the United Kingdom, though a rapid success has been achieved in securing a flow of North Sea gas adequate to meet the Gas Council's plans to market 40 Bcm by 1975, future rapid expansion of supplies now necessitates a change in policy towards production possibilities. To date, the relatively low "beach-price" paid by the Gas Council, as the only buyer of North Sea gas from a variety of producers, has enabled the gas industry to determine to convert the equipment of its 20 million or so consumers to natural gas use. But additional resources now need to go into the further exploration for more gas in the attractive southern parts of the North Sea.

In the now very likely absence of the considerable state investment that the Labour government had planned to make for finding new gas, the only alternative to achieve this desirable objective is to offer individual companies of the right to sell any new gas they find to the highest bidder – whether that be the Central Electricity Generating Board, anxious to diversify away from coal and now under pressure from rising oil prices, or industrial consumers in Germany and Belgium who are certainly in the market for additional gas supplies in order to supplement their imports from the Netherlands and the Soviet Union.

A European Gas Industry

In this way the production potential of the British sector of the North Sea could be quickly bid up in the interests of maximising the contribution of indigenous natural gas to the total energy supply of Western Europe. The possibility of British sector North Sea gas being sold in the mainland of Europe, could also represent a significant and desired further break though towards the Europeanisation of the natural gas industry. This is quickly becoming an essential development if the region's natural gas resources are going to be fully and rationally developed, for it is only within the context of an area with the size and population of Western Europe as a whole, rather than within the framework of individual nations, that this can happen.

As in the United States, where the gas resources of geographically limited regions have successfully found markets across the length and breadth of the country within the framework of ground rules determined by the Federal Power Commission, Western Europe will shortly come to need an equivalent body charged with sustaining the public interest in general, rather than have the interests of individual countries and companies emerging out of an unregulated situation, or out of a situation irrationally regulated by a series of national entities – as will be the case given a continuation of the existing policies by individual nations.

Europe's Oil Potential

The significant opportunities and challenges which face the development of the still youthful European natural gas industry, nevertheless appear relatively modest compared with the greater ones that face the indigenous oil producing industry. Indigenous oil has languished for years as a very minor contributor to the total oil needs of the continent, but it now stands on the threshold of becoming large, even by world standards, as a result of the recent proving of the middle North Sea basin as one with a very high production potential. Based on the present limited knowledge of the basin's structure, its potential annual production has been calculated at a minimum of 200 million tons – divided roughly equally between Norwegian and British sectors.

Beyond this there can, of course, still be no certainty about the maximum possible production rates on full exploitation of the whole basin, but extrapolating from experience from elsewhere in the world there appears to be a high probability of an ultimate reserve position capable of sustaining a long-term European annual output rate of between 300 and 400 million tons per annum. Such indigenous oil resources development would enable them to meet up to 25 per cent of the region's rising energy demand in the late 1970s. Roughly the same contribution that is, as has been suggested above, for indigenous supplies of natural gas.

Europe's Indigenous Hydrocarbons

Thus, when the potential for oil and gas resources' development in Western Europe are taken together, and added to the continent's continued production of indigenous coal and water power, plus its development of atomic energy, then the future contribution of imported oil need play only a minor part in the total energy supply. Even after allowing for the continued growth of Western European energy demand at a rate of over five per cent per year, the actual tonnage of foreign oil required could peak by 1973 or 1974 and thereafter suffer a slow, but steady decline. It is perhaps calculations similar to these that have already persuaded the Iranian National Oil Company and, reportedly, other traditional major oil producing nations, to seek a stake in North Sea oil production.

Such an advantageous development for Western Europe is

obviously contingent upon the rate of exploration and exploitation which can be achieved in the North Sea and other oil bearing regions. Though the exploitation of these resources is fraught with technical and other physical difficulties, these appear likely to be overcome by the application of existing and/or only slightly improved technology. It is much more likely to be various political and economic considerations which restrain the rate of growth in production levels, unless there are some big changes in existing attitudes and policies.

In the first place the continuation of narrow nationalistic attitudes towards the oil resources will have an adverse effect on the speed of development – in the same way that they have already slowed down the rate of development of Dutch and British gas resources. There is already evidence of this happening over oil in the failure to date of Norway to permit the large-scale exploitation of the Ekofisk field (the first giant oilfield to be discovered in the North Sea), because the producing consortium proposed to pipe the oil to shore elsewhere than to Norway. In that the achievement of a Norwegian land-fall for the pipeline from the field involves crossing an under-sea trench several times deeper than has previously been crossed anywhere else in the world; and in that, by contrast, lines to carry not only the oil, but also the associated gas from the field, could have been constructed without difficulty to the coast of Denmark, whence the oil and gas could have been piped over land to North Germany with immediate sales prospects in a rapidly expanding energy economy, one can readily see that nationalism in this case involves a significant loss for the European energy economy as a whole.

Neither is it really in the interests of Norway which stands to lose revenues and taxes from the delay in initiating large scale production from the field. Similarly, the recent history of the slower-than-necessary exploitation of the Netherlands' onshore gas resources is apparently being repeated in the lack of a quick and positive response to the potential for oil production from the Tenneco find in Dutch off-shore waters. In this case the stumbling block to rapid exploitation could be the government's inability to secure its ready incorporation into the country's oil economy which is dominated by the refineries of the major oil companies and which, in turn, exercise a very powerful control over decisions in this sector.

North Sea Regulations

Apart from these examples of individual nationalistic responses to the challenge and opportunities of North Sea oil potential, there is also the broader question of the existence of five different national sets of regulations for its exploitation, in a situation in which the countries concerned all have very much the same essential set of interests at stake and in which the product concerned can only be rationally exploited in a Western European context, rather than on a nation-by-nation basis. Given the over-riding common interests of the five countries in maximizing indigenous oil output as quickly as possible, in order to reduce their dependence on high-cost and insecure imported supplies, the need for a European approach to the regulation of this new industry would seem to be axiomatic.

The work of any such regulatory body, however, would need to be anything but negative in order to ensure against the second danger likely to limit the rate and extent of development of Western Europe's oil resources. This is the danger arising from the fact that the resources are being exploited either directly, or under the operational control of the very same oil companies which currently supply Western Europe with the great bulk of its oil requirements, under the terms of their agreements with the major oil exporting countries. It is thus not unlikely, partly as a result of profit maximization considerations by the companies themselves and partly because of assurances the companies have given to the producing countries guaranteeing them minimum levels of future oil production, that the companies' interests could best be served by their developing the North Sea and other Western European oil resources at less than the maximum possible rate.

If the impact of such considerations merely created too slow a rate of production from fields already discovered then it would be a relatively simple regulatory matter (given adequate technical expertise on the part of the regulatory body) to require the companies concerned to increase their rate of production to acceptable levels. Similarly, straightforward regulations could be imposed to ensure that Western European refineries are obliged to take appropriate quantities of European produced oil, irrespective of inter-company rivalries.

The Need for European Companies

A much more fundamental issue is, however, also involved. This is the need to create a situation in which much more of Western Europe's oil potential is controlled and developed by companies which are not already heavily committed by earlier investment and by concession agreements to very large-scale oil production elsewhere in the world. Very few such companies already exist, but the scope for profitable operations in oil activities in Western Europe, coupled with appropriate national and European policies, ought to bring them quickly into being.

This could conceivably be done through state entities, as in the decision of the Norwegian government to made state enterprise responsible for operations north of Latitude 62° or as in the now-aborted proposal of the previous British Labour government to have various existing state enterprises involved in all new off-shore oil and gas activities. Much more significant than such direct state involvement, however, could be a really serious effort to mobilize part of the savings of millions of affluent West Europeans who, so far, have had virtually no opportunity to participate directly in the exciting and risky, but nevertheless potentially highly profitable, exploitation of Europe's hydrocarbon resources.

A few British institutional investors – merchant banks, insurance companies and so on – are now tentatively dabbling in the possibilities, but they are, no doubt, constrained by the high risks involved, given their traditional investment attitudes. In addition a number of individuals have been invited to invest in venture firms and have, apparently, responded overwhelmingly to the opportunities. But no one has yet sought to open up any possibilities in this new investment field to the millions of relatively small savers throughout Europe. Many such would undoubtedly welcome the chance of taking a small direct stake in the production of indigenous oil and gas – in the same way as have millions of Americans in the United States industry.

Indeed, Norway's 3.8 million people reacted immediately to the opportunity offered to them of a personal stake in North Sea oil developments in Norwegian waters – and contributed the equivalent of almost £1 per head for every man, woman and child in the country to a risk capital venture (Norsk Oljeselskap). Is there any reason why Western Europe's 300 million people should react any differently to a similar offer? The chances seem to be high that they would react equally enthusiastically to the possibility of taking a stake in a multi-hundred million pound venture designed to exploit the continent's hydro-carbon resources as quickly and as effectively as possible. In doing so they would introduce the single most important element in the total investment required to fulfil this objective and consequently place the non-European controlled interests in hydro-carbons' exploitation in Europe into a more appropriate perspective.

Western Europe's indigenous capital availability, once mobilized, would be more than sufficient to enable the development to be run from Western Europe itself for the benefit of the continent's population and economy. The region's economy would thus, in the medium to long term, be relieved of its potentially dangerous dependence not only on foreign oil, but also on oil brought in by companies whose quite legitimate interests do not necessarily coincide with those of Western Europeans. Perhaps an affluent Western Europe also has the responsibility to take such an initiative – as a means of releasing the flow of traditional oil industry funding to those other parts of the world with expanding energy needs, but where local investment capital is not available on a large enough scale.

Chapter I – 5

The Western European Energy Economy*

Introduction

Our societies were built up on the basis of cheap and readily available energy. Industrialisation was closely associated geographically with the exploitation of coal resources found in relatively limited parts of Western Europe and North America so producing the industrial conurbations of northern Britain, the Ruhr, southern Belgium/northern France and those in the Appalachian region of the United States. Needless to say, neither the exploitation of the coal resources nor their manner of utilisation showed concern for the environmental impact. Indeed, in the search for the new standards of wealth that coal-based industrialisation could create the irrelevance and acceptability of the adverse environmental impart were summed up in the well-known Yorkshire saying 'Where there's muck, there's money': an evaluation which was as acceptable to the poorly-paid work-force, with its concern for the continued production and use of coal in order to provide the continuity of employment, as it was to the entrepreneurs and rentiers who could only earn their profits when there was a demand for the commodity in industry, transport and commerce.

Cheap coal thus provided the basis for development in the Western world and though it has now been substituted as the most

*** An edited version of the author's Lord Stamp Annual Memorial Lecture to the University of London, November, 1975, published by the Athlone Press, London, 1976**

important energy source in every part of the industrialised world, some continuing attributes of the energy sector clearly owe their origin to its former importance. As coal was cheap, there was little incentive to utilise its energy efficiently and thus we had the beginnings of the lack of attention to thermal insulation in homes and other buildings.

Later, an electricity production system was built up in which a fuel-use efficiency factor of only 25 per cent could be achieved and it was also acceptable to develop a transport system in which the main motive power was provided by inefficient steam engines. Our coal-based societies were thus generally energy inefficient in themselves, but, in addition, that traditional acceptance of inefficiency also pre-determined the relaxed attitudes to energy use in the recent period of rapid economic growth based on oil and gas.

Thus, cheaper oil replaced cheap coal – so that country after country became increasingly dependent on oil. Outside the United States the impact of low-cost oil was, however, restricted until the mid-1950s by the influence of the controlled international oil system, dominated by the major international oil companies. These companies, sometimes working together quite formally as an international cartel and at other times achieving their objectives of high prices and market stability through informal understanding of each other's position,[1] ensured that coal had a more protected position than would otherwise have been the case.

In Western Europe it was not until the late 1950s and the 1960s that the oil system became highly competitive, so enabling oil products to compete effectively with indigenous coal. The latter's use started to decline sharply, so undermining the coal industries of the countries concerned and, of course, the regions dependent on coal production. Western European countries welcomed this access to cheaper energy and gave little thought to the internal or external (strategic, military and economic) implications of dependence on oil. Cheap energy, indeed, became the watchword, with the strategy enthusiastically backed by oil companies anxious to prove that there was no problem concerning supply of oil that could not be overcome.

Given the expectation that this declining cost of energy in real terms would continue, there have been consequential developments in the ways of 'doing things' and of 'going places' which were little concerned with encouraging the careful use of energy. The 'waste' of energy emerging from such societal developments has been increasingly well documented over the last few years[2], but there are four main strands which demonstrate the hypothesis of 'energy careless societies'.

First, transport systems have been increasingly orientated to private means, so eliminating the advantage of economies of scale in the use of energy. Goods move overland by truck at an energy cost per ton/km which is at least four times greater than that involved in rail haulage. People move over long distances by air rather than by rail and/or by sea at a specific energy use per passenger kilometre some five or more times higher (depending on type of aeroplanes used) and, over shorter distances, by car, instead of by means of mass transit systems, with a consequential reduction of some 65% in energy use efficiency.

Second, the patterns of urbanisation with compact cities and the rational ordering of the spatial system of central places, have now given way to the dispersed city/city region. The earlier hierarchy of cities and towns required a near minimum use of transport facilities to interconnect them and to provide for the circulation of both goods and people, but within city regions the opportunities for the movement of goods and people often seem to encourage the maximisation of the intensity and the distance of exchange. As a result of this modern urban form, the use of energy 'necessary' to enable the system to function has moved up to a level which is an order of magnitude higher than in the pre-existing geographical ordering of society.

Third, the 'architecture' of the built environment has evolved into a form in which ignorance, carelessness, and aesthetic considerations, as well as the domination of interests concerned with first-costs rather than with running costs, have combined to produce high levels of energy inefficiency in most Western European urban areas. The failure properly to insulate houses, coupled with the usual preference in them for low capital cost heating systems (such as open fires and night storage electricity heaters), instead of thermally efficient systems based on fossil fuels or higher capital cost developments based on solar power and heat pumps etc., is a prime example of the lack of attention to energy considerations. Such failures in the domestic sector are, moreover, repeated in the inefficient energising of commercial and industrial premises, epitomised by the aesthetically-desirable all-glass, all-air conditioned office blocks of city centres. There, it often seems, conditions have been deliberately created for a maximum input of energy no matter what the external weather conditions.

Fourth, there is the issue of the production and use of electricity. In spite of continued technological developments, the efficiency with which our electricity authorities make their product available to consumers does not, at very best, exceed 35 per cent. In other words, the electricity systems as developed in modern Western economies (with a few exceptions) have to input about three times as much energy as they can output in the form of electricity. The rest goes up the chimneys of the power stations or is otherwise dispersed as 'waste' heat to atmosphere or to water, often with adverse environmental consequences. Whilst coal remained as the only possible input fuel for power stations, then the centralisation of electricity production in order to achieve economies of scale and of mechanisation and automation in the handling of the coal, was inevitable. Further, given societies' wishes that such 'monstrosities', with their environmental dangers, should not be built in proximity to the locations of electricity users, it was also inevitable that the heat produced in the power stations as an inherent part of the process of electricity production had to be dissipated to the atmosphere or to water.

However, once oil products and natural gas became available to produce electricity at prices comparable with those of coal, then the inevitabilities of the coal-based electricity producing system no long applied. These alternatives to coal can be utilised in a geographically dispersed pattern of electricity production so that the production of electricity can be related closely to the geography of the demand for heat. The combined marketing of both produces a conversion factor for the primary energy input of between 55 per cent and 80 per cent – implying an energy saving of at least 45 per cent over the centralised electricity production system.

It is sometimes still argued[3] that 'inefficiency' in the use of energy emerges out of 'natural' economic forces which give a strong preference to minimising the utilisation of capital and of time and that the systems which we now have, represent these optimal preferences. There is quite obviously some truth in this argument. For example, a car in the right place at the right time will 'save' over 90 per cent of the time required for the same journey by public transport, whilst the conversion of a coalfield-located power station to burning oil or gas is much less demanding of capital than changing the whole system of production of electricity. Yet, on looking at the energy-using systems as they have emerged, there seems to be much *prima facie* evidence that these rational elements are the exceptions rather than the rule. Economic adjustments to changing energy supply patterns and demand needs have been constrained by institutional factors and by non-rational behaviour on the part of consumers, sometimes even backed by government decisions which put obstacles in the way of rationality and change.

It is within this context of the lack of incentive for change that one needs to look a little more critically at the approach to the energy sector which has been accepted in recent decades. This is necessary because the rapid rate of increase in the demand for energy and the ways in which institutions and infrastructure have been built up to meet these demands, has been an essential component in the revolution of rising expectations. Essential, that is, to the achievement of the 'freedom' that is given by the use of private transport; to the expectation of a home in a low-density housing development with a private garden and in a suburb rather than in a city; to the convenience of being able to turn on energy for activating labour-saving household appliances; to the expectation that the use of energy would not only increase industrial and commercial productivity, but also serve to make work much less burdensome; and even to the extension and the diversification of leisure activities, in which indulgence became possible by increasing numbers of people as higher labour productivity was achieved through the increasing use of energy in industry and commerce.

Meanwhile, the consumer society and the generally accepted view that each succeeding year would, almost inevitably, be better than the previous one, necessarily implied the use of more and more energy just to keep the consumption of things and of services moving ahead. It also implied successful exhortation that old things should be discarded and substituted by new things, with 'inbuilt obsolescence' an important component in the degree of success; and it also implied the need to persuade people to travel further and more often at ever increasing speeds.

This is the context in which possible Western European responses to the problem over its future energy supplies must be evaluated. The problem, that is, which arises from the effective control now exercised by the oil producing and exporting countries over the supply and price of the sort of energy, viz. oil, on which the economies of most of the nations of Western Europe have come to depend so heavily.

The Politico-Economic Crisis over the Supply and Price of Oil

The low price and the ready availability of oil from those parts of the world in which large quantities of inherently low-cost-to-produce supplies were found by international oil companies[4] produced the rapid growth in Europe's post-1950 dependence on the commodity. Cheap energy policies based on competitively-traded low-cost oil became acceptable and accepted because of the expectation that the international oil companies would continue to be able to deliver the increasing quantities of oil demanded. Indeed, the oil companies encouraged this view of the world energy supply situation wholeheartedly until 1968 and with only slight reservations (in public) until as late as 1973.[5]

Thus, when OPEC finally took over effective power in the international oil system at the end of 1973, all Western European nations were still pursuing energy policies which served to enhance the role of the hitherto low-cost and low-priced oil resources of the Middle East and a few other very limited parts of the world. This was in spite of the many indications that all three parties concerned with the production of these resources, viz. the producing countries themselves, the international oil companies and the United States, were all anxious to limit their availability and to increase their prices.

From its initial formation in 1959, OPEC took over 10 years to achieve a consensus for collective action amongst its members. The consensus emerged at OPEC's Caracas meeting in December 1970. OPEC then became the formal Trade Association of the oil producing countries and over the next two years, it was gradually able to structure a strategy for increasing very considerably their share of the profits made by the producing companies on the exploitation of low-cost oil. In the changed circumstances of the time, notably the lower profit situation which had emerged out of the competition for markets in the oil industry between 1955 and 1970, this OPEC agreement was not only accepted but even welcomed by the oil companies which, a decade earlier, had declined to recognise the existence and significance of the Organisation. Their motivation was clear in that they saw an opportunity, arising from OPEC's role as an international cartel, of protecting their profits. They knew they would have to pay steadily increasing taxes to the producing countries, but saw OPEC as an 'excuse' for passing these on to their consumers. The companies could, in effect, turn OPEC to their advantage. Hitherto, the companies had been barred from collective action to protect their profits by United States' anti-trust legislation, but now they were able to persuade the Nixon administration that there was real danger of a crisis in world oil and that the best way to avoid this lay in concerted action by the companies in their negotiations with the producing governments. Thus, early in 1971 the companies secured American government permission to work together and thus enabled them, in due course, to reach collective agreements with the OPEC countries.

Though the negotiations between OPEC and the oil companies were presented publicly as a struggle between parties with divergent interests, it is clear that each of the parties recognised the need of the other in the establishment of a changed pattern of oil power, in general, and the re-distribution of benefits from the production and sale of oil products, in particular.

Thus, OPEC/oil companies co-operation became a fact of the oil-power system of the early 1970s with the positive encouragement of the United States which evaluated that it had much to gain, both politically and economically, from the development. On the political side, the United States wished for a successful *raprochement* between oil companies and oil producing countries as an element in a renewed effort to find a solution to the Middle East conflict. It argued that higher oil revenues and a greater degree of economic certainty would made it easier for the Arab oil exporting countries to accept a compromise with Israel which, in turn, would be more willing to modify its position, given the enhanced Arab economic and military strength. On the economic side, the United States was penalised by a situation in which the rest of the industrialised world enjoyed access to cheap energy from the oil exporting countries – an advantage which the U.S. did not share because of its decision to protect its domestic oil industry. The limited oil imports which were allowed were sold at the much higher domestic price. Thus, the United States had strong motivations to 'talk up' the price of international oil amongst the producing nations.[6] In as far as the U.S.A. itself would be affected by the higher priced crude oil imports (which were expected to grow rapidly in quantity after 1970 because of a lack of expansion in the domestic oil industry), these would be offset, or even more than offset, by the enhanced abilities of the U.S. oil companies to remit increased profits back to the United States, once the companies could base their oil supplies to consumers elsewhere in the world on the much higher prices they were able to charge in the transformed international situation.

Within this joint framework by the parties concerned as to how their separate interests could best be served, the scene was set for a reversion to the more usual pattern of producers' control over the supply and price of oil. Until 1973, however, it was intended and accepted that this would once again be under the leadership and direction of the major international oil companies which, in turn, recognised that the greater part of the enhanced profits to be made out of the retrained supply situation would flow to the producing countries, whose interests, the companies thought, would thus be served to the satisfaction of the countries concerned. The latter, for their part, had to continue to accept the idea that the major oil companies would play an essential role in the international oil industry, not only in respect of transporting, refining and marketing the oil, but also as decision takers on such fundamentally important matters as levels of production and the development of producing capacity in different countries. Given this *modus vivendi*, the companies could, moreover, still ensure that the supply of oil expanded more or less *pari passu* with the expected average 8 per cent per annum rate of growth in demand, so achieving an orderly and profitable international oil industry – in place of the 'chaos' of the previous fifteen years of intense competition between suppliers.

This grand strategy for an orderly world of oil, presented the prospect of a very significant change in the oil power situation. It was, however, quickly undermined when the producing countries decided that co-operation with the multinational oil companies was not necessary to protect their interests. Instead, they realised that they could take absolute control over decisions on price levels and on the levels of production. Straws in the wind had already indicated that the oil-producing nations were moving towards the assertion of control over supply – as, for example, in the expropriation of company assets in Libya and Algeria; in unilateral decisions such as that of Kuwait to fix maximum rates of offtake; in the willingness of Saudi Arabia to accept production expansion programmes scheduled by the companies; and in close national attention to the horse-trading between companies of their oilfield assets as, for example, in Abu Dhabi. By mid-1973 there was already an expectation that producing-country control over production and development decisions would gradually become the norm, as unilateral decision-taking by the oil-exporting countries replaced the 1971–72 agreed bilateralism between countries and companies.

The renewed outbreak of was between Israel and the Arab states in October 1973 and the latters' decision to use oil as an economic and political weapon in their struggle, provided the opportunity for accelerating this process. Within three months the world arrived at a state of imbalance between the available supply of oil and a potential demand which would, without the war and the Arab's use of the oil weapon, have taken up to three years to develop. Given three years, then there *might* have been time enough for adjustments to have been made to the structure of demand in response to the constrained supply position, though there was no evidence between 1971 and 1973 of any European realisation that such radical action was required. Even now there still appears to be no general acceptance that the world of oil has undergone a near-instant revolution so that there will be no return to unconstrained supplies of OPEC oil, if and when the politics of the Arab-Israeli dispute have been settled. The depth of the changes may be measured by the absolute control which the producing nations have taken over decisions on the level of posted prices, on the amount of oil to be produced and on the choice of customers with which to trade. As far as the Arab oil states are concerned there is no doubt, of course, that their actions reflect their decision to use oil as a political weapon in the struggle with Israel. However, the more fundamental nature of the change of attitude amongst the oil-producing countries is clearly seen in the decisions by the five non-Arab members of OPEC to take equally effective action to control the supply and hence the price of oil. It is, for example, Iran which has led the moves for price increases, whilst the first act of the newly elected President of Venezuela was to announce in January 1974 that his country had no interest in increasing oil production beyond the levels already reached. He has since indicated that reduced levels of production are more appropriate in order to ensure the continuation of price increases.

For all exporters the motivation and ability to keep the oil supply constrained has become stronger; firstly, because they have eliminated the power of the multinational oil companies and turned them simply into their agents for implementing the essential decisions over the supply and price of oil which the countries themselves have already taken; and, secondly, because their efforts massively to increase their return on each barrel of the limited supply of oil were so immediately and increasingly successful. OPEC oil has thus become very high cost energy indeed, but, behind this incontrovertible fact, there now lie strongly contrasting interpretations of the strength and motivations of the oil producing countries – and hence contrasting ideas about the strategies that should be followed in response to the very serious situation.

Many oil importing countries have been persuaded that the oil exporters are justified in their action and can be accommodated in the existing economic system. Strategy should, thus, they argue, be based on a package of measures designed to ensure that OPEC co-operates in getting the oil moving in the volumes which the importing world has calculated that it requires over the next five to seven years, at prices which the rich importing countries, even if not the Third World importers, can just about afford! This view also requires the further optimistic assumption that the oil producing lands can be persuaded to continue to circulate the oil revenues which they are now receiving and so provide the mechanism both for automatic adjustments of balance of payments' difficulties and the means whereby demand in Western economies can be kept at a high and expanding level. It further assumes that Western world inflation can be controlled and moderated, in spite of the unfavourable impact of continued rises in the price of oil, and then it conveniently glosses over the high probability that something will go wrong (albeit only accidentally) in this untried and fragile system which is subject to all kinds of economic and political pressures. The optimism appears to stem from an unwillingness to accept that there has been a revolutionary change in the world of oil power and to face the fact that, for the first time in some 400 years, there has been a loss of control by the Western world over an essential element in its system to a set of countries which have hitherto not been considered to be decision-taking entities within that system.

A contrasting view emerges from the recognition of this revolutionary change in the oil world. Hence, it is argued, the future of the world economy now depends on the policies and actions of a group of countries which have no hitherto had such power and responsibility. It is, moreover, impossible to be optimistic that the group will use its power and responsibility mainly to enable the member countries of the Western economic system to continue to enjoy their privileged position in the world. The lack of optimism arises from three considerations. *First*, the new decision-takers lack the experience and background adequately to gauge the results of their policies – in other words they could accidentally destroy the fabric of the system. *Second*, they may not even want it to survive, given the period of 400 years in which they consider that they have been the victims of Western economic exploitation. The undermining of the system could well be considered a prerequisite for a more appropriate world order. *Third*, as most OPEC countries give pride of place in their policy decisions to the elimination of, or to their control over, Israel, this brings them into confrontation with the Western world which, to the Arabs, is the mechanism whereby Israel was created and continues to exist. In the light of these three factors, the chances of the oil exporting countries moderating and tempering their demands sufficiently to enable the Western economic system to survive unchanged, appear to be small and are growing smaller as time goes by without any agreed Western response to the situation.

There is thus a three-fold threat to the system consisting of the following elements. *First*, there is the whole question of the availability and security of oil supplies from OPEC countries in a situation in which the Western economic system depends on a steadily increasing flow of oil to sustain the standards of living achieved and the accepted ways of life of their inhabitants. The near certainty that there will be interruptions in these supplies from time to time and/or a high probability that there will be decisions by the producing nations permanently to cut back the amount of oil exported to the West, serve to undermine the planning and the policies of all Western governments and so make it impossible for them to sustain the rising expectations of their populations for continued development.

Second, there is the hyper-inflation which has been generated by the traumatic increases in oil prices. In this respect there has been a widely accepted view that the increases in oil prices would simply create a once-and-for-all problem of readjustment to economies based on more expensive energy. These hopes have, however, been set back by the decision of the oil exporting countries eventually to index their oil prices to the inflation rate in the Western world: a decision which will, of course, further strengthen the inflationary trends in our economies, especially as so many internal costs and prices are also indexed to inflation, so opening up the real possibility of internally and externally generated cost and price spirals feeding on and sustaining each other.

And *third*, there are the difficulties arising from the large surpluses of foreign currencies which the oil producing countries are building up from their sales of high-price oil. In 1973 these countries earned less than \$5000 million – and spent most of them on buying goods and services from the West, so keeping the money flowing round the system. In 1974 their revenues surplus to needs were over \$30,000 million, in 1975 they seem likely to exceed \$50,000 million and then to continue to rise unless, in the meantime, the flow of oil is much reduced. In such circumstances the oil exporting countries can choose their monetary strategy and, if they so wish, bring the world's monetary system under severe pressure – and so lead to the further undermining of our highly money-orientated system!

Overall, there can be no doubt over the contemporary importance of oil power with all countries now feeling the impact of the recent quite fundamental changes in the international oil system – and with prospects of the problems becoming even more serious. In brief, as a consequence of power in the international oil system having been taken over well-nigh absolutely by the oil producing and exporting nations working together through OPEC and as, in the short term at least, they seem more likely than not to act irresponsibly, the fundamental question must be whether the international Western economic system can stand the strains.

Much though there is that is wrong and inadequate with this system, and with individual part of it, its demise is not automatically going to produce anything that is better, and most certainly not in the short term. Yet it seems to be the demise of this system that has now to be considered as a possibility as long as there is dependence on the OPEC countries. This implies an urgent need for energy policies which seek to reduce this dependence. In this context, a re-evaluation of the validity of accepted Western European views on energy supply/demand questions is required, together with a look at the possible establishment of a new framework for determining production and use patterns.

In this context, the question of the long-term availability of energy resources is essentially irrelevant. Indeed, the question of a sufficiency of energy for the mid-21st century is somewhat academic, if the control over the energy supply in the immediate future threatens the ability of modern, technologically-based civilisations to survive. Nevertheless, the concern that has recently been created over the long-term availability of resources does serve to create a willingness to accept energy conservation measures and the direction of society into less energy intensive activities. This issue of attention to demand must thus be the first element in the strategy required by Western Europe for tacking the problem created by the politico-economic control over oil supplies now being exercised by OPEC and its member countries.

Constraints in the Use of Energy

In as far as the demand for energy is derived from previous decisions on the organisation of society, as well as on previous purchases of goods and services, there are constraints on the degree to which there can be an automatic demand response to a changed supply and price situation. Indeed, demand response *per se* can only produce a limited curtailment of use – especially in a strongly inflationary situation when consumers' resistance to high prices is significantly weakened.

Over the period since the beginning of 1974 this limited potential demand response to the new supply situation has, moreover, been overshadowed by the effects of a much reduced level of economic activity in the industrialised world. Thus, energy savings over this period have been significant compared with previous years' use and even more so compared with what was expected to be used.[7] The impact of economic recession does, however, create unit-cost and/or cash-flow problems for the energy supply industries and for other industries, such as transportation and vehicle manufacturing, subject to high energy costs so that they find it necessary to try to stimulate demand in order to solve their own problems. But demand stimulation is, of course, contrary to the energy conservationist policy required in response to the politico-economic crisis over oil, as indeed are all steps designed to reflate the economy unless they are preceded by specific energy saving measures. In the absence of the latter, the inevitable impact such reflation has on increasing energy use makes it imperative that tools other than the use of the ordinary pricing mechanism and/or official exhortation are employed for limiting the growth of energy demand.

In the immediate future this seems likely to require the use of rationing and/or other physical controls on the use of energy, so producing sectoral and regional patterns of growth which will be different from those which might otherwise have been expected and necessitating consequential steps to avoid possible social distress and/or social unrest arising from such changes. Such short-term regulation of demand is, however, a crude approach to the problem and it must be seen only as an interim solution prior to a more radical longer-term approach to energy conservation.

Many years will be required to bring about changes in societal attitudes and organisation which produce more energy efficient systems without foregoing the continued economic growth still sought by the great majority of the populations of Western countries. As already indicated, there are two sectors of the economy where changes in societal attitudes would bring important energy savings, viz. transportation and residential use, and one sector, viz. the electricity supply industry, for which changes in the system of organisation and of the institutions involved in its production would generate even bigger energy savings.

In the case of transport, the essential requirement is the substitution of private facilities by public transport involving a combination of the deliberate encouragement of the latter, in geographical circumstances where it can be achieved without an extortionate investment requirement, and the restriction of the use of the private car through a joint pricing/non pricing allocatory mechanism for fuel. This mechanism would recognise the convenience of the private car for many journeys, but seek to penalise its unnecessary use. This implies the introduction of a system in which all families (or other units in society) are able to buy a defined quantity of relatively low-price gasoline to enable them to undertake those journeys which give them the greatest utility (or, if not car users, to sell on to others). Additional gasoline would be available at a much higher price so as to institute an effective limitation on the use of private transport and thus to stimulate the use of public transport facilities. And in order to emphasise the 'socially just' nature of the system there would, of course, have to be appropriately stringent rules to prevent the cost of the high-price gasoline being passed on to the customers of business users and to its cost being charged against tax obligations. For the transport of freight, the trend to the increased use of energy-intensive air and road services offers obvious scope for measures designed to get traffic back to water and rail facilities. Transport decisions have become too orientated to convenience and/or speed – to the exclusion of other considerations –

with a consequential heavy penalty in terms of energy use. The basic validity of this needs to be challenged for the sake of oil conservation, if for no other reason. If this is viewed as a "burden" on society, then the penalty constitutes part of the price which has to be paid to ensure the survival of most of the rest of the system.

Similarly, most of the population of Western Europe has embraced the idea of maximising convenience in their use of energy in the home. Again, the conventional wisdom that this is necessarily inevitable needs to be challenged and alternative strategies evaluated. While there can be no doubt that every household in a modern society does have the *right* of access to an appropriate amount of energy at reasonable prices, there seems to be no inherent reason why the rich, the lazy or the wasteful should enjoy the additional energy they use at lower unit prices than that paid for the essential quantity of energy needed in a household. To provide a means for redressing this quixotic situation, the calculation of an accepted 'norm' for the use of energy per household (dependent on variables like location, family size, types of infrastructure, etc) is an energy policy priority. Thereafter, tariff structures would reflect the need for relatively low-priced energy for the calculated norm for a given household, but for a supply of increasingly expensive energy for ail above-norm consumption, in order to diminish much luxurious or unthinking use of the commodity. Again the combination of a non-price based allocation mechanism with the use of the pricing mechanism for curbing the use of energy gives emphasis to the concept of a 'socially just' approach to the question.

Energy savings arising from changes in the system of allocating energy use in the transport and domestic sectors of modern economies are important, but are likely to be relatively insignificant when compared with the savings which are potentially achievable by a fundamental restructuring of the centralised electricity producing systems. Central thermal power stations and their associated high-voltage, long-distance transmission lines are energy wasteful – indeed, they waste about 65–70 per cent of the inherent calorific value of the input fuel. The need for such a system arose because, as shown above, it provided the only way in which coal could be used at a low enough cost and on a large enough scale to enable the rapidly growing demand for electricity in Europe's economies to be met. Unfortunately, the momentum generated by the coal-based system and the unwillingness or inability of management to recognise the viability of alternative approaches to the production of electricity, when oil and natural gas started to become important as the input fuels, has inhibited the development of alternative more energy-efficient systems. Because oil and gas can be easily transported and used in relatively small quantities, they are suitable for producing electricity in small plants in locations where the heat produced as a by-product of the process of electricity generation can be utilised in time-coincident demands for it – so increasing the efficiency of fuel use by 55 to 80 per cent; roughly 2 to 2½ times that of the centralised system of electricity production. The latter therefore should not be any further extended. Instead, there should be the installation of on-site systems of electricity generation with waste-heat recovery and utilisation[8] in all new factories, commercial and institutional developments which are technically and economically suited to such installations and also in existing large electricity and heat consuming premises as opportunities arise from them to be divorced from the centralised electricity producing system. These requirements would lead to a more energy-efficient electricity production system over the medium-term future, so making an important contribution to the objective of constraining the demand for energy.

Given such appropriate action *vis a vis* demand for energy in Western Europe, one can hypothesise a long-term rate of growth well below the 5½ per cent per annum rate of the 25 years since 1950 – the basis used, hitherto, for forecasting overall demand. Note that although demand-control does not imply "zero energy-growth", it would, nevertheless, make a striking difference to the quantities of energy used in Western Europe over the rest of the century, when compared with hitherto conventional expectations. Using figures for 1973 as the base, there would by 2000 be a 40% difference between the 'conventional' and the 'alternative' estimates; giving a 'saving' of more than 2000 million tonnes of coal equivalent in that year – much more, that is, than the total energy used in Western Europe in 1975!

In this context of a radically changed outlook for the future use of energy in Western Europe one can turn to examine a supply side option for energy autarky as both an appropriate and a practical alternative for the continent's energy economy for the period from the end of this decade through to at least the beginning of the twenty-first century.

The Case for Autarky in Energy Supply

Given the uncertainties in the world economic system, we would argue that, for a commodity as fundamental as energy to the ability of nations to continue to function and even to survive, it is unsuitable to be traded (on any other than a very modest scale) between nations or groups of nations which lack a communality of interest. This view is, indeed, already accepted by countries as dissimilar in their economic systems and attitudes as the United States and the Soviet Union[9] There now appears to be a need for Western Europe to adopt a similar attitude towards its energy sector.

Given that the continent's energy economy had to a high degree always been self-sufficient until the post-war period, it is interesting to ask why Western Europe so readily accepted the idea of an open, free-trading energy sector over the last 25 years. This was not essentially the result of a basic politico-economic belief in the inherent wisdom of free trade (though this has, of course, been a component in the arguments used since 1950 by the protagonists of a free energy market), but was due more to two other important considerations. First, because in the aftermath of the 1939–45 war and the consequential economic and social difficulties faced by the then dominant coal industry in Western Europe, there appeared to be a lack of indigenous resources capable of energising a rapidly expanding economy. Thus, there was no effective alternative to imported oil – and if there had to be imports, then opening up the market ensured their availability at the lowest possible prices. This first reason, however, is insufficient in itself to explain why Western Europe chose to allow its economy to develop in directions which were energy intensive and, as shown above, even energy wasteful, in a situation in which there was an increasing dependence on oil imports.

A reasonable explanation for this lies in the successful efforts of the international oil companies to persuade Western European countries that the companies could guarantee the supply of as much oil as would be required, no matter what happened to the availability of other energy sources and no matter to what degree the economy and society became energy intensive. Given that the oil companies were always interested in selling additional quantities of oil available at a low marginal cost as a means of reducing their average unit costs, their efforts to stimulate the use of oil products were unceasing. Moreover, they succeeded in servicing their customers efficiently and continuously and so demonstrated to governments and to users that reliance upon oil was, indeed, a 'risk-less' venture. Capacity to handle the increasing demand was always available on time (or could be organised within the framework of the companies' operations which were highly integrated at a European level), so that neither country nor individual customers suffered from a supply problem even in times of difficulties at the international oil industry level. Continued intense competition between the companies in their European marketing ensured that all this was achieved at falling real prices over the period up to 1970. It is, therefore, hardly surprising that Western Europe became convinced of the appropriateness of an open, free-trading energy economy!

By the late 60s, however, under the influence of the changes in the international oil industry noted above, these factors were beginning to become less relevant to decisions on energy policies in Western Europe[10]. By 1971 they were already entirely irrelevant, though this was not widely realised until the oil supply and price crisis of the end of 1973, thus leading to several wasted years before the need fundamentally to reappraise attitudes to the energy sector was recognised.

One other motivation for an open energy economy remained – and still serves to perpetuate the lack of serious consideration of the autarkic alternative. This is the belief that openness is inherently good: a belief, arising, in part, from the idea that free-trade automatically produces net benefits to all economies and especially to those of Western European countries which 'depend' to a high degree on trade with each other and with the rest of the world. In part, however, it is related to the French-inspired view that the Middle East, where, of course, most of the major oil producers and exporters are located, is a 'natural' extension of Western Europe in political-geographical terms. Recognition of this demands a policy which keeps Western European energy markets open to Middle East oil.

This motivation for open markets is difficult to sustain on either score. The advantages of free-trade imply rough equality in trading rules for the suppliers of goods and services on both sides. Yet whilst different countries and companies in the Western world compete intensively with each other for the export opportunities on offer in the oil producing countries, and so produce a buyer's market for the OPEC countries' demand for Western European goods, the supplies of oil to Western Europe are sold under the aegis of a successful cartel, with consequential super-normal profits for the sellers and high prices for the consuming nations. In such circumstances the validity of the idea that openness in the approach to energy/oil imports is appropriate seems to be highly questionable.

In political terms the concept of a self-evident mutuality of interests between the Middle East and Western European countries seems to be equally open to justifiable scepticism. It savours of the idea of 'special relationships' of the kind which France has endeavoured to develop with Arab countries over the last 15 years or so – and which in respect of oil brought France no advantages whatsoever[11]. It also implies that the oil-exporting countries can do better economically by accepting 'responsibility' for guaranteeing oil supplies to Europe, than they can out of their cartel-like arrangement for trade in oil. There is little chance of this being right, however, as they are already doing more than satisfactorily and certainly better than they expected or, indeed, need to do. Furthermore we have just gone through a period of three years (1971–74) in which the oil exporting countries have clearly demonstrated their unwillingness to hold to the successive agreements reached in negotiations with the oil companies in Teheran, Tripoli and Geneva etc.

These agreements have all been unilaterally broken by the OPEC countries. This recent precedent is hardly an inspiring one for the view that Western Europe should rely on trade agreements made with the same countries to ensure continuity in the supply of oil! The expectation of success for such agreements, within the framework of the power relationships as they presently exist between the oil producing and the oil consuming countries (i.e. a relationship in which the latter are dependent on the former for the essential well-being, whilst the former are no more than ephemerally dependent on the latter) implies the hypothesis that the oil producing countries have now secured all their objectives in changing the world order in their favour and that it is thus unnecessary to use oil as a 'weapon' in a continuing struggle for change in the system. Surely such an interpretation is self-evidently untenable for two reasons – first, because of the continuing use of oil as a weapon in the continuing struggle by the Arab countries for the success of their policies; and secondly, because many OPEC countries appear increasingly to be motivated by a wish fundamentally to change the international economic order and so break the power of the Western system within which Western Europe is the most vulnerable element – given both its internal dissensions and its position *vis-a-vis* the world of Soviet dominated communism[12]

In other words, the idea that Western Europe can continue to depend on a system in which energy flows in openly from the outside is based on little more than unsubstantiated hopes that international developments in the world of oil will be less unfavourable in the future than they have in the past five or six years. Thus, a policy of energy self-sufficiency not only becomes a desirable alternative to the openness of the region's energy economy of the period since the early 1950s, but could also be a matter of survival.

With such a powerful motivation for a fundamentally changed approach towards energy policy, the stimulation of the search for indigenous resources and the development of their potential must become central themes in policy making in the sector. Such stimulation of indigenous production potential must, however, overcome an element in policy making which has already become important in several Western European countries, viz. the idea that known indigenous resources be conserved for use in energising the continent sometime in the twenty-first century when global resources will, it is argued, have become scarce. A 21st century scarcity of energy is, however, by no means self-evident, particularly within the context of a responsible attitude towards consumption. An energy policy largely concerned with the conservation of indigenous resources for a low probability 'scarcity' in the twenty-first century is tantamount to keeping one's umbrella dry for the sort of rainy day which is unlikely ever to occur. In the meantime, surely the umbrella of Western Europe's indigenous resources' production ought to be used as a protection against the rainy day that is already very much with us and which seems likely to give our economic system a nasty cold, at best, or, at worst, a fatal dose of pneumonia?

The Development of Western Europe's Indigenous Resource Base

A combination of geological, geographical, economic and political factors are involved in determining the production of energy resources and it is perhaps the fact that so many disciplines are involved in the analysis of the potential for resource development which has caused such a delayed appreciation of the recent fundamental changes in the outlook for energy self-sufficiency in Western Europe.

This has emerged from the discovery of massive quantities of natural gas both on-shore and off-shore north-western Europe (as well as smaller amounts to date in other parts of the continent) and from the discovery that the North Sea basin is one of the world's largest oil provinces. Having previously been led to believe that Europe's resources of energy were very limited and that hydrocarbons are available only from far-away places which have nothing apart from oil, it is difficult to have to completely reverse one's perception of the position. Even when large, low-cost resources have been found and, indeed, brought into production – as with natural gas from the Groningen Field in the Netherlands and from the southern part of the British sector of the North Sea – there have been subsequent failures to re-evaluate the potential and significance of the developments in the light of changed politico-economic circumstances.

This has seriously undermined the short-term ability of Western Europe to diminish its dependence on imported energy at a critical period in its economic and political relations with the rest of the world. For example, the Netherlands is still pursuing a policy which aims to husband the country's known gas reserves so that production is less than that achievable. This represents a failure to appreciate the significance of additional production and an unwillingness to view reserves as a dynamic phenomenon, As a consequence of the constraints put on the production of the discovered natural gas, reserves are increasing more rapidly than demand, with the reserves/production ratio now conservatively estimated at 31 years[13]. Likewise, for the British South North Sea. There the failure of the British Gas Corporation and of succeeding Governments to provide an adequate financial incentive to the oil companies concerned to extend and develop their production facilities means that large resources of gas remain unexploited in a situation in which an additional 10 milliard m^3 of gas per year could have been produced at a lower resource cost than any other energy source in the U.K. Had this been produced, would it would have saved the country a net £240 million per year on its balance of payments.[14]

In Western Europe as a whole the potential for developing its known gas reserves indicates an annual production possibility in the 1980s which would be capable of meeting some 25 per cent of Western Europe's consumption of energy – within the framework of the constrained demand option previously discussed. The creation of this relationship between the indigenous supply of natural gas and the demand for energy implies a 15–20 year depletion period for the initially declared proven reserves of a field and relatively short lead-times for the construction of production and transport facilities. In economic terms there would be no difficulties in marketing the gas at prices related to the high price of the oligopolistically constrained supply of international oil – or even to an agreed lower price. This would bring some downward pressure to bear on the international oil price and so help to keep Western Europe more competitive with the United States where some energy prices remain divorced from the international oil price. A lower indigenous price for European gas, at an oil equivalent of $7 to $8 per barrel, would, however, still be so far ahead of the long-term supply price of known off-shore natural gas reserves as to make its maximum possible rate of production a commercially attractive proposition for the oil companies. It could still even give governments some cause for concern as to how they should deal with the super-normal profits of the companies concerned. This 'problem', however, should currently be of the least concern to government whose primary responsibility in the present energy situation should be to create an environment in which maximum possible quantities of gas can be produced and marketed as a substitute for dependence on oil from OPEC countries.

Similar politico-economic requirements exist for ensuring the development of the potential for oil from the North Sea and adjacent basins. Decisions are required which give all possible encouragement to the companies to maximize their exploration and exploitation efforts. This necessitates not only appropriate policy decisions to ensure the exploration and exploitation of all geological structures thought likely to contain oil, but also appropriate decisions to ensure the development of the necessary on-land infra-structure by way of supply and servicing bases and the priority allocation, if and when necessary, of resources of manpower and/or of goods and services to the industry.

There should also be an agreement within Western Europe for a preferential market for all the North Sea oil that can be produced. Without this, Britain and Norway could well develop a capacity to produce oil which their neighbours do not have to take because some or all of the OPEC producers decide to undermine the validity of North Sea production by reducing the market price of their own much lower-cost-to-produce oil. The producing companies themselves may also require such a guarantee in order to facilitate their decisions to invest the thousands of millions of £s needed to ensure the development of the potential for European off-shore oil: and they also need to be allowed to earn some degree of super-normal profits out of the European oil system in order to persuade them to concentrate their activities here, rather than elsewhere in the world. Multi-national corporations do have an option to take their exploration expenditures elsewhere. This has to be recognised as a fact of life, no matter how much the existence of the option may be deplored and/or thought to be counter to the well-being of the economic system.

In essence, therefore, the development of a European potential to produce oil in very large quantities in as short a period of time as possible is a function of a necessary and a mutually beneficial set of partnerships. In the first instance there must be a partnership between the 'have' and 'have not' nations of Western Europe. In this, the former, in exchange for agreeing to the rapid depletion of their resources and to receiving a lower price for each barrel sold than that which could be obtained short-term on the OPEC-controlled international market, could secure guaranteed long-term outlets and prices for their production irrespective of what happens internationally in the oil system. The 'have-not' nations for their part, in return for their long-term guarantees to use North Sea oil in preference to other oil and for their financial and industrial help in ensuring that sufficient finance and hardware becomes available for developing it, achieve a long-term security of supply at prices which will be lower in real terms than those they have to pay now for oil from the OPEC cartel.

Secondly, there must be an acceptable partnership between the countries with oil resource potential and the international oil companies in that a too severe limit on the latters' expected levels of profitability or too great a restriction on their exploration and development activities, consequent upon the reservation of such opportunities for state companies, will persuade the companies to demote the North Sea well down the international rank-ordering for their exploration and production investment. Recent legislation in both Norway and Britain designed to increase the states' direct participation in North Sea ventures and to enhance the states' ability to secure the bulk of the profits earned from oil production has eliminated the hitherto 'easy' conditions which the companies enjoyed (compared with the situation elsewhere in the world). What is now required is a degree of flexibility in the way in which country/company relationships are implemented so as to ensure that the companies remain committed to the most rapid possible exploitation of the reserves.

Finally, there is then the question of the financing of the industry and the provision of the hardware required for the developments. These are, however, essentially non-problems as neither the investment nor the hardware is required in such quantities as to cause any difficulty whatsoever, given the capacities of Western Europe's sophisticated industrial structure and its equally sophisticated financial system. Even at the currently estimated investment cost of up to £3000 for developing each barrel per day of production capacity, the annual investment required only amounts to £7 per European per year: and outside financing will, of course, only be required for a small percentage of this, given the ability of the oil industry to self-finance a large part of the continuing investment – especially with the guaranteed prices and the guaranteed off-take provisions defined as essential elements in the proposed autarkic European energy policy. And as far as the provision of hardware is concerned, both the learning and the tooling-up processes for the involvement of European firms in this new off-shore industry appear to be more rapid than even the optimists had suggested. The recession in traditional industrial activities has no doubt helped to general this response – but it is really nothing more than one should have expected within the framework of an entrepreneurial, competitive response to the alternatives to traditional activities offered by the opportunities of the offshore market.

Thus, geography, economics and politics each play an important role in opening up – or in limiting – the opportunities for the exploitation of Western Europe's potential off-shore hydrocarbon resources. The parameter that remains for evaluation is that of the resource base itself – a function of the geology of the very large area of continental shelf and continental slope which lies around the much indented and lengthy coast-line of Western Europe. For the short-to-medium-term this necessitates, first and foremost, an evaluation of the full oil and gas potential of the North Sea basin and its adjacent regions. For the longer term, the geographical options in Western Europe are very much wider and will be discussed below.

The development of the natural gas resources up to 1980 has already been described. Beyond that the dynamics of the reserves' situation in the 1980s depends, firstly, on the 'normal appreciation' of already discovered fields[15] and, secondly, on the expected continuation of discoveries from the further exploration work to which large numbers of oil companies are already committed by the terms of their concession agreements with Britain, Norway, the Netherlands, West Germany and Denmark. Much of the gas potential of the southern part of the North sea basin remains to be developed under an appropriate combination of political and economic factors. It is now, moreover, also accepted that the volume of non-associated gas in the northern part of the basin will be at least of the same order of magnitude as those in the southern part[16].

Given the time still required for building up the necessary infrastructure to exploit most of these latter resources, production will not start until the late 1970s and will then generate peak rates of production in the post-1985 period. Likewise for the associated gas reserves from the discovered and the still-to-be-discovered oil fields in the northern part of the North Sea. Such associated gas has traditionally had to be flared to the atmosphere in most oil-producing parts of the world, given the absence of local energy markets in which it could be utilised. This, of course, does not apply in the case of the North Sea basin which is surrounded by areas of intense energy consumption and in which there is an increasing familiarity with and, indeed, a preference for natural gas over all other sorts of energy including oil products. Thus, we may confidently assume that all the associated gas which has to be produced along with the production of oil will be pipelined ashore and incorporated into the energy economy of then W. European countries. One can thus conclude that the known, the probable and the inferred natural gas resources of the North Sea basin will be capable beyond 1985 of being developed at a rate which enables the production of the gas to keep pace with the rising demand for energy. Gas will thus continue to provide about one-quarter of Western Europe's energy demand through to the end of the century.

Indigenous oil will, moreover, also be able after 1985 to meet at least a quarter of energy demand – and so become three to four times more important than imported oil in the overall oil supply picture. Given that the total annual production of oil in the whole of Europe through to 1973 had not generally exceeded 20 million tons, then the potential build-up of oil output implied in these figures (to some 400 million tons by 1985 and to over 600 million tons by 2000) is formidable indeed. There is, however, now little doubt that the North Sea basin has the potential for putting the growth of Western Europe's indigenous oil supply on the trend required, given, of course, the establishment of the politico-economic framework as already described.

The Fig. I-5.1 shows the location and size-order of the oil fields which had been discovered and initially declared (in terms of recoverable reserves) to September 1975. Fourteen giant fields, each with at least a 50 per cent probability of reserves of over 135 million tons of oil, and each capable of an annual peak production of up to 10 million tons, have already been proven and are in the process of development. In addition, there have been more than 30 other discoveries which, though of more modest dimensions, are still large enough to justify investment in production and transportation facilities within the framework of the guaranteed demand and price system presented in this paper. Overall, the fields already discovered are capable in themselves of taking North Sea oil production potential very close to a 1985 level for indigenous production of some 400 million tons. Thus the medium-term future for European oil output as hypothesised here is already 90 per cent probable in terms of the availability of recoverable reserves and the technical abilities to produce these reserves in the time period under consideration.

As with gas, the post-1985 development of the indigenous oil supply potential becomes more a matter for reasoned speculation in light of the continued significant opportunities for exploiting the favourable geological conditions around the coast of Western Europe. One such speculation[17] – based on computer simulation techniques in which the main parameters used were those which appear to have been of importance in determining the scale and speed of petroleum exploitation in other parts of the world and with the variables calibrated in the light of the emerging picture on exploration and development results in the North Sea itself – indicates a potential unrestrained production curve which rises to a peak in the early 1990s. The mean result, on being

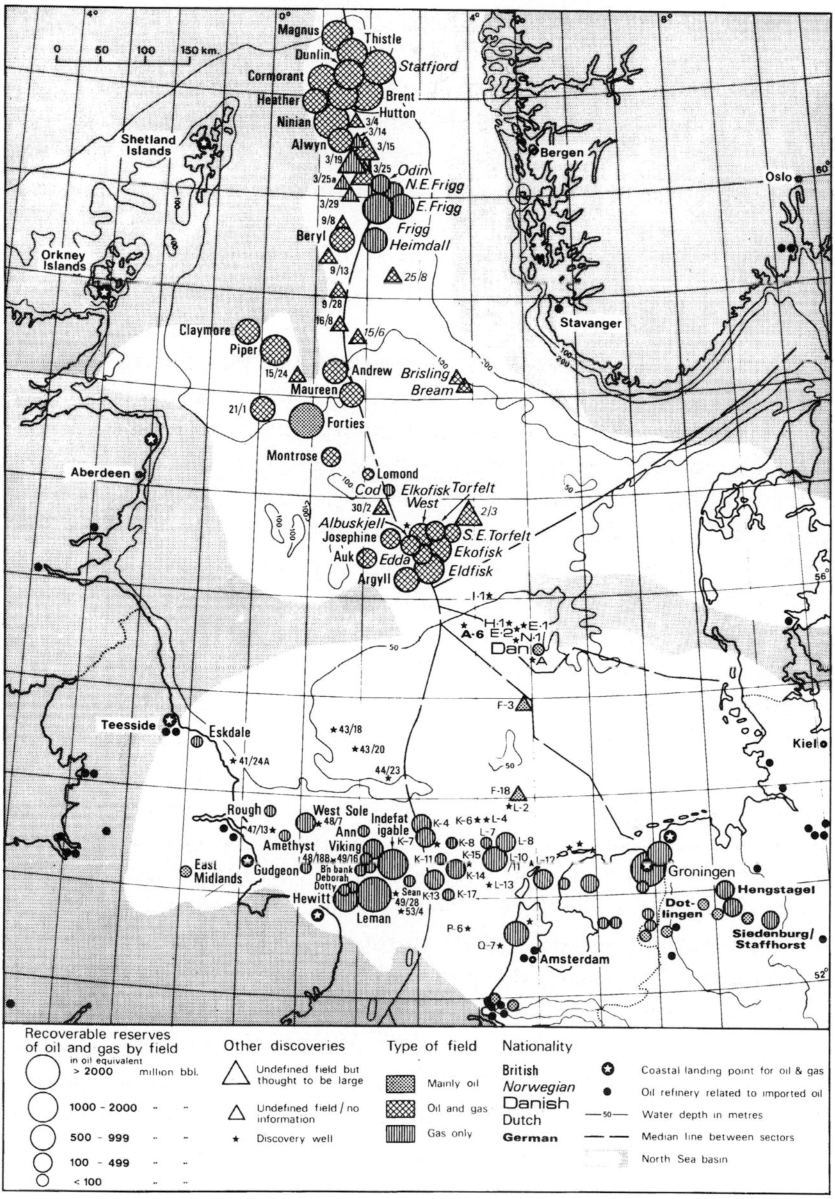

Fig I-5.1: The North Sea Oil and Gas Province: Discoveries by Size, Type and Location up to September 1975 also showing the year when production started or is planned to start

compared with the expected curve of 75 per cent of Western Europe's possible demand for oil up to the year 2000 shows that the North Sea basin alone may be more than able to meet this overwhelming part of the oil needs of the continent up to at least the end of the twentieth century. In brief, given an appreciation of the dynamics of the build-up of oil reserves in a province under continuing exploration over a long period and of their active depletion under the assumptions as indicated, it appears that the North Sea basin in itself offers the possibility of a long-term supply of indigenous oil which is capable of bringing down the dependence of Western Europe on foreign oil to an almost derisory level.

Should this reasoned speculation for the post-1985 oil supply potential from the North Sea be thought too optimistic, then it remains open to relate the need for, and the opportunity of Europe becoming, self-sufficient in energy to other possible developments of oil and gas producing potential from one or more of the large number of off-shore possibilities which exist around the continent. These are specified in Fig I-5.2 showing just how vast the total potential area for European offshore oil and gas is when compared with the very limited area of the North Sea. Norway, north of latitude 62°, has almost 950,000 sq.km of prospective areas, compared with less than 150,000 sq.km in its sector of the North Sea. Prospective areas to the west of the British Isles are about three times larger than the British sector of the North Sea and beyond this there are the prospective areas around the Faroes and Greenland in one direction, whilst in the other direction there are extensive areas off the west coast of France, Spain and Portugal as well as off-shore prospective areas along the Mediterranean coastline of Europe. Well before the end of the century it is near certain one or more of these areas will be contributing to Western Europe's oil (and gas) supply – to supplement, if necessary, the potential from the North Sea and so to provide a continuing ability by Western Europe to be self-sufficient or near self-sufficient in oil. In brief, the Western European oil and gas resource base potential appears highly unlikely to constitute a limiting factor in the development of an energy economy based on the concept of maximum possible self-sufficiency.

Conclusions

In the paper I have hypothesised a turning point in the development of Western Europe's economic and political relationships with parts of the Third World; on those parts, that is, on which we have hitherto been able to depend (or to exploit —depending on the viewpoint adopted) for supplies of resources essential to the mass-consumption way-of-life

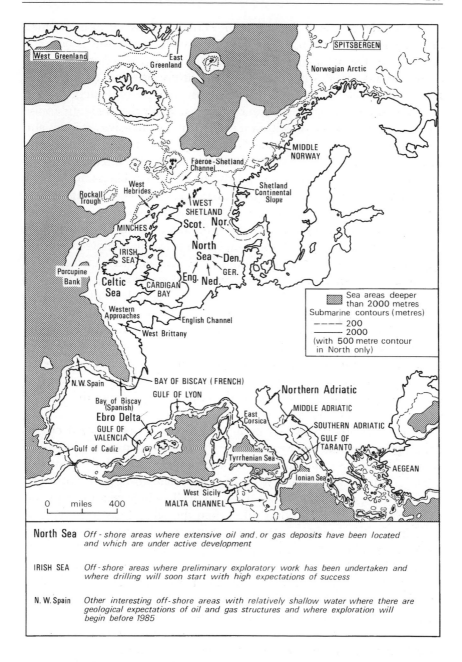

North Sea	*Off-shore areas where extensive oil and, or gas deposits have been located and which are under active development*
IRISH SEA	*Off-shore areas where preliminary exploratory work has been undertaken and where drilling will soon start with high expectations of success*
N. W. Spain	*Other interesting off-shore areas with relatively shallow water where there are geological expectations of oil and gas structures and where exploration will begin before 1985*

Fig I-5.2: Off-shore Western Europe showing the areas where oil and gas reserves have been discovered and are under development and other areas where exploration for oil and gas will be undertaken within the next decade

chosen by the people of the continent. There is no other case in which this changed relationship is as important as with energy; this is not only essential for the maintenance of our economic and social systems, it is also overwhelmingly important for their future prospects. I have thus argued, first, for a reappraisal of energy use patterns and suggested the introduction of a radically different attitude towards the 'freedom' to use energy and the gradual substitution of less energy intensive activities and more conservationist systems of 'doing things' and of 'going places'. And secondly, within the framework of a consequently much reduced rate of increase of demand, I have argued the need for a deliberately autarkic policy in respect of energy supply in a European-wide geographical framework. In this autarkic system indigenous coal production and water power seem unlikely to be able to fill more than a relatively modest part of the total supply needs (of the order of 15 per cent, compared with an approximate 25 per cent contribution in 1975). Neither can there be a realistic expectation that nuclear power will provide the answer. Its contribution may rise to 10–15 per cent in the post-1980 period, but it seems neither necessary nor, indeed, desirable for it to go above that level for at least the rest of the century. The resource costs involved in a nuclear based energy system are up to three times higher than those involved in a conventional (oil and gas) system and, in addition, there are considerations of the public concern for the safety of the reactors and for the long-term dangers of irradiated waste-material from the nuclear stations which serve to confirm this conclusion.

On the other hand, given an appropriate set of policy decisions designed to generate and to guarantee the necessary investments in off-shore oil and gas exploration and development around the coasts of Western Europe and a framework of appropriate co-operative attitudes between the governments concerned, as well as between governments and the international oil companies, then the oil and gas resource base capable of being developed offers an opportunity for near self-sufficiency in Western Europe's energy requirements. Indeed, indigenous oil and gas could supply an estimated 50 to 60 per cent contribution of the continent's energy use throughout the 1980s and the 1990s. It thus constitutes an opportunity for developing an energy system which will help ensure that the European way-of-life can, with some modest modifications, continue to thrive in what is becoming, in economic as well as in political terms, an increasingly hostile world.

There is one final point. The pursuit of energy autarky in Europe ought not to be presented or interpreted as an unfriendly act against the rest of the world or as an element in confrontation politics between 'rich' and 'poor'. Europe's choice of autarky in energy will certainly undermine the control that OPEC has over the supply and price of energy, but in doing so it will eliminate the ability of the member countries of that organisation to sell a cheaply producible commodity at a grossly inflated price (at a 100-fold 'mark-up' on costs!). Europe's energy independence could well provide the mechanism whereby the poor, non-oil producing parts of the world can, indeed, resume their much-needed progress towards economic growth as it would lead to an opportunity for their renewed access at a reasonable price to the world's considerable reserves of low-cost energy in the Middle East and elsewhere, given that these resources will not be required in Western Europe.

Until the time, however, that Western Europe clearly indicates its intention to seek to become self-sufficient in energy by actions designed to stimulate rather than to hinder the exploitation of its already considerable reserves of off-shore oil and gas, other parts of the world are going to continue to suffer from consequential problems and, as a result, create even greater dangers for the stability of the world's economic system.

An energy policy in Western Europe which seeks to solve these problems through the maximisation of its indigenous energy production and some curtailment of its rate of growth in energy demand is a more appropriate approach to the energy sector of its economy than one aimed simply at the conservation of indigenous resources for their possible utility to the well-being of its population in the middle of the 21st century. That time horizon is still too remote for us even to begin to guess what the issues might then be. Our task in the meantime is to ensure that we pursue policies which enable our societies to survive. In this respect, this approach to energy in Western Europe is important. The interpretation and the proposed solutions set out in this paper might appear radical in both economic and political terms – but they are not unrealistic. Their radical nature is simply an indication of the seriousness of the situation – a situation in which traditional solutions no longer seem to be appropriate.

References

1 J. Hartshorn, *Oil Companies and Governments*, Faber and Faber, London, 1967

2 See, for example, *Energy Conservation: Ways and Means* (J.A. Over and A.C. Sjoerdsma, Eds.), Stichting Toekomstbeeld der Techniek, Den Haag, 1974.

3 As, for example, in J. Maddox, *Beyond the Energy Crisis*, Hutchinson, London, 1975

4 See Chapter 3 of P.R. Odell, *Oil and World Power*, Penguin Books, London, 4th Edition, 1975, for further details

5 See, for example, issues of the influential oil industry journal, *Petroleum Press Service* (now *The Petroleum Economist*), through to 1973 for evidence of the oil industry's continued advocation of the free market economy/low energy prices as the desirable norm.

6 As, for example, in the speech made in 1972 at the Algeria meeting of the Organisation of Arab Petroleum Producing and Exporting Countries (OAPEC) by Mr. J. Akins, at the time the U.S. Department of State's adviser on oil questions.

7 In 1974 W. Europe used 1 per cent less energy than in 1973 – and 7 per cent less energy than had been expected to be used in that year. The figures for the first six months of 1975 show a further fall in use such that the rate of consumption is now almost 15 per cent less than hitherto expected.

8 Usually known as Total Energy Systems. See R.W.S. Mitchell and N. Gasparovic, 'Total Energy' in successive issues of *The Steam and Heating Engineer*, Vol. 41, 1974, for a full description

9 See P.R. Odell, *Oil and World Power*, op.cit., Chapters 2 and 3

10 It is ironic that the first real effort in the U.K. rationally to determine an energy policy, the results of which efforts were published in the 1967 *White Paper on Energy* (HMSO, London 1967), was made immediately prior to these changes in the international situation and so produced a study based on parameters which were very quickly to become outdated and irrelevant. See M. Posner, *Fuel Policy: a Study in Applied Economics*, MacMillan, London, 1973, for a presentation of the kind of analysis and results which emerged from a serious consideration of the conventional view of the energy sector.

11 See P.R. Odell, *Oil and World Power*, op.cit., Chapter 9

12 These issues are dealt with in P.R. Odell 'Oil Power in 1975', *The World Today*, Chatham House, London, June 1975.

13 Use of Dutch gas in 1974 was about 85 x $10^9 m^3$. Esso's estimates in early 1975 of the reserves remaining are 2,680 x $10^9 m^3$ giving a R/P ratio of 31.3 years.

14 Assuming oil at a net foreign exchange cost of $10 per barrel and the remittance of an additional $50 million of profits overseas by the oil company which produces the natural gas in the southern North Sea.

15 See the Annual Reports of the *Alberta Energy Resources Conservation Board* on 'Reserves of Crude Oil, Natural Gas etc.' for an explanation of the concept of the normal appreciation of reserves. Kommandeur and De Vries have shown that already-discovered gas reserves in Western

Europe appear to have appreciated by roughly the same factors as in Canada. See 'Natural Gas Reserves in W. Europe' *Energy Policy*, Vol.3, No. 1, March 1975

16 See W. J. George, 'Recent Activity in the British Sector', *Financial Times Conference on Scandinavia and the North Sea*, Oslo, 1974

17 For full details, see P.R. Odell and K.E. Rosing *The North Sea Oil Province; An attempt to Simulate its Exploitation and Development*, Kogan Page Ltd, London, 1975

Chapter I – 6

Europe and the Cost of Energy: Nuclear Power or Oil and Gas?

Western Europe is at a critical juncture in the development of its energy economy. From 1950 to 1973 the region became increasingly dependent on imported oil, in a period when the international oil companies successfully exploited the low-cost oil reserves of the Middle East. Post-1957, within the framework of an intensely competitive marketing situation they were able to make energy available to consumers in Western Europe in rapidly increasing quantities and at decreasing real prices. Events in the international world of oil between 1970 and 1974, including the creation of the OPEC cartel, have undermined the validity of this open energy economy and Western Europe is now faced with the need to reduce the degree of uncertainty to which an over-exposure to dependence on imported oil has placed its economy.

There are two medium-term alternatives (that is, for the period from 1980–2000) on the supply side[1] for reducing Western Europe's dependence on imported oil, viz. the rapid development of nuclear electricity on the one hand, on the other, the development of a large-scale indigenous production of oil and natural gas. These alternatives are here considered as real-world alternatives for the years 1980–2000 and ignore the following considerations:

- that there are uncertainties which exist in relation to the rapid expansion of nuclear capacity, not only technological problems, but also a high degree of unacceptability of the

* **Originally published in *Energy Policy*, Vol 4, No 2, pp 109-118, June 1976**

nuclear option on the part of the populations of Western countries.[2]

- that there are some remaining doubts about the existence of an oil and gas resource base in and around Western Europe capable not only of meeting the new demand for energy in Western Europe, but also of eliminating much of the need for imported oil.[3]

Apart from the necessary study of these uncertainties over the supply potential of the two alternatives, the main economic question affecting the choice between them involves the resource costs necessary to produce the energy required. In this respect, however, there is a great similarity between them; both nuclear power and oil and gas exploitation are capital intensive, and both involve relatively low running costs and employ little labour. For the purposes of this study the running costs and the labour costs involved in the development and use of the alternative systems are assumed to be equal and thus irrelevant to the decision in favour of one of the other – except that some allowance is made for the future capital expenditure which will be required in the oil and gas based system to sustain the productivity of the fields. However, as this is so far in the future (at least 5 years after peak production is first achieved) its present value is low, relative to the front-end loaded investment which has to go into the initial development of the capacity to produce both oil and gas and nuclear power.

The study is thus essentially a comparison of the investment costs required in the two possibilities which exist for expanding the Western European energy supply for the period after 1980.

The two alternatives are, first, the development of nuclear power based electricity production to the highest possible degree, with additional conventional (ie, non-nuclear) energy limited to those uses – such as transport – for which electricity cannot be used; and, second, the development of indigenous oil and gas resources to provide the necessary additional energy inputs to all sectors of the economy, including the electricity component.

Basic Assumptions

- The two alternatives are both essentially indigenous at a Western European level; their development does not involve the transfer of resources in or out of the Western European economy.[4]

- The expected growth of the Western European economy involves a steady quantitative increase in the use of energy, which in heat equivalent terms is equal to 45 million metric tons of coal per year.[5]

- The nuclear option implies the development of what is, as far as possible for the medium term, an 'all-electric' expansion of the energy economy. In this, 60% of the additional energy output is in the form of electricity, with only the remaining 40% provided directly by fuel for end-uses in which electricity cannot be used.

- The oil and gas option, however, is a minimum-possible[6] additional electricity option, so necessitating an expansion of the output of electricity which is equal to only 40% of that in the nuclear option. There is, consequently, a much greater direct use of oil and gas for the larger non-electric sector of the energy economy.

The Expanded Energy Economy based on Nuclear Electricity

This option assumes that 60% of the annual energy growth of 45 million tons coal equivalent (mtce) is to be produced as electricity from nuclear power stations. The remaining 18 mtce required in each year is assumed to be produced from indigenous oil.

The electricity component.

The output of electricity required (in kWh) is equal to the heat value of 27 million tons of standard coal. Given that 1 lb of coal yields 12,500 British Terminal Units (BTU) of heat and that 1 kWh of electricity equals 3,411 BTUs, then the kWh output of the required nuclear power stations is:

$$\frac{(27 \times 10^6) \times 2{,}240 \times 12{,}500}{3{,}411} = 222{,}000 \text{ million kWh}$$

An average power station load factor of 65% is assumed. Whilst this somewhat lower than that which is aimed for in a system in which all nuclear stations automatically go on to base load operation,[7] it is very much on the high side in relation to the development of an energy economy in which most of the additional energy is going to be provided by nuclear power, so that not all the stations can be used for base load purposes. The required installed capacity is now:

$$\frac{222{,}000 \times 10^6}{0.65 \times 8{,}760} = 39\text{million kW}$$

In terms of early 1976 prices in Western Europe, the overall average capital cost per kW of installed nuclear power appears likely to be as high as $1,000.[8] The total annual investment in the nuclear power stations is then $(39 \times 10^6) \times \$1{,}000 = \$39 \times 10^9$. This estimate of the required annual capital investment in nuclear power does not include transmission and distribution costs, which are specifically not included in this study.

The non-electricity component

An additional 18 mtce is required each year for the growing non-electricity demand in the 'all-electric' economy. This is for those uses for which electricity is not an appropriate source of energy for the medium-term future (eg, road, air, sea and much rail transport, some space and industrial heat loads etc). For ease of calculation it is assumed here that this non-electric energy will be made available entirely as oil products derived from North Sea crude oil output.

The quantity of oil products required is approximately 13.5 million tons, which in turn implies a requirement for the production of 14 million tons of crude oil as the input to the refineries (to allow for losses in processing) and thus the creation of 15.1 million tons per annum (= 302,000 barrels per day) of North Sea crude oil producing capacity, in order to allow for roughly 7½% down-time on the producing facilities.

This is a very modest rate of development of North Sea oil producing capacity (compared with the North Sea's overall potential for oil development) and so only necessitates the exploitation of some of the lowest cost fields. For these, an average investment cost of $4,000 per barrel day of producing capacity will be required in the production facilities and in the offshore-to-refinery delivery systems. In addition, there will be a capital investment need for 280,000b/d of refining capacity at a cost of $2,000 per b/d. The total annual investment in the non-electricity sector of the expanding energy economy is therefore:

$$\$(302{,}000 \times 4{,}000) + \$(280{,}000 \times 2{,}000) = \$1.75 \times 10^9$$

The annual required investment is therefore 39.0 + 1.75 = \$40.75 x 10^9. Thus, over the 20 year period from 1980–2000 a total investment of \$815 x 10^9 (in 1976 \$ terms and at early-1976 cost levels) will be required in order to create the nuclear and the non-nuclear capacity in a Western European economy in which the expanding energy needs are met largely by means of nuclear power. It should be noted again that this does not include the required investment in the transmission and the distribution of the energy, nor does it include the cost of the energy users' equipment.

The Expanded Energy Economy Based on Oil and Gas

This is an expanded energy economy in which the increasing availability of electricity is only two-fifths of that in the nuclear based economy. Moreover, in this option the electricity can be partly produced from oil or gas in traditional central power stations and partly in total energy systems, in which the on-site generation of electricity is combined with the recovery and use of the heat output. The remainder of the economy's additional energy needs are met by the direct use of oil and gas.

The electricity component

It is first necessary to calculate the fossil fuel inputs required in order to output 40% of the 222,000 million kWh of electricity produced in the nuclear economy, viz. 89,000 million kWh of electricity. We assume that 70% of this required electricity will be produced in combined power/heat systems (at a city, suburb or factory/office level) and the other 30% in traditional, large central power stations from which – for locational and/or load reasons – there is no recovery of the heat for any useful purposes in the economy.

The conventional centralised thermal power stations have to produce 26,600 million kWh of electricity. We will assume that these power stations operate at an average load factor of 60% (somewhat lower than our earlier assumption of 65% overall load factor for the nuclear power based system, in that the minimum-electric energy system seems likely to give a somewhat lower overall demand load factor in the use of the installed capacity). We assume that these power station swill be able to operate at an average fuel utilisation factor of 38%.

The fuel input and capacity needs can now be calculated[9]:

$$\text{Fuel input} = \frac{(26.6 \times 10^9) \times 3{,}411 \times 100}{12{,}500 \times 38 \times 2{,}240} = 8.5 \text{ mtce}$$

$$\text{Power station capacity} = \frac{26.6 \times 10^9}{0.60 \times 8760} = 5{,}061\text{MW}$$

The rest of the electricity (62,400 million kWh) has to be produced in combined power/heat systems for which we will assume an average load factor of only 55% and a fuel utilisation factor for electricity production of 30% (as it seems likely that such combined cycle systems will be a mixture of steam/back pressure turbines, gas/diesel engine and gas turbine based systems.[10]) The required fuel inputs and installed capacities of these systems are now:

$$\text{Fuel input} = \frac{(62.4 \times 10^9) \times 3{,}411 \times 100}{12{,}500 \times 30 \times 2{,}240} = 25.3 \text{ mtce}$$

$$\text{Installed capacity} = \frac{62.4 \times 10^9}{0.55 \times 8{,}760} = 12{,}950 \text{ MW}$$

In total, therefore, the electricity generating capacity required in this option to produce the electricity component in the system is 5,061 + 12,950 = 18,011 MW. Conventional power stations are currently running at a capital cost/kW of capacity which is roughly 60% of the nuclear stations' capital costs and thus, with a figure of $1,000/kW for the latter, a figure of $600/kW of installed capacity for the conventional capacity would seem to be appropriate.[11] In terms of 1976 $ and at 1976 prices, the annual investment cost in electricity generation equipment in the oil/gas based economy is then:

$$18{,}011{,}000 \times \$600 = \$10{,}807 \text{ million}$$

The non-electricity component

The required production of oil and gas for meeting the heat needs in this option is made up of two parts. First, the production of 18 mtce each year, as in the nuclear electricity economy, for the sorts of uses described in the section on the non-electricity component of the nuclear option. Secondly, the production of an amount of oil products and natural gas which will together provide the thermal equivalent of 60% of the electricity produced in the nuclear economy. Moreover, these requirements for oil and gas production will be additional to the oil and gas required for the production of conventional electricity, except in as far as the combined power/heat systems will produce utilisable heat-energy. This will be available instead of heat which would otherwise have to be produced from the burning of additional oil and gas.

60% of the heat value of the electricity output in the nuclear based economy must in this option be produced as heat-energy, ie. 60% of 222,000 million kWh = 133 x 10^9 kWh; the equivalent of:

$$\frac{(133 \times 10^9) \times 3{,}411}{12{,}500 \times 2{,}240} = 16.2 \text{ mtce}$$

But this amount of heat-energy will only be required as a separate output in as far as it cannot be recovered from the heat produced in the combined cycle systems. As shown in the preceding section, the fuel input to such systems is 25.3 mtce, of which 30% (=7.6 million tons) is converted to electricity. Of the balance of 17.3 mtce we may assume that 60% is recoverable as useable heat —to give a heat availability equivalent to 10.6 million tons of coal.

The only additional output of heat-energy required from conventional boiler systems is therefore 5.6 mtce (viz. 16.2–10.6 mtce). If this is produced in boiler systems with an average efficiency of 75% then the input of fuel must be 5.6 x 1.33 = 7.5 million tons coal equivalent.

The total coal equivalent input required in the oil and gas based economy can now be calculated as 18.0 mtce for transport use, etc. + 33.8 mtce for combined electricity and heat production and 5.6 mtce for supplementary heat production from conventional boiler systems = 57.4 mtce.

Oil and gas production to meet this demand

57.4 million tons of coal equivalent implies the supply of 38.3 million tons of oil equivalent per annum, which has to be produced from indigenous resources of oil and gas. For the purpose of calculating the investment required it is assumed that 20 million tons of oil products will be used, together with the natural gas equivalent of 18.3 million tons of oil.

For the production of 20 million tons of oil products, 21 million tons of crude oil will have to go to the refinery, in order to allow for refinery losses. In order to produce 21 million tons of crude oil, moreover, it will be necessary to develop some 22.5 million tons of producing capacity in order to allow for down-time on the producing facilities. The availability for use of 18.3 million tons oil equivalent of natural gas implies the development of a gas production capacity of 19 million tons of oil equivalent (mtce) also to allow for down-time on the facilities; this capacity is assumed to be split approximately 35:65 between the production of associated and non-associated gas. The total capacity needed will then be 420,000 b/d of refinery capacity, 450,000 b/d of crude oil producing capacity, 135,000 b/d oil equivalent of associated natural gas supply and 245,000 b/d oil equivalent of non-associated natural gas production facilities.

Investments in Oil and Gas Production

Refinery capacity. As indicated above, the investment cost per barrel per day in largely automated refineries, capable of producing a range of fuel products, is $2,000. The annual capital investment necessary to develop the required refining capacity is then 420,000 x 2,000 = $840 million.

Crude oil producing capacity. In July 1975, per daily barrel investment costs in North Sea oil field development and in the building of associated transport facilities (crude oil pipelines to on-shore landing points), ranged up to $5,000 according to the main companies concerned in the North Sea's development. Continued inflation in most costs since then suggest that the early 1976 investment cost could now range up to $6,000 per barrel per day of capacity. In this energy economy option it is necessary to develop all the discovered fields, rather than just the lower cost ones, and the larger fields must be developed to their fullest, rather than to a limited, extent. Thus, the higher figure of $6000 per barrel per day of producing capacity must be taken as the relevant capital resource cost for making indigenous oil available. To this figure, however, there must also be added *the present value equivalent* of future capital expenditure, which will be incurred over time in order to maintain the level of production from these fields for a period of 20 years. With a 20% discount rate this amounts to the equivalent of an additional initial investment of $2,000 per b/d of capacity. Thus to develop 450,000 b/d of capacity requires an investment of 450,000 x ($6,000 + $2,000) = $3,600 million.

Associated natural gas development. Associated natural gas, produced as a joint-product with oil production from oil and gas reservoirs, has a near zero resource cost at the point of production, as all the investment in developing the field has to be made in order to produce the oil. Such associated gas does, however, have to be pipelined ashore separately from the oil. For the 36-inch pipeline which would be needed to move this amount of gas, the investment cost would amount to some $400 million, on the assumption that the line would be about 350km in length from fields to shore.

Non-associated natural gas development. Development of these resources implies capital investment for exploitation as well as for transport facilities. The capital costs involved are, overall, of the same order of magnitude as those involved in oil production and its sea-to-land transportation, which would mean an investment of $6000 + $2000 per barrel per day of oil equivalent. For the equivalent of 245,000b/d of capacity the investment need is 245,000 x $8,000 = $1,960 million.

Summary of investments required in the oil and gas based economy

From the preceding paragraphs, the total investments required in this option can be summarized;

Electricity generating capacity	$10,810 million
Refining of crude oil	$840 million
Developing crude oil producing capacity	$3,600 million
Delivering associated natural gas	$400 million
Developing non-associated natural gas	$1,960 million
Total per year	$17,600 million

Over the 20 year period – at early 1976 prices and n 1976 $ values, the total investment required in the expanded oil and gas economy is thus about $ 350,000 million.

Contrasting Investment Costs in the two Alternative Systems

These calculations indicate that the 20 year investment cost involved (at 1976 prices) for developing the nuclear based, largely-electric economy is of the order of $815,000 million, while the investment costs involved in developing an expanded energy economy based on indigenous oil and gas are about $350,000 million, about 43% of the cost of the nuclear based option. This large cost difference between the options is such as to indicate that the oil/gas based energy system would be preferable in economic terms, but, before this preference can be seen to be justified, other costs inevitably involved in the two alternative energy systems must be examined, to ensure that these are not so overwhelmingly in favour of nuclear power as to reverse the economic case against it as shown above.

Several sets of costs have, indeed been omitted from the analysis, viz. the running costs of the two alternative systems, the costs involved in ensuring that the energy is available to consumers at the points of consumption (ie, the transmission, transport and distribution costs), and the users' investment costs, required to install equipment for consuming the energy supply made available to them.

Running costs

Both systems are highly capital intensive and relatively little labour is required for either nuclear power or oil and gas. There seems no reason to assume, moreover, that one system or the other will have maintenance costs which will be much above or below the alternative.[12] When comparing the economic viability of the two systems, differences between the variable costs are thus relatively unimportant. However, given the special characteristics of the nuclear system (in terms of safety and security arrangements for the producing units and in terms of the long period of surveillance and/or the high-cost reprocessing required for nuclear waste materials), the likelihood seems to be that its running costs will be higher than in an oil/gas system, in which a high degree of automation is already a very successful feature.

Transmission, transport and distribution costs

These are complex and significant for both systems but, in spite of this, they seem unlikely to be critical to the conclusions of this study. This is because the capital costs involved in transmitting and distributing electricity is several times greater than the capital costs of moving energy in any other form (except, possibly, coal by road transport). In effect, the higher transmission and distribution costs for the nuclear based energy economy accentuate its economic disadvantage compared with the alternative oil and gas based system.

Moreover, nuclear power stations must be remote – or relatively remote – from demand centres (for safety and/or public acceptability reasons) so that electricity transmission distances in a nuclear based economy are maximized. In the oil and gas economy, on the other hand, two-thirds of the electricity is produced at the point of consumption in combined power/heat systems, so that there are virtually no transmission costs. The remaining one-third of conventional electricity (equal, however, to only 13.3% of the amount of nuclear-based electricity), will also, on average, have to be moved less far than the nuclear produced electricity, as conventional power stations can be located relatively nearer to demand centres.

Oil products' use in the oil and gas option is about 50% higher than in the nuclear economy, accounting for over 40% of energy used compared with only 25%. The movement of oil products does not require any specific investment in transport infrastructure, as they can be, and are in the main, moved by means of water, rail and road facilities, which either exist or which will be built in an expanding economy, irrespective of the energy situation. The high running costs in such an energy transport system are unlikely to offset more than a small part of the capital costs savings, compared with the electricity transmission system, especially when they are evaluated in present value terms. Overall, therefore, the capital cost advantage in respect of taking the energy to the consumers lies very strongly with the oil and gas based systems – probably by a margin which is even wider than the 7:3 relationship of the capital costs in production.

Users' investment costs

Most components in the users' investment costs will be identical in the two alternative systems, and to this extent they are not a relevant variable in this study. Differences might emerge as a result of the greater use of fuel in the oil/gas economy, compared with the dominance of electricity use in the nuclear based economy. Specifically, these differences would result from any net additional investment costs which users have to meet, in order to enable them to use fuel rather than electricity. It is, however, not self-evident that there would be any net additional costs at all, especially when one notes that gas cookers cost less then electric ones, that gas boilers are no more capital intensive than boilers using electricity, that electrically powered compressors are at least as expensive as gas-powered compressors, and that internal combustion and diesel motors are cheaper in capital cost terms than motive power units based on electricity. In the all-electric nuclear economy, electricity has to be used for purposes for which the direct use of fuel is inherently cheaper. By contrast, in the oil and gas based economy, enough electricity is still available for those uses for which there is no alternative, or for use in sectors where electricity use instead of fuel use implies low capital investment cost in equipment. Intuitively, therefore, one can hypothesise that the indigenous oil and gas-based economy is, overall, a lower capital cost alternative, even as far as end-users' investment is concerned.

Conclusions

Western Europe currently has two medium term options open to it for the supply of energy which will be required by an expanding economy – a nuclear-based all-electric (or nearly all-electric) alternative, on the one hand; and, on the other, one based on the exploitation of the continent's known and likely reserves of off-shore and oil and gas. These reserves, when produced, are, in the second option, used only to make electricity for those purposes for which it is necessary and/or desirable, with, wherever possible, facilities for the combined production of electricity and useful heat – to avoid the great waste of energy inherent to a highly centralised electricity production system.

Both alternatives are high in cost (compared with the 1950–1975 option of using low capital cost foreign oil), but both offer Western Europe a much reduced degree of dependence on imported energy, so minimising the political and strategic, as well as the economic, dangers arising from a continued high ratio of imports to total energy demand. Both alternatives also have relatively low inputs (compared with the capital cost input) for labour and other running costs, and so their relative economic viability can be compared within the framework of an analysis of their comparative capital costs at 1976 prices and dollar values. This present exercise is relatively simple and unsophisticated, but it does compare like with like.[13] It clearly indicates that every dollar of investment in energy producing facilities in the indigenous oil and gas option is more than twice as productive as every dollar invested in the nuclear option.

Moreover, on looking, in a qualitative way, at the implications of the two options for related investment in energy transport and use, it is also shown that the additional resources required to get energy to the customer and to enable him to use it are also likely to be lower in the case of the oil and gas option – so orientating the economics of the two alternatives even more strongly against the development of nuclear power.

In brief, the expansion of the nuclear power-based all-electric economy in Western Europe appears to be highly unattractive on economic grounds,[14] and, if so, then the furious debate over its acceptability on security/safety reasons becomes gloriously irrelevant.

This paper has employed many simplifying assumptions on the variables, but these have been reasoned and/or based on the most recent information and there seems no reason to suspect that the assumptions unfairly distort the case. The next stage in the research is, however, to subject all the variables to appropriate sensitivity analysis to test both the depth and the strength of the conclusion. Such work would seem to be a relevant task for one or more of the many institutions which have to date worked exclusively on the nuclear power option in the mistaken – but still largely accepted – belief that we have no other way open to us for developing the energy economy over the medium term future without heavy dependence on high-price imported oil. It is this kind of unthinking acceptance of a particular line which has produced the situation in which almost all R and D funds in the UK go into nuclear power – at a cost so far of £600 million, compared with only £0.8 million on all other energy R and D in 1975[15].

Only very modest expenditure at national or European levels, is required to finance the economic, spatial and market research, coupled with applied technological investigations, on the potential for the oil and gas based option considered in this paper. Such investigations could indicate just how – and how far – the minimum electricity and maximum heat energy system with geographically dispersed producing facilities, based on the utilisation of offshore oil and gas, could be established in Western Europe for the period between 1980 and 2000. And as the potential for resource saving is up to a magnitude of $465,000 million over the period, such a modest investment in in-depth investiations would seem to be justified.

References

1 There are also possibilities on the demand side for reducing the degree of dependence. Demand limitation should be the first priority of energy policy made. Success in this respect would not alter the conclusions of this study – only the degree of investments required.

2 See A.J. Surrey, 'The future growth of nuclear power', *Energy Policy*, Vol 1 No 3 (December 1973), pp 208–224

3 These doubts are examined in P.R. Odell 'Indigenous oil and gas as alternatives to OPEC oil' in *Energy Policy Planning in the European Community*. (Ed. J.A.M. Alting von Geusau), Sijthoff, Leiden, 1975

4 This assumption is relatively more favourable to the nuclear power option for which, in addition to the required flow of American know-how and equipment (matched in the oil and gas option), there is also a continuing need to use foreign resources in importing natural or processed fuel for nuclear power stations.

5 After allowing for conversion (heat) losses in thermal electricity production in conventional central power stations this is a reasonable approximation to the annual amount of additional energy which, it has been estimated by the OECD, will be required in the Western European energy economy, viz. about 75 million tons of coal equivalent (mtce) per year. See OECD, *Energy Prospects in 1985*, Paris 1975, Vol 11, Tables 2D-10 to 12. pp 23–25

6 Note, however, that the assumed 27% electricity component in this oil and gas based option is still higher than the present (under 10%) share of electricity (measured in heat value terms) in the Western European energy economy. The option, in other words, does *not* mean a reversal of the trend towards more electrification.

7 Though note that the average load factor achieved in all nuclear power stations in non-communist countries in 1974 was only 58%. Only Canada, Spain and Switzerland (with 11 reactors out of the total of 130)

achieved a load factor of more than 61%. *Monthly Energy Review*, FEA Washington, December 1975. p35

8 Compare the optimistic estimate of $550 in the EEC publication, *A New Energy Strategy*, published in August 1974, and based on cost data from earlier in that year, with the February 1975 announcement by the British government that investment costs in the two new nuclear power stations it was planning would be over $800 per kW. Not only has inflation continued in 1975, but nuclear technology appears to have been particularly susceptible to price escalation – in part because of longer than expected building times and in part because of the need for a greater attention to security and safety considerations. A figure of $1000 thus seems eminently realistic for 1976 – even though it is on the high side compared with the figures which are still used for governmental planning purposes.

9 For conversion factors used in the calculation see second paragraph of the section on the nuclear option.

10 See R.W. Stuart Mitchell, 'Total energy' *Steam and Heating Engineer*, Vol 41, 1974

11 There is assumed to be no difference between the capital costs of producing electricity in the centralised and in the combined power/heat systems. Although the latter do have a somewhat higher capital cost per kWh, part of the investment in these systems can be offset against heat-raising capital costs which would otherwise have to be incurred. These, for the purposes of this study, are considered to be user-costs and thus are not included in the comparison here.

12 Except in the case of the oil/gas system in which a continued flow of capital into the producing fields is required, in order to ensure the ability of the original investment to continue to output oil and/or gas for the 15–20 years required in this analysis. This continuing capital cost component has been included – in present value terms – in the capital costs calculations.

13 That is, in terms of the way consumers can obtain their energy needs, given the contrasting technologies of the two options. Compare this energy systems' based approach with the usual biased approach of the nuclear protagonists who simply compare a nuclear power station's costs with an oil or coal fired power station's costs as though fuel-to-electricity component were the only variable! See, as a particularly good example of this approach, H. Mandel. 'Construction costs of nuclear power stations' *Energy Policy*, Vol 4, No 1 (March 1976), pp. 12–24.

14 And even more so in the specific cases of say, the UK, the Netherlands and Norway, where oil and gas resources already proven indicate clearly the absence of any need to consider the nuclear alternative for the period up to the year 2000. For other countries in Western Europe the question is complicated to some degree by the fact that some of their oil and gas may have to come from one of the three aforementioned

countries, so raising issues of transfer prices rather than of resource costs. However, as nuclear power also involves transfer prices (for technology, fuel and equipment etc, from the USA), a West European solution seems likely to be preferable.

15 Editorial, *Energy Policy*, Vol. 4 No 1, March 1976

Section II

The Exploitation of Western Europe's Indigenous Hydrocarbons

Chapter II – 1

Natural Gas in Western Europe: a Case Study in the Economic Geography of Energy Resources*

Introduction

Except in the U.S. and the U.S.S.R., it is only very recently that natural gas has come to have a meaningful existence, except as a commodity produced at various locations as an inevitable, and often unwanted, accompaniment of oil production. The geography of natural gas production has thus reflected the pattern of oil production, modified only by contrasts in the amounts of gas produced per unit of oil production at different locations. There was little effective demand for most natural gas produced as a by-product of oil development and its consumption, in a physical sense, thus took the form of flaring at well-head – the lowest-cost way of getting rid of this unwanted joint product.

Even in the United States, the geographical isolation of the main oil and gas producing areas from the main energy consuming regions meant that much gas had to be flared until the development of pipeline technology made its long distance transmission possible, thus enabling it to be sold in the energy consuming areas in competition with other fuels.

This ability of natural gas to overcome the friction of distance in the U.S. has been largely concentrated into the last 25 years or so. Similarly in the Soviet Union, where much the same phenomenon of the geographical separation of gas production and energy consumption

*** An edited version of the author's Inaugural Lecture on his appointment in 1968 to the Chair of Economic Geography at the Netherlands School of Economics, Rotterdam (De Erven F. Bohn, Haarlem, 1969)**

also occur. There, major natural gas transmission lines are still under construction to allow the substitution of gas for other forms of energy in increasingly wide areas of the country.

Meanwhile, in the rest of the oil producing world, oil's joint product – natural gas – continues, in the main, to create both costs and embarrassments, rather than revenues, for the producing companies. Only when the natural gas can be reinjected into the producing formations at costs low enough to be more than offset by the revenues arising from the consequential additional oil produced by this method of secondary recovery do the companies enjoy a return on the gas production. Otherwise, the producing company must, at best, incur costs merely to flare the gas or, at worst, as a result of governmental pressures which deplore the "waste" of the country's natural resources, incur even larger costs, unmatched by revenues, of putting the gas back underground for possible future exploitation or of making it available at low prices for use within the framework of the country's energy economy. The remoteness of the major oil and gas fields from the main centres of energy demand in all the major producing nations has the effect of making such developments generally unremunerative to the companies facing such obligations.

Natural gas production in the oil exporting nations thus largely remains a commercially unmarketable commodity as a result of its limited ability to overcome the friction of distance, particularly in competition with the lower-cost movements of crude oil and oil products. Thus, the geographical separation of natural gas production from potential markets has been a recurring theme in influencing the speed and degree of development of the resource.

It is in contrast with this limited utility of natural gas elsewhere that recent events in Western Europe appear to be of a significance which far exceeds the relatively low key in which gas discoveries have so far been played. The importance of the massive Groningen gas field in the Netherlands (the world's largest to date) lies in the fact that, for the first time in the history of major natural gas finds, it was possible to produce an immediately marketable product. There was no need to "hawk" the gas around looking for any sort of markets whereby a minor cash flow could be generated, and there was no need to close down the field as had been the case just a few years earlier when exploration for oil in the Algerian Sahara had embarrassingly discovered an equally big gas field.

Groningen's unique position was, in essence, a matter of geography – that is, a matter of the spatial juxtaposition and interrelationships of a very large potential supply of gas with the very large potential markets existing for it in the world's most geographically concentrated area of energy use – north-west Europe. Such spatial considerations are clearly of no small matter when viewed in the light of the weight of transport costs in the delivered prices of natural gas, arising not only as a function of the length of the transmission lines, but also as functions of their diameters and load factors. European scale distances, coupled with the geographical intensity of energy demand and the existence side by side of customers with contrasting temporal demands combined to produce a low transport cost element in the delivered price of Groningen gas, compared with the situations in the U.S. and the U.S.S.R, (see Figure II-1.1) and hence a new resource favourably placed to provide effective competition with existing energy supplies.

Given many producers as well as many consumers, one could have hypothesised a "scramble" for markets in which the output of the Groningen field was quickly bid up to its technically optimum rate and the price of the gas brought down to a level at which a customer could be found for the last m^3 of profitable production. Such conditions have not, however, prevailed in the case of Groningen which has been the responsibility of a single joint-producing company with only three component elements. These are Shell, Esso and the Dutch government with commercial or national interests which clearly indicated that a simulation of an open market situation in the producing and marketing operations would not be justified. Without getting unduly involved in the controversy that has already been generated by this situation, the available evidence and comparable empirical and theoretical studies of the gas industry elsewhere all point to a non-competitive production and pricing response to the development of Groningen.[1] To illustrate this conclusion, one can point to the existence of a major contrast between production levels which are possible and those that are planned, coupled with a pricing structure which enables "super-normal" profits to be earned.[2] The maximum production planned appears to be of the order of 50,000 million cubic metres annually,[3] whereas the known optimal production rate is at least twice this amount. The well-head price appears to be of the order of 4.3 Dutch cents per cubic metre (= approximately 35 U.S. cents per million Btu) as Groningen is known to have gas producing conditions of a degree of excellence which are virtually unmatched in the world outside the Soviet Union. In that these conditions are coupled with a rationalised producing operation of a scale and technical finesse such as is only possible when a large field can be worked by one highly competent operator (Shell), the costs at the levels of output achieved and planned certainly fall well below the average costs

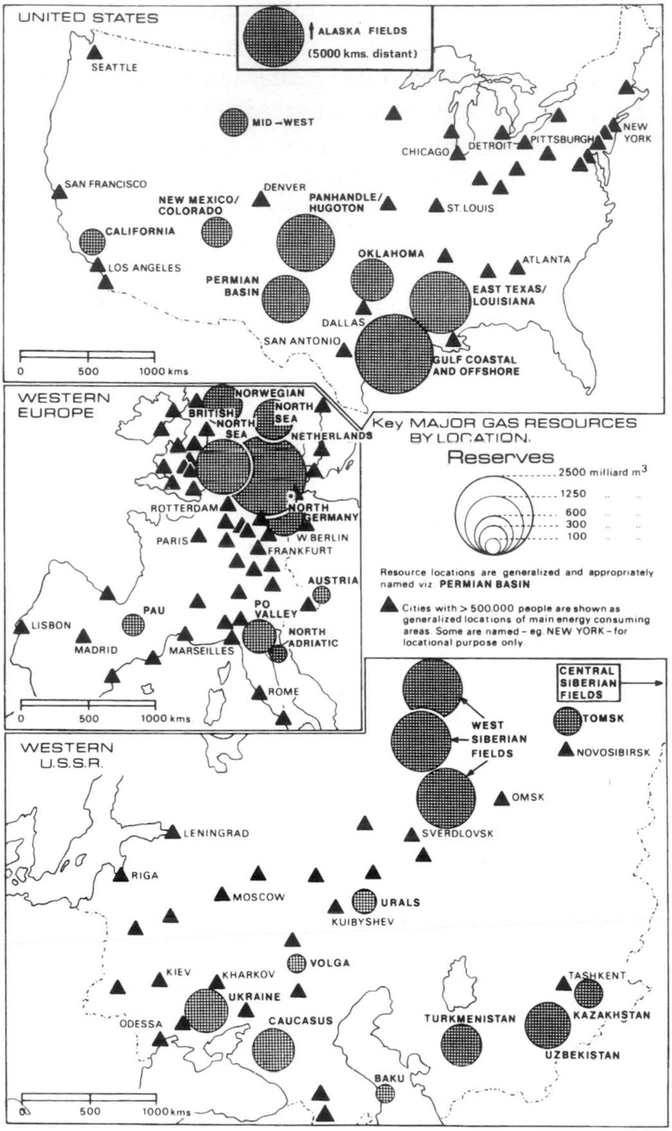

Figure II-1.1: The Geographical Relationship of principal Natural Gas Producing Locations with main Energy Consuming Areas in the United States, the Soviet Union and Western Europe.

In both the U.S.A. and the U.S.S.R. the main energy consuming areas are remote from natural gas supplies. By contrast the major supplies of natural gas within Western Europe are in the heart of the areas of heaviest energy consumption. Other things being equal, this situation should ensure the most rapid development and utilisation of Western European gas and consequent enhanced economic advantages for the continent's energy users

for gas production in the United States – yet there the average well-head price of gas (approximately 1.9 Dutch cents per cubic metre) in a profitable industry is well under 50% of the Groningen figure.[4]

Thus, there is good evidence to show that levels of output and the price of Groningen gas have been determined by other than "the competitive forces of the market place" and that neither the quantities to be produced nor the price have responded to the nature of this occurrence and the volume of the field's resources. One can thus argue that Shell and Esso, which have alone had the responsibility for selling the gas outside the Netherlands, were enabled to calculate the point at which such sales, taken in conjunction with the anticipated government controlled markets inside the countries concerned, maximised their profits. This would normally have involved a complex evaluation of the supply and demand characteristics of the field and of the markets respectively, but it seems likely with a field as large as Groningen and with its range of production possibilities, that the marginal cost curve could be assumed to be horizontal and the supply function thus ignored.

The exercise they had to undertake resolved itself essentially into a study in applied economic geography in that the marginal revenue curves arising from potential sales to different consumers in different locations had then to be "netted back" into well-head equivalents – i.e. excluding transport costs. But in that transport costs are a function not only of distance, but also of volumes to be moved along specific routes, the exercise demanded a high degree of knowledge of the geography of energy consumption within Western Europe. This availability-of-necessary-information, within the framework of the chosen ordered approach to marketing Groningen gas, thus appears to offer an adequate explanation for the decision to leave the gas export business in the hands of Shell and Esso. These business organizations were immediately able, through their international structures, to collect and process the immense amount of data required for the exercise. In parenthesis, one might note here the importance of a generally ignored factor which help to determine the geography of economic activities, viz. the nature and structure of the firm or enterprise. Had Groningen been discovered and developed by enterprises other than Shell and Esso, then the geography of natural gas in Western Europe would have evolved in a different way, not least because of a lower level of information that would have been available to others as a basis for decision taking.[5]

Shell and Esso's Market Evaluations

The results of Shell and Esso's complex market studies are, of course, closely guarded commercial secrets, but from the information that has been made publicly available[6] and from an examination of the nature of specifically located markets for energy in Western Europe, one can estimate that the optimal pattern of development for the foreign sales of Groningen gas emerged along lines similar to those shown in Figure II-1.2. Optimal in this context represents the nearest possible approach to a pattern producing, in present value terms at the time of the study, maximum profits for the companies concerned after taking into account the following constraints: the need to avoid frontier price discrimination in export sales in light of the Rome Treaty and other international obligations of the Dutch government; and the requirement to earn sufficient foreign exchange from sales abroad, readily to assure the Dutch government that the country's new resource was to play an effective role in strengthening the economy. Arising from these considerations, the optimally-possible market solution (additional to the estimated market for 20,000 million cubic metres of gas per year in the Netherlands by 1975), required the sale abroad of some 30,000 million cubic metres per year. This was destined mainly to relatively near neighbours, but some sales were anticipated to more distant markets, where opportunities existed for replacing or supplementing inadequate local supplies of gas (as in Austria and Italy) and where, therefore, imports would command prices sufficiently high to cover the greater transmission costs over the longer distances involved.

Planned annual output has thus been restricted in total to some 50,000 million cubic metres (at only about 50% of possible production) and, with the exception of sales within the Netherlands itself and to some degree in nearby Belgium, gas has been made available to distributors at prices which made resale possible only in "premium" uses, plus some short-term availability of interruptible gas to bulk energy users at lower prices so as to achieve earlier high load factors on the main pipelines. In particular, the optimum solution precluded any attempt to break into the low price, bulk energy market in West Germany or into the low-price markets provided by consumers such as the iron and steel industries of the Saar, Luxembourg and Eastern France.

At this point one must speculate just how far this optimal solution, within the framework of the north-west European spatial monopoly for natural gas that Groningen enjoyed, was also affected by another aspect of the structure of the enterprises involved. We have already seen how Shell and Esso, by virtue of their widespread and comprehensive

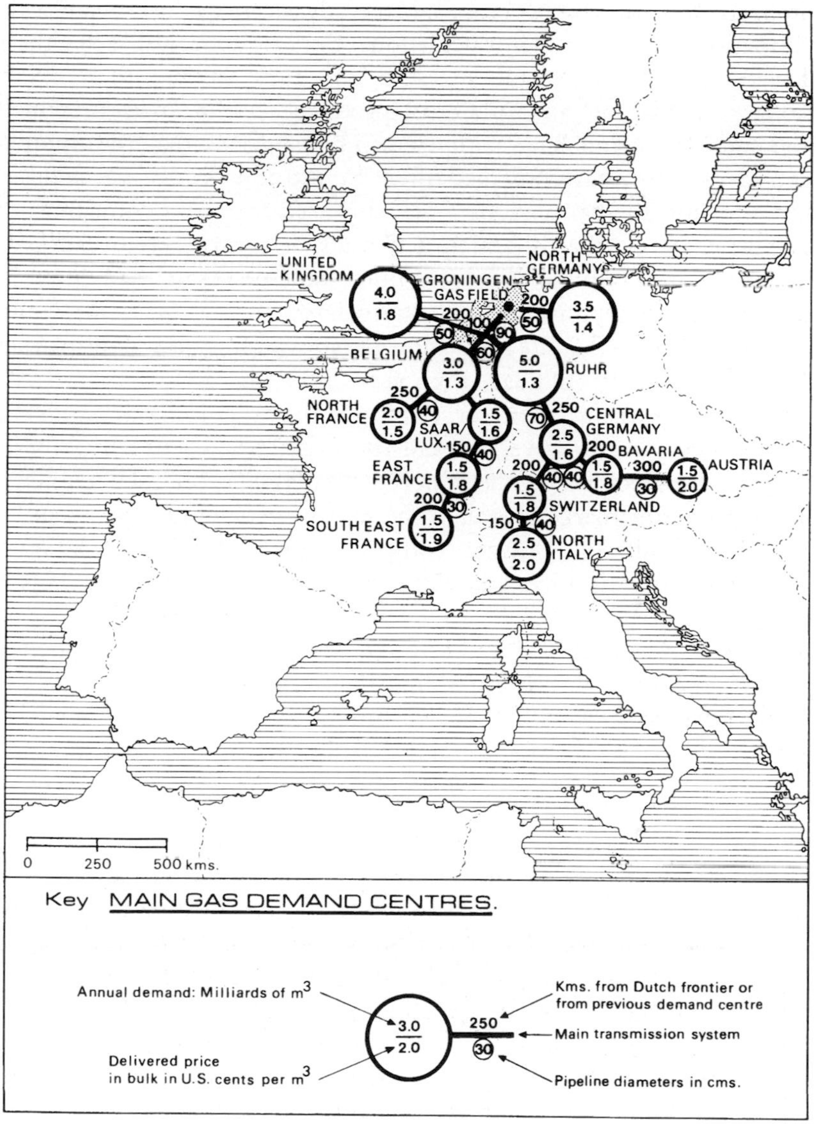

Figure II-1.2: Groningen Gas; a Reconstruction of NAM Gas Export's planned Western European Marketing Strategy for 1975.

Within a planned annual production of about 50 milliard cubic metres for the Groningen field, exports from the Netherlands were to reach slightly more than 30 milliard cubic metres annually. These estimates of the quantities planned to be sold in each main demand centre reflect the 1975 possibilities for the penetration of natural gas into the energy markets at the delivered prices indicated here. The delivered prices have been calculated on the basis of a frontier price of 4.3 Dutch cents per cubic metre plus transport costs over the distances indicated through pipelines of the diameter shown

European oil marketing operations, had ready access to the demand-side information needed for their calculations. But it is conceivable that their calculations were also made so as to take into account their major markets for fuel oil and other oil products in the areas in which Groningen gas could have competed. This suggests that the optimal solution may also have emerged after the introduction of yet another constraint into the calculations – viz. the need for these companies to maintain their previously anticipated rates of return on investments made in their oil refining and distribution systems. Again, had different enterprises been responsible for Groningen gas, such a constraint would not have applied so that the resultant geography of the natural gas market in western Europe would have been very different.

One can now turn to examine the repercussions of the development. First, however, it seems necessary to present a hypothesis of the way in which the market for "basic" energy (as opposed to energy sold in premium markets) over some part of geographic space can be divided geographically between competing fuels available at different locations at different prices and with contrasting transport costs. This is illustrated in Figure II-1.3. Natural gas available at location OG, with a supply price of SPG and with transport cost curves as shown (viz. by the upward sloping curves away from SPG reflecting the price of the commodity increasing with distance from the location OG), competes in the market AB only between c and f, with oil from OP providing an

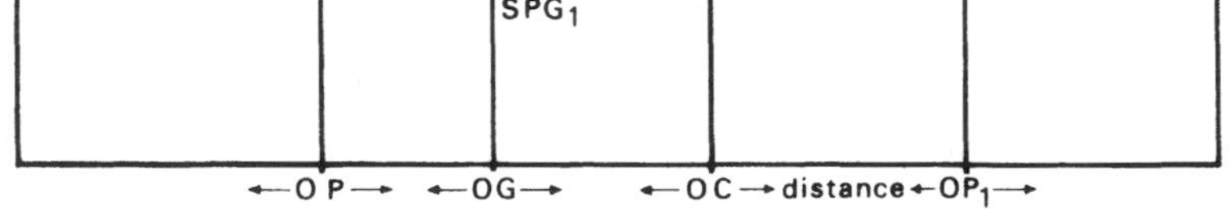

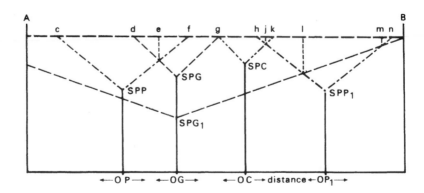

Figure II-1.3: Spatial Divisions of the Basic Energy Market
Between petroleum suppliers at OP and OP_1; a gas supplier at OG; and a coal supplier at OC

alternative source of energy between c and e. Gas from OG will, in theory, have an exclusive market only between e and f, and a steadily declining part of the market as one moves from e to c with oil taking over as the more important energy source at d.

This indicates how Groningen gas, available at well-head at just over 4 Dutch cents per cubic metre, can compete in the basic energy market of Western Europe. Its market penetration is limited to most of the market between e and f (viz. to much of the Netherlands itself): to some of the market from e to d (viz. to the rest of the Netherlands): and to a little of the market from d to c (viz. in parts of Belgium). However, if the price of gas at the point of origin is now reduced from SPG to SPG_1, then this lower well-head price, together with the now less steeply sloping transport cost curves (arising from the economies of scale that can be achieved through larger deliveries along the expanded pipeline systems), will now make gas competitive with alternative fuels originating at all other supply points. It thus now secures all the market except for the area between k and l, where oil from OP_1 can just compete. But for the producer of gas at OG this is sub-optimal behaviour, for with the much higher level of production required to meet the expanded market's demand, marginal costs rise above marginal revenues which of course decrease as market penetration is increased. This cuts into surplus profits and thus will only happen if and when there is an event which threatens to eliminate the spatial monopoly that the single gas supplier enjoys so that his output is pushed back well below the point at which profits are maximised.

Such events could have been responses from alternative energy suppliers at OP and OP_1 (oil) and at OC (coal), but these do not have enough incentive (or ability) to provoke a battle for the premium fuel markets and are content to take no action unless their shares of the basic energy market are threatened by gas. Thus, in the absence of external forces, there was no reason why Shell/Esso's profit maximisation model should not have worked. Indeed, until very recently, there seemed every possibility that the orderly marketing of Groningen gas would proceed as planned with a restricted output, selling at a price several times the long run supply of the gas, earning a large monopoly profit for the producers and severely limiting the geographical and economic impact of natural gas in Western Europe.

Politically, there was little reason for the Dutch government not to accept this situation for not only was the Netherlands getting an economic advantage from the gas sold locally in bulk at approximately fuel oil parity prices, it was also assured of continuing and reasonable foreign exchange earnings from exports. Moreover, France, Germany and, to a lesser degree, Italy were relatively content for they each had energy interests to protect, viz. high cost oil and Lacq gas in the case of France, Ruhr and Saar coal in the case of Germany and the price structure for Po Valley gas in the case of Italy. Only Belgium felt unduly grieved.[7]

But in spite of the combined effects of these political and economic factors, the carefully calculated strategy for marketing Groningen gas in West Europe has started to disintegrate under the impact of external forces – some unexpected, whilst others, though not excluded entirely from the calculations, have gathered strength at a rate considered improbable in the policy planning of Shell and Esso.

Countervailing Forces of Alternative Surplus

Figure II-1.4 repeats the spatially expected patterns of NAM gas export sales from Groningen but, in addition, it introduces the location and some indication of the potential availability of peripherally competing supplies of natural gas for which energy intensive, industrialised Western Europe provides attractive outlets – by virtue of the decision of NAM to produce Groningen gas only "up to the point beyond which any additional product offered would so lower the price that total profit would decrease"[8] with the result that no attempt was made to capture the basic energy market outside the Netherlands. The map shows that this decision of NAM has spurred competition from virtually all directions.

The North Sea

Gas is becoming available from the non-Dutch sectors of the North Sea basin and is potentially available from companies other than Shell/Esso in Dutch waters (assuming Gasunie will not be interested in this gas for Dutch markets). The 1968 price agreement on British gas (approximately 2.8 Dutch cents per cubic metre at well-head)[9] opened up a wide local gap with the price at which Groningen gas was on offer. The success achieved by the British Gas Council in its North Sea operations and in its gas price negotiations with the producing companies, eliminated the planned element in NAM Gas Export's strategy for sales to the U.K. Even more important, the negotiated price for UK North Sea gas is sufficiently below the Dutch price to enable the British gas to be delivered in the Netherlands (if this were politically possible) and elsewhere – notably in Belgium, France and North Germany – at border prices below the export price established by NAM.

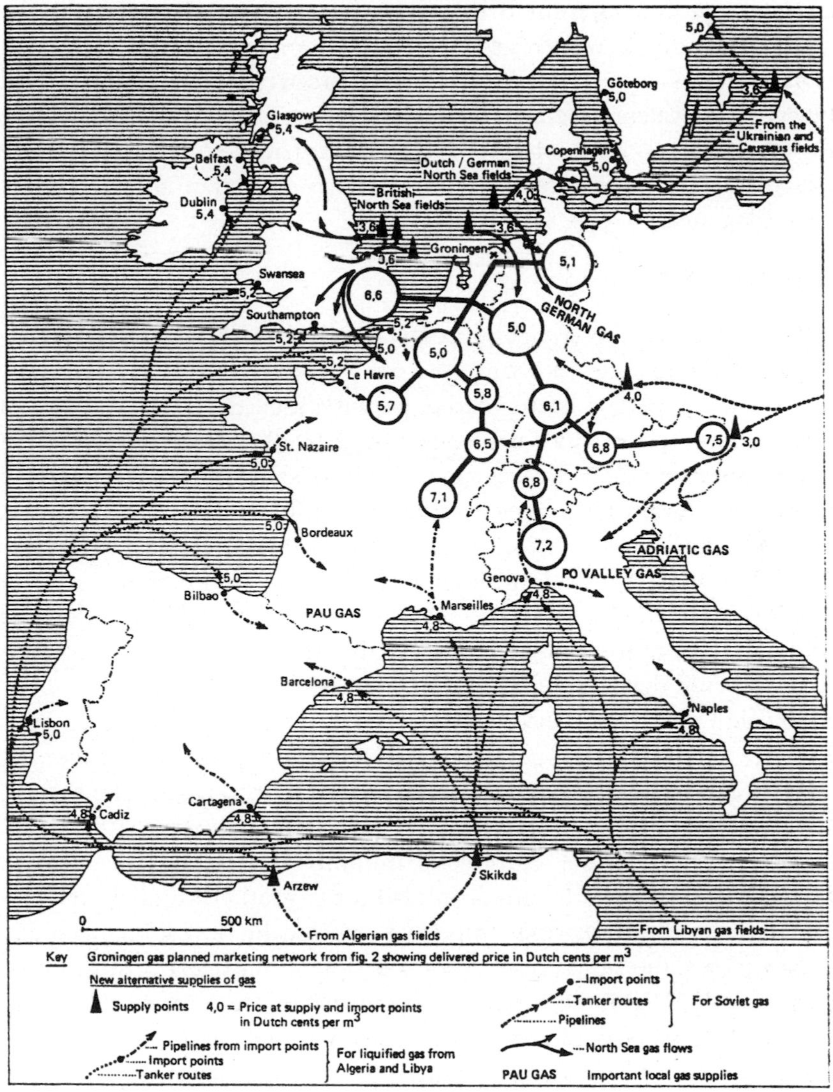

Figure II-1.4: Western Europe's Natural Gas Supplies in the 1970s: Competition for Groningen Gas

In this map, superimposed upon the reconstruction of the planned 1975 marketing strategy for Groningen Gas (see Figure II-1.2), are potential supply routes and supply points of North Sea, North African and Soviet Gas. Estimates are shown of the likely supply price of these alternative supplies at various import terminals. These prices may be compared with the prices which were anticipated in different markets for Groningen gas. The degree of contrast at different locations indicates the likely degree of competition which may be expected

Prospects of exports from Britain have been limited, however, by the national decision to use the gas in the large home energy market, but the availability of other North Sea gas – from German, Danish and Norwegian waters or from companies other than Shell/Esso in the Dutch sector – is anything but remote, such that the marketing strategy for Groningen gas is potentially undermined even close to the heart of its market area.

The Soviet Union

From the east the competition for Groningen gas is geographically more distant but, to offset this, one has to note that it is virtually unlimited (in relation to the market potential), except in the short-term when supply possibilities are constrained by the speed at which the required transmission systems can be built. Backing up the availability of this gas are the already proven reserves in the Soviet Union of over 9,000 milliard cubic metres – approximately four times Groningen's reserves (See Figure II-1.1). These give an annual production potential of over 400,000 million cubic metres to serve a current national Soviet demand of under 40% of this potential. And in that reserves are increasing even more quickly than the high rates of increase in consumption, a Soviet motivation to achieve the highest possible level of gas sales to West Europe as quickly as possible is clearly indicated. Moreover, to supplement its own resources, if necessary, and/or to reduce the costs of supplying the European markets that it secures, the Soviet Union can make use indirectly of the 10,000 million cubic metres per year of Iranian gas that it will be importing from the southern Iranian fields by 1970. By substituting this imported gas locally for Ukrainian and Caucasus gas, the latter can be made available to Europe at a substitution cost of only 2.2 Dutch cents per cubic metre: the cost to the Soviet Union of Iranian gas, plus transport costs to Western Europe. And as this gas can be moved for almost all the intervening distance through a gas transmission system being built primarily to serve the western parts of the U.S.S.R. and Eastern Europe, the Soviet economy will only need to recover the marginal costs of moving it, viz. only the difference between laying and operating, say, a pipeline of 102cm diameter, rather than the somewhat more modestly-sized line which would have been needed to make gas locally available. Thus, Soviet/Iranian gas could be offered for sale at the Western European frontier, in the quantities as large as it is possible to envisage by the mid-1970s, at a price as low as 3.0 Dutch cents per cubic metre.

In light of this favourable supply position, and in light of the fact that the Soviet Union has a strong economic motivation to sell its gas in this way, one can see quite clearly just how much of a geographical restraint Soviet gas places on the calculated marketing strategy for Groningen gas which is priced at the frontier some 40% higher than the alternative Soviet gas. Part of the outlet for Groningen gas anticipated in Austria has already gone to Soviet gas and, on delivered price comparisons alone, Groningen would lose its anticipated markets in Italy, Switzerland, Southern Germany and possibly even Eastern France where, however, the additional factor of Franco/Soviet cooperation indicates a preference in some circles for Russian as opposed to Dutch gas. Thus, potential, large-scale Soviet competition does much more than nibble away at the high cost fringes of Groningen's market area. The high export price established by NAM makes Russian gas attractive over areas within which no danger from alternative supplies was expected.

North Africa

There is, however, yet another major source of external gas which must be considered for its potential impact on the geography of the European industry. This is gas originating to the south, in the major producing fields of North Africa. Because of the absence of local markets and the existence of technological constraints on its transportation, this gas has so far remained almost without value, such that the associated gas from the Libyan oil fields has had to be flared, while the gas fields of Algeria have been largely unexploited.

Recently evolved techniques of liquefying the gas for transporting in specially designed ocean tankers have enabled a little gas to be moved to Europe. The likely availability of such supplies at about 7 Dutch cents per cubic metre, delivered and regasified at a point of import, had already persuaded NAM Gas Export that it would not be worthwhile to seek to compete in Spain, Southern France and most of Italy which were, therefore, (as shown in Fig. II-1.2) excluded from the planned market areas. However, the technical success of relatively small-scale liquefaction for the European market and a quickly emerging larger-scale technology from Japan's decisions to import LNG from Alaska and the Middle East, now suggest that North African gas can be viewed as a major, rather than a minor, potential source of Western Europe's energy. On the supply side, the problem is one only of completing facilities – viz. wells, pipelines, liquefaction plants etc – for the supply potential far exceeds the amount that can be placed. From Libya alone associated (near-zero cost) gas from the country's oilfields could currently support annual deliveries of some 30,000 million cubic metres (about 60% of the planned maximum output of the Groningen field), with the supply potential steadily increasing as oil output continues to rise rapidly. In Algeria, the Hassi R'Mel gas field could, on full development, turn out to be larger than Groningen, and thus capable of producing over 100,000 million cubic metres per annum. Other gas fields even deeper in the Sahara have not been evaluated, other than to show that they too are major fields by world standards and also likely to be capable of low-cost exploitation. Thus a North African equivalent of at least two Groningens is certain and the possibility of the equivalent of four by 1975 is not an unreasonable higher estimate. Such large amounts of gas could not find outlets in southern Europe alone, where the demand for energy is relatively limited and where there are existing local reserves of natural gas notably in south-western France at Pau – capable of a long-term production of over 5,000 million cubic metres annually – and in Italy, principally in the Po Valley where the present annual production, now of the order of 10,000 million cubic metres, seem capable of significant expansion on the basis of large discoveries immediately offshore in the Adriatic Sea. What now emerges, therefore, as a possibility in the developing pattern of natural gas supply to energy-intensive parts of Europe is one or more major northward running gas pipelines from the South European import terminals where the supply price of the regasified and purified methane will be a maximum of a little over 4 Dutch centres per cubic metre – or almost exactly the same as the frontier price of Groningen gas. Thus, North African gas could compete in markets well into western Europe, even taking into account its somewhat higher transmission costs, given the existence of the expensive-to-cross mountain barriers of the Alps and the need for more compression units on the lines south from Groningen.

Moreover, competition from North African gas will not be restricted to areas which can be served by pipelines from the Mediterranean coast. With the increasing economies of scale which will gradually be achieved through the use of larger LNG tankers, North African gas will be available to western and northern European import terminals at supply prices very little higher than those at southern European ports and will thus offer an alternative to Groningen gas, even in areas in closer geographical proximity to it. This could be of particular significance for some of north-west Europe's energy intensive industries – such as the steel works and chemical plants that are located on coastal or estuarine sites and are equipped with their own bulk importing terminals. Such large potential gas users might prefer to negotiate arrangements for the direct purchase of LNG, rather than buy gas through a state-owned or statutory-controlled distributor whose pricing policies do not necessarily reflect the economies of scale that can be achieved in providing gas to such large scale consumers.

Market Competition

Thus, not unexpectedly, NAM Gas Export's profit-maximising exploitation of its real monopoly has quickly had the effect of attracting potentially very large alternative supplies of natural gas to Western Europe. The "threat" posed to the carefully calculated and evaluated market strategy for Groningen by the availability in 1975 (given the completion of the required infrastructure of wells, lines and tankers etc) of some five or six times as much gas as Groningen is then scheduled to produce, seems very evident. The two companies concerned, however, still seem inclined to write down, at least in public, the dangers to their highly articulated policies.[10] Perhaps they are relying in this respect on the existence of long-term contracts, which tie certain customers (sometimes part-owned by the two companies) to them through pipelines which they also own or control, to save the situation. Though such vertical integration will undoubtedly help to maintain NAM Gas Export's position longer than would otherwise be the case, it seems unlikely to give other than very short-term protection, as one cannot envisage price discrimination of more than a fraction of a cent per cubic metre being politically acceptable in those countries where high price contract for Groningen gas will soon be outdated by events.

Thus, my geographical predictions for the natural gas industry in Western Europe by the mid-1970s indicate a more complex pattern of supply than has hitherto been indicated, coupled with the prospect for a greatly enhanced degree of penetration by natural gas into the European energy economy. This pattern seems likely to emerge in two distinct stages. The initial reaction will be to absorb "some" of the transport costs to more distant markets (a procedure which can be followed, given the producers' ownership of the main transmission lines) and to introduce small reductions in the frontier prices for new contracts – as, in fact, happened early in 1969 with the announcement of such a 5% price reduction. Such reactions are explicable in terms of an attempt to hold to the strategy of profit maximisation through limited sales to the premium markets only. But the attempt seems unlikely to succeed for two reasons: first, because the price reductions will not hold the more distant markets, or even some closer markets, against the competition indicated; and secondly, because it will lead to political difficulties as a result of the discriminatory pricing arrangement which it implies.

Thus, the second-stage reaction will be initiated involving a fundamental reappraisal of the strategy underlying the marketing of Groningen gas. Instead of a strategy based on its sale to premium markets, over a geographically extensive area and only up to the quantity where marginal revenue just equals marginal cost with a consequential monopoly profit, there will now have to be a strategy based on blanket sales in the energy markets (including the low priced, basic energy markets) over a geographically much more restricted area. With this new strategy, Groningen gas will become highly competitive and will actively seek out markets in competition with oil and coal as well as with natural gas from other sources. Output from Groningen will thus be expanded at least to the point where the marginal cost curve intersects the demand curve and the price will move down to eliminate the monopoly profit and to approximate to the long run supply price. The net result will be the optimum rate of production of the field – say 100,000 million cubic metres per year instead of 50,000 – with the gas available at an export price of roughly 50% of its present level for sale to markets almost entirely within a distance of no more than 300 kms from the Dutch frontiers. And what Groningen gas does in its geographically restricted market area, gas from the other sources described above will do in theirs, though some inevitable degree of overlap in the market areas will emerge as a result of individual decisions taken for other than economic reasons. My estimate of a possible result of this in terms of the economic geography of natural gas in Western Europe by 1975 – or thereabouts – is shown in Figure II-1.5.

By this time Western Europe will be consuming some 270,000 million cubic metres of natural gas per year. Of this, the Netherlands will supply about 43%, with Groningen itself producing up to its limit of 100,000 million cubic metres and the rest coming from other fields both on land and under the Dutch sector of the North Sea. Other indigenous Western European supplies will come from the British and Danish sections of the North Sea with some 30,000 million cubic metres from the former – whose production will, in fact, be constrained by a marketing bottleneck in the U.K. itself, coupled with a failure to develop export markets above a very low level – and about 5,000 million m^3 from the latter. German production – from fields on land mainly between the Dutch order and the River Weser and from North Sea discoveries – will amount to some 25,000 million cubic metres, while local production in France will be just under 10,000 million cubic metres and that from Italy

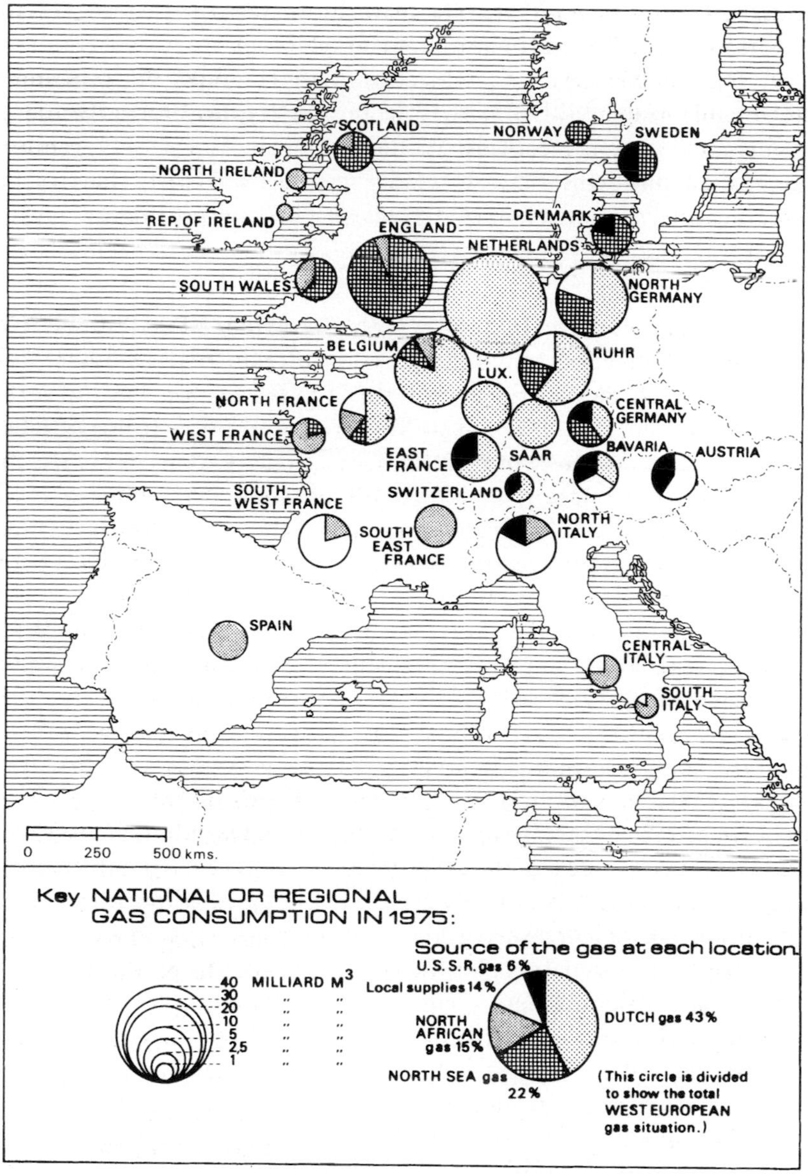

Figure II-5: Western Europe. Estimated Natural Gas Consumption in 1975.

Based on the scale and price effects of the competition of markets by the supplies of gas shown in Figure II-1.4, this map presents an estimate of the geographical pattern of consumption by the mid-1970s. Note that 79% of the consumption of 270 milliard m^3 *is indigenous gas, but that there is a trend to the emergence of clearly defined market areas arising from peripheral competition from imported supplies. This phenomenon is likely to develop further in the later 1970s as external gas becomes relatively more important*

up to twice this amount, mainly from the newly discovered Adriatic fields. In total, indigenous natural gas supplies will in 1975 account for nearly 80% of Western European consumption, whilst the remainder, split roughly in the ratio of 3:1 will be imported from North Africa and the Soviet Union, respectively. In the second half of the 1970s these imported supplies will increase rapidly to provide a larger share of the increasing total consumption of natural gas in Western Europe.

The European Gas Market in the Longer Term

Because doubts have been cast[11] on the ability of gas supplies to sustain the rate of growth in gas consumption envisaged here, it is necessary to introduce evidence of the supply position beyond 1975. Assuming a high 15% growth rate in gas consumption from 1975–80, which will have the effect of increasing gas' share of total energy used to about 36%, consumption in 1980 will be some 535 milliard cubic metres.[12] Pessimistically further assuming that West European gas production peaks at only a little above its 1975 level of 220 milliard cubic metres, then by 1980 gas import needs will be of the order of 300 milliard cubic metres – or the equivalent of three Groningens. But a supply potential of up to twice this requirement has already been shown as exploitable. On the basis of the same assumptions it will be 1984 before Western Europe's annual gas consumption reaches 5% of presently known, exploitable and, given the necessary infrastructure, already available supplies. The degree of probability of no new discoveries in the next 15 years in areas capable of providing West Europe with gas at prices not significantly different, in real terms, from those indicated in this paper is virtually zero. This, coupled with a major slow-down in the rate of increase in gas consumption once substitution of coal and oil has been completed, plus the possibilities of supplies from known sources not considered here – e.g. the Middle East – has the effect of extending well into the next century the period within which we can anticipate natural gas being a low-cost, readily available source of West Europe's energy: right through, that is, until atomic power in some from or another and the increasing direct use of the sun's energy become economically possible. With such a highly-prospective long-term future for natural gas in West Europe, we can return to an analysis of the 1975–80 position secure in the knowledge that future supply limitations need not affect policy decisions.

By 1975 Dutch natural gas consumption will be moderately higher than currently envisaged (as a result of prices here having to be adjusted to the lower export prices) and will account for over 50% of energy consumption, thus giving natural gas a more important role in the Dutch economy than its role in any other country in the world. It will also make average energy prices in the Netherlands lower than those in any other industrial nation. Natural gas will also have become a source of energy of some significance in every other West European nation, except for Norway, Portugal and the Republic of Ireland, and its use will be generally more intensive than is currently anticipated, though nowhere to the same degree as in West Germany where consumption will be 72,500 million cubic metres in 1975 – approximately one-quarter of the country's total energy use. For Western Europe as a whole the implications of these significant increases in the use of natural gas for the overall structure of the continent's energy economy are brought out in Table II-1.1. This contrasts the 1968 energy situation (as estimated from the latest available data) and the currently forecast 1975 and 1980 patterns[13] with the patterns that would emerge given the developments of the availability and use of natural gas as detailed in this paper.

TABLE II-1.1
West Europe's energy consumption 1968–80

Energy Source	1968 Estimate of approximate Consumption		1975 "Official" Predictions		1975 "New" Prediction with Increased Gas		1980 "Official" Predictions		1980 "New" Prediction with Increased Gas	
	Mmtce[1]	% of total	mmtce	%	mmtce	%	mmtce	%	mmtce	%
Natural Gas	50	4.5	170	11.3	365	24.4	265	13.2	730	36.5
Solid fuels[2]	425	38.6	325	21.6	250	16.6	300	15.0	150	7.5
Crude Petroleum	585	53.2	925	61.6	815	54.2	1235	61.8	970	48.5
Primary Electricity[3]	40	3.6	80	5.3	70	4.7	200	10.0	150	7.5
Total	1100		1500		1500		2000		2000	

Notes: 1. *mmtce = millions of metric tons of coal equivalent. Conversions made according to coefficients in U.N. Statistical Papers, Series J.*

2. *Includes coal, lignite and brown coal etc*

3. *Comprises hydro-electricity, nuclear electricity and geo-thermal electricity*

The Implications for Europe's Energy Economy

The effects of such a greatly enhanced role for natural gas in the energy economy of Western Europe would be of major importance, particularly as it is assumed that total energy demand will not increase as a result of the availability of cheaper natural gas – though this is possible in the longer term as new energy intensive industries and demands for higher standards of comfort start to develop. Here, however, it has been assumed that natural gas will achieve its larger role as a result of it being substituted for other sources of energy that would otherwise have been used. This does, of course, have important geographical implications.

Because of geographical propinquity, the great promise of natural gas as a low-cost energy source lies primarily in Western Europe's remaining geographical concentrations of coal production and consumption – the Ruhr, south Belgium, the Saar, Luxembourg, Lorraine and the Midlands of England. In these areas of intensive energy use, high unit-area gas demand will have the effect of bringing down unit transport costs because of the economies of scale that can be achieved on the main transmission lines and because of the low capital cost of providing distribution facilities to large consumers.

Though coal production in these areas has been adversely affected by competition from oil, the demise of the fields concerned remains unlikely, if only because of the restraints that continue to be built into the competitive processes for social and political reasons. But in the context of gas at the prices indicated in this paper even coal from the lowest-cost coalfields appears expensive[14] and thus seems unlikely to withstand the competition of an even more flexible and more easily handled source of energy than oil. And one, moreover, which in 1975, will be 80% indigenous to Western Europe, and, therefore, not subject to control in the interest of security of energy supply or balance of payments considerations to anything like the same degree as oil. Thus one can estimate that by 1975 about another 40% of Western Europe's already much diminished coal industry will have disappeared and by 1980 it seems not impossible that yet another 50% of the remainder will have gone, leaving coal only as a minor contributor to West Europe's energy needs – unless there are high-cost subsidies to coal producers and users.

The ability of coal in the U.S., which also has large supplies of natural gas, to have remained competitive is not valid evidence on which to be more hopeful about the European coal industry. In the U.S. coal has maintained its position largely on the basis of open-cast rather than deep-mined production and, even more significantly, as a result of its wide geographical separation from the regions with the largest natural gas supplies. Neither of these saving conditions applies to Western Europe, where its mainly deep-mined coal production is in the closest possible proximity to the major gas resources. Thus, in the absence of political restraints on competition a further major contraction in Europe's traditional coal industry seems highly probable.

But the rapidly increasing use of gas will also take the "steam" out of the markets for many oil products over many parts of the continent. Technically, gas for oil substitution is usually a relative easy one, such that industrialists and others who over the last 15 years converted from coal to oil at high cost (arising from the fact that new equipment was generally required), seem unlikely to hesitate very long before making the much lower-cost conversion to gas – once the appropriate competitive conditions of supply and price are established. Thus, the average annual rate of growth which can be envisaged for oil from 1968 to 1975 is only 5%, compared with 12% in the last seven years. This will produce a total incremental demand by 1975 of under 250 million tons, compared with over 300 million tons between 1961 and 1968 and the 350 million tons that has hitherto been forecast for the next seven years.

Western Europe is thus now approaching the time when the annual incremental demand for oil will level off (at about 50 million tons) and then, by the early 1970s, start to fall away as gas replaces it over wider and wider areas of Western Europe and in an increasing number of end-uses. By the later 1970s, oil's incremental market will be down to about 20 million tons per year (compared with a currently expected rate some four times higher) – mainly accounted for by oil's exclusive energy markets, mainly in the field of transport. This will require the rescheduling of refining operations – first in the main areas of competition from gas and later over most of the continent – designed to switch refineries away from fuel oil production towards the greater production of gasolines and diesel oils etc. Such investment for up-grading seems likely to become more important than this hitherto expected continuation of the post-1950 trend of large investments for commissioning massive new refinery capacity to meet Europe's basic energy needs. Many areas of Western Europe such as Rotterdam, which have become used to enjoying the results of the local multiplier effects of continuing refinery expansion may find their prospects becoming much less positive.

Finally, the expanding availability of natural gas as its lowest prices in most of the continent's most energy intensive areas, where interruptible gas will be available on a large scale for its most appropriate use in thermal power stations, will serve to curb the growth potential hitherto expected for nuclear power in the 1970s. Atomic power station developments would have been orientated to just these locations, but the nuclear power industry's promise of electricity generated in these stations at about 0.5 U.S. cents per kWh will now be much less attractive in the face of a challenge from new or converted natural gas-based thermal power stations. The latter's production costs of electricity, given the low gas prices envisaged in this paper, will be some 20–30% lower.

In addition, low-cost natural gas will stimulate the development of small scale, market located, on site-generation of electricity within the framework of total energy systems. Such systems will be developed for office blocks, shopping centres, blocks of flats and, eventually, even individual homes such that the requirement for a public electricity system may grow less rapidly than currently expected. This will further delay the use of nuclear power on a massive scale for the production of electricity.

In the meantime, natural gas seems destined to become and to remain a major and, eventually perhaps, even the dominant source of Western Europe's energy supply until sometime in the first quarter of the 21st century, when conventional hydro-carbons' based energy sources may at last start to give way to currently-non-conventional methods of energy production. Until then, the use of natural gas as the dominant or major energy source, has implications for the geographical patterns of secondary economic activities in Western Europe – the possibilities of which are briefly considered in this final part of this paper.

Energy and Europe's Regional Development

Until recently coal and manufacturing industry were in an almost symbiotic relationship, such that the location of the main industrial areas of the continent largely reflect the patterns of occurrence of highly localised coal deposits. These industrial conurbations have, moreover, remained as main centres of manufacturing activity even though their constituent industries have, for a decade or more, relied increasingly on alternative energy sources, especially oil. Oil itself has added a further contribution to Western Europe's spatial pattern of industrialisation through the growth of refinery complexes and immediately associated petrochemical developments – of which the continent's leading example by a very long margin is here in the Rotterdam area. However, transporting energy in the form of oil is particularly low-cost and hence spatial variations in its price in a continent as small and so well served by transportation facilities as Europe, have been relatively low, thus limiting oil's influence as a localising factor in industrialisation largely to the refinery/petrochemical complexes already mentioned.

By contrast, natural gas has much higher transport costs – some three times greater than oil on an energy-equivalent basis. It is also less perfectly mobile depending on a specialised infrastructure (pipelines and distribution lines) rather than, in large part, on the general transportation network (waterways, railways, roads) as with oil. In the United States these factors, coupled with institutional and political factors, have led to a considerable use of natural gas in industry in the areas of gas production In 1966 over 40% of all gas used in industry was used in the two main gas producing states (Texas and Louisiana) and 60% in the six most important gas producing states.[15] This does suggest the possibility of a new element in the geographical pattern of European industrialisation – viz. an element of localisation based on the continent's major gas resources with the development of complexes consisting of major energy using industries in which each industry secures external economies through an "over the fence" use of neighbours' products and "waste" materials. The outstanding existing example of this type of development is seen in the great industrial complex that has grown up alongside the Houston Channel in Texas.

In Western Europe, localisation based on gas use in this structural context does not necessarily mean the development of industries only in areas immediately adjacent to the well-heads – in the province of Groningen as opposed to South Holland, for example. Rather, it implies localisation on a somewhat broader geographical scale – viz. the development of such complexes in relatively nearby areas which are already attractive for industry – such as the western parts of the Netherlands, north-west Germany, parts of Eastern and central England and northern Belgium – as opposed to development in areas more distant from the centres of gas production. Political difficulties and constraints apart, is it stretching geographical prediction too far to suggest an industrial growth zone, utilising massive quantities of the locally available gas at lowest possible prices, stretching across the areas mentioned above. The large scale availability of gas from the external sources detailed earlier in this paper and the effect of these in ensuring that the Groningen reserves are used competitively and to their fullest extent in a market area of relatively limited geographical extent, will serve to enhance the attractiveness of the Rotterdam region for industrial development, by ensuring it of a large scale availability of very low cost natural gas (at about 2 Dutch cents per cubic metre) to add to the other considerable advantages it already enjoys.

References

1 Two books in particular viz. M.A. Adelman, *The Supply and Price of Natural Gas*, Oxford 1962, and P.W. MacAvoy, *Price Formation in Natural Gas Fields,* Yale U.P. 1962 set out the characteristics which would mark a monopolistic response. Such characteristics are clearly recognisable in the case of Groningen though the situation is made more complex by government intervention in the pricing decisions through the Central Plan Bureau, See, "*Energie in Perspectief*" (Nederland Economisch Institut, Rotterdam, 1966) pp. 77–83 for a discussion of the pricing process.

2 For a definition of "super-normal" profits see R.G. Lipsey, *An Introduction to Positive Economics*, London 1963, p. 205

3 Shell Briefing Service. *The Development of Groningen Gas*, London 1968

4 And in case it should be argued that this is an unfair comparison, because gas is often produced jointly with oil in the U.S. such that most of the costs can be set against the oil component, then one can also contrast the Groningen price with the price of gas out of the gas-only producing fields of the Permian Basin of Texas/New Mexico. There, in 1968 a Federal Power Commission rule established 2 Dutch cents (approx.) per cubic metre as a level high enough to produce an adequate return on investment. Though this price level is considered too low by the producing companies, they have been quite happy to sell at about 2.5 cents from these fields. Yet in neither a physical nor an organisational/managerial sense are the fields of the Permian basin as favourably placed to secure low cost production as gas from the Groningen field.

5 For a recent discussion of this point see G. Krumme, "Towards a Geography of Enterprise", *Economic Geography*, Vol. 45, January 1969.

6 Principally evidence gleaned from the pages of the journals *Petroleum Press Service* and *World Petroleum*, and from various published papers of the Shell Petroleum Company and its employees.

7 See J Martens, *La Politique Energetique en 1967 – 75*. A Report by the Director General of Energy Administration in the Ministry of Economic Affairs, Belgium.

8 M. Adelman, op. cit. p. 38.

9 80% of the figure of Dutch cents per cubic metre gives an approximate equivalent for the gas price in U.S. cents or British pence per therm.

10 See, for example M.E. Orlean, Manager of the Economics and Planning Department of Esso Europe's natural gas organisation, "Impact of Natural Gas on West European Energy" *World Petroleum*, April 1968, pp. 45–48 and Shell Briefing Service, *Development of Groningen Gas*, October 1968.

11 See, for example M.E. Orlean, op.cit.

12 To put this seemingly astronomical figure in perspective one should note that the United States used rather more natural gas than this in 1967.

13 As indicated in energy studies carried out in the last few years by organisations such as the E.E.C. and the O.E.C.D. and from information published by European energy companies such as Shell and Esso. The approximate figures in the Table represent a general consensus emerging from the separate studies which are in general agreement with each other.

14 Low cost British coal – the cheapest in Europe – will be at least 50% and could be up to 100% more expensive at source than the going price for gas envisaged in this paper.

15 U.S. Bureau of Mines, *Statistical Yearbook.* Vol. 1 & 2, 1966.

Chapter II – 2

Indigenous Oil and Gas Developments and Western Europe's Energy Policy Options*

Europe's Energy Policy Options – the Conventional View

Over the last twenty years, Western Europe's energy economy has gradually become orientated to the availability of oil imported from a small number of supply points most of which are politically unreliable with consequential dangers for the continuity of supply. However, because of significant cost differentials between imported oil and indigenous coal and because of strong consumer preference for the former over the latter for reasons apart from price, European governments came to accept a high degree of dependence on foreign oil as an essential component in the progress towards an expanded economy and improved living standards.[1]

Several factors temporarily combined to make such acceptance of dependence on imported oil an appropriate policy to follow. Most notable was the ready availability of a seemingly inexhaustible supply of low cost oil. This was provided within the framework of an international system in which there was competition for markets, not only by the different oil producing countries, but also between the increasing number of oil companies operating in Europe, partly as a result of their having had their planned markets for oil in the US cut off by the imposition there of import controls in 1959. At the same time, developments in oil refining and transport technologies – particularly the

* **Originally published in *Energy Policy*, Volume 1, No.1, pp.47–64, June 1973**

trend towards increasing scales of operation which led to greatly reduced unit costs in the industry – made possible the supply of greatly increased quantities of oil to Western Europe at prices which declined in real terms.

Given the consequential buyers' market in oil, no country saw any real need for co-operation with its neighbours and, in any case, such co-operation was made more difficult because of the conflict of interests between those countries with an indigenous coal industry to protect for economic and social reasons[2] and those without such a constraint on their energy policies.

Thus, by 1970 – and in spite of a number of earlier oil crises arising out of the interruption of supplies in the producing countries – Europe had happily allowed itself to become 60% dependent on imported oil and appeared to face a situation in which a move towards a 70% dependence was generally assumed to be unchangeable.[3] Since 1970 some fundamental changes in the world oil industry structure have demonstrated the dangers inherent in such an open energy economy. Most notably the supply and price of oil has been brought under control by the success of OPEC in constituting itself as a producers' cartel and by the decision of all the major international oil companies to pursue co-ordinated, orderly marketing policies. Europe now stands exposed to the dangers of supply limitation and of price escalation for its basic energy requirements[4]. As a result, energy policy options have had to be reviewed and the potential of coal and hydro-electricity, Europe's traditional energy sources, re-evaluated so that they can make a larger contribution to the future total energy needs, along with greatly expanded nuclear power production.[5] However, neither the attempt to stabilise – or even increase – coal output nor a crash programme for nuclear power expansion appear to represent more than a 'clutching at straws' solution to the oil import problem in that neither option can have much more than a marginal impact on the overall level of demand for other forms of energy over the next decade at least.

Europe's inattention – even irresponsibility – towards the role of natural resources in development has left it in a very exposed position both economically and politically. It is perhaps this unfamiliarity with the importance of resources which now inhibits Europe from taking a serious view of the potential it has to satisfy a very large part of its future estimated energy demand from its own resources of oil and gas. Previous disappointments over the significance of earlier oil and gas finds partly account for this attitude, but there has also been a too-ready acceptance of low estimates of reserves and production potential made by parties with particular vested interests at stake; for example, earlier attitudes towards the giant Groningen gasfield in the Netherlands.[6] Profound official pessimism over natural gas availabilities in Western Europe persist whilst, in exactly the same vein, there are concurrent attempts to minimise the importance of the continent's developing oil reserves – especially those from the developing North Sea province.[7]

As a result of these attitudes the future role of indigenous oil and gas in Europe's energy economy is officially considered to be a limited one which does not provide a basis for a radical reappraisal of the continent's energy policy options. In all these options the essential components remain appropriate combinations of economic, political and strategic/military measures designed to ensure the continuity of the required, expanding flow of foreign oil to Europe at fairly stable prices.[8] In other words, European energy policy options will increasingly become bound up with foreign political and economic policy options – such as decisions on the appropriate stance to adopt towards Middle East conflicts and on attitudes towards the oil exporting countries' influence on world monetary problems.[9] The known dangers and insecurities inherent to such policy options strongly suggest a pressing need for research into ways of curbing the rising demand for oil, including recommendations for changes in our societies to make them less energy intensive.[10] But all this is long-term, so the prospects for reducing oil imports remains bleak within the framework of the conventional analysis of Europe's energy policy options.

An Alternative hypothesis

A policy option based on a high degree of self sufficiency in oil has already emerged as a result of the first decade of exploration and development work in the North Sea oil and gas province. These fields rank behind only the Persian Gulf and West Siberia in terms of production potential and, in terms of general significance, they are more important than anything the oil world has known since the major oil and gas fields of the USA were first exploited half a century ago, changing the basis of that country's energy economy. There are no financial limitations on the rate of exploitation[11] with plenty of risk and development capital available. The necessary industrial and technical capacities and know-how also exist to provide the hardware and expertise required for the province's development within a relatively short period of time.[12]

TABLE II-2.1
Conventional and alternative views on Europe's energy supply 1971–85

	1971		1975 Estimates				1980 Estimates				1985 Estimates			
	Consumption		Conventional		Alternative		Conventional		Alternative		Conventional		Alternative	
	mtce*	%	mtce*	%	mtce*	%	mtce*	%	mtce*	%	mtce*	%	mtce*	%
Total	*1475*	*100*	*1700*	*100*	*1750*	*100*	*2250*	*100*	*2300*	*100*	*2750*	*100*	*2850*	*100*
of which														
Indigenous	*545*	*37*	*580*	*34*	*760*	*43*	*680*	*31*	*1350*	*60*	*1045*	*38*	*1825*	*65*
Primary Elec.	40	3	75	4	55	3	210	10	150	7	330	12	340	12
Coal etc.	385	26	315	19	250	14	205	9	200	9	220	8	200	8
Oil	25	2	40	2	100	6	50	2	500	22	195	7	660	23
Gas	95	6	150	9	355	20	215	10	500	22	300	11	625	22
Imported	*930*	*63*	*1120*	*66*	*990*	*57*	*1570*	*69*	*950*	*40*	*1705*	*62*	*1025*	*35*
Coal	45	3	55	3	50	3	75	3	50	2	90	3	60	2
Oil	880	60	1050	62	880	50	1450	64	775	33	1530	56	715	24
Gas	5	–	15	1	60	4	50	2	125	5	85	3	250	9

Sources: *OECD, EEC and various national estimates of future energy demand and the author's own alternative estimates.*

* *million tons of coal equivalent.*

Table II-2.1 shows both the current conventional estimates and the author's alternative estimates of the future of Western Europe's energy economy to 1985 by type and origin (imported/indigenous) of the different energy sources. These data clearly illustrate the fundamental change in the relationship between indigenous and imported energy which emerges from this alternative hypothesis. In particular – in terms of the requirement for imported oil – the current conventional view of a steadily rising growth to over 1500 mtce by 1985 is replaced by the idea that current import levels will not be exceeded in 1975 and, thereafter, will steadily decline until by 1985 they are only 715 mtce, representing 25% of Europe's total energy supply. This has implications for the world oil industry, in general, but these fall beyond the scope of this paper, the rest of which will be devoted to an examination of the validity of the assumptions made in establishing the hypothesis of a much expanded oil and gas production potential in Western Europe.

As an underlying assumption however, which we shall not attempt to justify, we assume that the overwhelming majority of interested parties are in favour of a policy option for as high a degree of self-sufficiency in energy as possible, provided that following such a policy will not made any great difference to the price at which energy is available to the consumer. This is not unreasonable, since it can be argued that an enhanced use of indigenous resources produced under a system of governmental or European level controls – including price controls – could ensure that Europe gets its energy cheaper than would otherwise be the case, given the great pressures by OPEC, the international oil companies and the US government to raise the going price of internationally traded oil over the next decade.[13] Thus, the option of the maximising the use of indigenous oil and gas now open to European energy policy makers lies within the context of a very different situation from that of the late 1950s and the 1960s, when the use of coal, a plentiful indigenous energy source, was reduced in the face of competition from very much lower cost imported alternatives. By contrast, an autarkic energy policy for Europe in the 1970s and the 1980s seems unlikely to bring penalties in the form of higher all-round energy prices for the majority of consumers.

The Size of the Relevant Oil and Gas Resource Base

The essential component in establishing the alternative hypothesis is the contention that the current official and/or company presentations of the scale of Western Europe's oil and gas resources and their production potential are unrealistic – either through ignorance or deliberate distortion on the part of some vested interests.

The ignorance stems from the failure of governments to place obligations on the companies to publish a comprehensive set of facts on their activities and then to make sure that they have adequate numbers of staff competent to collate and evaluate the flow of information, and so able to give valid advice on which to base policy decisions.[14] The exploring companies cannot, of course, be accused of ignorance, but their distortion – or witholding – of information appears to be the North Sea norm. This is, in part, understandable, given the needs of companies to protect the knowledge they have gained on oil and gas prospects through high risk expenditure. It is this which makes them able not only to take informed decisions about what new concessions to seek, but also to secure information which can be sold to companies with concessions in the vicinity of a discovered field or even a dry hole. In part, however, the reticence of at least some of the oil companies to talk any kind of sense about the success of these exploration efforts activities may be rationalised as a function of their international structure and their wish to evaluate internally, without public discussion, the impact of possible alternative North Sea strategies on their overall profitability.

All the major companies are currently involved in difficult and delicate negotiations with the present major oil exporting countries. The latter believe they have effective control over the energy supplies needed by the rest of the world and are thus seeking not only higher taxes per barrel of oil exported, but also direct equity participation in the activities of the oil companies.[15] Participation will not only give the countries more of the profits, but also a right to share in decisions by the companies – in order, for example, to prevent them from reducing their rates of production such as may well arise from a reduced demand for oil imports by Western Europe consequent upon North Sea expansion.

Given knowledge on their possible loss of future bargaining power, to say nothing of future revenues, the oil exporting countries could well choose to negotiate for a faster rate of build up of their participation in the companics, together with an acceleration in the rate of change in the tax position. All this would, at least, have the effect of reducing the economic viability of the international companies and, at worst, lead to the premature expropriation of their assets in the exporting countries.

Most companies accept this ultimate fate but, in the meantime, hope for a decade of more or less unrestricted opportunities to take out as much as possible of the oil they have found. They thus have every incentive not to take any action which will upset the governments of the oil exporting countries – and may also lack a sense of urgency in developing alternative resources, the use of which would oblige them to take less oil from the exporting countries. Many oil companies, therefore, do not wish to do anything more than the minimum possible with their discovered North Sea reserves, whilst continuing to supply Western Europe with as much overseas oil as they can possibly lift in their last decade of opportunity in this respect.

In brief, the commercial interests of the major oil companies need not necessarily be met by the rapid exploitation of indigenous Western European reserves of oil and gas and thus we have a positive reason as to why there should be distortion over and/or lack of information on the North Sea's resources and its producing potential.[16] Because of this danger, and because Europe, as opposed to the oil companies, has no motivation whatsoever for risking the hazards implied by continued dependence on the outside world for fuel supplies for a moment longer than is really necessary, the establishment of the validity of the minimum-imports option demands as rigorous as possible an evaluation of the resource base – utilising the little information that has been made available by the companies and through reasonable extrapolations from earlier experience in other major oil and gas producing provinces.

Natural Gas

Natural gas is almost always the preferred fuel, for environmental as well as economic reasons, and in Europe its greatly expanded use offers an opportunity to reduce the continent's dependence on fuel oil for electricity generation and industrial steam raising. This could swing the balance of oil products' demand more closely into line with the quantities and range of products that can most economically be made from the generally light North Sea crude oils. Table II-2.2 shows the likely development of indigenous natural gas production.

Current conventional estimates of possible production and consumption levels by 1980 – of 215 and 275 thousand million m^3 respectively[17] – are based on a low annual rate of depletion (about 4%) of the currently declared reserves of just over 5000 thousand million cubic metres. It is hypothesised that the annual production rate could be more than double present conventional estimates, in part as a result of more normal annual depletion rate of the reserves from the discovered fields. By 1980 these could all have reached their optimum production rates of about 7% of their known reserves, viz. some 350 thousand million m^3 per year. Beyond this, the higher total production will depend on discovered, but undeclared reserves being declared. We thus hypothesise

TABLE II-2
Western Europe: an estimate of its natural gas resources and production potential in the early 1980s

Region	Reserves As Declared in 1972 ($x\ 10^9m^3$)	Reserves As developed by 1980 ($x\ 10^9m^3$)	Annual Production potential by 1980 ($x\ 10^9m^3$)	Millions of tons of coal equivalent
On-shore Netherlands	2250	2750	135	150
South North Sea – British Sector	1000	1250	60	75
South North Sea – Other Sectors	250	1250	40	50
On-shore West Germany	400	650	35	40
Austria, France, Italy etc.	400	600	35	45
Middle North Sea Basin (Britain/Norway)	750	?*	80	90
North North Sea Basin (Britain/Norway)	–	?*	80	90
Rest of Continental Shelf	–	?*	15	20
Totals	5050	9500+	480	560

Source: 1972 Reserves from Oil and Gas Journal, December 15, 1972. Other figures are the author's estimates.

* *As these reserves are mainly associated with oil, the production potential is a function of the rate of oil off-take, rather than of the gas reserves. However, we can assume a level of oil production from these areas which will enable a total annual output of 175 x 10^9 m^3 of gas per year to be sustained (from an equivalent of about 3000 x 10^9m of reserves.)*

that gas discoveries already made, but not yet declared (eg, in fields in the North Sea, such as Heimdahl, Brent, Piper etc), combined with new discoveries from on-going exploration could lead to a total of not less than another 4000 million m^3 of gas reserves. These fields will not reach their peak production levels by 1980, but should thereafter be capable of producing some 140 thousand million m^3 a year.

This estimate of the development in the reserves position for already developed areas is, moreover, quite modest. The figures for the on-shore Netherlands and Western Germany imply no more than the continuing discovery of a series of small fields (small, that is, by comparison with Groningen) and some upward adjustment in the declaration of reserves from already discovered fields. This seems much more likely to understate rather than overstate the possibilities, given the unofficial view in the Netherlands that reserves there are up to double those currently declared. A similar conservative evaluation is made of the likely expansion of the gas resource base in the British sector of the southern North Sea. There the 25% increase in the estimates of recoverable gas implies only the continued slight upward appreciation of the reserves from existing discoveries, as production experience provides more information about the structures in a gas basin under active exploitation. In this sector there are also many minor structures which remain to be explored. With the price for gas rising over the decade, a series of small fields can be expected to be discovered and brought into production by linking them to delivery systems from the main fields.

Elsewhere in the southern sector of the North Sea we have been somewhat more optimistic over resource development in the expectation that we shall, by 1980, gain access to data on the 17 or more discoveries that have already been made in the Dutch sector. There the reserves so far declared refer only to one field, viz. the Placid field in Block L10, with about 150 thousand million m^3 of reserves. Given the likely 1000 thousand million m^3 of reserves in the Dutch sector, plus another 500 in Danish waters, it requires only very limited success in the recently allocated West German sector (where exploration is about to begin) and in additional areas in the Dutch sector to bring the total early-1980s reserves position to the 1250 thousand million m^3 shown in Table II-2.2. In spite of the continuing, and, unhappily, likely successful attempts to constrain the levels of production in the southern North Sea basin including Groningen, annual off-take rates by 1980 ought to be around 235 Bcm rather than the 150 m^3 currently conventionally estimated. Apart from the impact of general demand factors on production levels arising from the insatiable gas needs of customers lying around the fields, more positive EEC regulations could insist on maximum possible production levels, with the gas to be made available to member countries on a nationally non-discriminatory basis.[18]

Quantatively even more important over the next decade are the commonly ignored possibilities arising from the exploitation of the middle North Sea basins where the gas seems likely to be found mainly as an associated product in the oilfields now being discovered. There is already an agreement for the recovery and use of 10–15 Bcm a year of associated gas from the Ekofisk field and the use of gas from the joint British/Norwegian Frigg field is being negotiated. Plans for the recovery of natural gas from other fields already discovered are being made, but no decisions have yet been announced. Given the fact, however, that these early-discovered oil and gasfields are likely to be quickly brought on stream, the achievement of an early 1980s production potential of 80 Bcm per annum would represent only a modest scale of development from the 30–40 Bcm already sold or under negotiation.

Further north, one moves into a region where one can essay little more than intelligent speculation about the gas potential. However, given the initial occurrences in the northern basin of even larger oil and gas fields than have been found in the middle North Sea basins, it necessitates only a much lower than expected success rate in exploration and rather low gas/oil ratios in the discovered fields to sustain the estimate of associated gas reserves capable of an annual gas production of 80 Bcm. Finally, the estimates of 15 Bcm of gas from other parts of Europe's continental shelf is a speculative small element in the total figure, representing the high probability of a modest degree of success over the next few years in proving fields to the west of Scotland, between Ireland and England, and off-shore France and Italy too.

Thus, in overall terms, by taking nothing more than a modest view of the dynamics of reserves' appreciation in a province which has already proved to be gas rich by world standards (as evidenced by the Groningen field which is the largest gasfield in the world outside the USSR and by the Leman Bank field in British waters, which is the world's largest offshore field), it appears to be entirely unreasonable not to expect an appreciation in the reserves of existing fields and the discovery of new fields, both of gas and of oil and gas from which the gas will automatically become available as oil production is stimulated. Thus the rate of growth in gas production and use which Europe has enjoyed over the last few years seems likely to continue for the foreseeable future and so enhance the contribution of indigenous gas to Europe's energy economy to the level of 22% indicated in Table II-2.1. Thereafter – and in light of experiences from second and third decades' exploitation of reserves in other major provinces around the world – there seems to be no inherent reason why the development of production potential should not continue; at a rate at least as fast as the growth in energy use, as hypothesised in Table II-2.1 for the period after 1980.

Given the present very early stage of the exploration for natural gas in the large and still relatively little-tested sedimentary basins of Western Europe, it seems very unlikely that gas production in Europe will peak before the end of the century at the very earliest – unless the European gas provinces behave completely differently from those in all other parts of the world with longer periods of exploration and development history on which to draw.

Oil

If we accept that the build up in indigenous gas supplies will be much more rapid than expected, then, as shown in Table II-2.1, the Western European demand for oil will expand much more slowly than hitherto conventionally forecast. Substitution of oil by gas can become much more general in terms both of economic sectors and geographical regions: This has already happened in the Netherlands since 1969 when plentiful supplies of gas, available at oil-related prices, started to cut into markets for oil products. Total oil consumption is thus now little higher than it was then, while the use of fuel oil is down by a third in only three years. When this fact of consumer preference for gas over oil is linked with the virtual ending of the 1950–1970 phenomenon of the substitution of coal by oil (given, firstly, the near elimination of coal markets in some countries except in non-substitutable uses like the iron and steel industry and, secondly, the decision by Britain and West Germany, together accounting for over 80% of Western Europe's coal production in 1972, to give a high degree of protection of their remaining coal mines), then the future growth of oil demand must be much below the average 7% per annum which became the norm in the post-war period. Indeed, the alternative estimates in Table II-2.1 show only an average 3% per annum rate of growth in oil demand over the next decade.

This lower rate of growth means a reduction in the conventional forecasts of demand for oil in Western Europe by some 150 million tons per annum by the early 1980s and some 240 million tons by 1985. In percentage terms it reduces oil's contribution by 11% and 14% respectively, to a share below that to which it had risen by 1970. In terms of the infra-structure required for handling the future demand for oil, this means, on average, three fewer new refineries or refinery expansions per year and the use of up to 20 million tons less of crude oil carrying capacity than is conventionally estimated, on the basis of the extrapolation of the historic rate of growth in demand.

These hypothesised changes in oil are traumatic enough for they take the industry in Western Europe out of the ranks of the rapidly growing sectors of the economy to join those which grow at only about the same rate as the economy as a whole. But new companies are still moving into the market, whilst old established ones still work to infrastructure development plans related to the continuation of the historic rate of growth of the market. In this they are aided and abetted by government plans for very large scale investment in new marine and marine terminal facilities for greatly expanded crude oil imports within the framework of existing geographical patterns of delivery (as, for example, in the projected new mammoth tanker terminal at Foulness in the UK and the proposal to deepen the channel into Rotterdam sufficient to take tankers of up to 500,000 tons). Much of this investment seems highly unlikely to produce the optimistically calculated rates of return, given the failure of the investors – private and public – to recognise soon enough the changing trends in the patterns of oil demand.[19]

This consequential impact of the enhanced supply of indigenous natural gas on the appropriate policy decisions by companies and governments in the oil sector is, in itself, of great significance for energy policy makers, both in the immediate and the foreseeable future. However, these consequential changes must now be viewed along with others which arise from the dramatic and, until quite recently, generally unforeseen growth in indigenous oil producing capabilities in Western Europe.

The supply of indigenous oil could rise rapidly from its traditional 20 million tones a year to some 320 million tons a year by the early 1980s and to over 400 million tons by the middle 1980s (See Table II-2.1). Thereafter, on the basis of speculation over oil finds in areas known to be geologically interesting, but unlikely to be probed by the drill before the end of this decade, even further growth could be expected up to an ultimate level of 600 million tons per year in the 1990s. These possibilities certainly do not emerge out of the information which is made generally available by the exploring companies and the government organisations involved but, as already explained, these entities lack either the motivation or the capabilities to present other than a static interpretation of the potential as revealed from initial or early reserves declarations on discovered fields. A realistic evaluation involves the utilisation of experience from other parts of the world, where market located oil resources have already been developed under the stimulus of a more or less unlimited demand for the locally available crude oil. We assume the latter will be the case of the North Sea province on the understanding that sufficient incentives and/or requirements will be introduced by governments and/or the EEC to ensure maximum possible production in the shortest possible time.

Analogous and relevant experience – mainly from North America with its comparable demand-motivated situation – points clearly to the dangers of under-evaluating the level of recoverable oil reserves and the medium term (10–15 years) production potential during the early period of the discovery and development of a major oil province. For example, Halbouty states the general hypothesis as follows:

> *'We recognise that... reserve estimates are subject to considerable upward revision, particularly in the case of new fields that are not fully developed and thus lack adequate production history'*[20]

The appropriateness of this warning is already well demonstrated in the case of Groningen gas.[21] Elsewhere in the North Sea basin, one notes the initially very low recovery rate figure (16%) used to declare the Ekofisk field as having 1140 million barrels of recoverable oil. A three to three and a half fold increase in the evaluation of the amount of very light oil recoverable from the 7000 million barrels in place seems highly likely, given the experience from elsewhere of final recovery rates of between 45 and 55% from fields with oil of this quality.

It therefore appears reasonable to apply an appreciation factor to the initial information supplied by the discovering company on reserves in a discovered field. Only in this way can one arrive at a realistic estimate of resource availability on the basis of which to investigate different policy options for the energy economy. The question is – what appreciation factors should one use in this respect? The most systematised examination of this problem emerges out of the work of the Canadian Province of Albert's Energy Resources Conservation Board – to which comprehensive data series must be supplied by all oil and gas companies operating in the province's oil and gas basins. The result is a comprehensive annual publication including full data on reserves' development and the results of the Board's analytical work on the data.[22] The Board's work is based on the analysis of more than 20 years of data on the development of 128 separate oil fields and has enabled an Average Appreciation Curve to be evolved (Figure II-2.1). This indicates that initial reserves declarations have, on average, to be multiplied by 8.89 to get a realistic measure of the ultimately recoverable reserves from a field. Reserves as estimated after one year of development work have to be multiplied by 3.44 and declared reserves after three years by 2.31 and after ten years by 1.10.

In the absence of any alternative evaluations from the companies which are themselves exploring the North Sea province, we would tentatively suggest the application of aspects of such a methodology of future reserves' calculations to the presently declared reserves figures for the 15 or so North Sea oil fields about which anything was known by the end of 1972[23] (Figure II-2.2). Given the small number of fields involved and the uncertainty of the information (most of which is unofficial rather than official as for Alberta), it would not appear justifiable to divide up the discoveries by the year of their discovery and apply the appropriate appreciation factor to each set of fields. Instead, let us assume that we are dealing on average with reserves as declared at the end of the first year after discovery. In this case an average appreciation factor of 3.44 has to be applied to the 10 thousand million barrels of reserves declared in December 1972[24]. These, after appreciation, become 34,000 million barrels of ultimately recoverable reserves. From this figure one

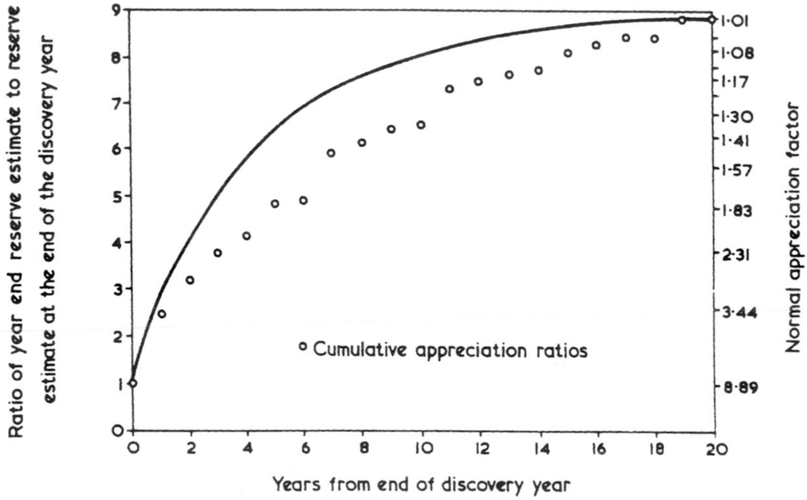

Figure II-2.1: Normal appreciation of initial recoverable crude oil reserves

This is reproduced from the Alberta Energy Resources Conservation's Board publication, Reserves of Crude Oil etc, December 31, 1971, pages V-31. The accompanying text in that publication reads as follows (pages V-2); 'Appreciation factors have been computed by aggregating, for all the reserves considered, the appreciation experience over successive years. The actual values have been used to construct a curve yielding smoothed appreciation factors, or estimates of average appreciation'. 128 oil pools were involved in the analysis which, as far as the author is aware, cannot be undertaken elsewhere, given the usual high degree of oil company secrecy over their reserves' position

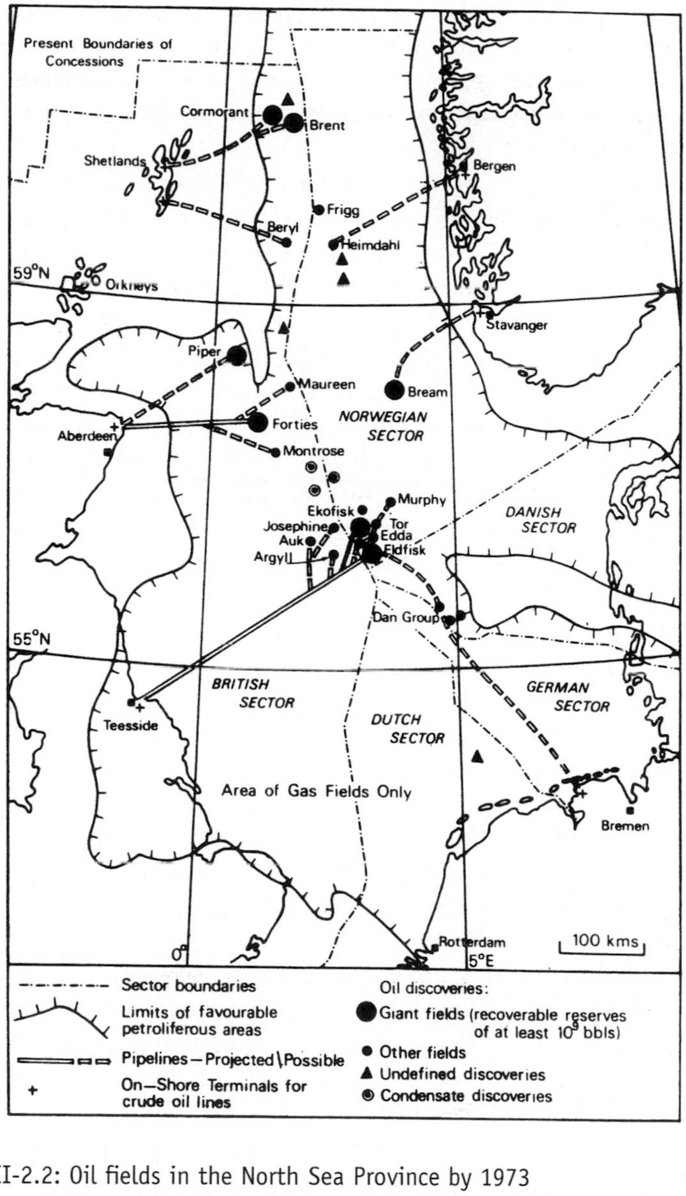

Figure II-2.2: Oil fields in the North Sea Province by 1973

The existence of seven giant fields, a dozen smaller fields and a set of as yet undefined discoveries is based on the limited amount of information made publicly available and on information freely circulating in oil industry circles. Their geographical distribution in three main groups is reflected in known or possible pipeline projects to Teeside/North Germany for the southernmost fields: to Eastern Scotland and/or Western Norway for a middle group: and to landing points in the Shetlands for the northern group. Intensive exploration activity continues and may be confidently expected to prove the existence of new fields in all three areas over the next two to three years requiring much additional pipeline capacity

can now build up the production rates achievable over the next decade – though in the second half of the decade this will, in part, tend to understate the probable expansion of capacity, a part of which will then depend upon yet-to-be-discovered fields being brought into their initial stages of production.

Beyond that one can think in terms of the possibilities for further discoveries from the 80% of the potential oil bearing structures in the province that remain to be drilled. If we assume very conservatively that drilling the remaining structures will lead to the discovery of ultimately producible reserves only twice as great as the potential from the first 20% of structures (implying, of course, a marked fall in both the success rate and in the average size of the fields discovered), then the province's ultimately producible reserves reach a speculative 100 thousand million barrels. This figure is, of course, highly speculative in much the same way as for Alaska where the current range of estimates of 20 to 100 thousand million barrels of ultimately producible oil is almost identical with the range of 12 to 100 thousand million suggested here for the North Sea.

European energy policy makers should note, however, that in the USA the upper end of the Alaskan estimate range is not ignored in official evaluations of the energy policy options open to that country over the medium to longer term. There would seem to be no good reason why Western Europe should not take the same stance over the higher North Sea reserves' estimates – especially as the technical and conservation constraints on getting the oil out of the ground are less stringent than in the case of Alaska, whilst the economies and geography of the regional supply/demand position are certainly more favourable for the North Sea than for Alaska.

Thus, the ultimate outcome needs to be considered in respect of longer-term European energy policy options for which the ability of the province to sustain over a long period the production levels which can be achieved in the shorter term are vitally important. The way in which this works is shown in Figure II-2.3. This has been derived from the simplified assumption of a set of fields with 10,000 million barrels of initially declared reserves and a potential for a 3.44 appreciation over the subsequent 20 years becoming available for production every five years. The following facts seem to justify this approach: viz. first, the very large size of the total sedimentary areas still to be explored in off-shore Western Europe and the large number of potential oil bearing structures already inferred from seismic work and geological interpretation; second, the allocation of concessions by the governments concerned in a sequential pattern, with requirements for continued and expanding exploration; and, third, the expectation of a firm market for as much oil production as can be achieved in the foreseeable future.

At the present time, however, the size of the ultimate recoverable oil resources, based on a speculative hypothesis of long-term supply and demand conditions, is largely irrelevant to the possible rates of build up of production over the next decade or so. These short-to-medium term possibilities, covering a critical period in the development of European energy policy, also depend on three sets of factors: first, on the full development of the initially declared and subsequently appreciated reserves from fields discovered by 1982 with a further assumption that customary time profiles of production from these field will be adopted; second, on the ability to install the producing and transporting facilities necessary to get the oil to the refineries; and, thirdly, on the availability of sufficient finance to enable the development to proceed.

A peak annual output of about 15 million tons is possible after five years or so of steadily increasing production from each thousand million

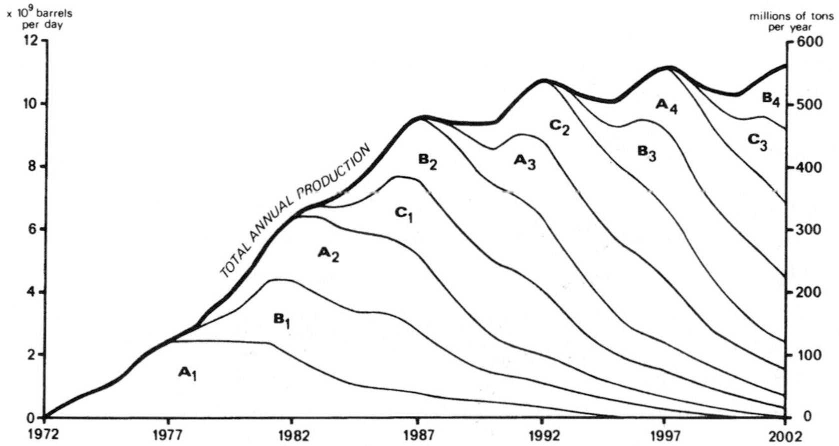

Figure II-2.3: A simple model of the build-up of oil production potential from north west European off-shore basins 1972–2002

A_1, A_2, A_3 etc – production from declared discoveries of 10 x 10^9 barrels of recoverable oil made in the preceding 6 years; B_1, B_2, B_3, etc – production from 8 x 10^9 barrels of appreciated recoverable oil from a 5 year period; C_1, C_2, C_3, etc – production from a further appreciation of 8 x 10^9 barrels in the recoverable reserves from the discoveries of earlier periods. (1, 2, 3,). Further assumptions are a 5-year build-up period, a 4-year peak production of 275,000 barrels per day for each 10^9 barrels of reserves and a 66% depletion of the reserves in the first 10 years of production

barrels of recoverable reserves. Over 10 years an appropriate profile of production would deplete about 66% of the initially declared reserves. On the assumption that the reserves of new fields and the appreciated reserves in previously developed fields are sequentially brought into production, then annual production would rise to about 350 million tons by 1982 and to a plateau of almost 475 million tons in 1987.

This rapid rate of resource exploitation will not be achieved as it will be 1975 at the earliest before all the early-discovered fields will come into production. An overall slower sequence of exploitation must be expected. Even so, given the foregoing assumptions, an annual rate of production of some 300 million tons per year could be achieved early in the 1980s. Thereafter, an annual production of between 350 and 400 million tons will be sustainable into the 1990s.[25] In brief, the West European 1980 and 1985 oil production potentials shown in the "alternative estimates" in Table II-2.1 are largely sustainable from the North Sea base alone – without there needing to be much concern for possible additional significant sources of supply from elsewhere in Europe – other than those already known to exist in parts of Germany, France, Austria etc. Between now and 1985, the supply of indigenous oil could increase from around 3% of demand to near 50% of the higher demand by the mid-1980s. This indigenous oil, together with the envisaged supply of indigenous gas could by then account for almost 40% of the continent's total energy supply.

Finance and Hardware for Indigenous Oil and Gas Production

This does, of course, pre-suppose the absence of problems in the supply of hardware and the availability of finance for developments. As the latter is mainly concerned with paying for the former, the two issues are closely linked. The problems in both cases can be wrongly magnified by treating the requirements in Western Europe as totally additional to what the oil industry has already considered as necessary for meeting the continent's growing demand for energy. This is not the case, since the industry's expectation of a 7% annual growth rate in the demand for oil – to be met by steadily increasing imports – already implied new tankers, new refineries and new pipelines etc. Some of the expected investment pattern will remain unchanged, for example, the strengthening of the Mediterranean oil supply systems and their associated pipeline developments up from southern Europe.[26] One consequence of large-scale oil production from the North Sea basin will, however, be a greatly diminished flow of oil by tankers around the coast of Western Europe to import terminals located on estuaries in the southern part of the North Sea. This change implies investment savings in tanker capacity and marine terminal facilities.

Outside Europe itself there will be a reduced demand for additional investment in the traditional oil exporting countries as new capacity for providing Europe's oil remains un-needed. Thus, the overall additional investment requirement for producing indigenous oil and gas is not the £1000 multiplied by as many daily barrels of oil or oil equivalent are producible from the North Sea, but only this amount *less* the investment saved by not building facilities elsewhere. Expected company expenditure involved in building up to a producing capacity of 6 million barrels per day in 10 years will thus be about £6000 million – say, an average of £600 million per year. We may modestly assume that the companies will provide at least 50% of this total investment requirements from their own resources (they are, in fact, used to providing up to 90%), so leaving a residual £300 million per year to be raised from European financial institutions and investors – or little more than an extremely modest £1 per head of the population. This figure is not, moreover, high by energy-sector-investment standards; for example, the UK electricity supply industry alone has over the past decade made an average annual investment of over £500 million.[27] Furthermore, the location of the development within Europe virtually guarantees the ready availability of whatever funds ultimately prove to be necessary, as many investors who have traditionally rejected investment in oil activities in overseas producing countries will be attracted to investment in Europe itself. Consequently, during the last few years – and in spite of stock market restraints on so-called speculative investment in resource exploitation activities[28] – a number of issues have been brought to the market and have been over-subscribed. Other privately launched ventures similarly seem to have had no trouble in raising their capital requirements. And, most important of all, a wide range of companies active in other – often entirely unrelated – fields have been easily persuaded that the North Sea offers the prospect for profitable investment and have provided risk capital for exploration work.

But the need for risk capital is a small part of the total investment required in North Sea oil and gas. Much more important are investment funds to develop discovered fields' producing and delivery systems. These, however, are low risk, bankable investment opportunities in which the oil in the ground constitutes a form of collateral against which money can be borrowed at the going rates, given that Western European oil markets guarantee the continuation of demand for the foreseeable future.

There is a similar position in respect of the supply of the hardware for the developments. Most of the requirements will come from firms which would otherwise have been supplying their products to other locations in the world where oil and gas exploitation was under way. A steadily growing part, however, will come from companies in Europe which, having recognised a major and continuing demand for oil industry products in their own backyard, will take the opportunity (or be so persuaded by governments) to diversify into types of production which would not otherwise have attracted their attention. The broad scope of Europe's industrial activities and the general and widespread familiarity with the basic, if not in the first instance the specialised, technologies required will ensure an availability of whatever range of supplies that the off-shore oil and gas industries demand: and in a timespan which is far shorter than would have been necessary with developments anywhere else in the world outside the United States.[29]

Conclusion

Western Europe offers the perfect location for the development of oil and gas activities: with not only a highly favourable geography of demand for whatever quantities of oil and gas can be produced, but also with a regional availability of investment funds, supplying industries and technical and managerial know-how literally targetting the area from which the oil and gas is waiting to be won. When one adds to this very powerful set of economic factors favouring the rapid development of the resources and the military, strategic and political concerns of most of Western Europe's governments for a significantly reduced dependence on imported oil, then a setting is created for a highly dynamic episode of oil and gas exploitation (even by the standards previously set by the industry elsewhere in the world). The challenges and the opportunities inherent in this developing situation also require the equally rapid evolution of a European energy policy which will not only recognise the fundamental nature of the changes, but also seek to ensure appropriate action by governmental and intergovernmental agencies to counter the impact of those forces which are less enthusiastic over the possibilities for a more autarckic energy sector in the European economy.

This necessitates much less apathy amongst European energy policy makers over the potential changes. There should also be a significantly improved information system in order to ensure that they are at all times sufficiently well informed about progress and the significance of events for new policy options. Europe's interest in incorporating as much indigenous oil and gas as possible into its energy economy must take precedence over matters such as imbalance between a company's crude oil production potential and its ownership of refineries and pipelines. The infra-structure of refineries and pipelines may have to be given a kind of common carrier status in order to avoid bottlenecks in absorbing North Sea production as quickly as possible.

The consequential effects of North Sea oil and gas developments on other sectors and on regional questions must also be taken into account. Transportations and regional planners need to be aware of likely reductions in the demand for marine facilities for handling large crude oil tankers, since most North Sea oil and gas will be used in north-west Europe.[30] And if oil and gas production develops as envisaged in this paper, there could be a dramatic effect on the economies of eastern Scotland and western Norway within a few years, removing the fears surrounding their present over-dependence on declining fishing industries and marginal agriculture[31].

Finally, reduced dependence on fuel from overseas could remove a major obstacle to freedom of political action for Western Europe in the Middle East. It may also cause Western Europe to question very seriously the current American proposals for a joint OECD policy, designed to help mitigate the worst results of the US self-inflicted energy crises.[32] Within a decade, Western Europe could become free of the political and economic pressures which have been imposed by the colluding oil producing and exporting countries with the co-operation of the major international oil companies. But this depends on Europe's energy planners ensuring that patterns of energy-use are the most efficient possible and that the indigenous oil and gas is rapidly incorporated into the overall energy supply. This much in policy terms is at stake as the world's most significant reserves of oil and gas outside the Middle East and Western Siberia are increasingly revealed by North Sea drilling.

References

1 For a general background to Western Europe's relationship with oil and the oil industry see P.R. Odell, *Oil and World Power*, Penguin Books 1972, Chapter 5.

2 See M. Posner, *Fuel Policy, A Study in Applied Economics*, (London, MacMillan, 1973) for a discussion of the policy constraints associated with the existence of a large indigenous coal industry

3 Compare this with the current worst possible forecast for US energy supplies. These estimate about a 40% US use of imported oil in a situation in which oil provides half of the country's total energy. With an additional import requirement for natural gas accounting for 2% of

total energy supply, the over all US dependence on imported energy is no more than 25%. Even so, this prospect has led to talk of 'crisis' and 'national emergency'.

4 See P.R. Odell, Europe and the Oil and Gas Industries in the 1970s, *Petroleum Times*, Vol. 76, nos. 1929 and 1930, January 1972 and M.A. Adelman, 'Is the Oil Shortage Real? Oil Companies as OPEC tax collectors' *Foreign Policy*, No 9, Winter 1972/73 for a discussion of these developments.

5 This has been done both at a national level, as in the British government's decision in February 1973 to increase its subsidy to the coal industry so as to prevent it declining below its existing level, and at a European level as seen in two recent basic energy policy documents approved by the EEC, 'Les Problemes et les Moyens de la Politique de l'Energie pour la Periode 1975–85' and 'Progres Necessaires de la Politique Energetique Communataire' – Com 72, 1200 and 1201 of 4 October 1972.

6 See, for example, *Problemes et Perspectives du gas natural*, EEC, Brussels 1965, and *Impact of Natural Gas on the Consumption of Energy in OECD European member Countries*, OECD, Paris 1969, as policy document studies which failed to question the validity of the assumptions concerning European natural gas reserves. This is treated at greater length in P.R. Odell, *Natural Gas in Western Europe: A Case Study in the Economic Geography of Energy Resources*, Haarlem, E.F. Bohn, 1969.

7 See T. White, ' The significance of the North Sea in the context of world oil reserves', *Northern Offshore*, Vol.1, No. 1, November/December, 1972

8 With, of course, appropriate stockpiling as insurance against temporary difficulties. See H. Roenneke, 'Crude oil and product stockpiling in member countries of the EEC', *Jahrbuch der Europaischen Erdol-industries (Annulaire de l'Europe Petroliere)*, Hamburg 1972

9 In 1972, the oil exporting countries collected about $15 thousand million in taxes on oil. Given present trends in tax rates and the production levels expected as a result of demand in Europe, the USA and Japan, their tax income is expected to rise to about $55 milliard in 1980, when according to Professor Adelman (*op cit.*), 'much of that wealth will be available to discupt the world monetary system and to promote armed conflict'.

10 For example, by substituting mass transit facilities for the use of motor cars in commuting, and long distance rail passenger and freight servies for the use of air and road transportation respectively. See P.R. Odell, 'The future of oil', *Geographical Journal*, June 1973, for consideration of such points

11 See the 1972 *North Sea Report* produced by Casenove and Co. (12, Token-house Yard, London EC2R 7AN), a major firm of stockbrokers, for evidence of the great interest of investing institutions in the North Sea opportunities.

12 See '*Study of potential benefits to British industry from offshore oil and gas developments*' by International Management and Engineering Group of Britain Limited, HMSO, London 1973

13 The views of the USA in this respect were expressed, eg, by the State Department's Oil Adviser, Mr J. Akins, to the Eighth Arab Petroleum Congress in Algiers, May–June 1972

14 As for example, in the case of the Venezuelan Ministry of Hydrocarbons or the Alberta Energy Resources Conservation Board whose annual reports of oil and gas activities enable an informed opinion to develop and so influcnce policy making. Some comparable sort of official Board is already an overdue requirement in Western Europe in order to avoid the continuation of the kinds of errors made by the UK administration over concessions and tax policies – see the report of the House of Commons Public Accounts Committee on this matter, HMSO, London, March 2, 1973.

15 For details of the participation agreements reached by early 1973, see *Petroleum Press Service*, Vol. XL, No. 9, February, 1973, pages 44–47

16 Circumstantial evidence of such commercial interests at work lies in the decision by Shell and Esso not to deplete the reserves of the Groningen gasfield at a rate exceeding 4% per annum and their refusal to produce any evidence on their reasons for this other than the somewhat meaningless assertion, given the non-availability of evidence against which to test it, that 'it is our conviction that the production level of the field must only be based on proven techical practices and observed reservoir behaviour.' The Managing Director of the Shell/Esso Company involved (Nederlandse Aardolie Mastschappij) continued, in a personal letter to the author (June 1972), 'we see no merit in discussing the geological, petrophysical, engineering and environmental factors involved' although their decision to limit the plateau level rate of output to as little as 4% is, to say the least, an outstandingly abnormal one, given the usual depletion rates 'in the range of 7 to 11% per annum for fields whose output is in demand' (Adelman, *op. cit*). This is certainly the case for Groningen which is surrounded by customers anxious to buy the gas.

17 *Petroleum Press Service*, Vol. XL, No. 2, February 1973, page 54

18 This would eliminate, for example, the present ability of the UK to prevent 'British' gas being sold outside the country – as happened in 1971 when the Conoco/NCB consortium was refused permission to sell Viking field gas in West Germany. It would also inhibit NAM (Shell/Esso) from restraining production as as to keep 20 thousand million m^3 of gas per annum off the West German and Belgium energy markets and stop the Dutch government from prohibiting the sale of its off-shore gas to Germany. These petty national and company interests are inhibiting the rational exploitation of the southern gas basin resources to a high degreee and must discappear in any renewed moves towards effective European integration.

19 See P..R. Odell, 'Europe's Oil' *National Westminster Bank Quarterly Review*, August 1972 pages 6–22, and his 'Import terminals and Europe's future oil supplies', Petroleum Times, Vol. 77, No. 1958, April, 1973, for more detailed discussions of these issues

20 M.T. Halbouty et al, 'World's giant oil and gas fields: geologic factors affecting their formation and basin classification', *American Association of Petroleum Geologists, Memo 14*, November 1970, page 509.

21 For several years following its discovery, the Groningen field was said to be an event of only modest proportions and, as already shown, even as late as 1965 it was still declared to be a field of only 1000 thousand million m^3, the optimum depletion of which enabled the operators to sell its limited production potential mainly as a premium fuel in European markets – at, of course, high prices. Now it is known to be over twice as large, a fact which if appreciated at the time of decision taking on Dutch gas policy would have opened up a much wider range of options for consideration.

22 Province of Alberta Energy Resources Conservation Board, *Reserves of Crude Oil, Gas, Natural Gas Liquids and Sulphur*, Calgary, Alberta, 1971

23 It should be noted, however, that by the end of 1972 about half as many fields again were known to exist in the North Sea, but no relevant information at all had by then become available about them.

24 From the paper given by W.J. George of British Petroleum to the Second North Sea Conference, London, December 1972.

25 These estimates are based on evaluations made by a group of international oil consultants in London on behalf of clients and emerge from discussions with 'most of the main groups now developing production' (from a personal letter to the author). They can, therefore, be viewed as representing 'backroom' opinion in the oil industry, as opposed to the public relations exercises in wich the companies have to indulge in order to 'justify' their policies.

26 For further treatment of this development and other elements in the geographical supply pattern in the early 1980s, see P.R. Odell, 'Europe's Oil' *op. cit.*

27 *Digest of United Kingdom Energy Statistics*, 1972. HMSO, London, Table 98

28 One is entitled to question the validity of the restraints in a situation in which the risk element in North Sea investment is probably less than the risk associated with investment in activities in the secondary and tertiary sectors of the economy trying to cope with labour unrest, inflation and price control at one and the same time.

29 See the recent British government sponsored report, *Study of Potential Benefits to British Industry from off-shore Oil and Gas Developments* – by International Management and Engineering Consultants, London, (HMSO, January 1973) – for a survey of the opportunities and

challenges opened up for British industry by the North Sea oil developments.

30 P.R. Odell, *Europe's Oil*, *op. cit.*

31 For a survey of the impact of developments on the East coast of Scotland see the 1973 report to the Church of Scotland Assembly by J. Francis and N. Swan, *A Social and Environmental Assessment of the Impact of North Sea Oil and Gas on Communities in the North of Scotland,* Church of Scotland, 1973, Edinburgh

32 See M.A. Adelman, *op. cit.*

Chapter II – 3

The North Sea Oil Province: a simulation model of its development*

The North Sea oil province, which is now generally accepted as a major one by world standards,[1] is unique in that it can be assumed that all the oil discovered will be produced in as short a time as constraints of hardware and technology will allow. The opportunities emerging from this form the basis on which this simulation model of its oil development has been built. Technically it has been constructed within the framework of a constrained random choice for the values of each of the variables involved.[2]

Calculating an Expected Rate of Discovery

Many hundreds of wildcats will have to be sunk to ensure full exploration of the province. Technology, hardware and manpower considerations imply the steady build-up of the exploration effort which began in 1969. We hypothesise a continuing intensification of the effort through to 1979 and a decline thereafter as the blocks allocated in the province are gradually drilled up. To build this model it was necessary to quantify the likely rate of build-up of exploratory drilling; then, from assumptions about the potential success rates, to calculate an average annual rate of discovery of initially declared recoverable reserves. This calculation defines the AM curve of such reserves (see Figure II-3.1); its

* **An edited version of the paper published in *Energy Policy* Vol. 2, no. 4, December 1974, pp 316–329. The joint author of this paper, Dr K.E. Rosing, has given his permission for its use in this book. This is gratefully acknowledged. That paper, in turn, was derived from a Monograph by the authors, *The North Sea Oil Province: an attempt to simulate its development and exploitation, 1969–2029*, Kogan Page, London, 1975, pp.74**

derivation is show in Tables II-3.1 and is briefly explained below. The number of wildcats for full exploration is calculated at 1675 on the assumption that the 365 'prime' blocks in the north Sea will require an average of 3.3 wells for full exploration and the 261 'fair' blocks an average of 1.8 wells.[3] These wildcats were then allocated by year and by size of structure being tested, ranging from Class I (the largest structures) to Class IV (the smallest ones). Overall annual success rates are now assumed for each year from 1969–1988 with an assumption that the highest rate of success has already been achieved and joint probabilities are attached to the likelihood in each year of finding fields of different sizes. A calculation of fields discovered was then made for each class size for each year and the matrix of Table II-3.1 is the result. From the expected number of fields in each size class in each year, it became possible to calculate the initially declared discovered reserves which can

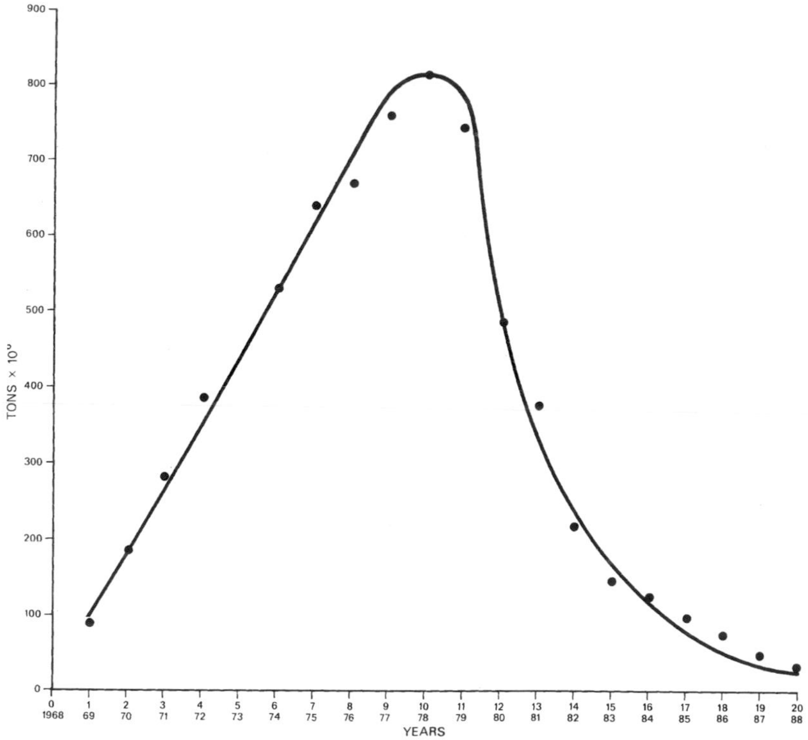

Figure II-3.1: Mean curve of initially declared recoverable reserves (derived from Table II-3.1)

TABLE II-3.1

Derivation of the mean curve of annually discovered reserves (in million tons)

Year	Class I No. of fields	Class I Reserves	Class II No. of fields	Class II Reserves	Class III No. of fields	Class III Reserves	Class IV No. of fields	Class IV Effective fields	Class IV Reserves	Totals for Each Year No. of fields	Totals for Each Year Reserves
1–1969	0.4	80	0.08	8	–		–			0.48	88
2–1970	0.74	148	0.38	38	–		–			1.12	186
3–1971	0.9	180	0.9	90	0.2	10	–			2.0	280
4–1972	0.99	198	1.74	174	0.26	13	0.13	0.09	2	3.12	387
5–1973	0.88	176	2.14	214	0.93	47	0.18	0.14	3	4.13	440
6–1974	0.67	134	2.65	265	2.45	123	0.54	0.41	8	6.31	530
7–1975	0.6	120	3.0	300	4.2	210	0.6	0.45	9	8.4	639
8–1976	0.62	124	2.22	222	5.58	279	2.99	2.24	45	11.41	670
9–1977	0.68	136	1.44	144	7.95	398	6.14	4.61	92	16.21	760
10–1978	0.8	160	1.2	120	8.01	401	8.9	6.7	134	18.91	815
11–1979	0.66	132	1.1	110	5.64	282	14.61	10.96	219	22.01	743
12–1980	0.6	120	0.6	60	2.38	119	12.42	9.32	186	16.0	485
13–1981	0.48	96	0.48	48	1.56	78	10.28	7.71	154	12.8	376
14–1982	0.12	24	0.36	36	1.15	58	6.77	5.08	102	8.4	220
15–1983	–		0.27	27	0.79	40	5.22	3.92	79	6.28	146
16–1984	–		0.38	38	0.75	38	3.4	2.55	51	4.53	127
17–1985	–		0.3	30	0.6	30	2.6	1.95	39	3.5	99
18–1986	–		0.25	25	0.5	25	1.65	1.24	25	2.4	75
19–1987	–		–		0.6	30	1.4	1.05	21	2.0	51
20–1988	–		–		0.38	19	0.87	0.65	13	1.25	32
Totals	*9.14*	*1828*	*19.49*	*1949*	*43.93*	*2200*	*78.70*	–	*1182*	*151.26*	*7149*

Notes: See the text for the derivation of;

a. the number of fields in each class in each year

b. the following average sizes have been assumed for fields in the different classes. Class 1: 200m tons; Class II: 100m tons; Class III: 50m tons; Class IV: 20m tons

c. only 75% of Class IV fields discovered are considered to be effective on the assumption that the others will be too small to be profitable to develop

be expected for each year in the 20-year exploration period. This is shown in the final column of Table II-3.1. These 20 annual values were now plotted and a smoothed curve fitted to the data points; a value for each year was read off this curve and the set of 20 annual values is now defined as the *AM* curve (see Figure II-3.1) and used as the basis on which the random choice of the simulated reserves discovered in each year is made.

A Random Choice of Annual Discovery Rates

Given the uncertainties of any exploration effort, coupled with the uncertainties of both companies' and governments' policies, it is obviously quite unreasonable to assume that the curve of reserves discovered over the 20-year exploration period will be smooth. The uncertainties which affect discovery rates from year to year are thus built into the model by means of a constrained random choice as to the volume of discovered reserves for each year. Briefly, this requires a choice within the framework of a normal distribution around the agreed value for a particular year – with a calculated standard deviation for each year providing the means of introducing uncertainty into the chosen value for the year. It has been further assumed that uncertainty will decrease over the first five years – as information about the oil province is accumulated. Thus over this period there will be an increasing probability that the values will have smaller variances from the mean. It is then assumed that the maximum probability of a minimum variance will continue for a further five years (as a result of good information and the search into the most likely prospects); and, thereafter, that the probability of achieving a result near the mean value as given in the *AM* curve will begin to fall – as uncertainty, following the beginning of the need to drill the less likely structures, starts to increase.

Appreciation of the Initially Discovered Recoverable Reserves

The initial declaration of the reserves recoverable from any oilfield is made on the basis of very limited information. Thus initial declarations are tentative and cautious both in respect of the size of the discovered fields and the proportion of the oil in them considered likely to be economically recoverable. In the light of the conservatism of initial declarations of producible reserves[4] and the general phenomenon of reserves' appreciation over time, it is considered appropriate and necessary to write into this model the possibility of the future appreciation of the initially declared recoverable North Sea reserves arising out of each year of exploration effort,[5] given that there is no reason why this oil province should behave any differently from all other major provinces.

The problem is how reasonably to write this inevitable process of reserves appreciation into the model. Appreciation is, of course, a continuing process but, to simplify this dynamic process, we here assume that there may be three discrete appreciations of each year's initially declared recoverable reserves. The size of the appreciations is determined at random within certain assumed limits which are related to the size of the initial reserves for a given year. The mean summed value of the three appreciations implies a doubling of the initial reserves but the random choice can be made within the range of 40%–160% appreciations, with the first, second and third appreciations as independently determined component parts of the total possibility. This is a simulated rate of appreciation which lies very much on the conservative size when compared with experience elsewhere. Such conservatism is in keeping with the general philosophy behind this model and it also allows for the claim of some oil companies that appreciation is becoming less important because of their increasing abilities to interpret the ultimate reserves position earlier in the exploration process.

The timing of the appreciations (an important factor influencing the rate of production) is also randomly chosen, within defined constraints, with the B reserves (the first appreciation) appearing between 1 and 3 years after the discovery year; the C reserves (the second appreciation) between 2 and 6 later than the B reserves; and the D reserves (the third appreciation) between 5 and 11 years later than the C reserves – thus giving a series of appreciations with generally increasing elapsed periods between successive revaluations of the fields. This has been done to simulate the observed real-world phenomenon that successively longer periods are required for the establishment of the finer upper limits on the size and the producibility of oilfields.

Total Recoverable Reserves

The elements in the model so far described have aimed to simulate the possible size of the total potentially recoverable reserves within the framework of a continuing 20-year period of exploration and discovery – and of associated appreciations of the initially discovered recoverable reserves over a total period of more than 30 years. The initial output from each iteration of the model is thus a series of figures for each of 80 sets of oil reserves of which each group of reserves (A,B,C,D) has 20 sets. Each individual value in the set varies according to the random choices made. Each set of reserves must then, of course, be produced within the framework of an appropriately timed and shaped depletion curve. Before proceeding to a discussion of the method used in the model for those aspects of the simulation, it is appropriate to look at the build-up of the reserves over time – both year by year and cumulative.

This is shown in Figures II-3.2 and 3, both of which present the results of multiple iterations of the reserves simulation procedures. Figure II-3.2 shows the total (of A,B,C and D reserves) declared in each year. It demonstrates, by use of the calculated mean of the values for each date, the rise through to the 11th year in the annual rate of discovery of

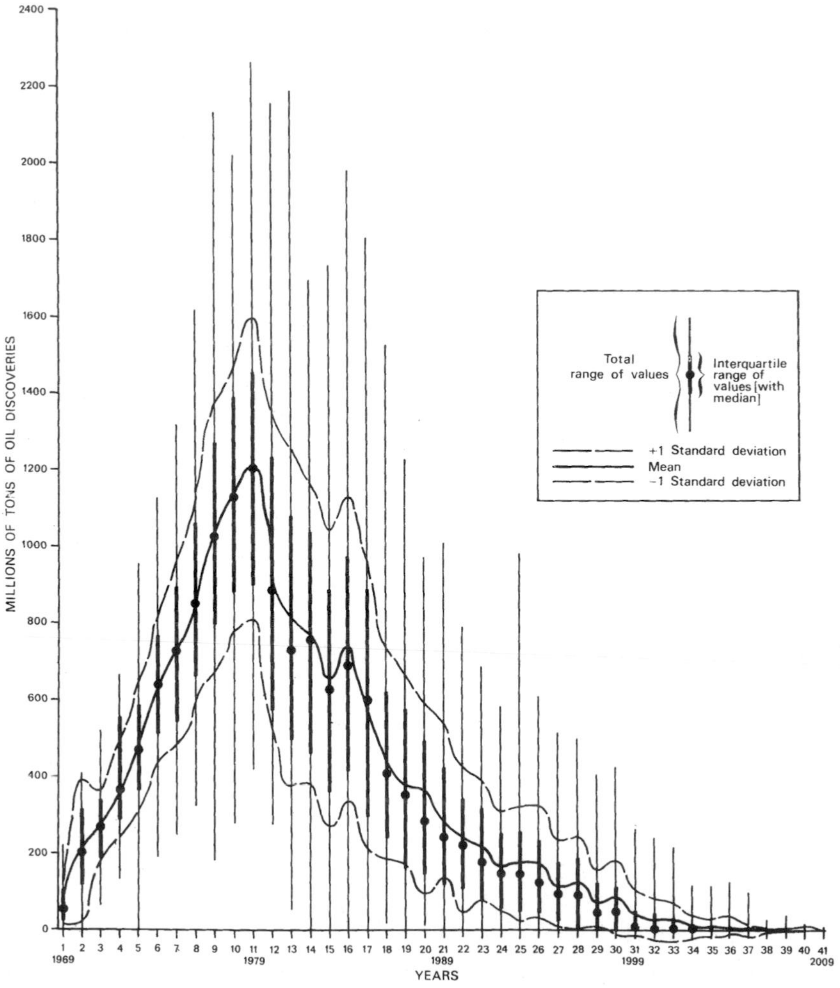

Figure II-3.2: Year-by-year discovery and appreciation of North Sea reserves from 100 iterations of the model

recoverable reserves and, beyond that, shows a general decline in the new reserves' situation arising from the fall-off in both discoveries and appreciations. However, the impact of the variability written into the model is clearly indicated by the great range of values which emerge for a given year from different iterations. The standard deviation curves around the mean and the range of values between the first and third

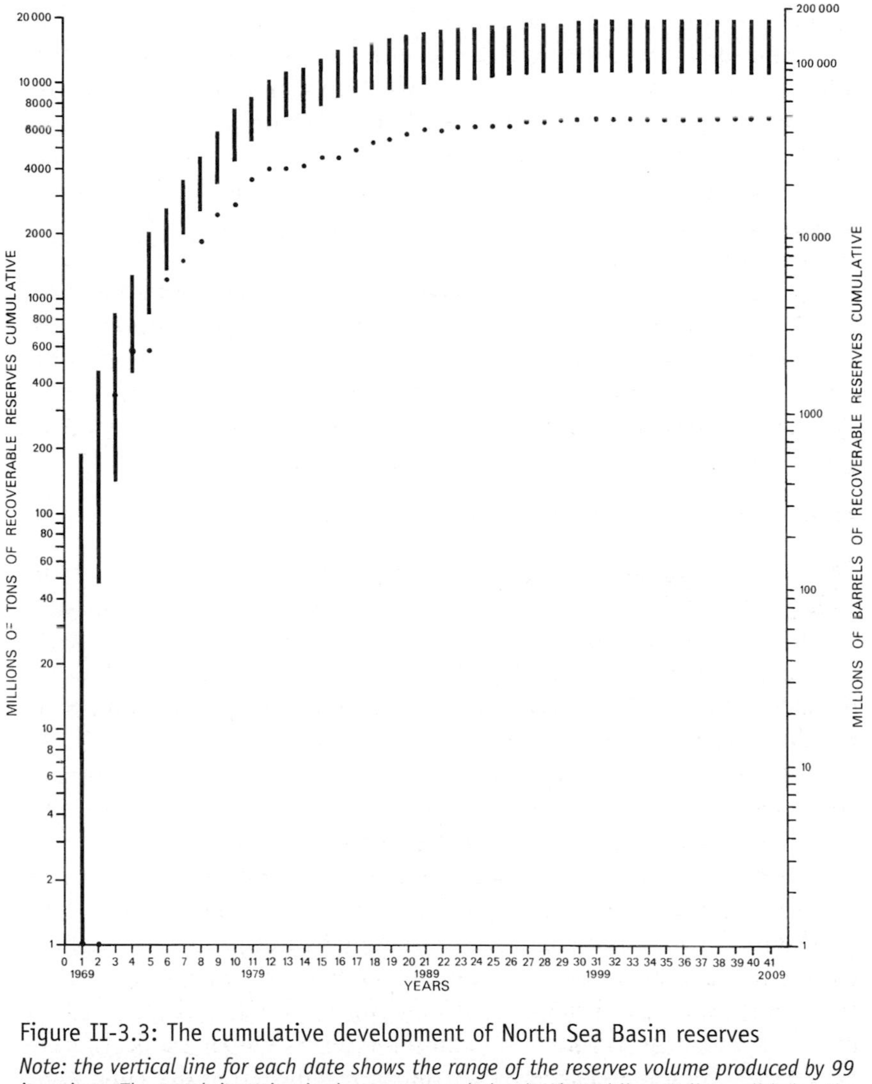

Figure II-3.3: The cumulative development of North Sea Basin reserves

Note: the vertical line for each date shows the range of the reserves volume produced by 99 iterations. The 100th iteration is shown separately by the dotted line. It lies well below the general range as it emerged from an iteration in which the early exploration efforts 'failed' and led to a diminution of exploration activities by the companies

quartiles for each date indicate a possible high/low interpretation for the expected annual reserves' position for the North Sea Basin over the remainder of the century.

It should be noted, however, that even with an assumption of no discoveries of new fields beyond 1989, new reserves are, in fact, still being added 22 years later – in 2010. This indicates the essential longevity of the reserves' build-up in major oil provinces – even when, as in the case of the North Sea, there is every incentive to explore and exploit with all possible haste. Figure II-3.3 clearly shows – on a log scale – the cumulative reserves' build-up over the period up to the year 2009. The range of total recoverable reserves is from a minimum of 10.8×10^9 to a maximum of 18.9×10^9 tons (equal to 79–138 milliard barrels). Even the lowest end of the range is over 50% higher than other present estimates of ultimately recoverable reserves. However, given the dynamics of the processes at work in the development of the basin, a static view of the situation is obviously inappropriate[6].

We would claim that the simulation model described in this paper enables the dynamic elements of rapidly developing situation to be introduced and so provides a much more reasonable base for the establishment of policy options on questions of appropriate annual production rates of oil from what is obviously a resource base in the process of a steady and lengthy period of expansion.

Timing and Rates of Depletion

All the recoverable reserves, as indicated in each iteration of the model, will ultimately be depleted given the assumption of a demand for oil in Europe which is at least equal to the amount that can be supplied from the North Sea at all times as soon as is technically possible. This involves variations in the start date of production and in the shape of the depletion curve for each set of reserves.

In the real world, of course, each field discovered and each newly-appreciated reserve emerging from any one year of continuing exploration and development effort in an oil province will start to be produced at a particular time depending upon the way the specific geological conditions affect production schedules, the way in which geographical conditions affect transportation facilities and in the light of particular company responses to the supply/demand situation. However, to avoid undue complications in the computing work, it is assumed for this model that the production of all the reserves which are declared to be discovered or appreciated in any one year will start at the same time. Reserves A will begin to be produced, according to the random choice made in the computing system, from two to five years after the discovery year; subsequently B,C, and D reserves will also begin to be produced in the second, third, fourth and fifth year after declaration – depending on the random choice of year – and will be fully depleted 20 years after commencement of production on a basis as set out in the next section.

The depletion curves for declared recoverable reserves are, of course, almost infinitely variable – even within the framework of a static economic situation (as assumed for this model)[7] – as a result of the influence of the same factors as mentioned above. Here again, however, for this simulation model, simplification is necessary and thus the following constraining parameters have been introduced to give limits to the shapes of the depletion curves:

- full depletion of any set of reserves to take place over 20 years;

- 66–75% of reserves to be depleted in production years 1–10 and the remaining 34–25% in years 11–20;

- the build-up period to the peak production rate to be between 4 and 7 years;

- the peak production rate itself to lie within the range of 6–10% of the original total reserves to be depleted.

A set of eight contrasting curves which encompass these parameters was constructed. For each set of reserves the computer randomly selects one of these eight curves for establishing its depletion over the 20-year period.

Simulation of Unforeseen Events

Oil exploration and development work and oil production are subject to risks of a physical character – particularly in locations as difficult as the North Sea where there is an important 'frontier of technology' component. In reality, there is some chance of something happening each year which will temporarily or even permanently inhibit the production as planned from any given discovered or appreciated reserves. To allow for the occurrence of these unforeseen events and their impact on developments as planned, an 0.02% probability is written into the model that an event will occur which will prevent any one of the 80 sets of reserves from being produced at all. When this random event occurs the production possibilities from any set of reserves so chosen – no matter how big or how small these reserves – are completely eliminated from that iteration of the model and the reserves concerned play no part in helping to determine the build-up of the North Sea oil province's production potential.

In addition to such 'unforeseen events' however, we must also recognise the possibility of changes in the expected behaviour of the exploring companies – arising from events in the North Sea itself. Thus the impact of a run of 'bad years' in terms of reserves discoveries will affect confidence – and influence earlier investment plans made on the basis of an average expectation of success. This is simulated in the model by means of noting a series of sub-normal values for volumes of discovered reserves: such values are penalised cumulatively over successive years. When the penalties reach a defined threshold level (roughly intended to represent a run of two 'very bad' years or three 'rather bad' ones), future values of the average initially declared reserves in the iteration are reduced by one-half – to represent the reduced effort which is likely to be put into the continued exploration of the province. A return to normal behaviour – and to the selection of the quantities of initially discovered reserves randomly based on 100% of the mean values for each year on the AM curve – can only be made if and when there is a succession of very good years in the rest of the iteration.

Individual and Multiple Iterations of Annual Production Rates 1969–2029

The computer program (see the Appendix) involves controlling the selection of random choices as defined earlier. It produces a print-out of results, diagnostic information and a graph of production over the period 1969–2029. An example of the graph plot of the build-up of production for a single iteration of the model is shown in Figure II-3.4. This appears as an onion-type structure in which the depletion curves of the 80 sets of evaluated reserves are superimposed one upon the other. In this particular instance it shows production from the province as starting in 1973, rising to a peak of 738 million tons (14.8 million b/d) by 1990 and then steadily declining to zero by 2028.

100 such separate iterations of the model have been run. The cumulated 100 top lines of the total production curve have been plotted in a superimposed format as shown in Figure II-3.5. This demonstrates that the general range of curves (that is, excluding the extreme values) gives peak values varying from 600 to 975 million tons per annum in years between 1987 and 1993. These superimposed curves thus give an impression of the potential production curve but, in order to give a clearer indication of the North Sea's potential contribution to Western

Europe's demand for energy Figure II-3.6 has been constructed. The lowest curve is of the 90% probability curve of production. The top curve shows the 10% probable set of values taken from Figure II-3.5.

Between those two curves is the calculated mean curve from the 90 curves. This we would define as our best estimate of the future shape of the North Sea's potential for oil production. This estimate is used in the final part of the article to compare potential production with Western Europe's future demand for oil.

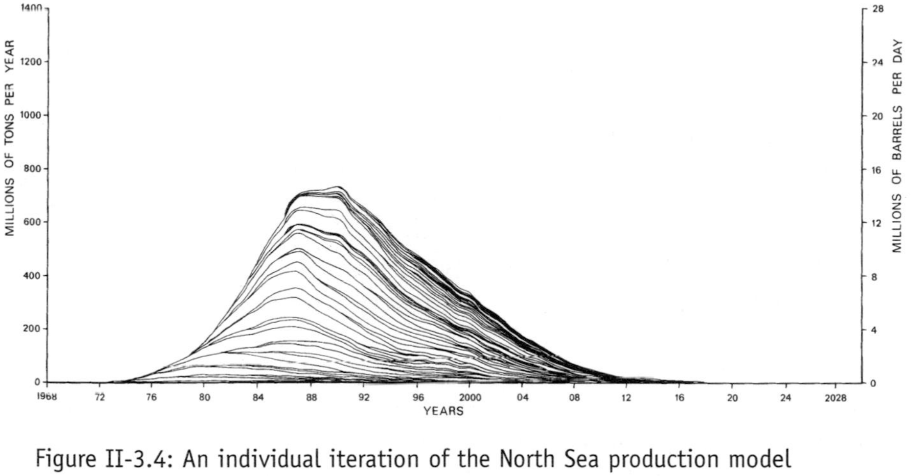

Figure II-3.4: An individual iteration of the North Sea production model

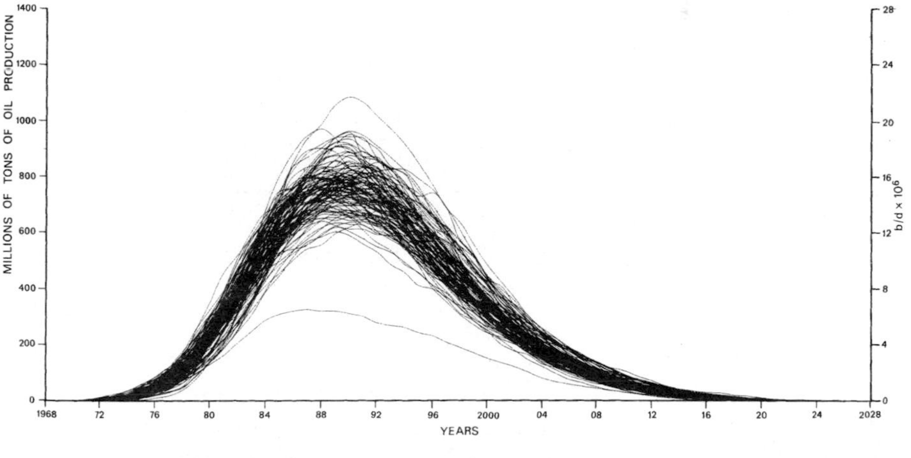

Figure II-3.5: 100 iterations of the production model

Note: The very separate production curve emerging from the 'failed' exploration efforts (see Figure II-3.3) also stands out clearly in this diagram – this, of course, represents the 1% probability of a curve of total production from the North Sea as low as this

Simulated Production and Demand

On Figure II-3.7 we have superimposed the mean production potential curve from Figure II-3.6 on a curve showing 75% of the expected demand for oil in Western Europe over the rest of this century.[8] This percentage of the total expected demand is taken on the assumption that no more than this could be met out of North Sea production either because of prior commitments made by various Western European countries to import oil on a long-term basis from other parts of the world; or because the North Sea is an inconvenient supply point for some parts of the continent; or because North Sea crude oils cannot meet the demand for the range of products required in Western Europe.[9]

The mean result from the simulation model on the long-term oil production potential from the North Sea Basin indicates that it can indeed contribute 75% to the total demand for oil for the period from 1982–1996. Thus some constraints on the rate of development of the Basin might be appropriate in order to keep the production potential much more closely related to the developing demand position and thus enable oil to be 'saved' for use in the first quarter of the 21st century.[10] Beyond that time the Western European economy may become orientated to the use of other cheaper and/or preferred energy sources available by then as a result of technological developments. In brief, the North Sea oil production potential may well be great enough – given full and appropriately timed development, as well as the efficient use of oil implied in the demand curve in Figure II-3.7 – to see the whole of

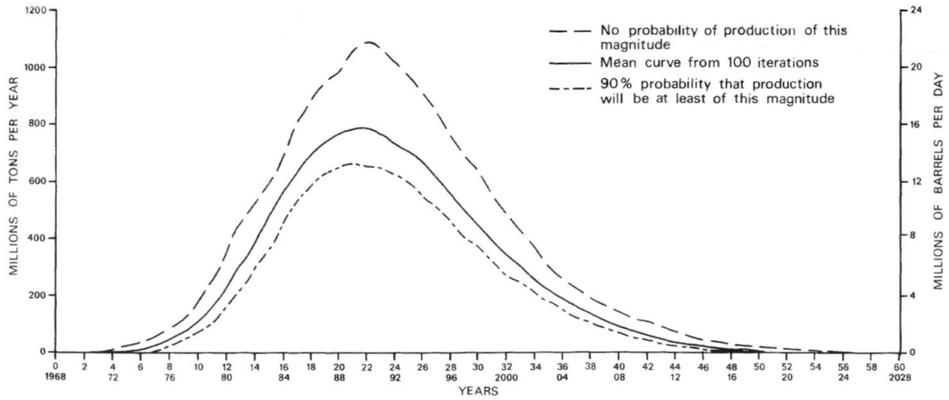

Figure II-3.6: Production potential from the North Sea Basin

Western Europe through into the post-oil age without any further undue dependence on supplies of foreign oil

Opting to go Ahead

The results from the simulation model thus clearly indicate that it would be inappropriate and, indeed, unwise for any single country sharing in the North Sea's potential riches to pursue a policy which seriously restricts the rate of development of the resources, for fear that inadequate supplies will be available over the long term for sustaining its own continued economic growth. This is a very obvious danger in the case of Norway, with its very small demand for oil – both now and in the future. It is even a danger in the case of the UK, where the future demand for oil is likely to increase only very slowly above 100 million tons per annum given, first, the continued and even increasing use of coal arising out of policy attitudes towards the coal industry and, second, the continued rapidly increasing contribution of natural gas to the country's energy economy. Like all major oil-producing countries, the UK will also become a mainly natural gas-using economy. Thus, even its next 30–40 years' demand for oil would deplete only a small part of the simulated potential from the North Sea Basin and so render inappropriate a policy directed at restraining development of the Basin to the country's expected demand for oil alone.

More positively, the long-term simulation of the potential from the North Sea points to the need for a European – or at least a north-west

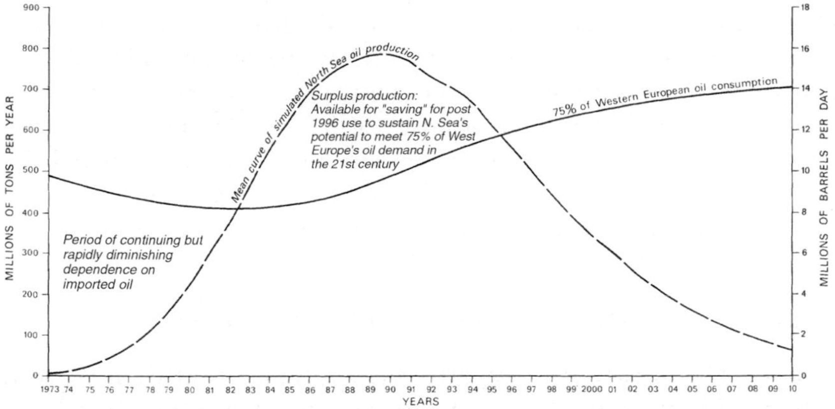

Figure II-3.7: Simulated North Sea production 1973–2010 compared with 75% of the future demand for oil in Western Europe

European – approach to the utilisation of the resources that appear likely to be delivered during the rest of the century. Within this political/economic framework, the UK and Norway (together with the Netherlands because of its role as a major gas producer) will obviously have the main part to play in determining that the resources are produced but, as the model indicates, it seems that they will be able to do this without any dangers whatsoever to their own long-term security of energy supplies.

In the meantime, their co-operation with the other oil-consuming countries of the region, notably France and Western Germany, will ensure the well-being of the whole of North-West Europe in respect of the availability of energy and so lead to a mutually advantageous development potential for the region.[11] It is this possibility which offers a broader range of energy policy options for Western Europe, where supply options have hitherto been seen only in terms of continued dependence on foreign oil. The only alternative seemed to be dependence on nuclear power with all its attendant security and environmental difficulties,[12] as well as the enormously high capital cost involved for the development and construction of a nuclear-based economy. The first requirement in looking at the broader range of options is an adequate international monitoring and evaluation system at a North-West European level for the effective development of the North Sea Basin. This would generate the information on which the shorter (5–10 years) policy options could then be determined. Evaluation would then indicate options for the next two generations. In this respect, the simulation model as presented in this article can, it is hoped, be used as the prototype from which to develop a more sophisticated and a more elegant approach to the longer-term potential from the province – not only for oil, but also for natural gas the availability of which will, of course, be increasing at the same time. The simulations of future possibilities can indicate limits against which actual developments can be examined for their significance, and provide a tool for examining policies which seek to control the rate of development of the resources of this major oil and gas province for political and economic reasons.

Appendix: The determination of parameter values

The model consists of a computer program implemented on an IBM 360/65. This program contains instructions supplementing the various assumptions outlined in the article, as well as the generation of the random choices for each component part in each of the 100 iterations of the model constructed for this study.

Annual Recoverable Reserves

A normally distributed random number[13] is generated to determine the size of each initially declared recoverable reserve. In order to generate a random number, GAUSS must be supplied with the mean and standard deviation of the normal probability field; one number is then returned from this field.

The means for GAUSS are defined in a data statement, changing for each discovery year. This A (discovery) mean (AM) curve, (see Figure II-3.1) was defined as stated in the article (p.185). A second curve, the A coefficient of variation (CV) curve, is also defined in a data statement. This curve represents the changing degree of uncertainty year by year with regard to the magnitude of initial recoverable reserves. The coefficient of variation[14] is a relative measure of the variability of a distribution. It is the standard deviation expressed as a percentage of the mean. When rewritten, one may calculate the standard deviation of a distribution given a mean value and a desired level of variation. The curve of standard deviation, SM, was then generated from the CV and AM curves. For each year of initial oil discoveries the appropriate mean from the AM curve and the appropriate standard deviation from the SM curve are supplied to GAUSS which returns a random number from a field of normally distributed random numbers. This random number is taken to represent the initial volume of declared recoverable reserves to be depleted.

Appreciation of Reserves

The first appreciation of reserves, viz. the B curves, are numbered 1–20 and each one corresponds to the same numbered A curve; that is, B1 is the first stage appreciation of the declared recoverable reserves of the 1969 discovery (represented as A1) and B2 is the first appreciation of the initial 1970 discovery curve, A2. The C and D appreciations are, respectively, the second and third appreciations of each A discovery. Thus, A1, B1, C1 and D1 reserves taken together constitute the fully appreciated volume of oil to be depleted from all fields originally discovered in the first year (1969); and A2, B2, C2 and D2 the reserves of the second year's (1970) discoveries.

The volume of each B, C and D set of reserves is related to the volume of the A reserves. RANDU[15] is used to generate a random number from a rectangular probability field which, in the case of B appreciations, is set at not less than 25% or more than 85% of the volume of the A discovery. This number is then taken to be the volume of the B appreciation. The limits for the C appreciation of reserves are from 15%–45% and D reserves from 0–30%.

Calculation of the declaration dates of the appreciations

The discovery year of the A reserves is, of course, fixed by an annual sequence of 20 years. The B, C, and D appreciations of these reserves are declared at a variable number of years after one another. For each discovery year, 1969–1988 (A reserves), the declaration date of the D reserves is dependent upon the declaration date of the C reserves whose declaration date is, in turn, dependent upon the declaration date of the B reserves. RANDU is used to select a random number from a rectangular probability field. The B reserves declaration date could be 1, 2 or 3 years after the discovery; the C reserves declaration date could be 2, 3, 4, 5, 6 or 7 years after the B reserves declaration; and the D reserves 5, 6, 7, 8, 9, 10 or 11 years after the C reserves declaration date. The A discovery year is 1969 and there can be no temporal variability. The corresponding B declaration date could be 1970, 1971 or 1972. If a B is declared in 1970, C could be declared in 1972, 1973 1974, 1975 or 1976. If a C of 1974 is randomly selected, then D could be declared in 1979, 1980, 1981, 1982, 1983, 1984 or 1985. If we calculate the corresponding dates for the 1988 discovery we find the D20 could be declared in any year between 1996 and 2009.

Timing of production from declared reserves

The date for the beginning of production from any set of reserves is calculated by generating a random number between 1 and 4 to give a minimum time-lag of 2 years and a maximum time-lag of 5 years between the discovery or a declaration of appreciation and the commencement of production. Using the example of D1 reserves, the production could begin as early as 1978 or as late as 1993. In the D20 example the corresponding dates would be 1998 to 2014.

Depletion of a set of reserves

A set of eight depletion curves, the DEP curves, were drawn and initialised in a data statement. These eight curves allow a variety of different possible depletion processes, within a series of constraints which are given in the text. The volume below the line of each curve was set equal to unity by making the summation of the Y values – that is, the values for each annual figure – equal to 1·0.

RANDU is used to generate a random integer between 1 and 8 from a rectangular probability field. This number is used to achieve a random choice of the depletion curve (DEP curve) for the production of any one set of reserves. The chosen curve is then multiplied by the randomly generated volume to scale, firstly, the total volume of the curve and, secondly, each of the 20 yearly production volumes.

Returning to the example of D1, total depletion could occur as early as 1997 or as late as 2012. In the case of the D20, depletion could occur as early as 2017 or as late as 2033.

Elimination of reserves from the production curve

(i) Since the A volume is generated from a normal probability field around a given mean there are cases where the volume chosen involves a negative number. In addition, the possibility of finding small fields which it is uneconomical to bring into production is recognised. Such occurrences have the highest probability in the later discovery years when the mean is small (near zero) and the coefficient of variation is large and so the standard deviation is relatively large. If the set of reserves for any discovery year is negative or less then 10 million tons, then it is considered impossible to produce these reserves, in that they must occur in fields which are too small to be economic, and thus the volume of the A discovery and also of the associated B, C and D appreciations are all set equal to zero.

(ii) There is some probability in all oil-seeking operations, particularly in the North Sea with its frontier of technology conditions, that some accident will make the recovery of a certain reserve impossible. A 0.02 probability was allowed for the elimination of any of the 80 sets of A, B, C and D reserves in each iteration. RANDU was used for this.

Changes in oil companies' expectations

A feedback from experience was included by scoring minus 1 penalty point for any year with a reserve discovery of less than 30% of the expected mean value; minus 2 points for less than 20%, and minus 3 points for less than 10%. In addition, if any unforeseen event prevented the production of a given reserve, minus 1 point was scored for the year it should have commenced production. Each year the previous 3 years experience was examined – if a total of minus 5 points had been reached the AM curve was divided by 2. Checking then began for both good and bad years. If any year was 125% above the mean value, plus 1 point was scored; if over 150%, plus 2; and for over 175%, plus 3. A total of plus 5 in the preceding 3 years was necessary to multiply the reduced AM curve by 2. At the same time checking for further reduction continued.

Build-up of a simulated production curve

After the random choice of the fixed parameters and the random selection of a volume, which together constitute each individual set of A, B, C or D reserves, the 80 individual curves[16] which represent these reserves must be summed to create the total reserves position. An array T is used to hold the total simulated production in each year, summed from the production of all curves producing in that year. The chosen and scaled depletion curve is added to the appropriate segment of the T array (as determined by the declaration date and the production date) and this segment of the T array is plotted. Thus, in the single iteration model (Figure II-3.4) the space between each two successive lines is equal to the total volume of the simulated reserve and the vertical space at each year, between each successive line, is equal to the volume of production in that year from a particular A, B, C or D set of reserves producing in that year. The single iteration model is not calculated or plotted in a temporal sequence. Instead, for ease of computing, the plotting is in the order of the A discovery date, ie A 1969 (A1), is first followed by its associated B 1969 (B1) then by C 1969 (C1), and by 1969 (D1). Next A 1970 (A2), B 1970 (B2), C 1970 (C2) and D 1970 (D2) are calculated and plotted. Plotting ends with A 1988 (A20), B 1988 (B20), C 1988 (C20) and D 1988 (D20).

Figure II-3.5 displays the final accumulated topmost curve for each of 100 iterations of the model. The determinations of each individual line is as shown in the single iteration model (see Figure II-3.4), except that the plotting of each individual A, B, C and D curve was suppressed. At the end of the time period for discovery, appreciation and depletion of reserves the total (T) curve was plotted. Each line thus indicates the total amount of oil produced in one iteration of the model for each year. That is, the vertical (Y) distance measured from the X axis in each year is the total production in that year arising from the commencement and continuation of production from one or more individual sets of reserves from each iteration.

In each iteration of the model the total production of oil (the final position of the T curve as displayed in Figure II-3.5) in each year was recorded on magnetic tape. After 100 iterations of the model this tape was used to construct Figure II-3.6 which displays the 90% probability minimum and the maximum production of oil in each year from 1970–2028, as well as the calculated mean value of the 100 curves.

References

1 See an earlier appreciation of its significance in P.R. Odell, 'Indigenous Oil and Gas and Western Europe's Energy Policy Options', *Energy Policy*, Vol 1, No 1, June 1973, pages 47 64

2 The methodology used in the determination of the parameter values was set out in the Appendix to the original paper.

3 Based on blocks adjusted to British block size. See the report, *The Outlook for Large mobile Drilling Rigs on the European Continental Shelf* prepared by the Investment Research Division of Kitcat and Aitken, 9 Bishopsgate, London, EC2

4 See P.R. Odell, 'The Future of Oil', *Geographical Journal*, Vol 139, no 3, October 1973.

5 The appropriateness of this is already being demonstrated by reserves revisions for North Sea fields. For example, the Ekofisk field, initially declared to have 1140 million barrels of recoverable reserves is now declared at 1800 million bbl. Brent, originally declared at 1000 million bbl is now, two years after discovery, declared at 2250 million bbl. The Thistle field, said to have 800 million bbl of recoverable reserves on discovery early in 1974, has now been appreciated to 1300 million bbl.

6 For a conventional estimate of the reserves' position see, for example, the most recent UK government estimate of reserves in the UK sector of the North Sea. This was published by the Department of Energy, *Production and Reserves of Oil and Gas in the United Kingdom,* HMSO, London, May 1974.

7 That is, an economic situation in which the supply of and demand for North Sea oil is not affected by energy price changes: as could be the case within the framework of an administered autarkic energy market in Western Europe.

8 For the derivation of the future demand for oil in Western Europe from 1973 to 1998, see P.R. Odell, '*European Alternatives to Oil imports from OPEC Countries; Oil and Gas as Indigenous Resources*'; a paper given to the Colloquium on Energy Policy Planning in the European Community at the J.F. Kennedy Centre for International Studies, Tilburg, The Netherlands, May 1974. This paper was published in the proceedings of the Colloquium in 1975.

9 This implies imports of other crude oils for refinery blending. However, it is unlikely to be as important as the present incompatibility

between light North Sea crudes and the large European demand for fuel oils would suggest. The increasing availability of natural gas from the North Sea Basin (from gas fields and from as production associated with oilfields) will substitute the hitherto expected rising demand for fuel oils and thus change the balance of product out-turn required from the refineries.

10 A potential 'excess' of North Sea oil over the European demand does, of course, offend the basic assumption of the model, ie. that all North Sea oil producible will have a market. While the result does not invalidate the model as such, it does indicate that the supply/demand situation will be more complex than assumed and so have a feedback effect on decisions about production levels. Decisions to delay production will emerge either out of governmental action or from downward revisions in investment levels in exploration and development work, thus extending the period over which the reserves will be produced.

11 For an examination of this particular issue in more depth, see P.R. Odell, '*Norway's Oil and Gas and Europe's Energy Needs – a Complementarity of Interests*', a paper presented to the Conference on Scandinavia and the North Sea in Oslo, April 1974 and published in the proceedings of the Conference by the *Financial Times* Business Enterprises Division, London, 1974.

12 See A.J. Surrey, 'The future growth of nuclear power. Part 2: Choices and obstacles'. *Energy Policy* Vol, No 3 1975, pp 208–224.

13 By Scientific Subroutine GAUSS, *IBM System 360 Scientific Subroutine Package Version III program manual*, No GH20-0202-4, (New York, 1970), describing program number 360A-CM-03X, p77.

14 K. A. Yeomans, *Statistics for the Social Scientist*, Vol 1 (Penguin Books, London 1968), pp 112–113.

15 Ibid, p. 77

16 Or a lesser number of curves in the case of iterations where sets of reserves have been eliminated on the basis of exclusions as described.

Chapter II – 4

The Economic Background to North Sea Oil and Gas Development*

Introduction

The international oil industry has, quite justifiably from the point of view of the international oil companies concerned, always endeavoured to integrate each new development of oil and gas producing capacity into a pre-existing pattern. The development of the North Sea basin – largely concentrated in the hands of these international companies – has been no exception to this general rule. We note, for example, how the exploitation of the first major discovery in the basin – the massive Groningen gas field in the Netherlands – was essentially planned by Shell and Esso, in co-operation with the Dutch Government, to ensure that the gas supplemented, rather than competed with, the marketing of imported oil.[1] The same companies – and others – fought hard to secure a similar opening for the gas they discovered a few years later in the southern part of the British offshore sector, but were thwarted by the Labour Government of the day which insisted on a supply price, rather than a market price, for the gas, and saw to it that the nationalized gas industry rather than the oil companies collected the economic rent.[2] That this was unsatisfactory to the companies, which saw it as a threat to their ability to integrate the new energy supply into their markets, quickly became apparent as they virtually ceased to explore further for gas in the British sector and also pretended that the

* **Reprinted from *The Political Implications of North Sea Oil and Gas* (M. Saeter and I. Smart, Eds.), I.P.C. Science and Technology Press, Guildford, 1975, pp. 51–80**

adjacent Dutch offshore sector was likely to contain no significant resources of gas. As a result, reserves of non-associated gas in the southern part of the basin failed to grow, even though the companies knew then that there was, in the words of a Shell spokesman at a recent Institute of Petroleum London Conference on the 'Geology of North West Europe's Offshore Areas', "a lot of gas still lying under the southern part of the North Sea which remains to be discovered".[3]

Western Europe's natural gas reserves – and the annual production of the commodity – have thus failed to grow as rapidly as would have been the case if either the companies had been allowed to do as they wished, or if governments had been better informed on the limitations of the companies' power and had been prepared to require the companies to adjust their policies.

Unfortunately, however, (in the light, that is, of the immense difference that an additional availability of indigenous natural gas would have made to Western Europe's current energy policy options), the Dutch and the British Governments – aided and abetted by indifference to the situation on the part of the West German and Danish Governments – chose not to challenge the companies' policies over the development of non-associated gas reserves and based their own policies on a firm belief in the near-future scarcity of these resources.[4] The policy of conservation can only be described as inappropriate based on a lack of understanding of the nature of the resource base and an unwillingness to question the motivations of the international oil companies in deciding to cease exploring for the remaining gas in the southern part of the basin.

At the end of the 1960s the companies had no interest in finding more gas than they could incorporate in their overall oil/gas supply systems for Western Europe[5] – particularly as they were at that time, through the work of the London Oil Policy Group, attempting to restructure the European energy market away from the intensely competitive situation that had developed after the late 1950s, under the pressure of oil supplies moving to Europe instead of to the United States, whose quota system had effectively excluded oil which had originally been discovered to serve the US market.[6]

North Sea oil potential, however, – as distinct from gas potential – represented a preferable option for the companies – for two reasons. Firstly, most European governments did not treat oil as a public utility (as was the case with gas), so that the companies had more commercial freedom in dealing with the commodity; and, secondly, North Sea oil could be incorporated into the existing infrastructure for foreign oil without any expected difficulties. It thus represented a safer investment opportunity for the international oil companies which had, in the meantime, secured the lion's share of the concessions awarded for oil exploration in British, Norwegian, and Danish waters.

Moreover, the incorporation of North Sea oil in the already-developed European oil economy was not foreseen as likely to create problems. There were two reasons for this; first, the expectation that the 7.8% annual rate of growth in the demand for oil since 1950 would continue at least until the mid-1980s.

Second, because there was a generally accepted view that the amount of oil which would be available from the North Sea would be small, relative to future Western European demand. Company spokesmen described the prospects as follows:

> *"North Sea oil will provide a small, but useful addition to Western Europe's indigenous energy supply, but it will not even provide the incremental amounts of oil which will be needed each year by the end of the 1970s and so the continent will remain heavily dependent on traditional supplies of oil from the Middle East and North Africa etc."*[7]

In this context the international oil companies had no cause to think other than that they would be welcomed as the producers of North Sea oil, given their already known and accepted European-wide ability to provide most of the energy which the continent required.

One should bear in mind that this was not simply what the companies expected: it was also what they wanted – given the way the international oil situation had developed by 1971. Prior to this, initial North Sea exploration – and the early discoveries – had taken place in the context of continuing low-cost oil in Western Europe where competition had ensured a falling real price of the commodity. By 1970 oil products were, on average, less than 50 per cent of their price, in real terms, of ten years previously. Oil from the North Sea was thus only interesting in economic terms in relation to imported oil readily available at refineries in Western Europe at under $3 per barrel. This minimized most companies' interests in too rapid exploration and exploitation of the North Sea's potential. Instead, at this time, they only wanted a build-up of knowledge of the North Sea potential, together with the opportunity to experiment with offshore drilling and field developments in deep water – in preparation for their expectation of an increasing need for such offshore oil later in the century.

In 1971, moreover, the international oil companies had yet another reason for adopting a relatively go-slow attitude towards North

Sea resources. This was a result of their success in achieving a *modus vivendi* with the Organization of Petroleum Exporting Countries (OPEC) following the decision of that organization at its Caracus meeting in December 1970 to seek a collective revision of the terms governing the production and export of its member countries' oil. The agreement achieved was highly favourable to the companies in the short-to-medium-term, given an assumption of the ability of companies to pass on price increases to oil consumers. This was, indeed, a requirement to which the companies in Europe, which they saw as the critical market in this respect, had already given their attention and where, indeed, they had decided to try to implement a strategy of orderly marketing, involving the voluntary elimination of competition between companies.

Thus, the marketing strategy agreement with OPEC ensured higher profits to the companies – within the framework of steadily rising prices. At the same time, it left them in control of essential decisions on the supply and development of oil production in most OPEC countries, for the agreement limited state participation to 25 per cent in the immediate future and, most important of all, to less than 50 per cent until 1982. In other words, the companies' agreement with OPEC not only left them in control of the international oil system for another decade, but also gave them a positive incentive to maximize the production of oil they had discovered in OPEC countries and to export it to their global markets in the period up to 1982.

This, of course, meant that their immediate interest in the North Sea – both in respect of a rapid development of its potential oil supplies and in respect of its potential for natural gas – in that such indigenous gas production would substitute imported oil – was even further diminished. Their commercial interests made them even more firmly in favour of the development of the potential after 1982, again implying a 1970s premium on securing knowledge and on gaining experience in offshore procedures. European governments either lacked or did not seek access to alternative intelligence and thus accepted the very strongly commercially-motivated presentations of the interested oil companies concerning the 'limited' potential for North Sea oil over the next decade.[8] There was thus a general expectation – at both individual governmental and at European organizational levels – that the North Sea was not very significant, either in terms of its likely impact on the domestic economy or in terms of its ability to change Western Europe's economic relations with the rest of the world – and, particularly, of course, those with the oil exporting countries.

The Revolution in World Oil Power and its Implication for the North Sea

OPEC/international oil companies co-operation, however, proved to be very short lived. By mid-1973 it was becoming increasingly clear to the companies that the oil producing countries were intent on taking control of the system – in terms, that is, of securing majority participation in oil producing operations – much sooner than 1982 and of securing the right to determine levels of production and prices. With the realisation of this impending change, the North Sea took on a new significance for the oil companies. This began to show in their revised appreciation of its medium-term supply potential as a hydrocarbon province from which much greater quantities of oil could be produced very much more quickly than hitherto planned. Thus, in the second half of 1973 all the companies with North Sea interests began a re-evaluation of the opportunities open to them from the concessions they had obtained. This re-evaluation became even more emphatic in the early months of 1974 when the unilateral quadrupling of international oil prices by OPEC members completely changed the economic outlook for large scale investment in the North Sea.

This was in spite of the fact that, by that time, the increased pressure of demand for, and the limited availability of equipment needed for offshore exploration were, together with the ravages of inflation, already bringing serious upward estimates to bear on the investment required to generate each barrel per day of producing capacity from the North Sea. This was, however, seen by the companies as but a relatively minor difficulty in a situation in which they could do their calculations of profitability against a price of oil delivered in Western Europe of upwards of $12 per barrel. Even smaller fields in deeper waters at increasingly remote locations became potential profit earners – and the companies were now, paradoxically, less concerned with the question of how to raise the capital necessary for such developments – for two reasons. First, given the rapid rise in their profit levels from 1971 through to 1974, they had arrived at a position in which they could internally generate much of the cash flow needed for North Sea work. Second, given the much higher value of discovered oil, the collateral against which they were able to borrow money from banks and other financial institutions had increased to the point where there was no cash-availability problem – at least for the larger, well-established international oil companies.[9]

Governments of the North Sea countries concerned have adjusted in one way only to this revised appreciation by the companies of their potential production and profit-earning capability. They have sought, in true oil producing country manner, to regulate the activities of the companies to ensure that surplus profit-earning opportunities are taxed away, in order to direct the economic rent to national exchequers. The debate between these parties on this issue soon resolved itself into a power conflict. On the one hand, the British and the Norwegian Governments threatened to exercise their sovereign rights while, on the other, the companies threatened to withdraw if they were left with insufficient incentive (from a profit-earning point of view) to make continued activities worthwhile. Though there was undoubtedly a large element of bluff in the companies' attitudes (for their commercial interests dictated that they had to made the situation appear worse than it really was), they could, in the final analysis, have decided to move elsewhere – most notably, of course, back to the United States, where a whole new range of profit-earning options had been opened up by 'Project Energy Independence'. This would have left Western Europe in the deepening clutch of the Middle East oil producing countries for its future supplies of essential energy, with all the many implications of such a position for the economic outlook for the continent.

Eventually, both Britain and Norway ultimately made tax proposals acceptable to the companies. Indeed, there ought never to have been any doubt over a solution, given the size of the economic rent under dispute, the essential reasonableness of both the governments and the companies concerned and finally the fact that the conditions under which the companies were having to operate elsewhere in the world – including the United States – in offshore oil developments were also becoming steadily tougher.

One of the reasons why such an unnecessarily risky outlook was allowed to develop at all was because governmental and intergovernmental attitudes towards North Sea oil and gas did not mature at anything like the same rate as that of the companies. Governments had neither the abilities – nor the inclination even to develop the abilities – to evaluate independently the gas resource base in the southern sector of the North Sea. They thus allowed policies to emerge from a less-than-fully-objective appreciation of the situation. This was true also of the oil resource base of the rest of the basin. In this case the increasing complexities of the many oil and gas 'plays' ensured that governmental policy makers were kept groping in the dark.

However, behind all the detailed argument and counter-argument lay two contrasting views of the North Sea's potential. On the one hand, there was the view that the basin would only provide the *additional* conventional energy that West Europe needed after 1978 and through the 1980s – though, it was argued, this would also require a restraint on the rate of growth in conventional energy by a crash programme of expansion of nuclear power. At best, according to this view, there would be no growth in the net demand for imported oil, but North Sea oil and gas were certainly not expected to substitute the need for continuing hydrocarbon imports as the level to which they had already developed. Thus, Western Europe would remain essentially dependent on OPEC oil for the whole of the foreseeable future.[10]

On the other hand, there was the alternative view that the North Sea basin had a potential to produce oil and gas with a rate of increase in supply far outstripping the growth rate in conventional energy demand, given the assumption of a reduced overall rate of growth in energy use as a consequence of higher energy prices and conservation measures of other kinds. And as the supply of indigenous coal would be kept up (by means of protectionist measures in Britain and West Germany), this would mean a sharp and continuing fall in the dependence of Western Europe on OPEC oil, the demand for which could, within a decade, become a small residual element (less than 10 per cent) in the continent's total energy supply – even without any kind of a crash programme of nuclear power development (See Table II-4.1).

The Limited View of North Sea Resources

The first view remains the one still held in most official circles. It thus forms the basis for the present official economic evaluation of the significance for Western Europe of North Sea oil and gas. It emerges out of a belief that oil and gas reserves must be *proven* to exist before they can be incorporated into energy policy plans. There is an unwillingness and/or an inability to accept reserves' figures which stem from methodologies such as the probability modelling of the dynamics of an oil and gas province's development. Though such probability approaches are considered entirely adequate for government policy-making in many other fields, they are not thought to provide a reasonable basis for medium-to-longer-term evaluations of oil and gas reserves' development and of production potential, in spite of the fact that there are analogies to which one can draw attention on the dynamic development of reserves and production in other oil and gas regions around the world.

Even more extraordinary is the simple statistical error that is made in presenting the total proven reserves of the North Sea as the arithmetic addition of the declared 90% proven reserves from each individual field instead of constructing a joint probability curve for the set of occurrences, so defining a correct 90% probability for the reserves of the multiplicity of fields in the basin.[11] Not doing this implies a significant under-evaluation of the medium-to-longer-term growth potential for oil and gas reserves with two consequential effects. First, it appears to 'prove' that Western Europe cannot change its energy policy towards one centred on the use of indigenous hydrocarbons. Second, it appears to indicate that the individual country within Europe which has been lucky enough to have oil and gas resources should conserve such 'limited' wealth to ensure its own future well-being.

Such inappropriate interpretations of oil and gas prospects produce, a best, a tendency to treat the hydrocarbon resources of the North Sea basin as a minor factor for the future energy economy of Western Europe: and, at worst, they produce a positively dangerous scenario. This may be stated as follows: *Because the amounts of indigenous oil and gas do not made a really fundamental difference to future options for the Western European energy economy, the continent will remain largely dependent on traditional oil imports from OPEC countries. This, of course, involves not only uncertainty over the supply, but also, dangers of inflation, monetary instability, and the transfer of resources and of assets to the oil exporting countries – to say nothing of being subject to political pressures from the OPEC countries. Thus, the future of Western Europe depends essentially on achieving a satisfactory modus vivendi with the Arab/OPEC world – and it is to this end, therefore, that European foreign policies will have to be orientated.*

The scenario requires, however, not only requires an entirely optimistic evaluation of the development of future relations with OPEC countries (implying a much greater act of faith by our policy-makers than that required for their acceptance of the validity of probability analysis for indicating the order of magnitude of European oil and gas reserves), but also creates a set of great uncertainties in the economic outlook for the continent. The most important of these are as follows:

i. The occurrence of oil resources in Europe which can be developed over the next decade lie mainly in British and Norwegian waters (plus the gas resources in the Netherlands) and this creates the possibility of a fundamental have/have not division in Western Europe. Two indications of this have already appeared. First, in Norway's refusal to join the International Energy Agency because membership implies sharing indigenous energy production potential; and second, in increasing the pressure in the Netherlands for a cut back in its exports of natural gas. Furthermore, a recent UK government declaration that there will be constraints on the levels of oil production post-1982 in the British sector also indicates a similar attitude.

ii. It provides an opportunity for greater Soviet influence to be exercised within the framework of a Western Europe divided by entirely nationalistically-determined energy policies – notably by separating Western Germany and Scandinavia from the rest through offers to help them 'defend' their energy policies by enhancing their imports of Soviet oil and gas.

iii. It allows OPEC – at any time of its own choosing – to undermine the viability of the whole North Sea development. It could do this by offering oil (at a relatively small cost to its member countries) at prices below those which *have* to be charged for the much higher-cost North Sea production and so severely diminish British and Norwegian production and lead to problems of readjustment for these countries.

The Alternative View of the Economic Significance of North Sea Resources

Thus the possible economic consequences of present attitudes towards the North Sea's resources are adverse. In particular, they heighten the already great uncertainties to which the European energy economy is exposed as a result of the international oil situation. The alternative view of the situation, on the other hand, offers a challenging, but positive, outlook based, first, on an effective and realistic long-term evaluation of the North Sea basin's total oil and gas potential and, second, on the willingness on the part of governments to take actions to ensure that this potential for development and production shall be realized as quickly as possible; with a requirement (if such be needed) for the companies to assist in achieving this aim.

The basic hypothesis on the size of the resource base emerges from a simulation model which has been developed in Rotterdam.[12] The results of this clearly indicate a high probability that the North Sea is capable of meeting a large part of Western Europe's energy demand for at least the last two decades of the present century – and possible for the first 20 years of the 21st century as well.[13]

The 'Alternative estimates' columns in Table II-4.1 show the impact of this re-evaluated indigenous oil and gas potential within the framework of the Western European energy supply/demand situation up to the end of the century. The model, however, showed only the very large potential for reserves' discovery and production. For this to be realised it is necessary that Western Europe pursues appropriate policies for the energy sector of the economy. This implies, as a primary requirement, the choice of a deliberately autarkic energy policy in which the aim is the minimization of imports. This gives a long-term guarantee of markets for all the North Sea oil and gas that can be produced, together, of course, with the same prospects for all other reasonably cheap-to-produce energy in the continent. Naturally, it also involves the development of an economic organizational/institutional structure which will make an autarkic policy possible. The following elements would be essential in such a development:

i. A collective agreement on a base price for energy between the nations of Western Europe – with the price reflecting the long-term supply price of the most costly indigenous energy which has to be produced. The price will be independent of the oligopolistic price of internationally traded oil. In 1975 dollar terms – and with costs at 1975 levels – this base price seems likely to be of the order of $6–7 per barrel of oil (or its equivalent for other energy sources). This price, of course, then becomes the highest price at which we are prepared to import foreign oil (or natural gas or coal). Potential external suppliers can be asked to bid for the right to supply energy to Western Europe, with this $6–7 boe as the ceiling price. Competition between many external suppliers for the very constrained Western European market for imported energy will, however, most likely drive prices well below this level. It should be noted, however, that the role of the external suppliers of the marginal requirements of Europe's energy market could be kept quite flexible, so as to introduce the possibility of more favoured treatment for particular external supplying nations which are considered to be appropriate to link the internal guaranteed supply and pricing system. Such favoured suppliers could, for example, include some or all of the populous oil producing countries as these might well prefer to negotiate guaranteed, long-term European outlets for their crude oil, at the $6–7 per barrel level, rather than remain dependent on the higher, but increasingly uncertain, oligopolistic OPEC price.

ii. An agreement within Western Europe for a co-operative effort to raise the investment required for the rapid development of the North Sea oil and gas potential. This would also require each member country of the agreement to give priority to the allocation of scarce resources of manpower, hardware, and finance for the North Sea's expansion, so as to ensure that the available investment funds do lead to the production of great volumes of oil and gas. This required flow of investment funds and/or of resources of men and machinery to the countries and/or regions within which the production potential is concentrated is, of course, the *quid pro quo* from the consuming nations of Europe for their now guaranteed supplies of oil and gas from the North Sea basin at the base price.

iii. An agreement that there shall be help from other European countries for any nation which seems likely to face special problems arising for its socio-economic system as a consequence of the rapid production of oil and natural gas. In light of the present evaluation of the circumstances this would be particularly important for Norway, but it could also apply to countries like Denmark and Ireland.

The economic implications arising from this approach to the development of the North Sea oil and gas reserves are formidable and radical compared with those we have become used to thinking about – and they are very different from the implications of the more conventional view of the North Sea situation. These are, however, by no means sufficient reasons for rejecting the radical alternative out of hand. Thus we need to give our attention to the following issues involved in the evolution of the elements which a policy based on the alternative requires, viz.

i. Energy prices in Western Europe will as indicated be related to the costs of indigenous production – rather than to 'market' or 'oligopoly' prices – and thus are predictable (except for general problems such as rates of inflation) at the level of $6 to 7 per barrel of oil/or oil equivalents for other energy sources. The economic system will thus not be related to cheap energy as it was in the 1960s but, to set against this loss, there will be the high degree of certainty over something as important as energy supply prices. For a modern, complex economy, with private and public decisions closely related, this is an important consideration and will enable Western Europe to avoid the high – and still increasing – degree of uncertainty which otherwise has to be faced in a situation of continuing dependence on traditional supply sources. It will also avert the fiscal and monetary problems – as well as the dangers of social unrest arising from them – which would undoubtedly arise from the further escalation of oil prices to be expected if OPEC continues to be successful. Finally, such an approach to energy pricing in Western Europe guarantees the high and continuing investment flow required for exploiting North Sea oil and gas (and other European energy sources which can be developed), especially in the event of the failure of OPEC (leading to a collapse in oil prices) and/or of decisions by its member countries which deliberately try to undermine the economic validity of Western Europe's search for energy independence based on the production of indigenous resources.

ii. It will produce a much higher degree of medium-to-longer-term confidence in the prospects for the UK economy and thus create conditions in which short-term problems can be overcome by help from other countries; given that the latter will, at a later stage, be the recipients of guaranteed quantities of North Sea oil and gas from British waters. This economic policy approach can be hypothesized on the basis of a UK hydrocarbon energy production (oil plus gas in oil equivalent terms) rising by the later 1980s to a level of some 300 million tons a year and then maintained at about that level for the rest of the century. Of this total output, up to a half will be available for markets in other parts of the continent.

iii. All this implies that increased interventionism in the energy economy with non-pricing allocation procedures will be required to complement the impact of 'market forces'. In the short-term, such procedures will be needed to constrain the rate of growth of demand for energy and also to ensure that the necessary resources of manpower, materials, and

> equipment etc, are available for North Sea developments. In the longer-term, further interventionism in the energy economy will be required in respect of the marketing of oil and gas production. This will obviously involve the subordination of individual companies' preferences for levels of production and in the refining and marketing of North Sea oil to considerations which are determined to be the most appropriate at national and European levels. A European Oil and Gas Authority would appear to be required to achieve these aims.

Success in implementing these elements of a radically different approach to Western Europe's energy policy would lead to the collective strengthening of the European power position *vis a vis* the rest of the world. As an economic policy it could obviously provide the mechanism for breaking the OPEC cartel and thus re-open the question of economic and other relationships between Europe and the OPEC countries. It would, indeed, put Europe in much the same position as the United States towards the energy sector of the economy, though we should note that the price of US energy independence will probably be higher than a similar policy for Western Europe. Although Western Europe has much further to go than the US in achieving independence – viz., from only 35 per cent compared with over 80 per cent – two factors stand in Europe's favour. Firstly, Europe is much less energy intensive and can, moreover, because of contrasting economic, social, and spatial structures in the organisation of its societies, economise more easily. Secondly, the US has already used most of its lowest-cost hydrocarbon resources, whereas Western Europe still has most of its potential oil and gas to develop.

In as far as Europe followed the US in going 'energy independent', the ability of the members of the OPEC cartel to enforce high oil prices would be undermined. Given the lack of demand for OPEC oil, some of its member countries might well seek to increase sales through lower prices for their exports. This would then help to create an availability of low-cost oil for Third World countries so that they could renew their steps towards economic development – and so stimulate the demand for European capital and consumer goods. Thus, paradoxically, Europe's apparent retreat from internationalism in respect of oil, in face of the cartel, could provide the mechanism for the longer-term revival of the world economy. The absence of such an autarkic energy policy in Western Europe and our failure fully to pursue the economic opportunity provided by the considerable oil and gas resources of the

North Sea basin seems likely, on the other hand, to confirm the success of OPEC and so prolong – and even exacerbate – the problems for the world economy arising from it.

References

1 See P.R. Odell, *Natural Gas in Western Europe; a Case Study in the Economic Geography of Resources* (Bohn, Haarlem, 1969) for an analysis of the exploitation of the Groningen gas field

2 G. Polanyi, *What Price North Sea Gas?* (Institute of Economic Affairs, London, 1967) discusses the background to and the implications of the dispute between the oil companies and the British Government over the North Sea gas price. It is treated at length in M.V. Posner, *Fuel Policy: a study in Applied Economics* (McMillan, London, 1973), in Chapters 10,11 and 12

3 J.P.P. Marie (Shell U.K. Exploration and Production Ltd) in the presentation of his paper 'Rotliegendes Stratigraphy and Diagenesis', given at the Istitute of Petroleum Conference on *Petroleum and the Continental Shelf of North West Europe*, 26 November 1974. To be published in 1975 by Applied Science Publications Ltd., London

4 See, for example, Dutch Government policy statements – culminating in the *Energienota* of November 1974 published by the Ministry of Economic Affairs. This argues for restraints in the use of gas specifically on the grounds that the resources as discovered to date otherwise indicate a scarcity before the mid-1980s. Electricity utilities and large industrial users have already been warned that their contracts for gas are unlikely to be renewed after the initial 10 year period is over in the late 1970s.

5 One additional factor at this time in limiting producing companies' interest in expanding the supply of natural gas was competition from the Soviet Union. It was prepared to match whatever price was asked for Dutch gas – in spite of the great distances over which Soviet gas had to be pipelined to markets in Western Europe. See P.R. Odell, op. cit., for a discussion of this factor.

6 The influence of new oil companies moving in to the European market to sell oil which they had originally planned to supply to the United States was the most important single factor in making the European oil system so competitive in the 1960s. This is discussed in P.R. Odell, *Oil and World Power* (3rd Edition, Penguin Books, London 1974). See Chapters 1 and 4.

7 This was the conventional wisdom about the potential of the North Sea until the latter part of 1973. Note, for example, the comment as late as 26 September 1973 by Sir Frank McFadzean, Managing Director of the Royal Dutch/Shell Group of companies, viz: 'Western Europe at present produces only 15 per cent of its combined needs of oil and gas and the realization of present hopes for the North Sea will not *materially*

change its dependence on outside sources'. From an address to the Foreign Affairs Club, London on 'Energy and Oil in the Decade Ahead'.

8 This was not because alternative intelligence and alternative interpretations of the North Sea potential were unavailable. It was generally accepted that much more was known in Houston than in London about the rapidly developing and the very exciting knowledge of the petroleum geology of the North Sea basin: and that by 1972–3 even the highest hopes in respect of field size, well productivity and the producibility of the oil bearing formations were being exceeded. It was already possible by early 1972 to speculate quite reasonably on an oil production potential of up to 300 million tons per annum by the early 1980s. See P.R. Odell, 'Europe and the International Oil and Gas Industries in the 1970s, *Petroleum Times*, January 1972. This article suggested that the full development of the North Sea oil and gas resources would 'enable them to meet up to 50 per cent of the regions rising demand for energy by the late 1970s', p.103.

9 This, of course, refers to the well-established international oil companies which enjoy high profits from selling low-cost oil from their concessions in the OPEC countries in the years between 1971 and 1974, when the oil prices increased almost fourfold over their levels in the 1960s. It does not refer to companies like Burmah Oil and other relative newcomers to large-scale oil production for which the high costs of offshore development constitute a heavy financial burden; in terms of both self-financing and in terms of borrowing.

10 See, for example, the forecasts in the EC's 'New Energy Policy Strategy for the European Community' (Brussels, June 1974, Doc. COM (74) 550 final). This still envisages a 75 per cent dependence on outside resources for oil supplies in 1985, giving an import requirement which in tonnage terms is only about 15 per cent less than the quantity imported in 1973.

11 Individual fields normally have their proven reserves declared at the 0.9 probability level, ie. there is a 90 per cent chance of a field having at least as many reserves as this figure indicates. Statistically, fields in a province may be viewed as independent occurrences and thus, when the reserves of two fields are summed, there is only an 0.1 x 0.1 chance of their having less than the total declared reserves of the two fields. With many fields involved there is a very small chance indeed (of no more than 1 per cent) that the total reserves will be as low as the arithmetically summed reserves of the individual fields. To calculate the normally accepted probability (0.9) of proven reserves in the province, it is necessary to construct a probability curve for the set of occurrences – this could produce a figure for the proven reserves of the province at the 90 per cent probability level that is a 30–50 per cent higher than the total given by simply adding together the declared reserves of the many fields.

12 This work was undertaken in the Economic Geography Institute of Erasmus University in Rotterdam. See P.R. Odell and K.E. Rosing, *The North Sea Oil Province: an Attempt to Simulate its Development and Exploitation 1969–2029*, (London, 1975)

13 This conclusion emerges from the results of the simulation model of the North Sea's potential (Odell and Rosing, ibid.). The implications of this are more fully examined by Odell in his paper, 'Indigenous Oil and Gas as Alternatives to Imports of OPEC Oil' in F.A.M. Alting von Geusau (Ed.), *Energy Policy Planning in the European Community* (Sijthoff International Publishers Ltd., Leiden, 1975)

Chapter II – 5

Optimising the Oil Pipeline System in the UK Sector of the North Sea*

Powers taken by the British government in its Petroleum and Submarine Pipelines Act[1] imply a search for rationalization in what is now to be a state controlled (although *not* a state owned) system of pipelines for oil and gas from fields in the British sector of the North Sea and for oil and gas coming to landing points in the U.K. from adjacent sectors. The complexity in the search for an optimal solution, from the point of view of the public interest - as opposed to the interests of the individual companies which have very good reasons for not wanting to do other than optimize their own production/transportation facilities - arises in the first instance from the large number of fields of varying sizes and in different locations. The geography of the fields discovered before 1 September 1975, are shown in Figure II-5.1

Additional complexities arise from a variety of reasons. There are alternative possibilities of choosing one or more landward terminals for the pipelines. There are also uncertainties in the timing of the beginning and the build-up of the rate of production from the fields involved. These rates will be partially dependent upon the initial uncertainty in the quantity of recoverable oil in these fields and the subsequent high probability of appreciation of reserves. Such appreciation means that peak production rates would be maintained over a longer period of time[2] than originally expected and thus necessitate additional pipeline capacity

* **Originally published in *Energy Policy*, Vol. 4 No. 1, March 1976, pp. 50–55. This paper was jointly authored by Dr K.E. Rosing and the late Drs. H. Beke-Vogelaar. Dr Rosing has given his permission for this reprint and this is gratefully acknowledged.**

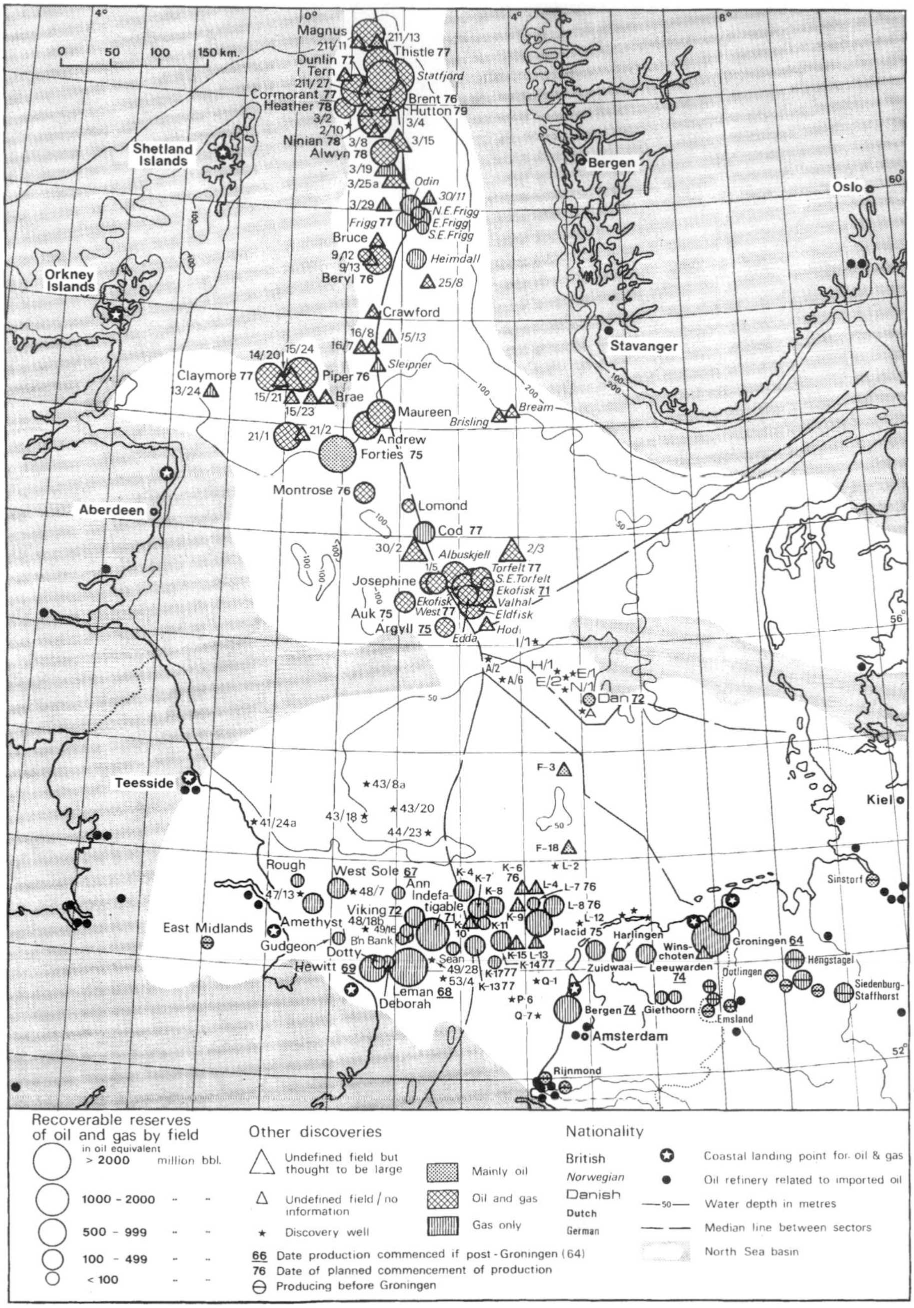

Figure II-5.1: North Sea discoveries to September 1975

from time to time if oil from newly discovered fields is also to be transported. The pipeline optimization problem is thus a complex one - as is always the case when there are both spatial and temporal variables to incorporate into a model.

Work is now proceeding on this problem in the Economic Geography Institute of Erasmus University, Rotterdam as a part of the programme of research on the North Sea oil and gas province.[3] The problem may be divided into two parts: firstly, the modelling of the possible pipeline networks in the North Sea and secondly, the development of mathematical programming techniques capable of identifying the optimal solution.

The objective of the research is to minimize the total joint cost of laying pipelines to:

(i) collect oil from some number of spatially discrete fields (n) to a number of seaward terminals (p)

(ii) transport that oil to (r) landward terminals in the United Kingdom.

The geography of the North Sea (see Figure II-5.1) dictates that the 'collector' lines will connect each of the n fields with the seaward terminus of one of the p 'trunk' lines, each of which will deliver oil to a specific landward terminal in the United Kingdom.

The model then involves costing the various lines and this too also has to be divided into two parts. First, the cost of the collector lines from the n fields to the p seaward termini; and second, the cost of laying the p trunk lines from these termini to the r landward terminals.

Linear programming is the most widely known and used method of solving this type of problem, but restrictions on the formulation of the linear programming model make its application to the pipeline problem impossible. The same is true of the associated techniques of dynamic programming and geometric programming[4]. Thus for the problem with which this paper is conceived it is necessary to utilize another sub-field of mathematical programming, viz. integer programming. This, however, suffers from the handicap of generally requiring a new code to be specifically written for each problem. Nevertheless, previous work in this field[5] by an American geographer has disclosed a group of closely related types of locational problems. His analysis in this field has enabled the identification of the exact characteristics of the collector line problem.

This is the so-called 's-median problem' which entails the identification of the p points of a network such that each of the n nodes (oilfields) is connected to one of the *p* facilities (seaward trunk pipeline termini), minimizing the total distance summed throughout. The resultant solution consists of two parts: first, the designation of *p* facility locations and second, the assignment of each node to one and only one facility for service. The *p*-median problem has previously been intractable of solution because of the sheer size of the problem (see Table II-5.1), given that feasible solutions include all those in which oil from many fields can be collected at one or other of the seaward termini of the pipelines.

TABLE II-5.1
Number of feasible solutions for combinations of nodes and facilities (oilfield and pipelines)

Oilfields (n)	25	25	25	25	25	25	25
Pipelines (p)	2	3	4	5	6	7	8
Feasible solutions $(^{n}_{p})$	300	2300	12650	53130	177100	480700	1081570

Theoretical work in the Economic Geography Institute over the last year has resulted in a branch and bound algorithm which is efficient in locating the optimal solution to such problems.[6] Branch and bound, also called partial enumeration, requires the examination of only a small fraction of the total possibilities and guarantees the optimality of the solution. This basic algorithm then, however, had to be modified to deal with the much more difficult situation posed by the oil pipeline problem. This involves a two level solution, not just the minimization of the cost of collector lines, but also the minimization of the cost of trunk lines, further increasing the solution space beyond that indicated in Table II-5.1. The evolved solution method has now been applied to the somewhat simplified model specified below - as a precursor to work on a model more closely corresponding to the complex, real-world situation.

Twenty-five fields in the British sector of the North Sea (including 40 per cent of the declared reserves of Stratfjord) were included in the analysis. Their recoverable reserves were taken from the latest published figures, or, in the cases of discoveries which have been designated as 'large' or 'significant' and for which there are no such figures, a nominal 500 million barrels of recoverable reserves was assumed. Peak recovery rates were then assumed at 27,500 barrels per day for each 100 million barrels of reserves.

Four possible landward terminals in the UK at the locations shown in Figure II-5.1 were assumed. Three of these will be recognized as actual terminal locations; the fourth, just to the south of the Firth of Forth is not. This location was chosen for convenience in the cartographic presentation. Trunk lines from seaward to landward terminals and collector lines were assumed to be shortest straight line paths and the existence of built, partly built and projected or planned lines was ignored.

Costs of collector lines were estimated on the basis of a cost of £300,000 per kilometre irrespective of line capacity, plus a variable cost of £0.625 per kilometre per barrel per day of throughput. These values were based on data published from the Brent system pipeline to Sullom Voe[7] and resulted in a square non-symmetrical matrix with zeros on the principal diagonal which defined the cost of connecting each field to

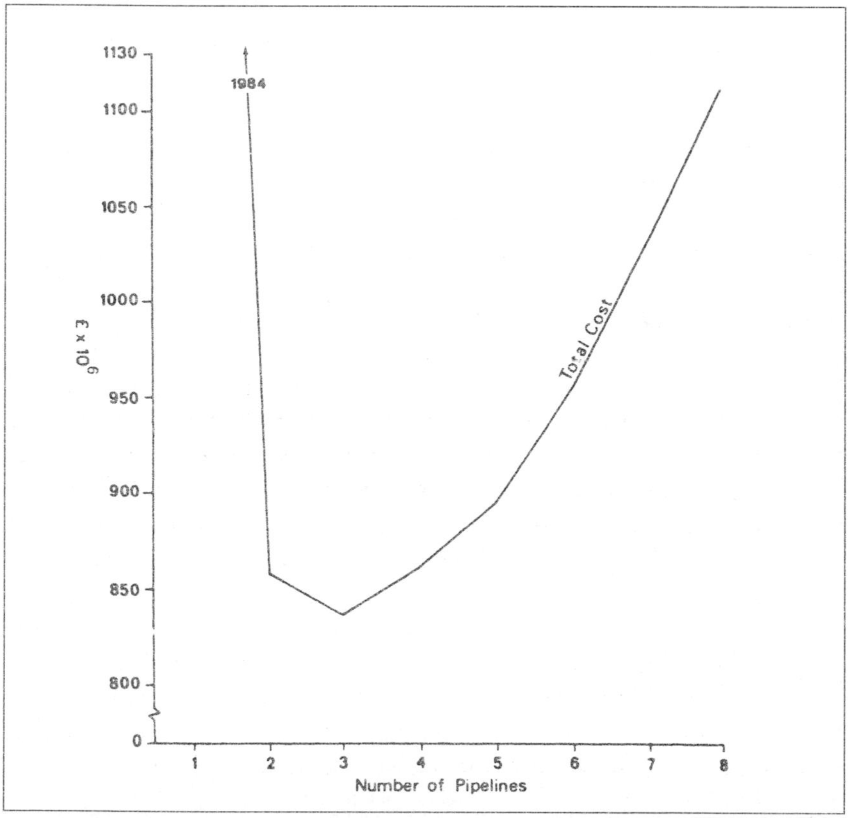

Figure II-5.2: Inter-field and sea-to-shore pipeline costs for the eight solutions

each other field. Costs of connecting the p seaward termini to the four shore terminals (r) were estimated on the basis of Brent line costs,[8] adjusted for the distances involved.

The modified branch and bound programme[9] was then used so as to minimize simultaneously the cost of the interfield collecting lines and the p trunk lines from the seaward termini to the land terminals. Eight different solutions were obtained (p = 1-8). Figure II-5.2 shows first, the slope of the curve for the costs of the connecting line (these generally decrease as the number of seaward terminals increase); secondly, the upward slope of the cost curve for the trunk lines as the number of these increase. Figure II-5.3 shows the total cost curve indicating that a three pipeline system provides the 'optimal' solution. The solutions, in

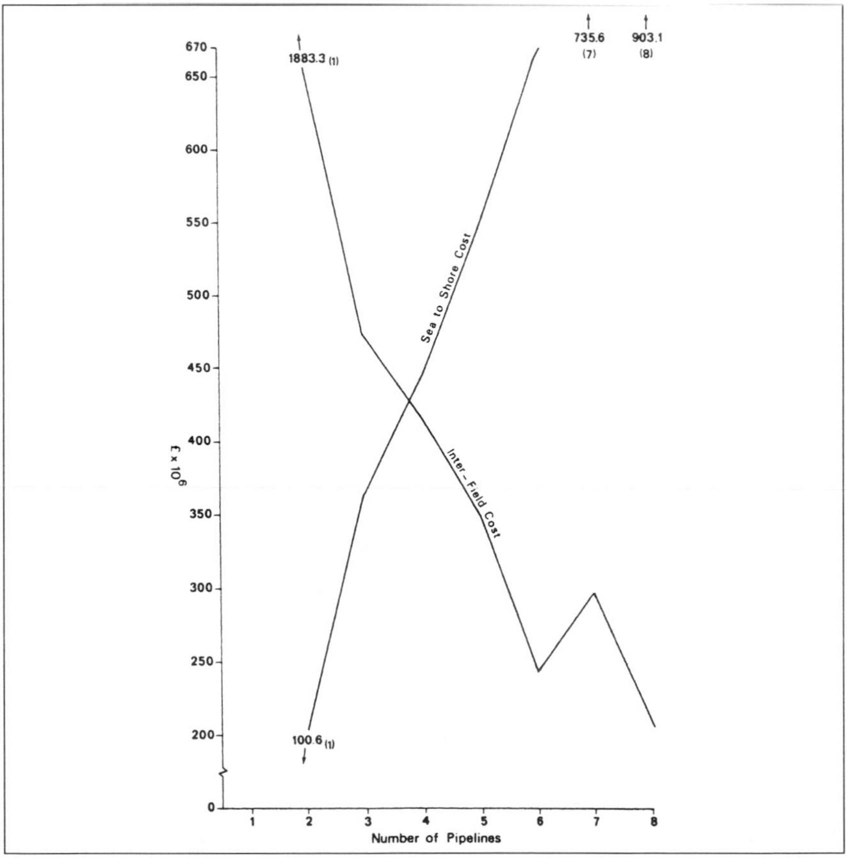

Figure II-5.3: Total costs of pipelines for the eight solutions

geographical terms, for one to eight trunk lines, are shown in Figure II-5.4. The capacity required in each trunk line is also shown in the set of maps, on the assumption - unrealistic, of course - that all the fields connected to the line concerned achieve simultaneous peak production rates.

At this stage the results are of little more than academic interest as a result of the over-simplification of the variables in the model. It has been proved, however, that the modified branch and bound programme can satisfactorily solve the problems formulated, with a degree of efficiency which is as high, or even higher than the unmodified programme performs on theoretical distributions. It would therefore appear that this programme provides a tool which is able to resolve the problems created by a more fully specified model. Much more work, however, remains to be completed before a model is developed which corresponds sufficiently to reality to provide realistic answers. Modifications to the model must include:

- A more precise specification of the cost functions, relating them to the specific interfield paths which might be used and the varying geographical and geological conditions of these paths.

- Relating the through-put capacities to costs in the form of a power function to include economies of scale.

- The designation of minimum and maximum sizes for the lines.

- The introduction of the existing and planned pipelines and on field tanker loading facilities.

The results will then be optimal *inclusive* of the utilization of the facilities which were built or under construction before the United Kingdom government introduced its control requirements, rather than, as in this paper, unrealistically ignoring their existence - and their capabilities for moving the oil.[10]

None of these modifications appear to present any undue difficulties, but, in one other respect, the modification of the programme and the collection of the necessary information inputs do pose a considerable problem. Projected production schedules for individual fields must, of course, be brought into the analysis so that account can be taken of variable flows of oil from the different fields over time, since the

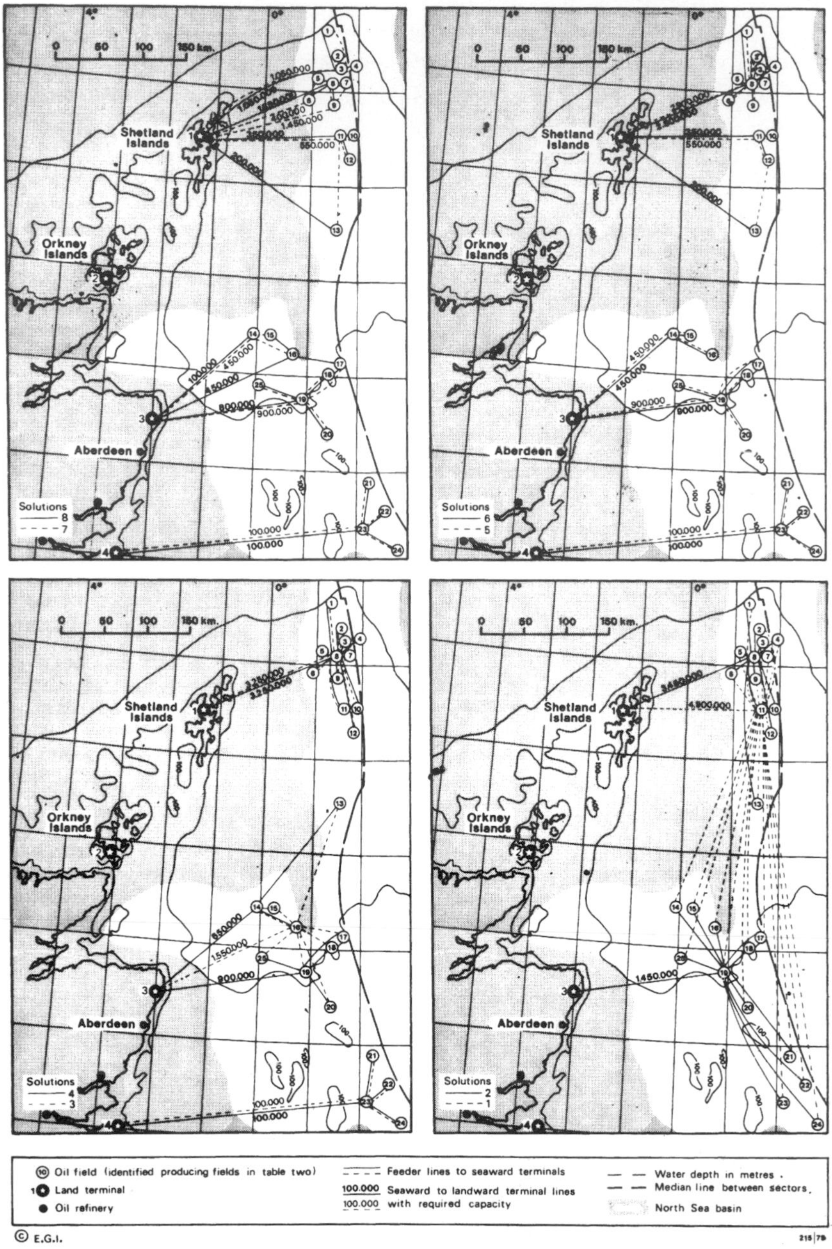

Figure II-5.4: The geography of the eight solutions. Each of the maps shows the collector lines and seaward/landward trunk lines for two solutions

timing of the necessary connecting links may influence the structure of both the collecting network and the trunk pipeline components of the system. The inclusion of this variable could also provide the basis for proposals to re-schedule production from one or more of the fields in order to avoid possible over-construction of pipeline capacity simply to meet the fortuitous and short-lived conjunction of peak flow rates from a number of fields. Such a possible production re-scheduling control has already been taken by the British government in its measures to constrain the amount of oil produced for national economic or conservation-of-resources reasons. The pipeline problem which has now also been accepted as a governmental responsibility offers another possible reason for national control over production levels, a development which will be viewed either as a threat or a promise by the groups with contrasting interests in the North Sea oil development.

The oil industry is proud, with some justification, of its record in seeking optimal solutions in its international activities - in terms, for example, of the allocation of production to different supply points and in the organization of its refining and transportation facilities. Any government which seeks to over-ride the companies' efforts at the optimization of the use of resources by substituting national for company interests must be able to employ methodologies which are equally sophisticated. The North Sea pipelining problem is complex and deserves an appropriate methodology. We have suggested one possible approach, but there will undoubtedly be others.

References

1 *Petroleum and Submarine Pipelines Bill*, Part III, Submarine Pipelines. Presented to Parliament by the UK Government, Bill 127, 1975.

2 See P.R. Odell and K.E. Rosing, *The North Sea Oil Province: An Attempt to Simulate its Development and Exploitation*, (London: Kogan Page, 1975), pp 29-35

3 Previous publications include that on the simulation model of the North Sea's production potential (see previous footnote) and P.R. Odell's contribution 'The economic implications of North Sea oil', to the recently published joint British/Norwegian Institute of International Affairs book, *The Political Implications of North Sea Oil and Gas*, (ed M. Saeter and I. Smart), (IPC Science and Technology Press Ltd, Guildford, Surrey).

4 This point is considered at greater length in: K.E. Rosing, 'Notes Towards the Optimization of Oil Pipelines' *Series A, Working Paper Number 75/14*, Rotterdam: Economic Geography Institute, Erasmus University, Rotterdam, 1975).

5 R.L. Church and C.S. Revelle 'Theoretical and Computational Links Between the p-Median, Locational Set Covering and Maximal Covering Location Problem', *Geographical Analysis*, (forthcoming).

6 C.S. ReVelle and K.E. Rosing, 'An Efficient Branch and Bound Algorithm for the p-Median Problem', *Series A, Working Paper Number 75/6*, (Rotterdam: Economic Geography Institute, Erasmus University, Rotterdam, 1975).

7 J. O'Donnell, 'International pipelining". *Offshore*, Vol 58, No 8, (July 1975), p61.

8 Ibid.

9 Revelle and Rosing, op. cit.

10 In this respect it is interesting to note that none of the solutions indicated that a terminal in the Orkneys is an appropriate development. The existence of the Occidental Oil Company's line to Flotta must, however, obviously be an element in the new government planning of an optimum North Sea system.

Chapter II – 6

Oil and Gas Exploration and Exploitation in the North Sea by 1977*

Introduction

The North Sea has become the world's most active region of offshore oil and gas developments – with the possible exception of the Gulf of Mexico. Even that would certainly be pushed into second place if one takes note of the speed of exploitation, the degree of success achieved in terms of the discovery of oil and gas reserves and the amount of technological innovation that has been engendered by the exploration and production efforts in the deep water and adverse weather conditions of the North Sea.[1]

What has happened to date in the North Sea over the period from the first offshore activity there in 1964 depends essentially on the international oil companies, for they provide the dominant element in the exploration for, and the production of, the province's oil and gas resources. Indeed, they have been given – directly and indirectly – well over 70 percent of all North Sea acreage allocated[2] (including an even higher percentage of the best blocks). They have discovered all the major fields found so far except two, together with most of the minor fields; and they are responsible for almost all the fields in production or under development toward production.

This is so in spite of the interest shown in the North Sea both by new oil companies formed with the North Sea potential specifically in

* **Originally published in the *Ocean Yearbook* 1 (Eds. E. Mann Borgese and N. Ginsburg), The University of Chicago Press, Chicago, 1978, pp. 137–159**

mind and by non-oil companies in Britain and Norway deciding to diversify into a new field of endeavour. Even the rapid growth of state oil companies – particularly the British National Oil Corporation and Norway's Statoil – seems unlikely to diminish *very much* the role of the international oil companies in determining the manner and speed with which the oil and gas province will be developed. Thus, the motivations of the international oil companies in respect of the North Sea opportunities – together with these companies' responses to the petroleum legislation of the countries surrounding the North Sea – have largely determined the progress towards, and still condition the prospects for, the development of this major new oil and gas province.

The Resource Base

Progress and prospects in the North Sea's oil and gas exploitation, though dependent in the final analysis on the "behaviour" of the companies (and of the governments), depend in the first instance on the size and wealth of the resource base. Knowledge of this is, in the initial period of exploration, limited and speculative. Thus hypotheses on the likely occurrence of oil and gas are based largely on geological analogies. The size and complexity of the North Sea basin constituted a major difficulty in the early evaluations. Size, though a simple concept, was consistently under-rated such that the North Sea came to be seen as a kind of European backyard in which prospects for exploitation were limited and about which the main concern was that false hopes should not be raised over its potential.[3] However, what has happened in terms of successful exploration and what remains to be done in this respect (and this adds up to an exploration effort that will be several times the magnitude of that already made) should be seen in the context of a potentially petroliferous North Sea province which is almost exactly the same size as the petroliferous region of the Persian Gulf – a comparison which is illustrated in Figure II-6.1. As in the Persian Gulf, where there are many different oil and gas "plays" (i.e. many potentially productive horizons in the underlying geological strata), so also with the North Sea where the first decade of exploration has also clearly demonstrated that there are many sorts of potential reservoir rocks which are worth of investigation. This opens up not only the likelihood of exploration in parts of the province which have, hitherto, not been considered worthwhile, but also the desirability of deeper exploration in areas previously investigated only by shallower wells.

Thus, both geographical scale and geological complexities underly the great success to date in the exploration efforts in the North Sea and they also provide the basis for continuing efforts which should certainly stretch out over at least the next 25 years. Success to date is pinpointed in Table II-6.1, which shows the number of finds of oil and/or gas that had been made in various classes up to mid-1976. The essential proof of the prolific nature of the basin lies in the fact of 208 discoveries to date, of which 83% have already been designated as gas or oil and gas fields. Of this number, 45 per cent have had reserve figures declared for them. Moreover, of the 79 fields with declared reserves, 31 have more than 500 million barrels of oil (or the oil equivalent thereof as natural gas), so making them international "giant" fields in usual North American oil industry parlance.

TABLE II-6.1
North sea oil and gas province: offshore discoveries to June 1977

	Oil and Gas	Gas Only	Not Known	Total
Number of discoveries – total	75	99	34	208
In southern basin	7	88	–	95
In northern basin	68	11	34	113
Designation of the discoveries – total	75	99	34	208
Fields *with* declared reserves	41	38	–	79
Fields *without* reserves declaration	34	16	–	50
Discovery wells (no other information)	–	45	34	79
Size distribution of declared fields – total	41	38	–	79
More than 2 x 10^9 bbl oil or equivalent	4	1	–	5
1–2 x 10^9 bbl oil or equivalent	9	2	–	11
05–10 x 10^9 bbl oil or equivalent	12	3	–	15
Smaller fields	16	32	–	48
Production and production plans:				
Number of fields with declared reserves	41	38	–	79
In production	8	9	–	17
Production plans	16	7	–	23
No plans for production	17	22	–	39

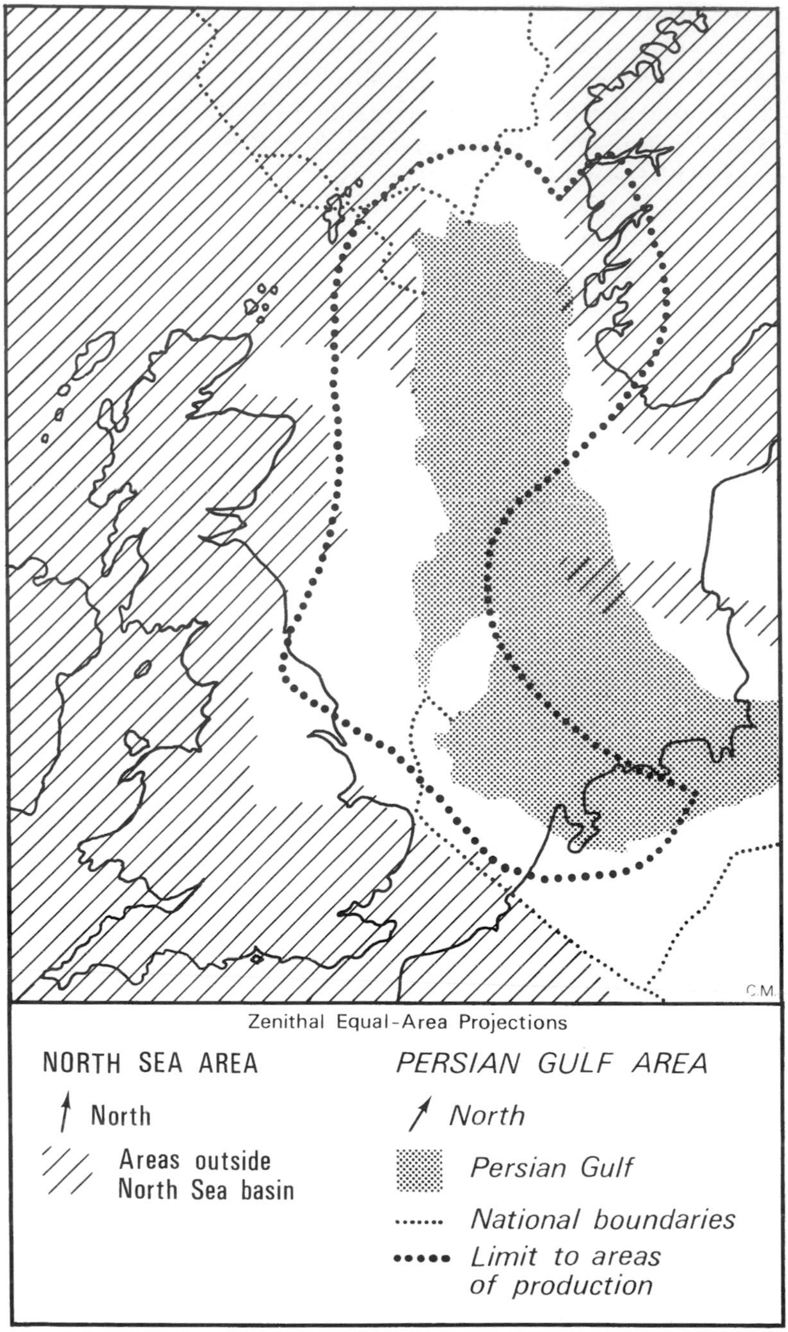

Figure II-6.1: A comparison of the size of the potentially petroliferous areas of the Persian Gulf and the North Sea

Natural Gas

What the discoveries mean in terms of effective overall recoverable reserves is especially difficult to evaluate in the case of natural gas. For example, many of the gas-only fields lie in the Dutch sector of the North Sea (see Table II-6.2). In this sector there is, unfortunately, no obligation on the part of either company or government to give any field and reserve information at all to the general public, so eliminating the possibility of depletion policymaking (and, indeed, energy policymaking) which can be seen to be justified by the rate of reserve discovery. There are at least 40 offshore discoveries[4] about which virtually nothing has been made known to the public in terms, for example, of reserve figures for individual fields. The Dutch government has merely given a total figure for the sector's proven reserves without indicating what is meant by "proven" or even the number of fields whose

TABLE II-6.2
Natural gas resources of the Southern North Sea basin, excluding associated gas

	Dutch Sector	British Sector	German/ Danish Sector
Total number of gas discoveries	51	31	3
Number of discoveries declared as gas fields	11	15	–
Governments' declarations of remaining proven gas reserves (10^9m3)	367*	552*	–
Number of fields in production	5	7	–
Other fields with announced production plans	5	–	–
Current (1977) annual production (10^9m^3)	ca. 3	40	–
Estimate of 1980 production from fields currently on production or in development (10^9m^3)	ca. 10	ca. 42	–
Likely remaining reserves for all fields in each sector at summed 90% probability (10^9m^3)†	1,000+	1050+	50+
1980 production potential with full exploitation‡ of the already discovered reserves (10^9m^3)	40+	50+	2–3

Notes

* *Arithmetic total of 'proven' reserves of declared fields*

† *Based on all discoveries made and not just on declared fields*

‡ *Based on 20–25 year depletion periods for the fields*

reserves are included in the total! One can conclude, however, that the officially declared reserves grossly understate the actual position. The Placid field (L10/11) alone has been forward sold (to Gasunie and to German customers) to the extent of 150 x 10^9 m^3; the Ameland field has been officially announced as containing 55 x 10^9 m^3 (in response to a parliamentary question which asked if the field was a second Groningen with over 2,000 x 10^9 m^3!); and each of the 10 other fields in production or being developed for production must have at least 25 x 10^9 m^3 (a minimum exploitable size) while their average size, based on hints dropped by the companies concerned, seems to be of the 40 x 10^9 m^3.

Thus, these 12 fields alone – even on a simple arithmetic addition of their individual reserves contain between 450 and 600 x 10^9 m^3 of natural gas. However, as shown in Table II-6.2, there are at least 40 other gas fields and discoveries in Dutch waters ranging in size from a few milliard up to more than 25 x 10^9 m^3. Assuming, conservatively, an average size of 5 x 10^9 m^3, then there is another 200 x 10^9 m^3 of natural gas to add to the 450–600 x 10^9 m^3 defined above – to give an arithmetic total of up to 800 x 10^9 m^3 of reserves. The simple arithmetic sum of the 90 per cent probable reserves from a large number of fields is not correct, as the answer derived gives a 98–99 per cent probability of their being at least that amount. To return to a reasonable 90 per cent probability figure for the overall reserves of the group of fields it is necessary to collate the individual probability curves for the size of each field. Though such curves are confidential in the Dutch information system, a conservative estimate of their shape and a subsequent recalculation of the overall 90 per cent probability of reserves from all the Dutch-sector fields combined indicates a total of at least 1,000 x 10^9 m^3.

More information, is, fortunately, available on non-associated gas reserves in the British part of the southern North Sea basin, though there is no evidence to indicate that all finds made have been announced or that the total figure of remaining reserves of just over 500 x 10^9 m^3 has been adjusted to give an overall 90% probability of recovery. Elsewhere in the North Sea, the reserves as announced for the Danish sector (by the DUC, the operating company) are 24–42 x 10^9 m^3. A much higher reserve figure of 62 x 10^9 m^3, has been estimated for three fields only by consultants appointed by the Danish government, again suggesting an understatement of reserves by the companies which have made the discoveries. Overall, the understatement of Western Europe's offshore gas reserves continues the tradition established by Shell and Esso in seriously understating for many years the likely recoverable reserves of the Groningen field, now known to be the largest gas field in the non-communist world.

TABLE II-6.3

Western Europe: Its currently "proven" and possible natural gas reserves and an estimate of their development potential by the early 1980s

	Remaining Recoverable Reserves				
	Declared "Probable" by 1977 ($x10^9m^3$)	Probable plus Possible ($x10^9m^3$)	As Likely by Early 1980s ($x10^9m^3$)	Mid-1980s Annual Production Potential ($x10^9m^3$)	Millions of Tons of Coal Equivalent* (Approximate)
Onshore Netherlands	2,030	2,150	2,100	105	120
South North Sea, British sector	550	725	1,050	50	70
South North Sea, other sectors	440	850	1,250	55	75
Onshore West Germany	310	515	450	25	30
Austria, France, Italy, etc	420	490	600	35	45
Northern North Sea basin, UK/Norway	900	1,500	2,500	115	135
Rest of European continental shelf (Ireland, Spain, etc)	50	150	350	20	25
Total	*4,700*	*6,380*	*8,300*	*405*	*500*

Source: *For 1977 various national and EEC/OECD estimates. Estimates for the 1980s are the author's own.*

* *Conversion to coal equivalent based on known or estimated calorific values of the various gas supply sources*

In Table II-6.3, therefore, we clearly see the contrast between the proven gas reserves position as it is officially presented and the situation as it would be if there were a more rational and open approach to the calculation of reserves. If discoveries of new associated and non-associated gas reserves continue at the level achieved over the last few years, then the estimate of over 8,000 x 10^9 m^3 of remaining recoverable reserves of gas by the early 1980s in Western Europe – mainly in the North Sea – could now be considered as the *minimum likely* rather than the maximum possible[5]. Moreover, even beyond that date one can also be confident that new discoveries – and additional reserves in earlier discovered fields – will continue to enhance the available amounts of recoverable natural gas. Even without the discovery of another major offshore gas province somewhere around the much-indented coastline of Western Europe, an ultimate natural gas resource base amounting to at least 20,000 x 10^9 m^3 does not seem to be unduly optimistic.

Given this adequate gas resource base and the proximity of a readily available market for as much gas as can be produced, one would expect (company and government behaviour permitting), a continued rapid contribution of natural gas to the Western European energy economy, particularly, of course, in the countries surrounding the North Sea from which the bulk of the reserves will be produced over the forseeable future.

Oil

Progress to date in the development of North Sea hydrocarbons has thus been related more to natural gas than to oil. More recent developments (post-1970) in the exploration for oil have, however, produced even more exciting results in terms of the resource base potential[6]. As a result, the prospects for the medium-to-longer-term future of oil production from the North Sea now exceed all earlier forecasts and expectations.

Table II-6.4 presents the minimum likely situation on reserves in already discovered North Sea oil fields. It shows ultimately recoverable reserves of at least 45,000 million barrels after making modest allowances for the appreciation of fields with already declared reserves and taking into account fields which have been been declared proven, but about which no information on reserves has yet been given. It does not, however, make any allowance whatsoever for the more than 60 additional discovery wells which have been drilled (see Table II-6.1). Some of these will eventually prove to have recoverable reserves of oil and so push the sum total of discovered reserves to date above the 45 x 10^9 barrels figure.

TABLE II-6.4
North Sea oil reserves, at the end of 1976

	Million barrels
1. As declared	
– By simple addition of the declared reserves figures for 43 fields	ca. 23,500
– After adjusting to a summed 90% probability (from the 43 fields, assuming each field is declared at a 90% probability)	31,000+
2. On extrapolation	
With upward revision following production from the 43 fields (average 15%)	36,000+
Addition of reserves from the so-far undeclared 35 discovered fields* with an assumption that these fields are on average two-thirds smaller than the declared fields	9,500
3. Minimum total* of North Sea oil reserves discovered to date	45,000+

* *In addition, there have been as shown in Table II-6.1, over 60 oil discovery wells many of which will eventually be declared as oil fields. The 45 x 10 barrels of oil shown in this Table as already having been discovered is thus a minimum likely figure and not even the most reasonable estimate. See below in Table II-7.5 for more detailed estimates of North Sea oil reserves.*

Table II-6.5 then puts these real-world reserves to date in the perspective of some of the estimates which have been made over the last few years about North Sea oil. Shell and B.P. both gave earlier estimates for 1976 and 1980 reserves which are 50 per cent or less of what can now be expected. Even these most recent estimates of the province's "ultimate" reserves range from only 35 to 50 x 10^9 barrels; figures, that is, for ultimate reserves which are lower than those that are currently known to exist or the existence of which can be confidently extrapolated from present evidence. This is in spite of the fact that most of the exploration work in the North Sea still remains to be done – including exploration work on some of the largest structures in the Norwegian sector in blocks which have only just been allocated; or even not yet allocated to companies for such work, in accordance with Norwegian conservation policies. Of the oil companies' present estimates on the North Sea's ultimate reserves, only Conoco's 1975 estimate of up to 67,000 million barrels remains credible. One is thus forced to the conclusion that the oil industry's conservatism over the question of evaluating reserves has once more led it to inappropriate pronouncements for the resource potential of a region whose future

TABLE II-6.5
North Sea oil reserves: estimates and forecasts compared

	By 1976	By 1980	Ultimate
1. As declared:			
– By simple addition of declared reserves for 43 fields	23.5	–	–
– After adjusting to a summed 90% probability (from 43 fields, assuming each field is itself declared at a 90% probability)	31+	–	–
2. Declared plus additional discovered reserves not yet declared:			
– From upward revision of reserves in declared fields	–	ca. 36	–
– From fields discovered but not yet declared (35 fields)	–	9–10	–
3. New discoveries from 1976 to 1980	–	ca. 10	–
4. As predicted by EGI simulation model (50% probable)	17.8	–	–
5. As hypothesized by EGI model for 1976 reserves after appreciation plus discoveries 1976–80 (50% probable)*	–	48.6	–
6. As forecast by EGI model on full development of the province:*			
– 90% probability	–	–	78
– 50% probability	–	–	109
7. As forecast by oil companies:			
– By Shell in 1972†	10–12	17.5	–
– By B.P. in 1973‡	–	–	38
– By B.P. in 1974§	16–18	24–30	44
– By Conoco in 1975¶	–	–	45–67
– By Shell in 1976#	23	–	35
– By B.P in 1976**	–	–	±50
– By Shell in 1977††	–	–	±50

Sources of Estimates

★ *PR Odell and KE Rosing,* The North Sea Oil Province: An Attempt to Simulate its Development and Exploitation, 1969–2029 *(London: Kogan Page, 1975)*

† *At E.I.U. International Oil Symposium, London, October 1972 (A. Hols, head, Production Division of Royal/Shell Exploitation and Production Coordination);.*

‡ *At a Financial Times North Sea Conference, London, December 1973 (Dr J Birks, Director, B.P. Trading)*

§ *At the Annual Conference of the Society for Underwater Technology, Eastbourne, April 1974 (H. Warman, Exploration Manager, British Petroleum Company).*

¶ *At the Conference on the Political Implication of North Sea Oil and Gas, Tønsberg, February 1975 (T.D. Eames, Oil Exploration Division, Conoco North Sea, Ltd.)*

Shell Briefing Service, Offshore Oil and Gas, North West Europe *(London: Shell International Petroleum Co., July 1976).*

★★ *At a meeting in London of the European Atlantic Group, December 1976 (P.I. Walters, a managing director of B.P.), reported in* Noroil, *vol.5, no.1 (January 1977).*

†† *At a meeting of the Empire Club of Canada, March 1977 (P.B. Baxendell, a managing director of Shell).*

economic and even political survival could well depend on a realistic evaluation of its indigenously available oil.

In this respect it should be noted that Table II-6.5 also shows that our simulation model of the North Sea oil reserves[7] – a model which was generally criticised in the industry for its highly "optimistic" conclusions – is also under-predicting the rate of development of reserves of the province. This, however, does not surprise us, as we described it as a conservative model in which the probabilities of discovery, of size of field, and of recoverability of oil, etc., were oriented to minimum, rather than to the most reasonable expectations. The mistakes seem, however, to relate to the simulated *timing* of the exploration and development efforts. If so, then, the model's results will within a few years, "catch up" with the real world developments, so that its 90 per cent probability figure of 78,000 million barrels of ultimately recoverable oil from the North Sea province remains a reasonable expectation in the context of the more than 45,000 million barrels which can already be shown to be recoverable from known fields and given another 60 discoveries still awaiting evaluation.

The Contrasts in the Oil and Gas Reserves Evaluations

Why should "officialdom" (governments and the oil companies) have so seriously underestimated the resource potential? It seems to be due largely to their unwillingness to try to quantify the future potential availability of the so-far undiscovered recoverable resources of the North Sea province – an unwillingness, that is, to do for Western Europe an exercise that is commonplace in other parts of the world, and most notably North America. There the USGS-directed evaluations of undiscovered recoverable oil and gas are used as prime inputs for estimating the US' future potential levels of oil and gas production. Yet this reasonable procedure is even more important for Western Europe, where most of the potential hydrocarbon resources still remain undiscovered, given that most of the petroliferous regions are round the long and much-indented coastline of the continent in off-shore locations in which oil and gas exploration and exploitation has only recently become possible. Figure II-6.2 shows the vast extent of these petroliferous offshore regions. Within the total extent of these regions the North Sea constitutes but a small part.

Within this context, the concept of total potential resource base is very different from the way in which it has been conceptualized at the official European level, where planners seem not to have succeeded in reaching out beyond the severe limitations of "proven" reserves. The

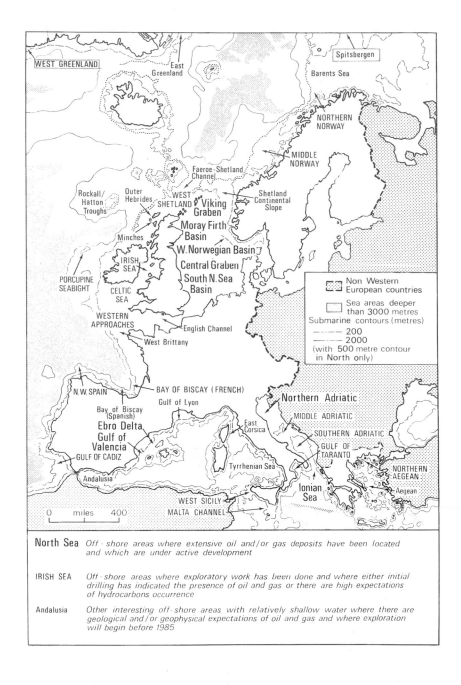

North Sea *Off-shore areas where extensive oil and/or gas deposits have been located and which are under active development*

IRISH SEA *Off-shore areas where exploratory work has been done and where either initial drilling has indicated the presence of oil and gas or there are high expectations of hydrocarbons occurrence*

Andalusia *Other interesting off-shore areas with relatively shallow water where there are geological and/or geophysical expectations of oil and gas and where exploration will begin before 1985*

Figure II-6.2: Western European regions of offshore oil and gas potential

familiar resource diagram – Figure II-6.3 – perhaps helps to clarify the issues involved as it is especially important at the present stage in the cycle of petroleum exploration and exploitation in Western Europe. Its proven and inferred reserves, which are currently considered to be economic to produce are very small relative to the total potential which could be developed as knowledge increases, technology improves, politics change, and economics give more encouragement to indigenous resource development.

To return to the North Sea, however, we can simplify the evaluation which is necessary to try to understand how its oil and gas resources are developing. This evaluation is presented in Figure II-6.4, with it *x*- and *y*-axes representing "discovery" and "recovery" – the complementary aspects of resource development.

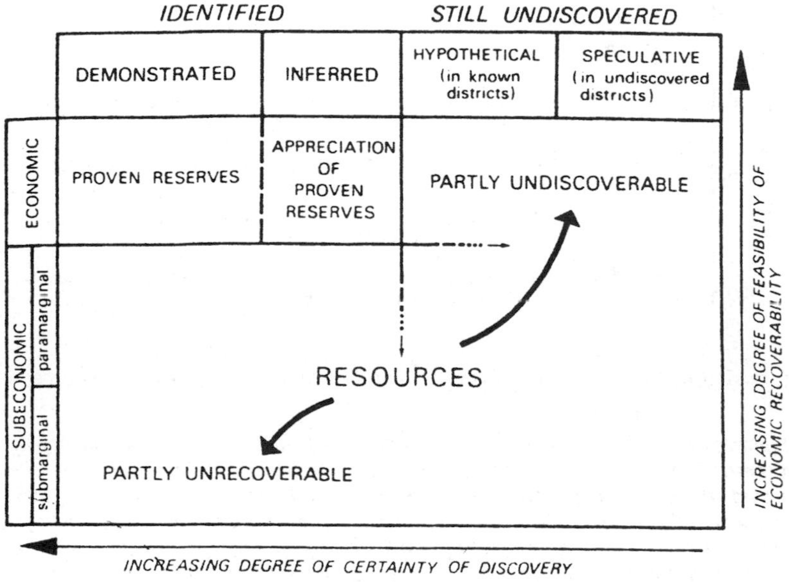

Figure II-6.3: A resource diagram

This illustrates the categories into which resources of oil and gas may be divided and the way in which geological and economic factors influence the relationships between the categories. In light of levels of knowledge, technology, and price, the dividing lines between the categories vary over space and time. The proven and inferred reserves of oil and gas in Western Europe at present are relatively small, but conditions have worked against their development to date. With changed conditions – of knowledge, technology, economics, and politics – the resource base can be more effectively explored and developed

The size of the resource base depends, in the first instance, on the number of fields and on their size. Equally obviously, in any province, there are a finite number of fields and how many of these are discovered is a function of the size of the investment in the exploration effort. The more fields that are discovered, the greater the quantity of reserves in the province – a development which is illustrated on the y-axis in Figure II-

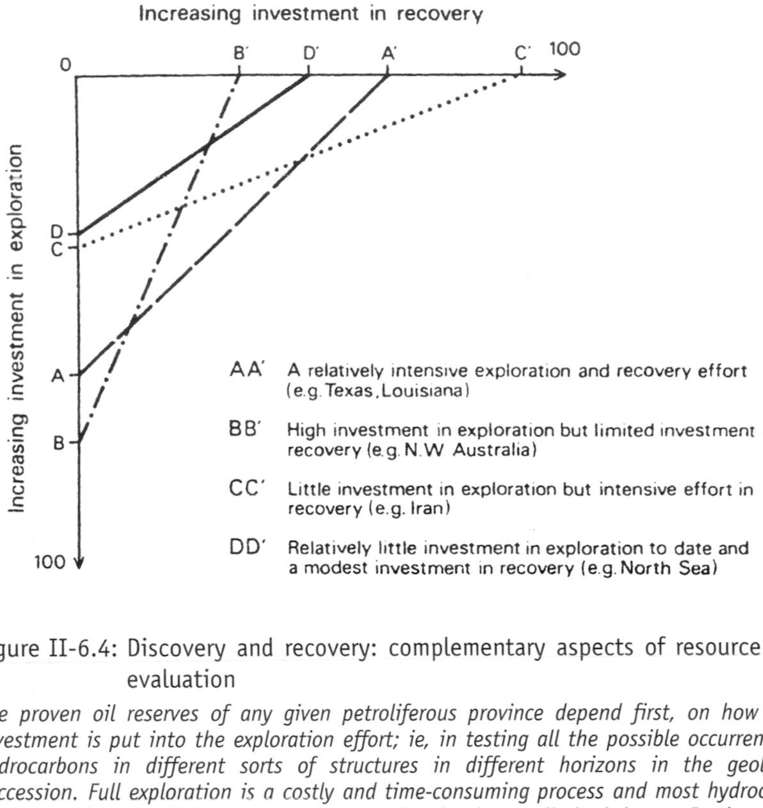

Figure II-6.4: Discovery and recovery: complementary aspects of resource base evaluation

The proven oil reserves of any given petroliferous province depend first, on how much investment is put into the exploration effort; ie, in testing all the possible occurrences of hydrocarbons in different sorts of structures in different horizons in the geological succession. Full exploration is a costly and time-consuming process and most hydrocarbon provinces of the world have, as yet, been explored only to a limited degree. In the case of the North Sea, the exploration effort to date, as shown on the diagram covers no more than one-third, at most, of the total exploration which is necessary fully to explore the province; this, if it happens at all, will be spread out over at least the next two decades. Second, proven recoverable reserves depend on investment in the facilities needed to recover the discovered oils. Fields may be creamed of their lowest-cost resources (with costs represented almost entirely by investment costs) or developed more intensively so as to push recovery toward the "limit" set by technology. The technology itself, of course, changes over time, as does the commercial viability of recovering more or less oil from a field or group of fields. In this diagram North Sea investment in recovery is shown to be relatively higher than the current level of investment in exploration (line DD). It is, however, still modest compared with what could be done.

6.4. Increases in reserves figures thus depend first, on the continuation of an exploration effort in the province over a long period of time during which the results of the continuing effort are expected to yield knowledge of successively smaller fields. However, in light of specific economic and political conditions, it is possible that the exploration effort will be terminated before all the prospects have been tested so that resources which are discoverable in a technological sense will remain undiscovered so keeping the reserves figure of the province at a lower level than would otherwise have been the case.

The degree to which this phenomenon is likely to occur in respect of North Sea exploration remains uncertain. Many companies, however, have indicated that much increased physical costs of developing North Sea fields, coupled with the higher share of the revenues which now has to be paid to governments, following the re-negotiation of the terms of the concession arrangements (and which the companies see as constituting an additional set of costs affecting the viability of their operations), constitute good reasons why exploration should cease when all the larger structures have been tested. This might then exclude the search for all fields which are expected ultimately to yield less than 200 million barrels of oil (or the equivalent thereof as associated or non-associated natural gas). However, as such reservoirs are unlikely to contain much more than 15–20 per cent of the total reserves of the province (based on the usual statistical distribution of reserves by field size in a petroleum province), then the effect of this component in accounting for the difference between the simulated availability of reserves and the quantities expected by the oil companies will be relatively small, unless, of course, other factors, such as government-imposed limits on the amount of exploration (as in Norway) intervene to inhibit discoveries.

Apart from this factor, however, the much more important component in determining contrasts in estimates of reserves emerges from the other axis (the x-axis) in Figure II-6.4. This is a component which represents variations in the degree to which the resources of oil and gas which have been found in a set of fields are actually exploited. Of the oil and gas in place in any reservoir a certain percentage will be recoverable with a given technology over an economically relevant time period. This percentage figure will be a function of the level of investment made in the oil recovery system such that the more money that is spent, the more oil will be recovered. Moreover, technology also improves over time and so increases the percentage recoverability of oil from discoverable fields, so enabling more oil to be recovered with

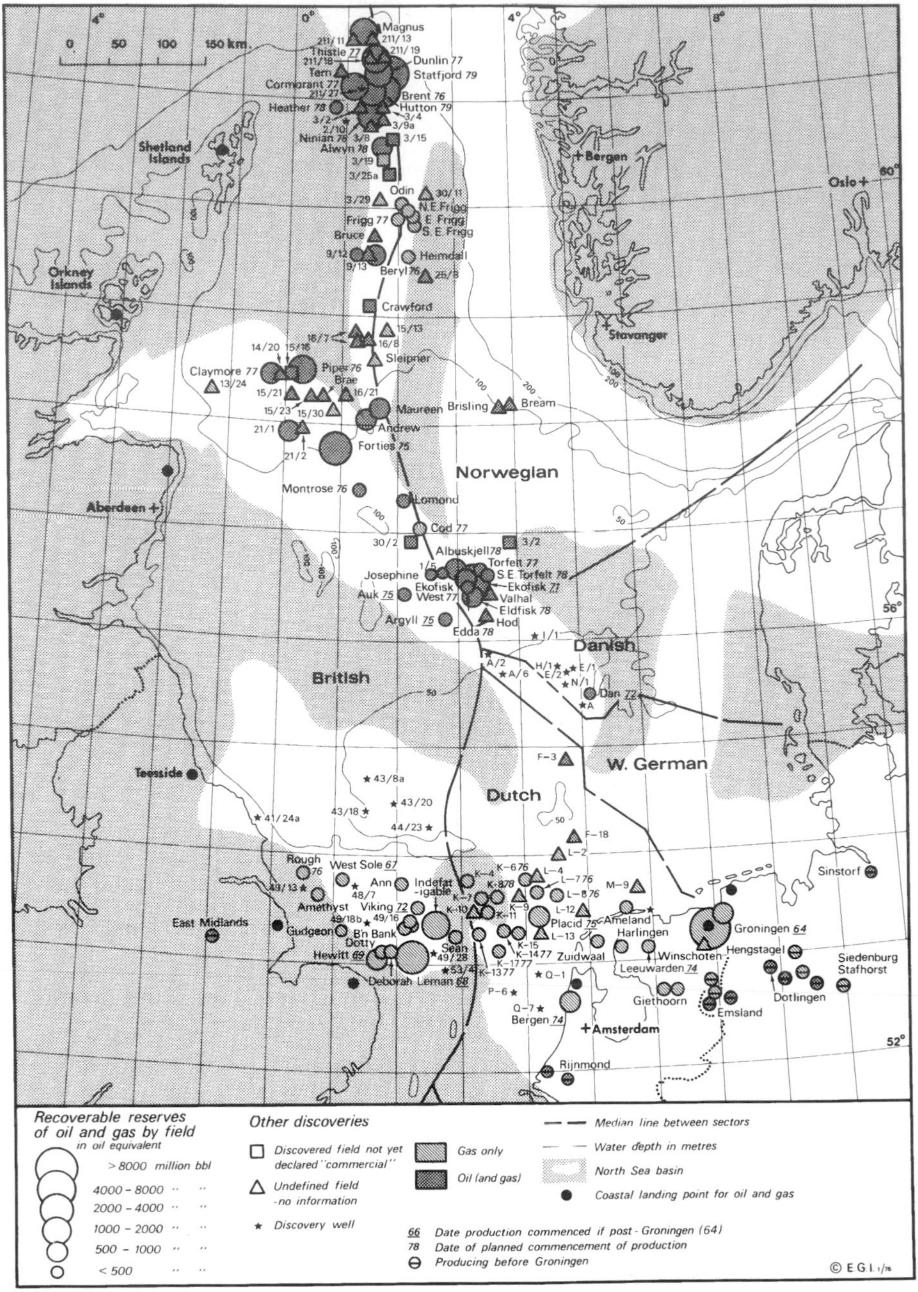

Figure II-6.5: Oil and Gas fields in the North Sea by the end of 1976

additional expenditure on the development systems. Finally, technological improvements also require investment whereby the recoverability of oil from a field can be enhanced. Overall, therefore, one can hypothesize that the more investment that is made to develop an offshore field more intensively and extensively, the more oil that can be produced, so pushing out to the right in Figure II-6.4 the recoverability component. By this means, too, total reserves are increased.

The major difference between our simulation model's predictions of the reserves of the North Sea and the estimates of "officialdom" appears to emerge from the way in which this component in reserves' evaluation is treated. The difference, in essence, boils down to a contrast between the *possible* and the likely recoverability of oil from a discovered field. This question – with both technical and economic aspects – is on which has not yet been adequately explored, in spite of its important policy implications for both companies and governments.

As shown previously (see Table II-6.1) a large number of oil and gas fields have already been found in the North Sea province. Their distribution, by size and by type at the end of 1976, is illustrated in Figure II-6.5. There are almost 30 "giant" fields (using American parlance, in which a "giant" is a field with over 500 million barrels of oil or oil equivalent). In the case of each of these fields some part of their technically recoverable reserves will be recovered on the basis of an installed production system, the decision on which will depend essentially on the operating company's evaluation as to how it can earn, at best, maximum and, at worst, sufficient profits. In other words, the declared recoverable reserves of an offshore field are not a fixed quantity. On the contrary, they are a highly variable element and are essentially a function of the investment decision which the operating company takes. The initial investment decision is, moreover, one which could, in the unique circumstances of the North Sea's physical environment, determine more or less once and for all what percentage of the technically recoverable reserves shall be recovered over the full production life cycle of the field. This is because, given the size, the shape and the deep-water location of the fields, the initial decision on the number of platforms to be put on the field and the number of wells to be associated with them is the critical variable for defining the quantity of the reserves which will be recovered – that is, for defining the size *of the field* in terms of recoverable reserves.[8]

How this works out in practice is illustrated in Figure II-6.6 showing how, on an hypothetical field, one, two or three platforms can be located to deplete the reservoir, or, rather, the oil reserves of part (or parts) of the reservoir. In economic terms, moreover, each additional platform is less productive than the previous one (successive platforms produce decreasing quantities of oil, but do not cost any less to install or to run) and so, as shown in Figure II-6.7, there is a rising average unit investment cost curve as a field is more extensively and/or intensively developed.

Thus, the location and the geographical extent and shape of any North Sea field together with its reservoir characteristics and the economics of the different production systems which can be built to deplete it have important consequences for the field's unit production costs. Moreover, as the unit revenue curve can be taken to be horizontal (as it is not affected by the production decision), then there are also consequences for the unit profitability of production.

In such circumstances one can argue that there must be a high propensity on the part of the operating companies to play safe and thus to take exploitation decisions which mean that the fields are "creamed" of their lowest-cost-to-produce reserves. Thus, some, or even most, of

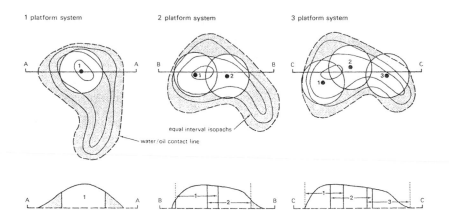

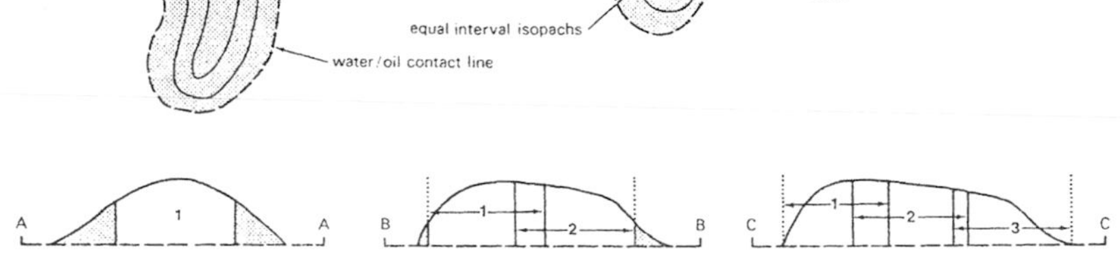

Figure II-6.6: Reorganization of platform location with increasing system size

A hypothetical oil field is shown with a one-, a two-, and a three-platform system, repectively, on the three maps of the field. Below each map is an appropriate cross-section diagram. If we compare the one- with the two-platform system and the cross section A-A with the cross section B-B, we can see that the introduction of the second platform has caused the relocation of the first, resulting in the deepest oil-bearing sands being shared between the two platforms for production purposes. The first platform now produces less oil than the one platform in the one-platform system. Comparing the two- with the three-platform system and sections B-B and C-C, we can see the same phenomena in respect to the location and productivity of platforms 1 and 2.

the technically recoverable reserves of a field are defined as uneconomic to produce in the context of companies' opportunities for investing in other prospects in the North Sea or even in oil-producing activities elsewhere in the world.

It is this set of conditions that constitutes the principal reason for the contrast between the simulated level of North Sea reserves (in which simulation a primary assumption was that all reserves which are discovered and technically producible will be produced) and the much lower levels which are being defined by the companies. The problem has been fully defined and analyzed in a recently completed study[9]. Here,

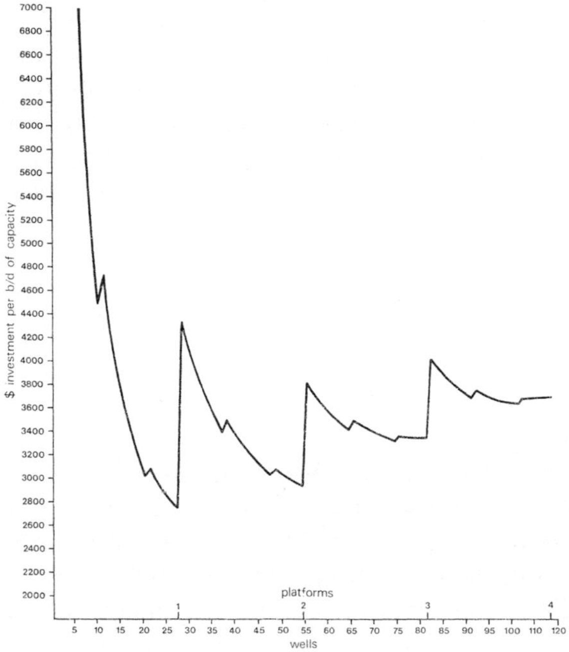

Figure II-6.7: Average investment per barrel per day of capacity with a four-platform system

For the installation of this system on the field, four separate platforms each with an ultimate capacity for handling 27 wells are required. Within each platform there is a falling unit cost curve as average productivity increases with the increasing number of wells – except that costs of expanded platform facilities at 12 and 20 wells per platform create the upward kinks in the curves. The installation of each additional platform, however, reduces overall productivity and there is a jump back to a higher unit investment cost – eg, from the minimum $3,350 when full productivity is achieved from the third platform to $4,000 as the first well on the fourth platform comes into production. Note the steadily increasing unit cost of investment from the most productive levels for platforms 1 to 4, respectively, viz. from $2,750 to $2,950 to $3,350 to $3,700 with the four platforms

Table II-6.6 offers conclusions of the study in respect of three North Sea fields (Forties, Piper, and Montrose). These were analyzed in detail in respect of possible alternative production systems with calculations for each system of the profit levels, revenues payable to government and their contributions to foreign exchange earnings. The figures in the Table demonstrate unequivocally the great difference between the reserves the companies have decided to produce from these three fields to satisfy their required rate of return on investments and the reserves which might have been produced if alternative production systems had been installed on the fields with the objective of achieving production levels to maximize the economic returns to the United Kingdom. The difference in terms of the reserves figures is of the order of 31.0 per cent; the differences from the point of view of the benefits to the U.K. economy are just as significant, ranging from 23.9 per cent in terms of the present value of the future flows of tax revenues, to over 50% in respect of the present value of investments in the production systems. This latter factor is of great significance for job creation in the U.K. economy in general and in its oil-producing areas in particular.

Conclusions

Offshore oil and gas exploitation in the North Sea has created a technological frontier – not only in terms of the development of the exploration and production facilities which have been considered in this paper, but also in terms of undersea pipelining and maintenance developments[10]. But this still leaves equally important issues unexamined; issues such as the safety of the installations themselves and of the additional difficulties created for navigation in a busy area; the working conditions and the safety of the men employed on the rigs, the platforms, and the servicing vessels – especially those in the deep-sea diving industry which has had to expand rapidly to cope with the demand for underwater inspections in waters of up to 250 metres in depth and which requirements have already caused a significant loss of life (over 30 deaths to the end of 1976); and, of course, the question of the impact of exploration and exploitation on the marine environments as well as the impact of major onshore installations (servicing facilities and terminals especially) in adjacent coastal regions, hitherto largely undisturbed by large-scale industrial activity. Such technological and environmental issues are obviously important and raise many hitherto unasked questions concerning the use of the marine environment for oil and gas production.[11]

In this article, we have tried to show that the developments of the North Sea for oil and gas has also created new frontiers for political and

TABLE II-6.6
The Forties, Montrose and Piper North Sea oil fields: Contrasts in the benefits for the UK emerging from company orientated and country required optimal developments

Characteristic	Forties 4 Platform Company Optimum	Forties 6 Platform Govt. Optimum	Montrose 1 Platform Company Optimum	Montrose 3 Platform Govt. Optimum	Piper 1 Platform Company Optimum	Piper 3 Platform Govt. Optimum	Total of the Three Fields Companies' Optimum	Total of the Three Fields Country's Optimum	Total of the Three Fields % Difference
Quantity (million barrels) of recoverable oil in economically relevant time period	1,935	2,315	159	333	625	903	2,719	3,561	31.0
Production of oil in 1980* (million bbl)	221	255	21	42	71	100	313	397	26.9
Peak year production of oil	1980/81	1981	1979/82	1981/82	1979/82	1981/82	–	–	–
Flow of Government revenues in 1980* (million $)	2,138	2,310	140	282	738	983	3,016	3,575	18.4
Foreign Exchange value of 1980* oil (million $)	3,210	3,697	320	628	1,071	1,494	4,601	5,819	28.6
Present value of all future Government revenues (million $)	7,040	8,070	460	710	2,450	3,540	9,950	12,330	23.9
Present value of total volume of oil produced (million $)	13,100	15,100	1,340	2,670	4,820	6,820	19,260	24,590	27.7
Capital investment (million $):									
1974	367	367	42	85	93	113	502	565	12.5
1975	397	443	98	213	217	279	712	935	31.3
1976	361	500	84	215	186	274	632	988	56.5
1977	277	599	28	162	62	191	367	952	159.4
Number of Platforms:									
Built by 1976	4	4	1	2	1	2	6	8	33.3
Under construction in 1976	0	2	0	1	0	1	0	4	

⋆ *1980 has been selected for illustrating the contrasts in Government benefits to show the relatively near-future importance of the differences arising from the systems and not because the differences reach this peak in 1980. Indeed, relatively, the gap continues to widen throughout the 1980s but, in terms of the absolute difference between the size of the benefits, the peak is reached in 1984/85.*

economic issues. These issues remain less than fully understood and, to date, have not been put into the context of the relationship between national governments anxious for security of energy supply and the oil companies whose justifiable commercial interests seem to indicate a pattern and a type of development of the oil and gas reserves which does not meet the needs of the communities involved.

References

1 Descriptions of these conditions can be found in K. Chapman, *North Sea Oil and Gas* (North Promfret, Vt: David & Charles, 1976), and I.L. White et al., *North Sea Oil and Gas*: A Study Sponsored by the Council on Environmental Quality (Norman: University of Oklahoma Press, 1973).

2 In all sectors of the North Sea (See Fig II-1.5), concessions for exploration and/or production are allocated on a discretionary basis to companies which apply for them. They are not auctioned by the states concerned to the highest bidders. See K.W. Dam, *Oil Resources: Who Gets What How?* (Chicago: University of Chicago Press, 1976), for a description and analysis of the procedures involved.

3 P.R. Odell "The Economic Background to North Sea Oil and Gas Development" in *The Political Implications of North Sea Oil and Gas* (M. Saeter and I. Smart, Eds.), IPC Science and Technology Press, Guildford, 1975 pp 51–80 (See above Chapter II-4)

4 Ministry of Economic Affairs, *National Gas and Oil in the Netherlands and Its Offshore Area, 1976* (The Hague, 1977)

5 In addition, by then, about 2,250 x 10^9 m^3 of natural gas produced in Western Europe will have been used.

6 This is partly because the oil companies are more interested in finding oil than gas. In almost all Western European countries gas distribution and sale is a state monopoly and thus subject to more control than the oil companies like.

7 P.R. Odell and K.E. Rosing, *The North Sea Oil Province: An Attempt to Simulate Its Development and Exploitation*, 1969–2029 (London: Kogan Page 1975). See Chapter II-3 above.

8 P.R. Odell and K.E. Rosing *The Optimal Development of the North Sea's Oil Fields*, Kogan Page, London, 1976

9 P.R. Odell and K.E. Rosing, ibid.

10 See Chapter II-5 above, for an analysis of the optimization of the UK's pipeline system

11 For a survey of these issues – and for references to the specialized literature – see E. de Keyser, (Ed.), *The European Offshore Oil and Gas Yearbook* (London: Kogan Page, 1976).

Chapter II – 7

The exploitation of Western Europe's hydrocarbon resources by the late '80s: the politico-economic framework and prospects*

Western Europe's energy resources: an 'embarras de richesse'

Indigenous oil and gas

One of the important results of the two oil price shocks and the associated deterioration in the security of supply of Western Europe's almost entirely imported oil needs has been the stimulus given to the indigenous production of oil and gas. In 1967 the total Western European production of natural gas and oil was a mere 40 million tons of oil equivalent. This came partly from a number of scattered and only locally important production centres in West Germany, Italy and France, partly from the initial exploitation of a new gas field in the Netherlands (the Slochteren Field, later to be renamed the Groningen Field) and partly from the first gas fields discovered in the southern sector of Britain's part of the offshore North Sea province (see Fig II-7.1). By 1972, gas production of the Netherlands and the UK had already grown significantly, but even so, total Western European hydrocarbon output was still only 130 million tons of oil equivalent. This then represented no more than 11% of Western Europe's energy use and it was still much less than the declining annual output of coal and lignite in Western Europe which, in 1972, still accounted for 230 million tons of oil equivalent.

* **An edited version of a paper presented at the opening session of the 1990 Joint Annual Conference of the European Associations of Petroleum Engineers and Geologists held in Copenhagen. Originally published in *First Break*, Vol. 8, No. 10, October 1990, pp. 361–374.**

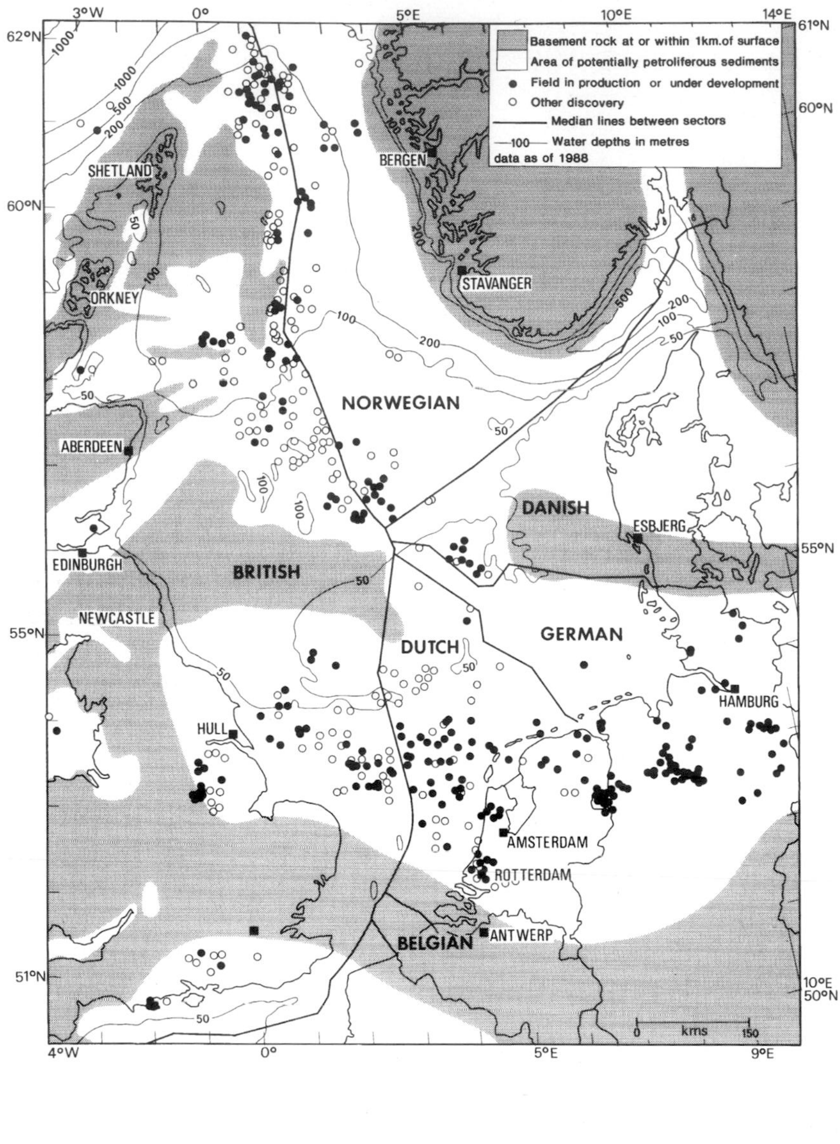

Figure II-7.1: The North Sea Basin: oil and gas fields and discoveries to December 1988

The North Sea Basin

Moreover, even amongst those who were aware of the geological significance of the Groningen and North Sea finds of gas and oil – in terms, that is, of what they indicated about the potentially petroliferous nature of the North Sea Basin stretching northward along the axis of the North Sea from Groningen as far as the latitude of the Shetland Islands (see Fig. II-7.1) – there was a continuing tendency to discount their importance. The potential for North Sea Basin oil and gas was, indeed, generally portrayed as very limited. At best, it was widely argued, even by spokesmen of the leading oil companies, the North Sea could be the means whereby Western Europe's incremental demand for oil and gas might be met for some years, but it would certainly not be a phenomenon which would allow large volumes of the then high and growing level of energy imports to Western Europe to be substituted.

However, the enhanced motivation for oil and gas exploration that was generated by the international oil price shocks of the 1970s soon showed that the North Sea Basin was a large, complex and potentially highly productive hydrocarbon province, the ultimate resources of which would take decades, rather than years, to be revealed in their entirety, even if there were a continuing intensive exploration effort in all the national sectors.

On the basis of the limited data available by the mid-1970s we were then able to show, with the help of a simulation model that there was a 90% probability of almost 80,000 million barrels of oil equivalent of recoverable reserves (= about 11,000 million tons of oil) and a 50% probability of reserves of more than 105,000 million barrels (= about 15,000 million tons of oil). Though the results of our study were generally disbelieved, it is, nevertheless, proving to be an understatement, rather than an exaggeration of the prospects. By the end of 1987, as seen in Table II-7.1, the hydrocarbon (oil plus gas) reserves already discovered were almost 100,000 million barrels of oil equivalent – and this is only counting the reserves of the fields that were then already in production, in development, or for which an indication of development potential had been indicated by the companies concerned. There are additional, but still unknown, volumes of reserves in the many other fields which have been discovered, but for which development plans have not yet been announced. Meanwhile, the search for additional fields continues, and this continuing exploration effort is still giving a high success rate when measured against international norms. The average size of new fields is not, moreover, falling away as quickly as had been generally expected. There is thus a high probability that significant additional reserves will continue to be discovered for many years into the future.

Simultaneously, the phenomenon of the 'appreciation' (that is, the up-grading of estimates over time) of the declared reserves of fields which are in production continues, so that production from most of these fields will go on for much longer than was anticipated in earlier evaluation of their potential. One prime example of this appreciation process under way is the case of the Dutch Groningen gas field (Fig II-7.2).

Four years after its discovery, at the time of the initial production of its gas in 1963, its recoverable reserves were declared as 1125 x 10^9 m^3. Since then, more than 1300 x 10^9 m^3 have been produced from the field, but the field's reserves which still remain to be recovered are now declared at 1324 x 10^9 m^3. This gives a latest estimate of the original

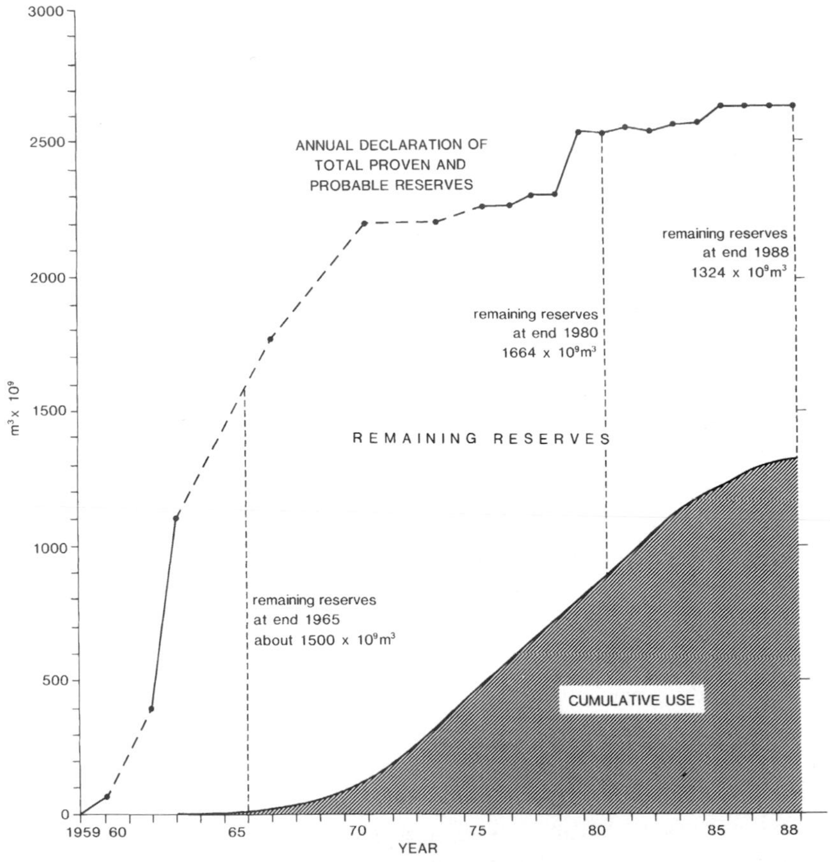

Figure II-7.2: The Groningen Gas Field: the evolution of reserves and use

TABLE II-7.1
North-west Europe's hydrocarbon production to 1987, recoverable reserves remaining and the total reserves declared in developed/developable fields by 2000
(barrels x 10^9 oil equivalent)

	Cumulative production to end 1987[1]			Reserves remaining: In existing fields[2]			Reserves remaining: In fields to be developed[3]			Reserves remaining: Total			Recoverable reserves declared to end 1987[4]		
Country	Oil	Gas	Total	Oil	Gas	Total	Oil	Gas	Total	Oil	Gas	Total	Oil	Gas	Total
Norway	2.2	1.53	3.37	7.5	2.65	10.15	6.75	12.1	18.85	14.25	14.75	29.0	16.45	16.28	32.73
UK	7.85	3.82	11.67	6.37	3.97	10.34	3.68	3.09	6.77	10.05	7.06	17.11	17.9	10.88	28.78
Netherlands	0.53	8.7	9.23	0.55	11.5	12.05	0.25	0.45	0.7	0.8	11.95	12.75	1.33	20.65	21.98
W Germany	0.45	1.9	2.35	0.73	1.55	2.28	0.2	0.74	0.94	0.93	2.29	3.22	1.38	4.19	5.57
Denmark	0.13	0.01	0.14	0.4	0.38	0.45	0.83	0.78	0.83	0.78	0.83	1.61	0.91	0.84	1.75
Ireland	0.0	0.07	0.07	0.0	0.3	0.3	0.2	0.18	0.38	0.2	0.48	0.68	0.2	0.55	0.75
Totals	*11.16*	*16.03*	*27.19*	*15.55*	*20.35*	*35.9*	*11.46*	*17.01*	*28.47*	*27.01*	*37.36*	*64.37*	*38.17*	*53.39*	*91.56*

Notes on Sources: (1) *Cumulative production from national/EEC/IEA statistics on annual production*

(2) *Remaining reserves in existing fields from national data on oil and gas reserves updated to end 1987 from the technical press*

(3) *Reserves in fields to be developed from the survey of development projects and prospects 1988–89 based on operators' replies to questionnaires, Offshore Engineer, January 1988, pp. 26–44*

(4) *Recoverable reserves declared to end 1987 = the sum of cumulative production plus total reserves remaining at end 1987. This is not an indication of ultimately recoverable reserves or even of reserves in all discovered fields. Many of the latter do not yet have development prospects or are not even fully appraised.*

recoverable reserves in place of almost 2700 x 10^9 m^3, an appreciation of the initial reserves declaration of 140%; and the process is not yet complete.

It has been a similar story with the oil fields brought into production in the British sector of the North Sea prior to 1980. The reserves declared for these 14 fields on the initiation of production totalled 1050 million tons. Collectively they have already produced some 900 million tons, indicating that only about 150 tons should now remain in the fields. The exploiting companies' current expectation of the volume of oil that remains to be recovered is, however, over three times as much, at more than 450 million tons; and again, in the case of this set of fields, the process of reserves' appreciation has by no means run its full course. Elsewhere in the oil world, the phenomenon of appreciation of oil reserves has been noted as occurring over a minimum of twenty years. It would thus be surprising if the recoverable reserves of this set of fields does not eventually turn out to be at least double those that were originally declared, so that significant production from them will be sustained for much longer than originally expected and planned.

The complexity of hydrocarbon development decisions

The continuity of oil and gas reserves' discovery and of their production is, of course, dependent on very much more than the mere presence of resources: though the establishment of the fact of their existence and the recognition of the producibility of their oil and/or gas – not only by governments, which made the decisions to awards concessions and production licences, but also by the companies whose investment funds have to be committed to their exploitation – is the first essential part of the development process. In addition, there has to be evidence that adequate overall profitability can be secured from the volumes which can be produced. This is a prospect which depends on appropriate technology, on markets, on prices and on tax regimes. These variables are individually difficult to evaluate and collectively they constitute a very complex background to the development decision-taking task for the companies concerned, especially as the variables are all subject to significant changes in value over time.

For the development of the supply of North Sea hydrocarbons, in the heart of an energy-intensive industrial Europe, markets may perhaps be assumed, though, as will be shown below, given that Western Europe is generally an open energy market, competition from alternative supplies in a period of easy supply/demand relationships cannot be excluded. The required technological inputs may also be assumed, especially given the high degree to which the technology of offshore oil and gas has been developed in Western Europe. Moreover, after a period in which European governments wrongly assumed that oil and gas production activities provide a 'golden goose' phenomenon in respect of guaranteed revenues, there is now a general recognition that taxes can only be collected when an economic rent (or a super-normal profit) can be earned on a project. The achievement of such rent depends, of course, on the market price of oil and gas and thus, in an open economy such as Western Europe, on the international price of energy. Earlier widely held expectations of continuing high, and even still rising, prices for oil have now been replaced by the recognition that prices are more likely to be much lower than prices of the late 1970s/early 1980s for the rest of the century – and beyond.

Such uncertainty over price does, of course, raise doubts over the economic viability of continued oil and gas exploitation offshore north-west Europe. But with levels of development costs already declining in real terms as technology improves, and with the continuing adjustment of tax regimes, then, providing international oil prices do not fall below about $15 per barrel (for other than short periods), there is not likely to be much price-imposed constraint on potential production levels for the remainder of the century. Nevertheless, lower prices will slow down the geographical extension of oil and gas exploration and exploitation into other areas which show hydrocarbon potential around Western Europe; notably in respect of the Arctic waters to the north of Norway and in the deeper waters to the west of the British Isles including Ireland (See Figure II-7.3).

Uneconomic Coal Resources

In contrast with the continuing prospects for further expansion of indigenous oil and gas production, little of the rump which remains of Western Europe's formerly wide-spread and productive deep-mined coal industry can be profitably exploited under existing or prospective costs, prices and technological conditions. The maintenance of most 170 million tons of coal production depends on the continuation of formal guarantees of markets and/or on continuing production-cost subsidies for European coal. The already widely-taken alternative of using inherently lower cost international-traded coal from competing overseas suppliers will become increasingly attractive, not only for coal consumers, but also from the standpoint of national economic considerations.

Until recently, concern for the environmental problems of coal use was at a lower level, and there was thus strong official encouragement for an increased use of coal – related to the widely held

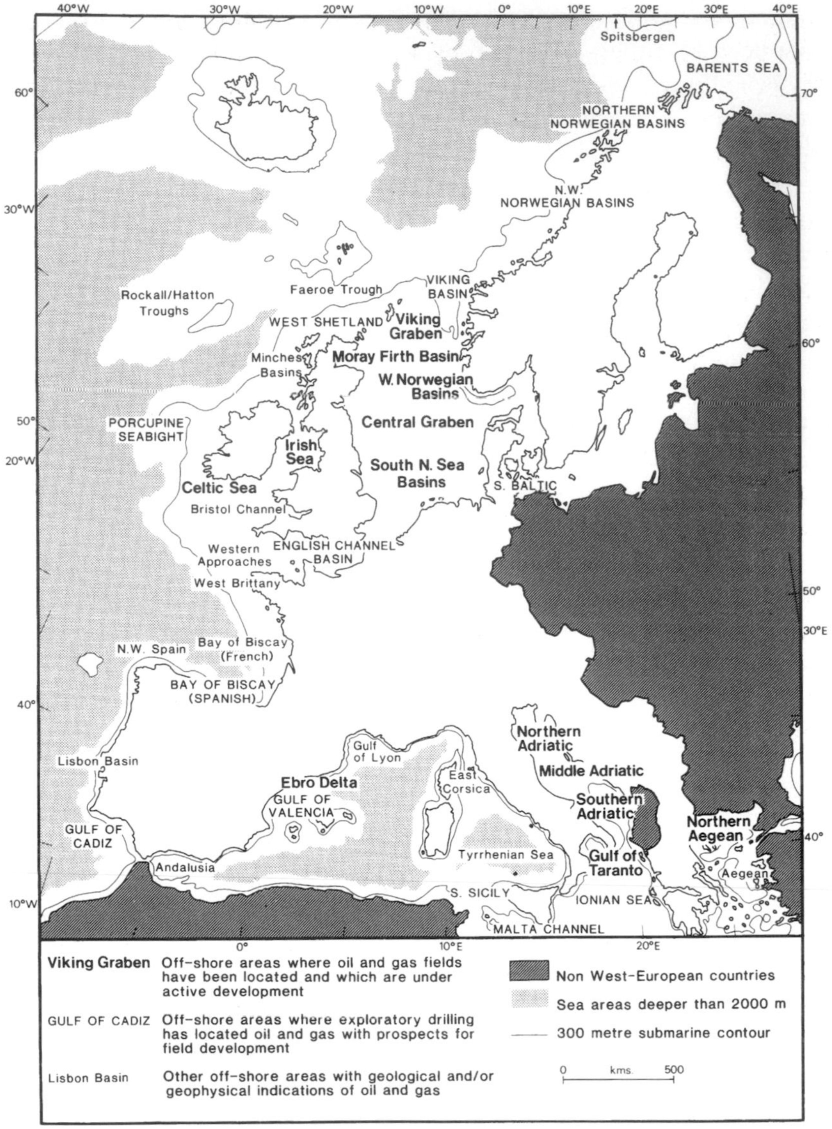

Figure II-7.3: Western Europe's Petroliferous and Potentially Petroliferous Offshore Areas

(though mistaken) belief that hydrocarbons (oil and gas) were inherently scarce, so that energy policies should be orientated to their replacement by coal. It was thus anticipated that imports of coal would supplement a slowly growing availability of indigenous production. As the mistaken belief in hydrocarbon scarcity has now been largely, though not yet entirely, discounted and as it is now generally accepted that oil and gas prices will not return to the early 1980s' levels, then in the context of a now greatly enhanced concern for the environmental dangers of coal use, less emphasis seems likely to be given to encouraging coal use. Thus, there will not only be less encouragement for coal imports, but also a much reduced willingness by governments – and the European Community – to subsidize indigenous coal production. The share of coal in Western Europe's energy supply thus seems set to fall below its current contribution of about 20 per cent.

Competition between external suppliers for Western Europe's limited markets

Import options

Within the context of the now expected limited expansion of Europe's energy demand, and given the continuing availability of indigenous hydrocarbons in significant volumes, it is evident that potential exporters of oil and gas to Western Europe will be obliged to compete for the relatively limited markets available.

Indeed, Western Europe has a number of options in respect of its relatively much more limited energy import needs than is generally indicated in official forecasts. The choice between the options is, moreover, of high-level geo-political significance, thus carrying implications for the EC's external policies well beyond energy supply and supply security considerations alone. The options involve three potential large-scale oil and/or natural gas supplying regions, viz. North Africa, the Middle East and the Soviet Union, with all three of which, for one reason or another, Western Europe needs to maintain close trading and economic relations.

Mediterranean basin prospects

With the expansion of the EC to include Portugal, Spain and Greece, the centre of gravity of the Community's interests is shifting southwards, so heightening the political importance of the Mediterranean Basin. This of course, involves trans-Mediterranean relationships between the EC member countries on the northern side of the Mediterranean and the North African Arab nations on the southern side. Three of the latter (Egypt, Libya and Algeria) depend mainly, or to an important degree, on their exports of energy for their foreign exchange earnings, and some formal arrangements in respect of energy trade with their close European neighbours on the north side of the basin are already in place. These include the Algerian to Italy gas pipeline (see Figure II-7.4) through which supplies, soon to be increased, have been contracted for 25 years and also specific Libyan/Italian oil supply arrangements, including Libyan interests in the refining industry in Italy.

In theory, the hydrocarbon resources of North Africa could supply most of the EC's medium-term energy import needs, while Western European markets constitute the only ones in which North African oil and gas exports have a transport cost advantage over supplies from elsewhere. In spite of current political difficulties, the potential mutual benefits from the development of EC/North African energy interests seems likely to be increasingly recognized in the coming decade, and thus lead to attempts to expand Mediterranean Basin interdependence in the energy field.

Imports from the Middle East

The oil and gas rich countries of the Middle East proper (around the Persian or Arabian Gulf) are geographically more distant, but in terms of general historical, cultural and political contacts, as well as specific oil interests going back to the early part of the century, there exists a powerful relationship with Western Europe. Recent EC energy policy, with its central emphasis on minimizing dependence on oil – and especially dependence on oil imports, most of which in the 25 years of rapid demand growth to 1973 originated from the oil-exporting countries of the Middle East – has been at odds with this background. Nevertheless, EC/Middle East relationships in respect of oil have recovered markedly from their low point at the time of the Arab/Israeli war of 1973, when there was an Arab embargo on oil supplies to the Netherlands and the threat of reduced supplies for all other countries.

EC/GCC (Gulf Cooperation Council) and EC/OAPEC (Organization of Arab Petroleum Exporting Countries) discussions on matters economic are now well established and have already led to proposals intended to meet the perceived threat to Europe's refining and petrochemical industries from the low-cost availability of oil products and petrochemicals from the newly developed facilities in the Middle East oil and gas producing countries.

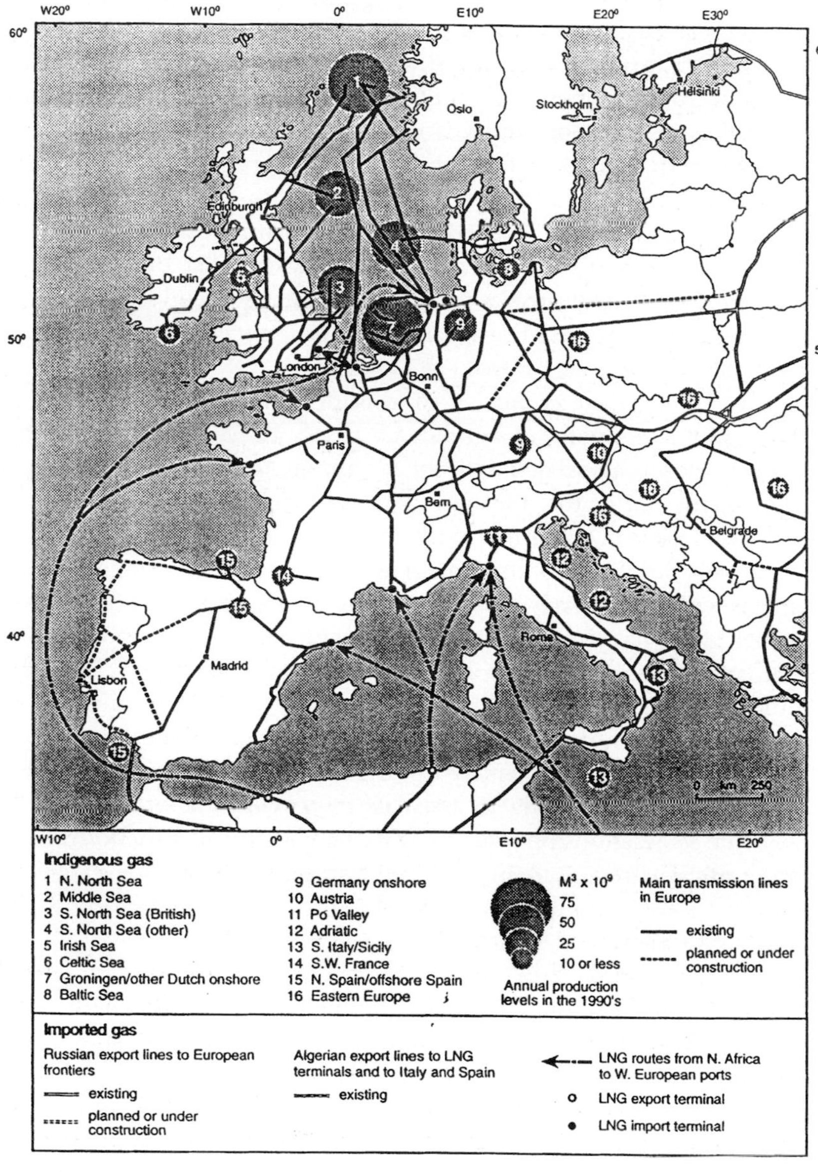

Figure II-7.4: Western Europe's gas production potential in the 1990s and the transmission system for both indigenous gas and for imports

Meanwhile, all European-based oil companies reached agreements over the valuation of their assets expropriated by Iran and other Middle East countries, in contrast with the pursuit of what appear to be unrealistic claims for compensation by many US oil companies. Since then, Saudi Arabia and, even more noticeably, Kuwait have invested in downstream oil activities in a number of Western European countries. Meanwhile, oil transport links from the Middle East, orientated to serving Western European markets, have been strengthened in recent years by the construction of new pipelines from the main oil-producing areas of the Gulf to the Eastern Mediterranean coast and to newly developed export terminals on the eastern coast of the Red Sea. Such routes, specifically dedicated to securing and easing the oil supply routes to Europe, serve to reduce European importers' perception of dangers of dependence on shipping oil from the Gulf.

The key transit countries in this development are Egypt and Turkey and it is not without significance that both have indicated a wish to be more closely associated with the Common Market. Membership of the EC seems out of the question as far as Egypt is concerned, but its wishes for closer relations with the EC can, nevertheless, be appropriately developed in its context as one of the Mediterranean Basin countries with hydro-carbons resources (as discussed above). Turkey, on the other hand, has already submitted an application for membership of the EC and, though its accession will not be a near-future development, it could well become a formal part of the European Community before the end of the century. It thus has a strong and heightening motivation to enhance its role as the country through which Gulf oil is transported to Western Europe in order to assist its own economic growth and to consolidate its relationship with the EC. Its role in this function would be the critical component in enabling natural gas from the extensive resources of the Middle East (notably from Qatar and Iran) to be pipelined to Eastern and Western Europe.

Turkey could thus make itself central to the potential redevelopment of Middle East/European hydrocarbon trade. Moreover, from the European importers' view-point, Turkey's membership of the EC would extend Europe right up to the frontiers of the world's main oil producing regions – both currently and for the foreseeable future – so reducing the present strong perception of the inherent insecurity of supplies from the Middle East. For the oil exporters, Europe would become a near neighbour with large energy import demands for the foreseeable future. It could thus offer potentially guaranteed off-takes of oil and gas. Such a development would, in turn, help to undermine the validity of the minimum Middle East oil policies of the EC, whereby prospects for the oil exporters have been – and currently remain – so adversely affected.

The Soviet option

Finally, the vast (and in effect, relative to prospective demand, near infinite) natural gas resources of the Soviet Union (Figure II-7.5) represent not simply a major new energy resource for long-term use in Europe, but also a means whereby relations more generally with the Soviet Union could be fundamentally changed. In essence, in order to achieve a more efficient, productive and acceptable economic system, the Soviet Union needs to import large volumes of consumer, capital and investment goods and services over the coming decade. The only means that it has to pay for these requirements (apart from borrowing) is by the export of large volumes of natural gas for which the only possible markets for this gas are in Western Europe.

Here, in return, a wide range of manufacturing industry and suppliers of other goods and services could be offered guaranteed outlets in the Soviet Union – without the difficulties of having to compete with alternative suppliers from other parts of the world. The economic bargain which could be struck between the USSR and the EC is thus a powerful one, with immense potential benefits to both sides and one that appears to override the hitherto well-rehearsed arguments of strategic dangers from reliance on Soviet energy supplies: particularly in the context of the broader geopolitical issues that are involved. One of these arises from the fact that the gas involved has to flow through the countries of Eastern Europe. Conceivably, therefore, the Soviet-gas-for-Western-European-goods-and-services bargain could, over time, help to sustain the economics of the newly democratic Eastern European countries, so helping them to become more closely involved with the rest of Europe. This potential political aspect to the massive export of Soviet gas to Western Europe could prove even more fundamental than the purely economic considerations. The combined economic/political changes which could be set in train by Western Europe's willingness to import several times more Soviet gas than the approximately 40×10^9 m^3 which are currently traded, clearly have a geo-political importance which would be at least the equal of the potential for change derived from Mediterranean Basin or Middle Eastern initiatives relating to oil and gas developments.

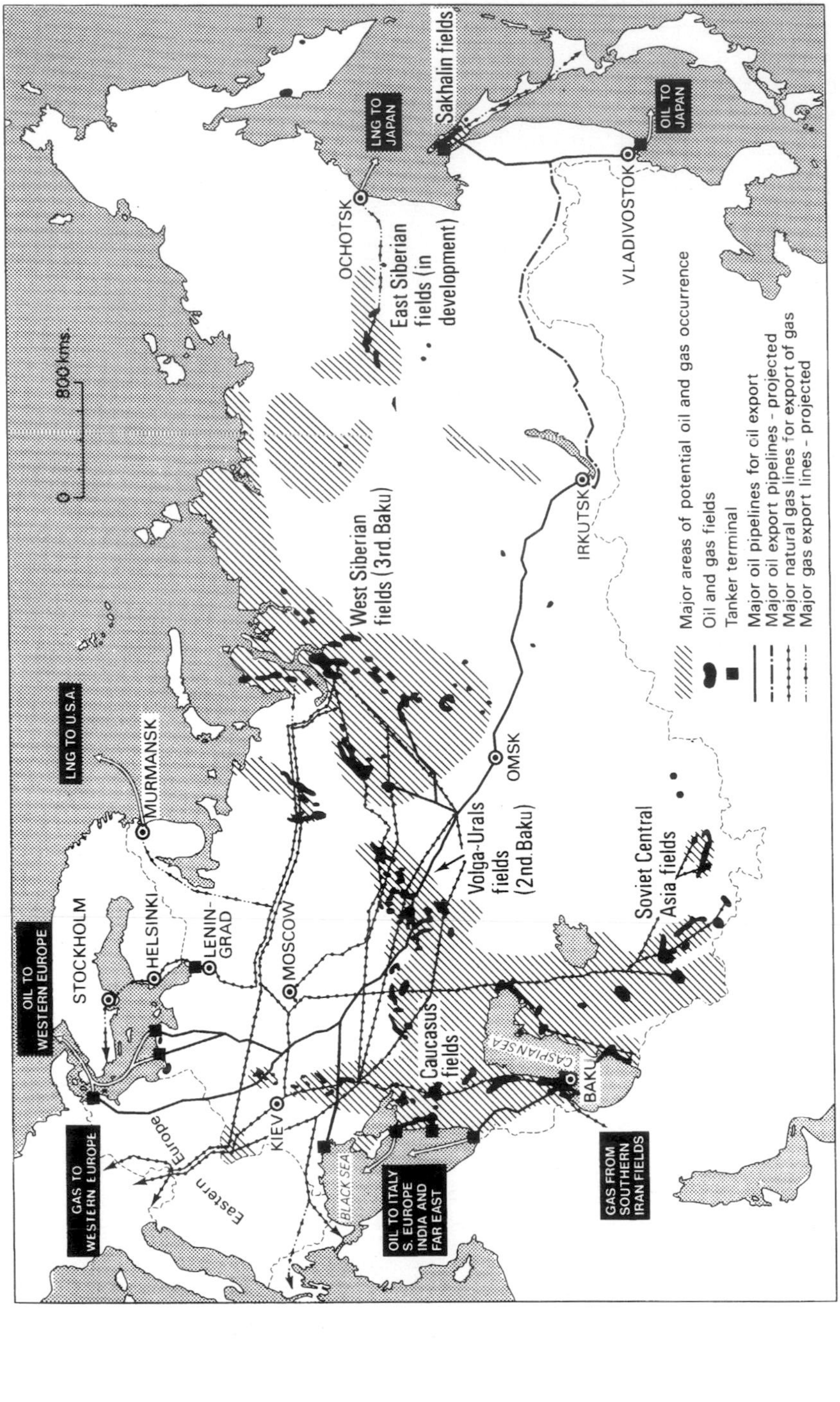

Figure II-7.5: The main oil and gas basins in the USSR showing export facilities (including potential developments

The choices

Clearly, in the context of a demand for energy in Western Europe which will grow only slowly, and in the light of an indigenous potential to produce large volumes of hydrocarbons for decades into the future, Western Europe's policy makers may well be obliged to exercise a series of Solomon-type judgments in their search for external energy relationships which made the most long-term sense. Unhappily, little thought has been given to date to the alternative options, and their significance. This is largely because Europe's energy sector policy makers, in spite of all the evidence to the contrary, remain generally besotted with the concept of supply scarcity and thus believe it to be more important to continue working and planning mainly for ways in which to protect the EC against future oil supply crises. This is perhaps a clear example of the way in which policy makers show that they are rather like the proverbial generals who concentrate their efforts on fighting the last war!

Institutional and technological aspects of Western Europe's energy outlook

The Single Market and competition

The changed energy demand outlook and the wide range of options which now exists for supplying Western Europe's oil and gas together represent a significant challenge. The dynamics of the energy sector will, however, be yet further accelerated in the years immediately ahead by both political/legal changes within Western Europe itself and by technological developments. The impact of the latter will, moreover, be heightened in the more competitive environment which the political/legal changes will create. These changes arise not only from the implementation of the Single Europe Act (theoretically scheduled for the end of 1992 but, in practice, likely to become gradually effective through the first half of the 1990s), but also from various national moves to denationalize and/or to deregulate industries which have traditionally been in the public sector and thus subject to strong political control.

Reduced protection for coal

The West European coal industry mainly falls into this category. As shown previously, most of the industry which has survived competition from oil and gas has done so as a result of direct or indirect state support, usually in the context of state ownership. Such support emerged out of post-war rehabilitation difficulties for the industry and as a consequence of declining oil prices between the mid-50s and the early 70s. However, it continued to be necessary even in the period of the high energy prices from 1974 to 1985. Since then, the sharp fall in the oil price has caused coal's already weak competitive position to deteriorate still further.

Thus, most of the small remaining Belgian and French coal industries are being closed, while in West Germany increasing opposition is being expressed to the rising cost, payable partly by the Federal government and partly by electricity consumers, of the early 1980s' agreement to sustain the use of 70 million tons of indigenous hard coal annually in power generation until the year 2000. In the UK the number of mine closured has been increased since 1985 and this process is expected to continue. Meanwhile, plans are now being made to sell off the remaining coal industry in bits and pieces so that only a small number of efficient collieries in favoured areas (both geologically and in relation to demand centers) seem likely to remain in production. Unhappily, no technological breakthroughs (such as the successful underground gasification of coal) appear to be in sight for the European coal industry, whereby it could hope to survive the increased competition from natural gas and from more freely imported and low-cost foreign coal. And an already bad situation and outlook for Europe's coal industry is made even worse by the developing opposition to the use of coal because of its high specific contribution (compared with gas and oil) to the atmospheric carbon-dioxide problem and its perceived greenhouse effect on the world's climate.

Technological changes favour gas

Meanwhile there continue to be restraints imposed on the use of gas, as shown in Table II-7.2. These are a result of the attitudes of the monopolistic and/or the protected gas transmission and distribution companies in most of Western Europe, coupled with government imposed minimum price controls and other regulations which have also succeeded in limiting gas demand, particularly the EC's and some national restraints on using gas in power generation. These actions have, moreover, also served to inhibit the implementation of technological developments in gas-using technology.

The most important of these by far has been the failure to expand gas use in high-efficiency power generation; either through combined-cycle technology, in which the 50+% efficiency at which electricity is produced is almost half as good again as the average 35% efficiency achieved in coal and oil-based condensing plant production; or in combined heat and power systems, in which heat recovered from the system is used along with the power generated to give overall thermal efficiencies in excess of 65%. In Japan a 2000 MW natural gas-fuelled, combined cycle power plant is already operational, together with several other somewhat smaller units. By contrast, the largest in operation anywhere in Western Europe is still only 230 MW, under the influence of a lack of interest until very recently in their development by the gas suppliers, and in the context of highly centralized and usually monopolistic electricity supply companies with a strong predilection to expand their systems by means of large centralized coal-fired and/or nuclear stations.

The liberalization of the electricity and gas supply industries from monopolistic control seems to be a necessary first step to a situation in which the cost and the environmental advantages of gas for electricity

TABLE II-7.2
Natural gas use from 1976–88 in present member countries of the EC
(in $m^3 \times 10^9$)

Country	1976		1980		1984		1988
Belgium/Luxembourg	11.3		11.4		9.4		8.8
Denmark	–		–		0.1		1.5
France	20.9		26.0		25.9		26.0
Greece	–		–		0.1		0.1
Republic of Ireland	–		0.6		1.9		1.3
Italy	24.2		25.2		29.2		36.6
Netherlands	36.3		33.5		34.3		32.7
Portugal	–		–		–		–
Spain	1.6		2.0		2.2		3.7
United Kingdom	38.1		45.5		49.6		52.6
West Germany	39.9		48.8		45.2		47.4
TOTAL	*172.3*		*193.0*		*197.9*		*210.7*
% increase in successive four-year periods		12.2%		2.5%		6.5%	

Average annual rate of growth, 1976–88 = 1.7%
Average annual rate of growth, 1980–88 = 1.1%

Source: BP's *Statistical Review of World Energy, 1989 and earlier years' issues of BP's Statistical Review of the World Oil Industry*

production could begin to be effective. The ultimate result could be a much more geographically dispersed pattern of power production closer to centres of demand, with the long distance transport element being that of the gas input, rather than the electric output. This would give an overall much more cost-effective and environmentally acceptable system than the separate gas and electricity systems to which Europe has become used over recent decades.

Conclusions

The dominance of fossil fuels

Since the end of the Second World War, Western Europe's energy system has undergone two dramatic changes; and a third is under way. The first was in the 1950s and the 1960s, when imported oil replaced indigenous coal and when there was a continuing rapid rate of increase in energy use. The second was between 1973 and the mid-1980s when much higher energy prices curbed demand growth to a near-zero rate and when there was a massive development of Western Europe's own oil and gas resources.

Nevertheless, in spite of the traumas associated with the first two changes, Western Europe's dependence today on fossil fuels is very little less than it was in 1945. For thirty years nuclear power has promised much, but its overall contribution to total energy supply is still only about 5% (when the electricity produced in nuclear power stations is appropriately measured in terms of its heat value equivalent). It has only just succeeded in becoming more important than hydroelectricity which continues to contribute about 3.5% to total supply. In spite of efforts over the last decade to develop additional renewable energy sources, hydroelectricity remains the only source of benign (atmospherically non-polluting) energy in large scale production. The other sources, such as wind, tide and wave power and the direct use of the sun's energy, still contribute only a fraction of one per cent of Western Europe's energy supply.

The apparent similarity between yesterday's energy sector situation and that of today is, however, superficial. Underneath, there have been important changes in the relative contributions of the three fossil fuels, together with changes in their sources of supply. (Table II-7.3). These changes have, moreover, been accompanied by the evolution of energy sector institutional and organizational structures which have reflected the rapid expansion of energy demand and the belief in 'bigness' and in economies of scale on the supply side. The energy sector has thus been dominated by large public and private corporate entities, with a high degree of centralization and, in some cases (particularly in respect of electricity and gas) with monopolistic controls over energy supply.

Future choices between fossil fuels

In the future too any apparent similarity in the energy sector, as fossil fuels continue to dominate the supply of energy, will also be superficial. Changing economic considerations, political ideas and technological developments will jointly create a significant potential for the establishment of a much greater variety of organizational structures and for changed market relationships between suppliers and users. In addition, as already indicated and as set out in Table II-7.3, there will be a continuation of the changing fortunes of the fossil fuels. Indigenous coal production will certainly continue to decline in importance but, in the event of a reduction in West Germany's high level of support for its coal industry and/or as a consequence of the privatization of the UK's coal industry, the European coal industry could be down to little more than half it present size by 2000. Coal imports will continue to grow, though by no means as rapidly as has been expected and is still widely assumed. Increasing concern for environmental issues (or, put in another way, the costs of achieving acceptably low pollution levels from coal use) will slow down the expansion of the market for imported coal.

Natural gas will be the major growth element in Western Europe's energy supply for the foreseeable future – for a mélange of easy supply, favourable environmental and new technology reasons, in the context of a much more open and competitive situation for incorporating gas in a broader range of end uses. In particular, natural gas use in high-efficiency generating plants will largely eliminate the present continuing level of fuel oil use in power station and industrial use, and will substitute what would otherwise have been nuclear – and coal – based capacity in most countries. By 2020 natural gas seems likely to be Western Europe's single most important energy source, on the assumption that, in the meantime, compressed natural gas (CNG) becomes an alternative automotive fuel.

Prospects for alternative energy sources

Non-fossil fuel energy sources will continue to be dominated by hydroelectricity (for which slow capacity growth will continue) and by nuclear power which, barring accidents, will grow slowly. The downside risk for the latter is, however, high, given that another major accident anywhere would produce an even stronger anti-nuclear

TABLE II-7.3

Evolution of Western Europe's energy supply by sources from 1964 to 1988 (in million tons oil equivalent)

Energy Source	1964 Total	1964 Indig-enous	1964 Imports	1972 Total	1972 Indig-enous	1972 Imports	1980 Total	1980 Indig-enous	1980 Imports	1984 Total	1984 Indig-enous	1984 Imports	1988 Total	1988 Indig-enous	1988 Imports
Solid fuels	381	350	31	261	230	31	264	222	42	240	162	78	264	187	77
(%)	(52.9)			(22.7)			(23.6)			(22.4)			(23.3)		
Oil	293	10	283	738	20	718	621	121	500	576	184	392	594	199	395
(%)	(40.6)			(64.2)			(55.5)			(53.8)			(51.9)		
Natural gas	15	15	–	119	113	5.5	182	160	22	186	154	32	199	150	49
(%)	(2.1)			(10.3)			(16.3)			(17.4)			(17.5)		
Nuclear power*	1	1	–	5.5	5.5	–	17	17	–	34	34	–	47	47	–
(%)	–			(0.5)			(1.5)			(3.2)			(4.1)		
Other sources*	30	30	–	27.5	27.5	–	35	35	–	36	36	–	39	39	–
(%)	(4.2)			(2.4)			(3.1)			(3.3)			(3.2)		
TOTALS	*720*	*406*	*314*	*1150*	*396*	*754*	*1119*	*555*	*564*	*1072*	*570*	*502*	*1143*	*622*	*521*
(%)	*(100)*	*(56.4)*	*(43.6)*	*(100)*	*(34.4)*	*(65.6)*	*(100)*	*(49.6)*	*(50.4)*	*(100)*	*(53.1)*	*(46.9)*	*(100)*	*(54.4)*	*(45.6)*

Notes: * *Nuclear power and the primary electricity component (e.g. hydroelectricity) in 'Other sources' are converted to m.t.o.e. at the heat value equivalent of the electricity produced.*

Sources: OECD and IEA energy policy studies and reviews for 1964, 1972 and 1980. 1984 and 1988 from BP's Annual Statistical Review of World Energy.

response than over Chernobyl, so leading to the cancellation of the few new facilities planned, as well as the closure of some existing stations. An accident in Western Europe itself, at one of the stations of the PWR type that constitutes most of the existing capacity, would seem likely to lead to the near instant close-down of the industry with severe implications not only for electricity supply over much of the continent, but also for the economies of the two countries (France and Belgium) which depend (to the extent of more than 70 and 60%, respectively) on nuclear electricity.

Meanwhile, there will be continued, albeit modest, expansion of wind-powered generating capacity, and a few large-scale schemes for harnessing tidal or wave power could well be under way by the turn of the century. Nevertheless, in spite of formal EC support, as well as national programmes already under way, for the promotion of renewable energy sources' development, major contributions from such new systems of power generation will be delayed until well into the twenty-first century, unless there is unequivocal and dramatic evidence in the meantime of a real medium-term threat of climate change as a result of the continued combustion of fossil fuels in increasing amounts. By then, fusion power and/or power based on the underground gasification of coal may conceivably be competitors for meeting the demand required, especially as the price of oil – and of fossil fuels more generally – may well at last be starting to go back up to the levels which were temporarily achieved in the early 1980s. Until then, Western Europe's energy users seem likely to have a ready choice of alternative energy resources available to meet their demands.

Security of supply

Political decisions will be required from time to time to determine how the choices shall be made; first, in the light of a broad range of economic and other considerations which Western Europe will have to take into account in its dealings with the main other potential energy supplying regions, viz. North Africa, the Middle East and the Soviet Union. Secondly, the decisions will need to pay attention to the extent to which it is considered appropriate to give preference to the production and use of indigenous oil and gas, the continued development of which may otherwise be thwarted by lower costs imports. Such a continuity of interest in Europe's security of energy supply seems highly likely in a world which will remain replete with problems of an international and inter-regional significance, particularly as Western Europe will continue to be second only to Japan as the most energy intensive using part of the

TABLE II-7.4

Alternative potential Western European energy supply patterns in 2000 under: (a) continued constraints on indigenous oil and production: and (b) encouragement of their exploitation

(in millions of tons of oil equivalent)

	Energy Supply Pattern 1988			Alternative Patterns in 2000 A			B		
	Total	Indigenous	Imports	Total	Indigenous	Imports	Total	Indigenous	Imports
Natural gas	199	150	49	220	130	90	335	205	130
%	17.1	17.9	26.7	17.9	23.8	13.1	26.7	31.8	21.3
Oil	594	199	395	595	125	470	580	200	380
%	51.2	48.3	46.2	48.4	22.9	68.6	46.2	37.0	62.2
Coal etc.	264	187	77	275	150	125	220	120	100
%	22.8	22.4	17.5	22.4	27.5	18.3	18.3	18.6	16.4
Primary elec.*	102	102	–	140	140	–	120	120	–
%	8.9	11.4	9.6	11.4	25.7	–	9.6	18.6	–
Total Energy Supply	*1159*	*638*	*521*	*1230*	*545*	*685*	*1255*	*645*	*610*
%	*100*	*55.1*	*45.0*	*100*	*44.3*	*55.7*	*100*	*51.4*	*48.6*

Notes: * *Primary electricity (including nuclear power) converted to m.t.o.e. on the basis of the heat value of the electricity produced.*

Source: *1988 data from BP's* Statistical Review of World Energy, *1989. Alternative patterns in 2000 based on the author's research.*

world through into the first quarter, at least, of the twenty-first century. Thus, in the year 2000, dependence on imported energy to a degree of 'only' 48%, could be significantly different for Western Europe's economic and political standing at that time, then an import dependence of more than 55%. These are the alternatives which emerge, as seen in Table II-7.4, from the impact of contrasting policies over the next decades towards the exploitation of Western Europe's indigenous oil and gas – with only 255 mtoe under policies of constraint compared with 405 mtoe with policies which encourage their production.

Chapter II – 8

North Sea Oil and Gas: The Exploitation of Britain's Resources: Retrospect and Prospect*

The North Sea's Resources: A Retrospect

A lack of appreciation of the potential size and longevity of Britain's hydrocarbons resources has been a recurring theme over the history of the North Sea. Initially, in the 1960s, this was from a lack of knowledge, experience and familiarity with the nature of the oil and gas discovery processes: but later, it was more the result of a widely-held belief in an inherent scarcity of oil and gas at the global level and thus, by definition, also at the national level. (Odell and Rosing, 1975 and 1980)

Thus, in the 1960s – after the initial motivation to bring the earliest gas fields quickly into production to serve the country's residential markets – North Sea gas came to be seen as only able to make a limited contribution to the country's energy economy. Sales of gas for power generation were banned and were priced out of other bulk energy markets (Odell 1978). This was subsequently formalised into a policy based on the concept as "gas as a premium fuel" and thus too "valuable" to be sold in competition with coal, the protection of which was a government priority at the time (Posner, 1973). As a result, the Gas Council defined its ultimate market as no more than 40 Bcm per year: so undermining its earlier motivation to create demand-pressure conditions under which supply developments would have been stimulated. Even gas reserves already discovered in the North Sea remained unexploited, while those which

* **Published in *The UK Energy Experience: a Model or a Warning?* (Eds. G. MacKerron and P. Pearson), Imperial College Press, London 1996, pp. 123–133**

were produced were (in the context of high oil prices after the first oil price shock in 1973/4) under-priced. Thus, UK gas production peaked in 1977, so that both direct and indirect benefits for Britain of more gas production at higher unit revenues were foregone: while, ironically, Norwegian Frigg field gas was imported by the BGC at prices much higher than it paid UK offshore producers under the terms of a contract signed as a consequence of the "belief in scarcity" (Odell, 1978).

Meanwhile, from the mid-1970s, UK North Sea oil began to flow from the recently discovered giant fields in the context of low development costs, very modest taxes on production and agreed field development plans (Annex Bs) which reflected company preferences (Odell, 1979). Government intervention in the development of the industry remained modest through to the late 1970s. The government could, on a relatively limited scale, stop things happening, but it was unable to cause things to happen. Even the creation of BNOC in 1977 and the rights it was given in the UK's offshore oil system which had already been developed made little essential difference. For example, defined and agreed work programmes on the concessions could not be enforced: the government could not intervene to require successful discovery wells to be developed into fields; field development programmes prepared by the operators were monitored and subjected to minor amendments, but were not fundamentally altered (so as to secure higher rates of recovery through higher investments) by means of government intervention (Odell and Rosing, 1976); sub-optimal transport infrastructure plans (viewed nationally) could not be re-ordered to serve national interests (Odell, Rosing and Beke-Vogelaar, 1976); and the essentially unfettered disposal of British oil was never seriously challenged by government on grounds of national interest (Odell, 1979). Legislation in the later-1970s finally secured part-government ownership of the exploitation process and a tax regime which was designed to bite, but these changes were soon overtaken by a radical change of government (in 1979) and by the second oil price shock (in 1980/1). There was thus no opportunity to see how a partly state-owned, a more regulated and a highly taxed industry might have worked under the conditions of the 1980s.

The second oil price shock made British North Sea oil an inherently much more attractive proposition for both state and companies: yet, paradoxically, it was followed by a retrenchment of activities, by a declining rate of growth and, eventually, a reversal in the volume of production of oil; and to the failure of policy to secure the highest possible returns from temporarily very high-priced oil. First, from 1981 specific taxes on oil production were pushed beyond their reasonable ceiling, so that investment and activities slowed down; and, second, there was a formal governmental move towards the establishment of a so-called depletion policy (more correctly defined as a restraint on production policy (Dept. of Energy, 1981).

The objective of government policy on oil production was announced as "self sufficiency plus 5 million tons for export" (Dept. of Energy, 1981); either to 'save' a scarce resource or to delay its production in the expectation of higher prices in the future (Dept. of Energy, 1979). Thereafter, as the largest fields brought on stream in the mid-to late-70s reached their plateau rates of production and were not joined by other fields which would otherwise have been coming into production, so the overall rate of growth in output declined and then, in 1985, peaked. By the time of the price collapse in 1986, it seems likely that the combination of the 1976/9 confrontational policy by the Labour government and the "conservationist" approach of the succeeding Conservative government, had cost the UK up to $25 billion of 'lost' output (750 million barrels at an average of $33 per barrel). Surely, in this way at least, the UK provided an energy experience which was a 'warning' rather than a 'model'?[1]

Meanwhile, not even the privatisation of British Gas in 1982 and the nominal end to its monopsony in respect of North Sea gas purchases and to its monopoly in the end-user markets, led to any significant and continuing expansion of the rate of exploitation of the country's North Sea gas resources. British Gas' preference for enhanced Norwegian imports (Odell, 1978: Stern, 1984) undermined development activities in the UK North Sea, at least until the decision by Norway in 1981 to market its new availability of gas from the Sleipner field into the higher priced mainland European market (Estrada et al., 1988). The indigenous supply response to this – and also to the higher prices which eventually came to be paid for UK North Sea gas – began to show through only in 1985 (when annual gas production returned to over 40 Bcm for the first time since 1977). It was not, however, until the subsequent liberalisation of the gas market and the privatisation of the electricity industry (with the privatised companies having a greater interest in gas, rather than in nuclear power, for electricity generation), that a sufficient motivation was generated for new investments to be made in the further exploitation

1 The "lost" oil remained, of course, available for production later; and it has, of course, since been produced, given the subsequent decision to abandon the idea of a depletion policy. Its value when eventually produced was, unhappily, down by two-thirds to no more than $11 per barrel (in early 1980s dollars).

of North Sea gas supplies (Stoppard, 1994). This was initially in the context of a perceived potential shortage of gas against the expected enhanced demand (so leading to a price expectation of over 20p per therm), but, more recently, as supplies have built up rapidly, it is now at prices generally down to only two-thirds of the previously expected level: and, at the margin, at under one-half the earlier expectations.

Thus, the exploitation of the UK's North Sea gas reserves is now running at an all-time high (with a 1995 production of some 70 Bcm). This is similarly now the case with oil, for reasons which will be examined below. Oil and gas output in 1995 together totalled over 200 million tons of oil equivalent with 65% as oil and 35% as gas. Cumulatively, total oil and gas production over the period since the beginning of oil output in 1976 has now reached about 2750 m.t.o.e.(=20 billion barrels of oil). If one places this in the perspective of Britain's total energy production over this period, it compares with a much lower coal output of about 1150 m.t.o.e. and with a primary electricity supply (hydro and nuclear) of only 120 m.t.o.e. (measured in terms of the heat value of the electricity produced).

Oil and gas' massive development has, moreover, been achieved in spite of large government tax-takes from production revenues for much of the 20 year period. By contrast, there have been continuing subsidies for the competing energy sources, coal and nuclear power. In addition, there have, from time to time, been regulatory constraints or end-user taxes on the consumption of natural gas and oil as a means of inhibiting their use in power generation, thus further protecting coal and nuclear power's markets (Manners, 1981).

The dominance which has been achieved by indigenous oil and gas in the country's energy supply over the past 20 years is self-evident from these data: and the degree of dominance is still increasing. 1996's 200+ m.t.o.e. of oil and natural gas production represents about 85% of total UK energy output and it is also almost equal to the country's total energy consumption.

The Prospects

a. The Reserves Question

Prospects for North Sea oil and gas depend primarily on the continuing year by year success of the industry in replacing the reserves which are produced. The publication each year (BP Annual Review) of proven reserves figures for UK oil and gas appears to indicate relatively near-future exhaustion (viz. the 1995 Review shows a minuscule 4.8 year reserves to production ratio for oil and only a somewhat better 9.6 ratio for natural gas). In spite of the highly constrained definition of the proven reserves and the statistically inappropriate arithmetic summation, rather than the *monte carlo* addition, of the 90% probable reserves of individual fields that are reported (initially in DTI's annual Energy Report), many analysts of the UK's oil and gas prospects use the figures to question the ability of the industry to continue to produce at existing rates for more than another few years (Rutledge and Wright, 1995).

The prominence given to such reserves declarations and the apparently very short life of the production process which is derived from them is unfortunate, given the near zero significance which they have in determining the continuity of North Sea production by the oil and gas industry. Neither, for that matter, do the estimates of additional probable and possible discovered recoverable reserves (D.T.I. Annual Report) have much greater significance in assessing the longevity of oil and gas production. These give a range of what is presented as a "low" and a "high" for "total oil and gas reserves remaining". For oil this range is currently reported as 1475–2075 million tons and for gas, 1515–1915 Bcm (= 1350 to 1700 m.t.o.e.) Taken together, this suggests a maximum possible future supply of offshore hydrocarbons of 3800 m.t.o.e.: or roughly 19 years at the 1995 rate of extraction of about 200 million tons. It thus appears to indicate an inevitable decline from the plateau peak production rate within no more than about 8 years.

This view of the UK's offshore oil and gas prospects reflects only present day knowledge on fields found-to-date in the context of current technology. It makes no allowance for the evolution of either knowledge or technology. But, as Adelman has pointed out, "ultimate reserves are unknown, unknowable and unimportant" so that declarations of what they might seem to be at any particular moment in time are not necessarily helpful in looking at future production potential (Adelman, 1994). What really matters in an examination of the prospects of the future production performance of an oil and gas province is the ability and the success of the industry in continually replacing reserves that are used. As long as such replacement is achieved, production can at least remain stable (at or close to present levels). It can, of course, increase above contemporary levels, providing the identified reserves' levels grow appropriately and the politico/economic conditions in which exploitation is taking place do not change significantly for the worse.

Looking back over the last 20 years' history of UK North Sea reserves' developments (DTI Energy Report 1995), one notes only five years in which the oil used has not been fully replaced by additional reserves. Three of these years were in the early 1980s when, as shown above, exploration and appraisal was undermined by the much higher taxes imposed by successive governments as well as by the government's 1981 declaration of its intention to impose production constraints because of "scarcity" fears. The other two years were 1986 and 1987 when the industry was striving to adjust its activities to the net revenue consequences of the international oil price collapse. Thus, all occurrences of a failure to renew reserves by at least the volume of a year's use were at times of adverse politico-economic circumstances for the industry. Since 1988 oil reserves additions each year have exceeded use.

For natural gas reserves' development the history is very similar. There have been only four years in the past twenty when reserves added have not exceeded reserves used, viz. 1978, 1982, 1987 and 1988. The former two were, as with oil, years when government policies (on tax rates and/or depletion) were considered to be disadvantageous by the companies concerned. The other two years were at the end of the period during which British Gas continued to enjoy its *de facto* monopolistic role as the gas buyer. It was prepared to pay only low prices for the gas and thus inhibited companies' interests in the development of more reserves.

Since 1988, in the case of oil, and 1989 for natural gas, there has been no break in the run of annual successes of the industry in adding reserves in excess of those used in each year. This, moreover, has been achieved in spite of the rising levels of production of both commodities over the six or seven years involved. In other words, the concern by governments and others (Rutledge and Wright, 1995) for the premature diminution of reserves available for production and use in the future, should the current level of production not be curtailed, can be shown to be groundless. Reserves' creation is, in reality, a function of demand as it is the latter from which revenues are generated; and it is from the revenues that the investments required to ensure the recoverability of more reserves in an oil and gas province are secured. This remains true, providing only that the province has not already been extensively and intensively explored from the shallowest potential reservoir formation down to the basement rock in all areas of potential hydrocarbons' occurrence (Brooks and Glennie, 1987). Except over very limited areas of the North Sea, such a totality of the ultimate exploration effort remains far from being achieved. Thus, as long as the politico-economic conditions remain acceptable, North Sea exploration will remain worthwhile, discoveries will continue to be made, additions to recoverable reserves will be defined and, in the light of the perceived demands for oil and gas in the relevant market areas, new fields' developments will continue to be financed (Kemp and Rose, 1993; UKOOA, 1994).

It requires a continuing sequence of several years during which there are no significant politico-economic changes and in which efforts to find reserves are relatively unsuccessful, so that reserves additions' fail to match production, before the issue of reducing or even terminating, the exploration efforts needs to be faced. If this were to happen, then fears for relatively near-term future levels of production would be justified. For the UK sector of the North Sea this eventuality currently seems to be at least five years in the future – and it could be fifty: depending on the industry and the results of exploration efforts in the meantime. Only if and when this eventuality occurs could policy makers at last be justified in seeking to regulate rates of production; so as to "save" some of the then known ultimately recoverable resources of the UK continental shelf for serving either national interests in the longer term future – or in the interests of inter-generational equity.

b. The Required Politico-Economic Conditions

The smooth, continuing and high-profile processes of discovering and extending reserves as a function of demand over the long-term, requires, as indicated above, the creation and maintenance of national politico-economic conditions which are basically attractive in the international arena in which the oil and gas industries operate. Such attractive conditions for the UK North Sea, in respect of both national and international considerations, have been in place for the last five or six years. In terms of national considerations, the post-1987 reforms of the oil tax system in favour of the companies, coupled with the industry's success in reducing costs by an estimated average of over 30% (UKOOA, 1994), has increased the post-tax return on investment in UK offshore operation to one of the highest in the world (Kemp and Rose, 1993). Given the physical attributes of the UK North Sea oil and gas occurrences, in terms of the success rate for exploration drilling and the relatively large average size of the reservoirs discovered, then the high returns are also accompanied by low risk – compared with the situations in alternative locations in which the companies could invest their funds. And, as always, the political stability of the British system (compared with elsewhere in the world) and the nearby-located and virtually guaranteed European markets for all the oil that can be produced, add yet other favourable components to the country's North Sea prospects.

Even gas supply developments, for so long constrained by limited markets, have now been broken open by competition in the UK itself (Stoppard 1994). There is, moreover, no longer any significant "threat" to additional UK North Sea supplies from new Norwegian imports;[2] while the projected gas line "interconnector" from Bacton to Zeebrugge, to take British gas into mainland European markets, currently seems more likely than not to go ahead. This will give market access for UK gas in relatively high gas-cost countries (Odell, 1995a). Some such customers in The Netherlands and Germany, are, indeed, already "lining up" to secure access to UK gas exports, in the expectation of buying at prices lower than those charged by their monopoly suppliers. With UK natural gas supplies in surplus for the short-term, from take-or-pay contract gas which is instantly available; for the medium-term, from already discovered and partly developed gasfields, the production of which can be enhanced over a two-to-three period; and for the longer-term, from discovered, but so far undeveloped, gasfields which can eventually be brought into production), the prospect for a UK gas output of 90 Bcm by 2000 and of 115 Bcm by 2025 is a reasonable prospective (Odell, 1995b).

Meanwhile, oil output will, over the next year or two, be maintained at its present +125 million tons per year level. Thereafter, the continuing depletion of the thirteen largest fields discovered in the 1970s will soon impose the difficult task for the industry of fully replacing those fields' much diminished future output (compared with their earlier plateau levels of production). For the UK's North Sea sector on its own, holding output above 100 million tons per year will necessitate increased investments in many, relatively very much smaller, newly-discovered fields. The sustainability of annual production above this level will necessarily depend on the successful exploitation of the West Shetland province on a growing scale. Given the still very recent start to the exploitation process in this area, the upward slope of its development curve is still highly uncertain, but it seems likely at least to be able to extend the plateau production rate of 100+ million tons per year well into the first decade of the 21st century. By then, however, natural gas seems most likely to be the more important component in the country's total hydrocarbons production (Stoppard, 1994; Odell, 1995b), so that total UK hydrocarbons output will remain ahead of the country's energy demand. The UK will thus continue to be a net exporter of hydrocarbons beyond the 50th anniversary of the first gas production

2 A proposal in 1994 by a group of large gas users – notably private electricity generators – to import gas directly from Norway was rejected by the UK government.

from the North Sea in late 1965.

European and some global considerations will, however, continue to form an essential part of the politico-economic backdrop to the evolution of the UK's offshore hydrocarbons industry. There are, however, only two specific considerations which are of central importance to the future of Britain's oil and gas industries themselves (as opposed to more general economic and political factors which will influence the prospects for the country's economy overall, including oil and gas). The first of these is one which affects the prospects for both oil and gas, viz. the evolution of the price of international oil. The potential evolution of supply, as indicated in the previous section, for the UK's oil and gas output depends on the maintenance (in real terms) of the current international price of oil – at some $17 to $18 per barrel – through to 2010; and on a real increase in the price beyond that date, so that there will then still be scope for profitable investment in higher-cost North Sea and west of Shetland oil and gas. Anything more than a modest and temporary fall in the international oil price would reduce the industry's propensity to invest in the North Sea and so lead to less additional production than predicated in this paper. In more extreme low-price circumstances, even a decline rate in output could occur at an early date. Note that this would apply as much to gas as to oil production prospects by virtue of the linkages between the prices of the two commodities throughout most of Europe (Odell, 1995b).

The second central consideration relates solely to gas, viz. the evolution of the gas market in Europe, either by actions or inaction which serve to restrict access for UK gas (as for example, through the maintenance of the transmission and distribution monopolies in most mainland European countries, if EU policies are not applied): or from the possibility that external gas supplies from outside Europe, notably from countries of the former Soviet Union and/or Algeria, will flood the markets and bring border prices down to under $2 per Mmbtu (approximately 13.5 pence per them). Such supply-side developments would undermine the viability of the potential for some of the British gas exports which are foreseen above as an essential component for the growth of UK gas production to 90 Bcm by 2000 and 115 Bcm by 2025.

Neither of these developments is impossible: but the joint probability for their occurrence is under 25%. Europe could, however, for security of supply reasons, choose to protect the prospective levels of its indigenous hydrocarbons' production from the impact on a collapsed international oil price and/or depressed gas price and thus establish a means of ensuring that the UK's production reaches the levels indicated above. The UK, however, currently seems unlikely to agree to this – even as part of a deal, for example, over other matters of dispute between the UK and the other member countries of the EU. Such a UK involvement would seem to be contrary to the principles on which present British policy in international economic affairs in general, and to international trade issues in particular, is based.

Potential protection for British oil and gas within a European Union market can thus, for the time being at least, be excluded from any realistic appraisal of the prospects for the North Sea. Longer term, however, such protection may well become welcome, if not essential: particularly if it proves to be British gas exports to the mainland of Europe which provide the catalyst for the emergence there of an intensely competitive market. In this development, potentially very large supplies of gas from Russia, other former Soviet republics, Algeria and elsewhere would not only threaten the viability of the UK's offshore oil and gas industry, but also raise the question as to the wisdom of having built the inter-connector in the first place!

References

Adelman, M.A., 1994, *The Economics of Petroleum Supply*, MIT Press, Cambridge

BP Annual, *Statistical Review of World Energy*, London

BP Policy Review Unit, 1979, *Oil Crisis... Again?*, London

Brooks, J. and Glennie, K.W. (eds), 1987, *Petroleum Geology of North West Europe*, Graham and Trotman, London

Department of Energy, 1979, *Energy Policy Green Paper; Energy Policy – a Consultative Document*, HMSO, London

Department of Energy, 1981, *Development of the Oil and Gas Resources of the United Kingdom*, HMSO, London, Appendix 9

Department of Trade and Industry, Annual *The Energy Report, Oil and Gas Resources of the United Kingdom*, HMSO, London

Estrada, J. et al, 1988, *Natural gas in Europe Markets Organisation and Politics*, Pinter Publishers, London

Kemp, A. and Rose, D., 1993, "Fiscal Aspects of Investment Opportunities in the UKCS and Other Countries" *North Sea Study Occasional Paper, No. 43*, University of Aberdeen, Aberdeen

Manners, G., 1981, *Coal in Britain: an Uncertain Future*, G. Allen and Unwin, Hemel Hempstead.

Odell P.R., 1978, "Constraints on the Development of Western Europe's Natural Gas Producing Potential in the 1980s", *United Nations Economic*

Commission for Europe Symposium on Natural Gas in the E.C.E. Region, Evian.

Odell P.R. 1979, *British Oil Policy: a Radical Alternative*, Kogan Page, London

Odell P.R. 1995a, "European Gas Markets to 2020" in *Die Energiemarkte Deutschlands im Zuzammenwachsenden Europa – Perspektiven bis zum Jahr 2020*, Bundesministeriums fur Wirtscahft, Bonn

Odell P.R. 1995b, "The Cost of Longer Run Gas Supply to Europe" in the proceedings of *European Natural Gas Economics in a Global Context*, SNS Energy Day, 1995, Stockholm

Odell P.R. and Rosing, K.E. 1975, *The North Sea Oil Province*, Kogan Page, London

Odell P.R. and Rosing, K.E. 1976, *Optimal Development of the North Sea Oilfields*, Kogan Page, London

Odell P.R. Rosing K.E. and Beke-Vogelaar, H., 'Optimising the Oil Pipeline System in the UK Sector of the North Sea' *Energy Policy*, Vol.4, No.1

Odell P.R. and Rosing, K.E. 1980, *The Future of Oil*, Kogan Page, London

Posner, M.V., 1973, *Fuel Policy: a Study in Applied Economics*, Macmillan, London

Rutledge, I. And Wright, P., 1995, "Taxing the Second North Sea Oil Boom: a Fair Deal or a Raw Deal?, *B.I.E.E. Energy Economics Conference*, Warwick, December 1995.

Stern, J., 1984, *International Gas Trade in Europe: the Policies of Exporting and Importing Countries*, Gower, Aldershot

Stoppard, M., 1994, *The Resurgence of UK Gas Production*, O.I.E.S., Oxford

UKOOA, 1994, *Cost Reduction Initiative for the New Era*, Institute of Petroleum, London.

Chapter II – 9

Four Decades of Groningen Gas Production and Pricing Policies*

Introduction

The Groningen gas field has been an extraordinary phenomenon since its discovery in 1959. This is true, first, in terms of its status as a super-giant gasfield and, second, in terms of its mode and pattern of exploitation. Our concern in this paper is with the politico-economic aspects of the latter, though it also involves the issue of the public presentation of the former, with particular respect to the size and productivity of the field and of the costs of exploiting it.

Groningen's Past

Three years after the discovery of the large, but then modestly defined, Groningen gas field in 1959, the Dutch Minister of Economic Affairs, set out the main principles proposed by the government for Dutch gas policy in the *Nota inzake het aardgas* (Kamerstukken II, 1961–1962, nr. 6767). First, in order to generate maximum revenues for the state and the holder of the concession, the *Nederlandse Aardolie Maatschappij* (NAM, a 50/50 joint-venture of Shell and Exxon), the Minister indicated that the "market-value" principle should be used as the basis on which the gas would be marketed. Thus,

*** A revised version of a paper jointly authored by Dr. A.F. Correljé of Erasmus University Rotterdam, presented at the May, 1999 conference, "Groningen: catalyst for the north-west European Gas Industry" and subsequently published in the *Netherlands Journal of Geosciences*, Vol.80, No.1, April 2001, pp.137–144. Dr Correljé's permission to publish the paper in this book is gratefully acknowledged.**

the price at which gas was to be sold to various types of consumers was linked to the price of the alternative fuel most likely to be substituted, viz. gas oil for small-scale users and fuel oil for large-scale users[1]. Accordingly, consumers would never have to pay *more* for gas than for alternative fuels, but the market value principle also ensured that they would not pay *less*. The application of this principle enabled the concession holders, Shell and Exxon, and the Dutch state to secure much higher revenues than they would have achieved from a pricing process in which the consumer price was related to the low production costs of gas from the Groningen field. An essential precondition for maintaining the 'market value' principle was, of course, that no alternative supplies of low priced gas could reach the market – a condition which was fulfilled until the early 1970s in Europe (Odell 1969; 1973) and,until recently, in the Netherlands itself.

Secondly, the *Nota* stated that the exploitation of the Dutch gas resources should proceed "in harmony" with the level of sales of gas achieved, in order to avoid disruptions of the energy market. Control over the supply of gas was thus seen as a government responsibility. Yet, the *Nota* also stated that the exploitation and marketing of the gas reserves should be undertaken by the private concession owners, Shell and Exxon, in order to enable the country speedily to benefit from their knowledge, experience and financial resources. In 1963, the Dutch government and the two companies agreed upon an organisational structure on which these principles could be implemented (see Figure II-9.1).

Thus, while direct state ownership/control of Groningen gas was avoided, its influence over the financial flows emerging the gas production was assured. The management of the state's interest by DSM established, moreover, an effective *arm's length* relationship between the public and the private interests[2]. The state's revenues were secured in a number of ways; first, through corporate taxes (48%) on the profits of the Maatschap, Gasunie and DSM; secondly, through an additional 10% government surcharge on the profits of the Maatschap, and thirdly, through the dividends and the 'state profit share' paid to the state by, respectively, Gasunie and DSM. From the early 1970s onwards, a 'state profit share' was also applied on the profits of the Maatschap (see

1 These principles had been suggested by Exxon, drawing upon its experience with the exploitation of gas reserves in the US (see Correljé 1998, Heren 1999).

2 See Peebles (1980, 1999), Stern (1984), Kielich (1988), Kort (1991), Ausems (1996) and Correljé (1998) for detailed accounts of the development of the institutional structure and the government's policy.

Wieleman 1982). The Ministry of Economic Affairs also formally confined its responsibilities for approving decisions taken by DSM and Gasunie in respect of prices, production, national and international trade volume and the construction of transport and storage facilities. In practice, it was always consulted on strategic issues and could initiate

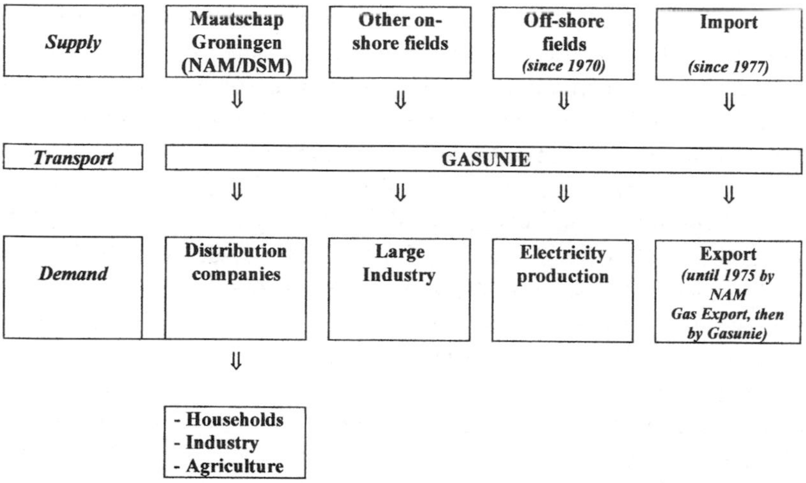

Figure II-9.1: The structure of Dutch gas sector

Note: Gasunie was also given the right of first refusal *to purchase all gas produced by other on-shore fields (since 1962) and off-shore Dutch gas fields (since 1970). These fields were exploited under regimes that differed from that for the Groningen field, but their production became part of Gasunie's monopolistic gas supply system (see Correljé 1998).*

- *The holder of the Groningen concession, the* Nederlandse Aardolie Maatschappij BV *(NAM), a 50/50 joint venture of Shell and Exxon, was exclusively to undertake the production activities.*
- *The state, via the Staatsmijnen (later Dutch State Mines or DSM), was to participate in the costs of exploiting gas from the Groningen field and in the flow of revenues through a financing partnership, known as the* Maatschap *(40% DSM, 60% NAM).*
- *Gasunie was established as a joint venture owned by the Dutch State Mines (DSM) (40%), the Dutch State directly (10%) and Exxon (25%) and Shell (25%). Gasunie was given the exclusive responsibility to co-ordinate the commercialization of the Dutch natural gas resources on behalf of the State and the concession-holder, NAM. NAM Gas Export was to be responsible for the sales of gas abroad.*
- *In 1972, in response to the increasing number of participants in the industry a separate entity was established:* DSM Aardgas BV. *In 1989,* Energie Beheer Nederland BV *(EBN) replaced DSM Aardgas when DSM was partly privatized. EBN has remained a part of DSM, surrounded by a so-called Chinese Wall. The state pays DSM a management fee.*

discussions for any changes it thought to be necessary in the national interest[3].

Given Groningen's discovery within a unified concession (except, of course, for the small part of the field underlying German territory) operated by a single company (NAM) plus a state interest, the arrangements agreed and implemented created a near-perfect monopoly, especially in the context of the absence at the time of its early development of any significant alternative sources of gas within a radius of a thousand kilometres (Odell 1969). Its creation as a monopoly was thus never in question and this, in combination with the state-supported position of the NAM in the Netherlands, established the framework for the ordered and profitable supply side of its exploitation (Odell 1969). On the demand side, the conditions for market developments were equally propitious. Groningen was located close to the centre of one of the world's most energy intensive using regions within which the two most important suppliers of hydrocarbons were the same two companies, viz. Shell and Exxon, which had discovered the field.

Under these conditions, the existence of a set of national or regional monopolistic transporters and another set of monopsonistic distributors and marketers of the potential gas supply from Groningen was axiomatic. Monopoly plus monopsony was thus imposed and implemented on the emerging European natural gas industry. This was achieved in a systematically planned way – as shown in Figure II-1.2 (p.139). Within this designated area, the price of the gas offered (the volume of which was initially restricted to well below the competitive optimum level) was related to the opportunity costs of exploiting the markets[4]. At the time of gas' extensive introduction to the markets (see Fig. II-1.2), Europe consisted of a set of relatively high-priced energy markets arising from the combination of highly protected indigenous coal and a less than fully competitive supply of oil products (Odell 1975).

Thus, the net-back value of the Groningen gas marketed was a multiple of the very low long-run supply price of the commodity; leading to consequential super-normal profits for the producers and generous revenues for the state. Energy consumers were, of course, the losers, but this was largely unrecognised by them in the context of the ordered system imposed. They were, in any case, as shown above, very used to high energy prices (Odell 1975) so that gas could, in effect, be offered on a take-it or leave-it basis as a so-called premium fuel at so-

3 See Correljé (1998) for an overview of the adjustments made to the Dutch gas regime.

4 As was also stated by the Head of the Gasunie Dept. of Marketing Research, (Portegies, 1969).

called "competitive" prices, in a market in which no gas at lower prices was available. Monopoly power was effectively exercised (Odell 1969).

Monopolies, however, are inherently fragile and, as shown in Figure II-1.4 (p.143), alternative suppliers emerged relatively quickly on to the European gas scene. In the case of the UK, with its newly found offshore gas fields, this happened even before the planned Dutch exports could be implemented[5]. Elsewhere on the mainland of Europe, initial imports of pipelined Soviet gas and of North African LNG were made available at locations (see Fig. II-1.4) where prices could be competitive with those for Groningen gas. Such competitive forces were, however, soon restrained by the much enhanced opportunities for marketing gas which were created by an important external development, viz. the oil price shock of 1973/4. Gas consumers' hopes that gas would be offered as an alternative lower-priced energy source were quickly thwarted as the steeply upward sloping demand curve – in the context of fears over the security of oil supplies – led to a tightening of the oil price/gas price links, so enabling gas producers' profits and host-governments' revenues to be boosted (Odell 1979). The ordered system originally established to market Groningen gas was thus made more extensive and intensive across much of the continent – to the great advantage not only of the various suppliers now acting oligopolistically, but also to that of regional monopsonistic distributors which were able to secure high sales margins.

But not even these structured restraints on the pace of gas developments satisfied the Dutch gas supply industry. Another factor was now introduced to curb demand growth and to add to upstream profits, viz. the "declaration" of gas to be a scarce commodity (Odell 1973). This eminently and demonstrably false vision of the future of gas availabilities (even at the time, let alone in retrospect) led to actions – both at various national and the European levels – to constrain gas demand, while Dutch policy deliberately sought to constrain supply (Odell 1984).

Whilst demand growth constraint was successfully achieved post-1976 – as can be clearly seen from the data in Table II-9.1 – gas suppliers, except for the Dutch, sought to expand their markets. This involved

5 This sequence of events, whereby the planned Dutch gas exports to the UK market in 1965 were excluded, led to the physical separation of the British and mainland European systems for 30 years. It also meant that the nature of the British gas supply system developed in a completely different way from that generated elsewhere in Europe by the Groningen phenomenon. Had events been otherwise, the privatisation and liberalisation of the British market may not have occurred and thus not have had the strong demonstration effect on reorganisation of the gas industry in the rest of Europe, as now in progress.

much enhanced Soviet exports – with Ruhrgas and SNAM as the main facilitators and beneficiaries; increased Norwegian investments in both fields' development (notably the new super-giant field, Troll) and in gas delivery systems to the mainland of Europe; the expansion of Algerian export potential through the Trans-Mediterranean pipeline to Italy and as LNG to market areas hitherto entirely dependent on Dutch gas (most notably Belgium); and, finally, the belated decision by the UK to interconnect its hitherto isolated gas system with that of mainland Europe (Odell 1988).

These supply-side reactions to the Dutch decisions to limit production as a matter of policy served to undermine not only the dominance in the European gas system previously enjoyed by Groningen gas, but also the role of NAM/Gasunie as the leading actors in the gas supply system. In terms of the Dutch contribution to mainland Western Europe's demand (outside the Netherlands itself), the decline was steep and continuing – as shown in Table II-9.2 – from almost 54% in 1971, to 36% in 1981, to under 19% in 1991 and to only 16% by 1998. Groningen, in particular, in effect a supplier of last resort (given the preference extended by the Dutch government to the exploitation of its small gas fields), was even more significantly affected: so much so, in fact, that it can be described as having lost its role as the principal catalyst in the European gas system. In spite of its ultra-low costs of production, it became (as with Saudi Arabia, the lowest cost oil producer in the

TABLE II-9.1
Natural gas demand in Western Europe, 1961–1998

Year	No. of Countries Using Gas	Total Use m^3x10^9	Total Use % increase	Gas Use as % of Energy Use
1961	5	16	–	1.8
1966	7	25	77.8%	3.3
1971	8	106	416.0%	9.7
1976	10	183	72.6%	13.4
1981	11	200	9.3%	14.7
1986	15	218	9.0%	15.2
1991	15	270	23.9%	19.4
1996	17	344	27.4%	21.5
1998	18	356	3.5%	22.0

Source: BP, Review of the World Gas, *various years to date.*

international oil system) the swing producer in the market, responding to, rather than initiating, changes in the European gas system.

Eventually, though already much too late to restore the *status quo ante* for Dutch gas supplies to the market, the fallacy of gas scarcity was finally discarded. As shown in Figure II-9.2, the cumulative use of Groningen gas has only slowly drawn down the volumes of declared remaining recoverable reserves. This is, in part, because of the factors noted above, but it has also been exacerbated by the guarantees of volume and price extended to gas production from Dutch fields other than Groningen, in a policy which, yet again, was deliberately sub-optimal in economic terms (Correljé 1998).

Thus, the recent attempts to revive the flagging fortunes of Dutch gas – through belated offers of additional gas to foreign customers – have achieved only modest success. A recovery from the 1991 low level of exports was temporarily secured (see Table II-9.2), but the Dutch share of the relevant European market has – after a temporary resurgence to about 46 bcm in 1996 – since again fallen to the level of the mid-1980s. The national proven reserves to production ratio, meanwhile, remains at a generous 28 years, while technological advances and new investments (Schweppe 1999) appear to remain capable of extending the productive life of the Groningen field beyond the end of the first quarter of the 21st century: and to a total ultimate recovery of upwards of 3000×10^9 m^3, compared with the reserves declaration in 1964, the first year of

TABLE II-9.2
Dutch gas exports to mainland Western European markets, 1971–1998

Year	Dutch Gas Exports ($m^3 \times 10^9$)	Demand in the Markets ($m^3 \times 10^9$)	% of Dutch Gas to the Markets
1971	25.2	46.7	54.0
1976	39.4	93.8	42.0
1981	43.0	118.0	36.4
1986	35.3	131.7	26.8
1991	30.9	166.3	18.6
1996	45.8	212.1	21.6
1997	40.1	210.3	19.1
1998	36.4	225.2	16.2

Source: BP (op.cit – see Table II-9.1)

production, of only some 1100x10⁹ m³ (see Figure II-9.2). It was perhaps the eventual recognition by the mid-1990s of the economically sub-optimal results from the country's exploitation of its gas resources – and the irrational combination of a limited market development, with the simultaneous existence of a resource "overhang" (Odell 1995 and 1997) – that really motivated the fundamental reappraisal of the organisation and structure of the Dutch gas industry then undertaken, rather than merely the threat of EU pressures for the eventual liberation of the sector.

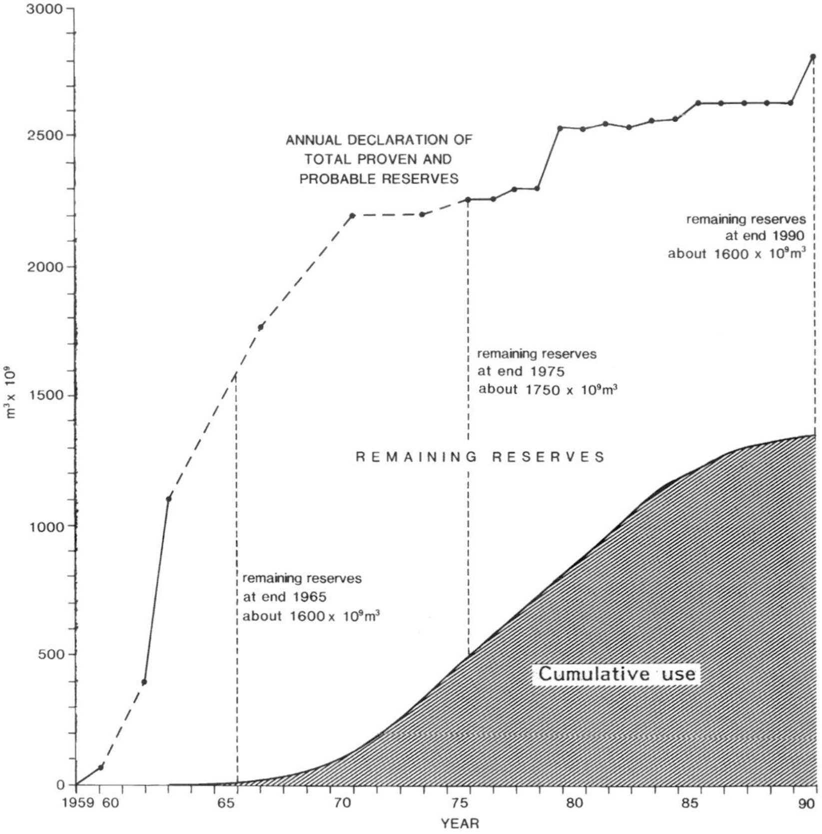

Figure II-9.2: The Groningen Gasfield: Trends in its Cumulative Production and the expansion of its Ultimately Recoverable Reserves from 1960 to date

Source: Odell (1991). Revised data from MEZ (various years); Brouwer and Coenen (1968).

Liberalisation

Much to the surprise of most of the European energy establishment in 1995, the Dutch Government's *Third White Paper on Energy* (MEZ 1995) contemplated a fundamental change to the traditional organisation and the operation of the Dutch gas industry. The 1998 draft Gas Law (MEZ, 1998a) based on that report provided that customers would have free choice regarding their gas supplier(s). Large consumers, accounting for around 46% of Gasunie's home market sales, were explicitly to be allowed to seek alternative suppliers immediately. In 2002, medium sized users, representing 16% of the market, are to follow. Eventually, in 2007, it is suggested that the small users will be allowed to shop around freely.

Gasunie's right of first refusal for all Dutch gas produced had already been withdrawn by the government in 1994, when the EU Hydrocarbons Directive was accepted (RL 94/22/EG, PB 1994, L 164). Under the proposed new Gas Law, other suppliers and traders have also been given the right of negotiated access to the transport and distribution systems. Gasunie and the distribution companies are all required, moreover, to establish *Chinese walls* between their trade and transport activities and to publish separate indicative prices for the services provided.

The basic structure of the industry, with a key role for Gasunie and De Maatschap/NAM – including the cross shareholdings – is, however, maintained. This, it is argued, is because it provides scale and organisational advantages for the most important parts of the Dutch gas industry and it also makes possible the continued co-ordination of gas sales and purchases from the Groningen field, on the one hand, and the small fields, on the other (MEZ 1998a, pp. 18–22). This raises the question whether and how the changes proposed for the trade regime can be reconciled with the core elements of current Dutch policy, viz., first, the maintenance of short term (daily), as well as longer term security of supply through the small fields policy, the balancing function of the Groningen field and the national depletion policy; second, the pricing of gas according to the principle of opportunity costs (or the 'market value' principle); and, third, the maintenance of state and private revenues (the *aardgasbaten*) (see Algemene Rekenkamer 1999, p. 7).

To appreciate the consequences of the proposals for the new Gas Law, it is essential to take note of the institutional and financial framework of the Dutch gas industry. Gasunie is obliged by statute to 'pass on' most of its sales revenues from sales (amounting to Dfl. 16,102.4 million in 1998) to its suppliers, viz. NAM and the other gas producers. Its operations' costs are less than 1 Dutch cent per cubic meter and its shareholders receive a fixed annual dividend of only Dfl 80 million. In 1998, Gasunie indicated its total sales were 79.8 Bcm (84 Bcm in 1997), of which about 4.3 Bcm (4.2 in 1997) were imported, 22.7 Bcm (29.4 in 1997) were purchased from Groningen, while the remaining 52.8 Bcm (50.4 in 1997) came from the smaller Dutch gas fields (Gasunie 1997 and 1998)[6]. Gasunie is obliged by government decision to give priority in its gas purchases to buying gas produced by the Netherlands' smaller fields, in relatively constant volumes and at an 'adequate' price, as determined by negotiation; and with the approval of the Ministry of Economic Affairs. The daily, seasonal and annual balances of requirements are met by purchases from the Groningen field and/or, more recently, by withdrawal from underground storage facilities. This necessarily means that a fall in Gasunie's overall volume of sales produces a near-instantaneous decline in production of Groningen field gas (Correljé, 1998, pp. 145–158).

As the production costs of Groningen gas are estimated at under 1 Dutch cent per cubic meter, while the costs of production at the larger off-shore and the on-shore fields are about 7 and 4 cents, respectively (Kunneke et al 1998), this implies that both the profits of the NAM partners and the revenues for the state on Groningen gas production are much larger than those on production elsewhere. Thus, the parties involved have an enormous commercial interest in maintaining output at Groningen. This would be true even if the unit price paid by Gasunie for all its gas purchases were the same, but in fact, the price paid per cubic meter by Gasunie to the Maatschap Groningen is higher than that paid for gas from other fields[7]. This is so because the price which Gasunie pays for Groningen gas (owned by the same parties as the Maatschap) is the item whereby Gasunie's books can be balanced so as to show its statutory required annual profit of Dfl 80 million. Under these conditions, the central element in Gasunie's strategy must be to protect its sales volume at as high a level possible above the level of gas purchases from the other fields. As a consequence, the declared overall Gasunie volumetric objective for its sales has been increased to around 80 Bcm annually right through to 2010 (MEZ 1995, p. 141).

6 These data are volumetric quantities, unadjusted, it seems, for varying calorific values of the gas purchased and sold (notably between low calorific Groningen gas and high calorific gas from most other fields).

7 This is derived from calculations over the period 1985–1993, based on Gasunie's reported costs of purchased gas (Annual Reports Gasunie) and the volumes purchased from the several sources; the 'Typical Gasunie Gas Purchase Price to Producers' (NOGEPA 1993); and the costs of imported gas (CBS Statistics of Foreign Trade).

Nevertheless, under the country's liberalization regime for gas the first supply contracts to Dutch consumers without Gasunie's involvement have been signed and become operational. Trading affiliates of Dutch distribution companies are thus importing UK gas, through the UK\Zeebrugge interconnector line and thence through their self-owned *Zebra* pipeline to the Netherlands, rather than transporting it through the Gasunie system. Gasunie anticipates a loss of around 20% of its sales over the mid-term, but a former Minister of Economic Affairs, Hans Wijers, indicated his confidence that Gasunie would remain able to underbid alternative suppliers of gas so as to keep a higher percentage of future Dutch markets. This is unrealistically optimistic, given that Gasunie has already lost about 10% of its hitherto anticipated sales for 1999[8] and given certain prospects for increasing competition from additional new suppliers in the near future years. It is also, as shown above (see Table II-9.2), having to cope with highly competitive markets abroad (see also Peebles, 1999).

These developments seem likely to necessitate a major revision of Gasunie's marketing strategy in which compensation for falling sales and revenues in the Dutch market itself is sought through an increase in Dutch gas exports. Indeed, the Dutch home market for gas is, by and large, already largely saturated and it may even decline in the near and medium term as prices continue to be increased by additional ecotaxes[9]. Thus, it is only the expected expansion of gas demand elsewhere in Europe that can provide Gasunie with an opportunity at least to stabilise its overall sales volume. To date, however, Dutch gas exports remain statutorily restricted by the requirement to maintain sufficient proven reserves of gas in the Dutch fields to satisfy the country's anticipated consumption over a 'rolling' 25 years' period (MEZ, 1995, p. 133; MEZ 1997, p. 18; Algemene Energieraad 1998, p. 10; Algemene Rekenkamer 1999, pp. 20–22).

Gradually, however, the fact that this reserves' requirement is excessive is at last becoming more widely accepted, as we have long argued (see Odell 1973, 1979, 1984, 1987, 1993 and 1997). Even Gasunie has finally recognized the illogicality of such a restraint on production[10]. The proposed new Gas Law states that Gasunie can no longer be held

8 As reported in the Dutch media, viz. *De Volkskrant* 8/5/1996; *Gas*, Juni 1997, 13; *NRC-Handelsblad*, 5/1/1999; *Het Financiele Dagblad*, 3/11/1998.

9 Steep increases in such taxes for domestic gas users were imposed on January 1, 1999 and are to be further accentuated on January 1, 2000 when the ecotax will rise by 26% from 19 to 24 Dutch cents per cubic metre. Thereafter, taxes (including VAT) will constitute 40% of the consumer price (European Energy Pricewatch, August 1999; *NRC Handelsblad Weekeditie*, 5 October 1999).

10 See the comments of Mr. G. Verberg, the managing director of Gasunie, as reported in *Het Financiele Dagblad* 3/11/1998.

responsible for the *long-term security of supply for importing consumers*. The Law is also even more explicit in the statement that "Dutch gas cannot be reserved primarily for Dutch consumption" (MEZ 1998a, pp. 20). Hardly surprisingly, this revised view has produced an adverse reaction from the Dutch public which has been brought up for more than 25 years with the idea that Groningen gas is a scarce asset and that it should be reserved primarily to Dutch consumers (Odell 1987; Witteveen 1998; Algemene Rekenkamer 1999).

Notwithstanding this background, an enhanced supply position for Gasunie in non-Dutch markets can be seen and should be accepted as an important component in the desirable further penetration of gas in the European energy market, given its associated advantages on environmental grounds and for enhanced efficiency in energy use. Such a change in policy can perhaps be sold on these grounds to the environmentally-conscious Dutch parliament and public. Flexible supplies of additional Dutch gas should, indeed, also stimulate the creation of additional markets in Europe for long distance, high load-factor gas from Norway, Russia and elsewhere[11]. The large underground gas storage facilities, with a total capacity of 35.1 Bcm, now being created by Gasunie/NAM in depleted gas and oil reservoirs in the Netherlands – at Alkmaar (jointly with BP/Amoco), Langelo and Grijpskerk, together with the highly flexible production regime which is technically possible from the expensively up-graded facilities which are being installed on the Groningen field, will be important instruments for achieving the objective of increased Dutch exports (Energie Nederland 8/12/1998).

Conclusion

We thus conclude that the liberalising steps taken to date by the Dutch government and the changes that can be observed in Gasunie/NAM's strategies, could eventually do justice to the continuing magnificent low-cost supply potential of the Groningen field (see Correljé and Odell 1996). The revision of the present limiting requirement on the volume of long-term supply contracts with foreign customers is essential in this respect[12]. Instead of being restricted to exporting limited volumes of gas at high prices, Gasunie will again have

11 Including supplies by the foreign affiliates of Gasunie's own shareholders, Shell and Exxon, in North-Germany, Norway and the UK.

12 Gas exports that have to be compensated by imports yield relatively moderate profits, because the crucial up-stream profits flow to foreign producers and governments and not to the operators of Dutch fields and the Dutch State.

to be allowed to exploit the wider European market as an unconstrained active and long-term strategic player, as it was until the mid-1970s[13].

The extent and speed of the liberalisation of the Dutch gas industry – which remains an important element in the Dutch economy and which has a significant competitive advantage over other suppliers – whereby competition for the smaller and medium-sized Dutch consumers can be incorporated into the system, will depend on the manner in which the new Gas and Mining Laws (MEZ 1998a and 1998b) and the Dutch competition authority (NMa) choose to enforce third party access rights to transportation, storage and handling facilities. Yet, it is precisely through its control over these storage facilities that Gasunie remains strongly placed – if so allowed – to protect its home market (Energie Nederland 1999). By inhibiting the access of alternative gas suppliers to these essential facilities for maintaining supplies to highly seasonally domestic and small-commercial variable end-users of gas, the greater part of the total Dutch gas market will remain essentially incontestable. If so, then the smaller and medium sized consumers in the Netherlands itself could largely be denied the potential benefits of liberalisation.

Eventually, however, the extension of the benefits of liberalization to these smaller users may turn out to be a political necessity. Indeed, if it is not done, it will be extremely difficult to justify to the population of Dutch residential consumers (which is virtually the same as the total electorate) the newly accepted 'non-scarcity' paradigm and the policy-led expansion of Gasunie's international supply role. The consumers – and voters – have long been given a deliberately fostered perception of the need for – and even the right of – the nation's users to be supplied with indigenous natural gas far into the next century (Odell 1987).

References

Algemene Energieraad (1998), *Liberalisatie van de gassector: Advies aan de Minister van Economische Zaken. Vastgesteld op 15 januari 1998*, Den Haag.

Ausems, A.W.M. (1996), 'Nota de Pous 1962 en het aardgasgebouw', *Handboek Energie en Milieu*, Samsom, Alphen aan den Rijn:.

BP (various years) *Review of World Gas*, London (to 1995), and *Exploration – Global Gas International*, London (1996–1998)

13 Provisions in the recent Energierapport 1999 (MEZ, 1999) imply, however, that the Ministry of Economic Affairs will remain in charge of depletion policy. The current production objective of 80×10^9 m^3 annually will be maintained, of which 25 to 30 milliard m^3 will be taken from the Groningen field. Export prices will have to be market conformist. This suggests that past depletion and export policies will continue virtually unchanged and that Gasunie will thus remain constrained in its commercial policy.

Brouwer, G.C., Coenen, M.J. (1968), *Nederland = Aardgasland*, A. Roelofs van Goor, Amersfoort.

Correljé, A.F. (1998), *Hollands Welvaren: De geschiedenis van een Nederlandse Bodemschat*, TeleacNOT, Hilversum.

Correljé, A.F., Odell, P.R. (1996) 'The politics of European Gas: Dutch play for a return to Centre-Stage', *Geopolitics of Energy*, Issue 18, Number 9, September 1996, pp. 7–10.

Energie Nederland (1999) 'Inkooppositie energiebedrijven zwak in nieuwe aardgaswet', *Energie Nederland*, Jaargang 2, No. 4, p. 9.

Gasunie (1997), *Jaarverslag Gasunie*, Groningen, Gasunie.

Gasunie (1998), *Jaarverslag Gasunie*, Groningen, Gasunie.

Heren, P. (1999), 'Removing the government from European gas', *Energy Policy*, 27, pp. 3–8.

Kielich, W. (1988), *Ondergronds Rijk, 25 jaar Gasunie en Aardgas*, Uniepers/Gasunie, Amsterdam/Groningen.

Kort, C.J.M. (1991), 'Milieu en Europa van grote invoed op Nederlands aardgasbeleid', Staatscourant, 25 September, p. 7.

Kunneke, R.W., Arentsen, M.J., Manders, A.M.P., Plettenburg, L.A. (1998) *Marktwerking in de gasmarkt*, No. 19, Beleidsstudies Energie, Den Haag, DGE, Ministerie van Economische Zaken, p. 105.

MEZ (Ministerie van Economische Zaken) (various years), *Olie en gas in Nederland*, Den Haag,.

MEZ (1995), *Derde Energienota* 1996, Den Haag, Sdu Uitgeverij.

MEZ (1997), *Gasstromen: discussienota*, Minsterie van Economische Zaken, Directoraat-Generaal voor Energie, Den Haag.

MEZ (1998a) *Regels omtrent het transport en de levering van gas (Gaswet)*: Memorie Van Toelichting en Voorstel Van Wet.

MEZ (1998b) *Regels met betrekking tot het onderzoek naar en het winnen van delfstoffen en met betrekking tot met de mijnbouw verwante activiteiten (Mijnbouwwet)*: Memorie Van Toelichting enVoorstel Van Wet.

NOGEPA (Netherlands Oil and gas Exploration and Production Association) (1993) *Exploration and Development of Marginal Fields in the Netherlands and on the Dutch Continental Shelf*, Marginal FieldsTask Force, Den Haag.

Odell, P.R. (1969) *Natural Gas in Western Europe: a Case Study in the Economic Geography of Resources*, De Erven F. Bohn, Haarlem.

Odell, P.R. (1973), "Het Nederlandse aardgastekort: onbeantwoorde vragen en een alternatieve hypothese", *Economisch Statistische Berichten*, 58e Jaargang, No.2900, pp.422–426

Odell, P.R. (1975) *The Western European Energy Economy, Challenges and Opportunities*, Athlone Press, London.

Odell, P.R. (1979), "Constraints on the Development of Western Europe's Natural Gas Producing Potential in the 1980s", *OilGas*, Vol.5, No.3. pp.21–30.

Odell, P.R. (1984), "Constraints on the Development of Natural Gas resources with Special Reference to Western Europe", in C. Campbell Watkins and L. Waverman (Eds), *Adapting to Changing Energy Prices*, Oelgaschlager, Gunn and Hain, Cambridge.

Odell, P.R. (1987), Aardgas en de Nederlandse Samenleving, *De Gids*, nr. 2/3, pp. 121–128.

Odell, P.R. (1988), "The West European Gas Market: the Current Position and Alternative Prospects", *Energy Policy*, Vol 16, No.5, pp.480–493.

Odell (1991) *Global and regional energy supplies: recent fictions and fallacies revisited*, Centre for International Energy Studies, Erasmus University, Rotterdam

Odell, P.R. (1992), "Prospects for Natural Gas in Europe", *The Energy Journal*, Vol.13, No.3, pp.41–62.

Odell, P.R. (1995), "The Cost of Longer-Run Gas Supply to Europe," *Energy Studies Review*, Vol.7, No.2, pp.94–108.

Odell, P.R. (1997), "Europe's Gas Consumption and Imports to Increase with Adequate Low Cost Supplies," *Energy Exploration and Exploitation*, Vol.15, No.1, pp.35–54.

Portegies, M.J., (1969), 'Marktstrategie van het Nederlandse aardgas', *Economisch Statistische Berichten*, 25 Juni, , pp. 650–655.

Peebles, M.W.H. (1980), *Evolution of the Gas Industry*, London, The Macmillan Press Ltd.

Peebles, M. (1999), 'Dutch Gas: Its role in the Western European Gas Markets', *Gas to Europe* (eds. Robert Mabro and Ian Wybrew-Bond), Oxford University Press, The Oxford Institute for Energy Studies, Oxford.

Schweppe, F. (1999), "Groningen Long Term: project van lange adem," *PetroChem*, No.2 Februari, pp.16–21.

Stern, J.P. (1984), *International Gas Trade in Europe: The Policies of Exporting and Importing Countries*, London, Heinemann Educational Books Ltd.

Wieleman, F.G.M. (1982a), *De economische betekenis van het Nederlands aardgas*. Institute for Economic Research, discussion paper series 8218/G, Erasmus University, Rotterdam.

Witteveen-Hevinga, T., (1998), *Gas: winnen of verliezen. Een aanzet tot een herbezinning op het aardgasbeleid in het licht van de volgende eeuw*, PvdA, Den Haag.

Section III

The Evolution of Europe's Gas Markets

Chapter III – 1

Natural Gas' Early Years: Geographical and Structural Constraints

A. The Exploitation of Groningen Gas – cui bono?*

a. An ordered development of the Dutch market

The natural resources of its European neighbours had long been somewhat envied by the Dutch, particularly as these had differentiated between the industrialisation of Britain, Belgium, Germany, etc., on the one hand, and by the continued agricultural base of the Netherlands' economy, on the other. Attitudes occasioned by this situation have, however, been much softened by the increasing importance of oil in Europe and by the choice of Rotterdam in the 1950s as the prime centre for refining in Europe – by virtue of its favourable location for serving extensive markets.

Simultaneously, relatively low labour costs were attracting other industry to the Netherlands, but neither development had stimulated the national consciousness out of the somewhat pessimistic mood which followed the disastrous floods of 1953. These floods had made the long-talked about Rhine/Maas Delta Scheme an imperative, rather than just a possibility. Its enormous costs thus emerged as a short and long-term burden on the national exchequer and on taxpayers. All this, combined with the problems of living space in a small country with the highest rate

*** An edited version of the article published in the *Petroleum Times*, Vol 73 No. 73 November 21, 1969, pp. 1519–1526**

of population growth in north-west Europe, indicated a somewhat sombre outlook for the Dutch economy.

Some of the Dutch enthusiasm for the Common Market emerged from this background. This, it was argued, would at least provide ample opportunities for the efficient Dutch farmers and thus help to maintain employment in agriculture and associated activities; and jobs for the growing population.

However, at much the same time something much closer at hand and much more tangible entered into Dutch economic and other calculations, viz. the discovery and initial evaluation of the Groningen gasfield. Here at last was a natural resource, the large size of which was known officially long before it was revealed to the man-in-the-street or, indeed, to the energy world outside Shell and Esso. The gas placed the Netherlands alongside other energy-rich nations and its exploitation would be planned to change the national energy economy. Not only would it provide the basis for attracting new industry, but the revenues to be earned from the gas would finance both the Delta Scheme and the increasingly comprehensive social services being introduced into the country.

Such an attitude to Groningen implied a deliberate, a rational and hence a state controlled or directed exploitation of the resource. The state thus took a direct 40 per cent interest in the producing operation in the name of the State (Coal) Mines, while transportation and marketing were to be organised through a company (Gasunie) in which the State had a 50 per cent share. Most actual sales were, moreover, to be made through pre-existing municipal or provincial-owned or operated utilities which, given the Netherlands system of national control over local purse-strings, could also be considered as quasi-state operations.

Perhaps even more important than the statutory apparatus involved in exploiting Groningen gas was the key role given to the Central Plan Bureau in decisions about the incorporation of gas into the nation's economy. This Bureau had, under the directorship of the recent Nobel prize winner in economics, Professor Tinbergen, built up for itself a formidable reputation for shaping economic policy making, and thus the "national welfare" optimisation model which it developed for the exploitation and use of Groningen gas became the main tool which determined the 1962–63 decision on gas strategy.

As the model was able to assume the absence of supply limitation it could be used to determine welfare maximisation on the basis of appropriate rates of incorporation of gas into the energy economy. This involved consideration of variables such as price, capital investment requirements, government revenues, substitution rates for other energy sources and changes in employment in the energy sectors. It should be noted, however, that the need to simplify the model prevented the evaluation of secondary and succeeding changes in the economy likely to be produced by the introduction of gas.

The model indicated that welfare would be maximised by a pricing/offtake policy which would restrict gas' rate of introduction into the economy so as not unduly to upset the existing energy structure. The wholesale price (viz. the price to local distributing companies and to very large consumers) was thus set at approximately a fuel oil parity price, so protecting oil outlets, but not those for Limburg coal the cost of which had become too high to enable it to compete with oil.

By and large the use of gas in the national energy economy has emerged as planned (see Fig. III-1.1). The task of building an appropriately sized nation-wide transmission and distribution system was pursued with vigour and is now virtually complete: though, as expected demand goes on rising, there will have to be expansion and duplication of facilities and, of course, their extension into new areas of rapid urbanisation and industrialisation.

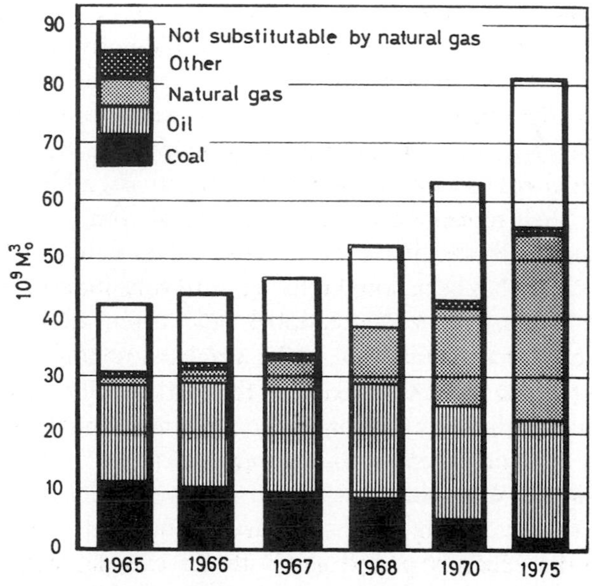

Figure III-1.1: Estimated Dutch Energy Consumption Substitutable by Natural Gas – in $m^3 x 10^9$

Demand is, however, tending to go ahead at a somewhat faster rate than expected for two main reasons. First, because a price gap in favour of gas opened up following the surcharge on oil products after the 1967 Middle East war and the closure of the Suez Canal. And second, because even with parity prices gas has clearly emerged as the preferred energy source. By 1969 it has secured nearly 25 per cent of the energy market, compared with the 22% forecast (Fig III-1.1).

Its impacts on coal and oil markets have, as planned, been in marked contrast overall. Oil has not suffered any reduction in its total markets as the fast rate of national economic growth has stimulated the demand for oil in those markets where oil products are immune from gas competition, so more than offsetting the fall in demand for heating/burning oils. These have not only failed to achieve their traditional shares of their markets in new outlets, but have also been substituted in existing outlets as customers re-equipped with gas-burning equipment.

For the coal industry the additional competition from gas – complemented with that from oil – has further reduced demand. The Netherlands coal production is already down to under 50 per cent of its level of 10 years ago and not even a relatively optimistic government report envisages the continued production of coal beyond 1975 when the last mine is now planned to close.

The final demise of the coal industry could be even earlier, not as a result only of falling demand, but also of a labour shortage, as miners that can leave the industry before being made redundant, do so. In that the government is providing alternative employment in the Limburg coal producing area by financial assistance to a car assembly plant, a joint State Mines/Shell refinery and a petro-chemical complex, this process has been encouraged.

So far there has been some reluctance to substitute natural gas for other fuels in thermal power stations, but current plans for public electricity generation envisage it as the single most important fuel by 1975. Nevertheless, about 40 per cent of the market will remain for other sources including the country's proposed first atomic power station. In a situation of an under-use of Groningen gas (by normal rates of depletion) and in a situation in which other gasfields are still being discovered, there is much to be said for turning electricity production over entirely to specially priced gas both for existing stations (where practicable) and for new stations to be built in the 1970s. Such a policy would give the Netherlands not only the important advantage of Europe's cheapest electricity (as with Texas and other major gas producing states in the USA), but also the lowest possible capital cost for new power generation capacity – an important consideration in a country whose capital resources are heavily committed to the Delta Project and to the infrastructure requirements of a rapidly growing population.

There is a second controversy over the use of gas. This arises from the implementation of plans for the so-called "Ministerial" gas: gas, that is, which the Minister of Economics is empowered to make available at specially low prices (reportedly at one-third off the normal quantity price) to new industries (for the Netherlands) which are located in the less prosperous parts of the country. So far, partly because of problems in defining a new industry, this provision has been used very sparingly; viz. for an aluminium smelter in Delfzijl and for a chemical complex in Zeeland. The province of Groningen – itself one of the less prosperous parts of the country – has been upset by this, particularly as consumers there were given only a small price advantage of about 5%, compared with the general price levels established for gas. Gas transport costs were, in effect, averaged for the whole country so that the producing region was denied the locational advantage of its own gas.

There is thus pressure for a revision of pricing policy so that a significant differential is opened up between prices in the Groningen region and those elsewhere. In this way it is hoped to stimulate the development of large scale energy intensive industry in the north of the country in conjunction with the plans for greatly expanding port facilities at Delfzijl. This would be not only advantageous for the Groningen economy, it would simultaneously take pressure off the rapidly developing Rotterdam area. This, together with a greater use of gas for power generation, would justify a less restrictive attitude to the exploitation of the Netherlands' main "natural" resource.

b.The regulation of Dutch gas exports

In the Netherlands itself natural gas has thus come to dominate the energy scene – in spite of a relatively high prices and constraints on gas use. A similar approach has, however, more seriously inhibited the ability of the major new energy source to secure high level sales in the very energy intensive using regions which literally surround the Netherlands. On the basis of the optimising model referred to above – and in light of advice given by Shell and Esso from their knowledge of European energy markets – a decision was taken to sell gas aboard only in the so-called premium markets. The maintenance of a fixed frontier price for exports meant that transport for export costs would act to gradually restrict the markets that could be won as distances from the

Netherlands increased. Given this strategy, planned exports were to amount to about 30 milliard m^3 annually and supplies were to be guaranteed for at least 25 years.

In this context a series of export agreements have been made since 1964. As with the national plan for gas, the strategy has on the whole evolved as it was planned by Shell and Esso – the joint exporters of Dutch gas. These companies have also become involved, either as owners of or participants in the companies concerned, for both the transportation arrangements and in local distribution networks. This degree of vertical integration in the natural gas industry, from production through transportation and into actual marketing, is the sort of development that has been prevented in the United States – through bodies such as the Federal Power Commission – in the interests of the ultimate consumers. Western Europe's lack of familiarity with natural gas and of the dangers arising from vertical integration can perhaps be held responsible for the absence of similar intervention here. Only the French were not prepared to accept such arrangements and insisted, instead, that Gaz de France should exclusively handle the gas itself. This now puts France in a much stronger position to re-negotiate price and other aspects of the gas supply agreement as a fundamental change in Western Europe's natural gas economy approaches.

The success of the Dutch planned gas export strategy was a function of the maintenance of the monopoly position enjoyed by Groningen gas. It needed time to evolve fully, with 1970 as the target date for the completion of the system. Thereafter, another five years was thus required to build up to the ultimate capacities.

Events, however, have moved faster than this. The first "casualty" to the planned strategy was the sale of gas to the UK as the British Gas Council achieved success too quickly in securing gas from both other companies and its own North Sea operations, so undermining its interest in imports of Groningen gas from 1967.

Meanwhile, the philosophy of the strategy itself was working to ensure its eventual and inevitable collapse. Not to put too fine a point on the matter, other parties elsewhere in Europe considered that the Nederlandse Aardolie Maatschaapij (NAM) has established a too profitable pattern of operation and they determined instead to seek part of the markets – and of the profits – for themselves. Over the past few months competitors have clearly demonstrated the speed with which they will move to challenge NAM's strategy for marketing gas in Western Europe. Thus, Austria has opted for Soviet gas rather than Dutch and France has contracted to take North African gas at a delivered price no greater than the border price of Dutch gas. Meanwhile negotiations by other nations for North African supplies continue.

Other companies that have discovered gas in the eastern part of the North Sea are busy negotiating sales to Denmark and northern Germany, where large scale possibilities emerged following the reported comment of a spokesman for Dow Chemicals that they were not going to pay "fancy prices" for Dutch gas for their new complex near Hamburg. Even a second Dutch gas export operation awaits only government approval following Amoco's agreement to sell gas out of its Bergen (N.H.) field directly, by a pipeline independent of the NAM and Gasunie lines, to a large electricity company in the Ruhr. Though the price apparently agreed for this large contract (up to 2.5 milliard m^3 per year) appears to be in line with the NAM Gas Export price, it is significant that this is a sale directly to a customer in the basic, rather than the premium, sector for primary fuels. It thus goes well beyond anything that NAM had in mind for Groningen gas in West Germany.

The example of V.E.W. of Dortmund in securing access to natural gas directly from a producer for power generation, instead of "going nuclear", will not be lost on many other similar customers in North Germany, the Ruhr, Belgium and Luxembourg to which areas transport costs on Dutch gas are relatively low. In these developments – and in the possibility of North Sea gas going directly to the Hamburg area for large scale industrial use – we see evidence of the breakthrough of natural gas into the base energy market. As this market is gradually forced open, natural gas will quickly achieve a role as a third primary energy source in Western Europe's most heavily industrialised region. The revaluation of the Deutschmark will help in this respect as far as West Germany is concerned, for it will enhance the competitive position of Dutch (and other) gas against both local coal and nuclear power, the costs of which will not be reduced by the revaluation. It will also help gas against oil to some degree in that only the crude oil component cost in the total cost of oil products will be reduced by the currency revaluation.

Of course, NAM could ignore all these developments by sitting tight on its Groningen reserves and remaining content with lower revenues and much lower profits as a result of the anticipated sales it will not now make. However, such an attitude is not tenable except in the very short term even with the events and developments mentioned above. But in addition to all these, there is a further factor, viz. the reality of Soviet competition which was recently dismissed as remote. "Not a single country in Western Europe", Gasunie's spokesman said, "is thinking seriously of importing substantial quantities of gas from

Russia... It must also be regarded as practically out of the question that West Germany will make itself dependent on the Eastern bloc for a substantial part of its natural gas supplies".

Now, just a few months later, Soviet competition has emerged into the reality of an almost signed agreement with the West Germans, into negotiations with the French whose markets in the iron and steel areas of the eastern part of the country will be served through an extension of the pipeline through Germany and in the reactivation of the Russo-Italian negotiations for supplies to the Po Valley to supplement local gas.

Soviet interest in Western European markets – and the ready response by West European governments to the opportunities for buying Soviet gas at prices at the frontier of Western Europe which are already below the Dutch gas frontier price – implies access to supplies which, with those of the Netherlands, the North Sea and North Africa, mean that natural gas in Western Europe in the 1970s will, as in Britain, become an energy source available for all markets where it can technically be used, instead of merely for the premium markets. In such a rapidly developing situation, only belatedly recognised by the OECD whose forthcoming study on gas will significantly "up" its earlier pessimistic evaluations of gas' role in the Western European energy economy, the question of a response by the owners and operators of the Groningen field – which is technically and economically capable of moving to a level of production almost twice that hitherto contemplated – is but a matter of time.

The time that will be required by Shell and Esso will mainly be that needed to persuade the Dutch government that the old "welfare maximisation" model is no longer relevant in an open and competitive market situation. Once the persuasion is successful, however, then export prospects for Dutch gas will lie in the achievement of higher level, lower unit-profit sales to all customers within easy reach. When that happens, neither the Russians nor the North Sea gas operators will have such an easy time in making their sales. Dutch gas will then emerge by the mid-1970s as the largest single source of Western Europe's energy requirements. By that time gas could account for over 20 per cent of total energy supply – with the sources of supply and the geographical pattern of consumption as shown in Fig.II-1.4 and 5 (pp.143 and 149).

B. From Monopoly to Competition in the Marketing of Natural Gas in Western Europe*

a. Introduction

The geographical separation of natural gas production from potential markets has been a recurring adverse influence on the speed and degree of development of gas resources. The impact of such separation is most apparent in the embarrassment that the flaring of unmarketable associated gas generates in the world's major oil producing countries where the phenomenon is a conspicuous reminder of the lack of economic value attached to this gas. When major gas fields have been discovered in such countries – as, for example, in the Algerian Sahara – then the struggle to market even a small part of the potential production has, at least until very recently, been a very unrewarding use of time and investment.

Even in the United States, the ability of natural gas to overcome the geographical separation of the main producing areas from the main energy consuming areas has been largely concentrated into the last 25 years or so, as the development of pipeline technology made gas' long distance transportation possible at costs which brought its delivered price down to levels competitive with both locally produced and also more cheaply transportable energy sources. The Soviet Union, too, shows many similarities in this respect. Major gas transmission lines are now being built to allow the substitution of gas for other forms of energy in increasingly wide areas of the country and also to enable the production of gas from areas even more remote from energy demand centres than the fields of West Siberia.

The situations in the U.S. and the U.S.S.R are illustrated in Figure II-1.1 (p. 136) in which the highly contrasting position for Western Europe is also presented. As shown, the massive Groningen gas field contained an immediately marketable product as a function of the spatial juxtaposition of its very large gas reserves with the large potential gas markets which already existed in the world's most geographically concentrated area of energy use. Such spatial attributes of supply and demand possibilities are of no small importance when viewed in light of the incidence of transport costs in the delivered prices of natural gas,

*** An edited version of a paper given at the U.N. Economic Commission for Europe's Symposium on *The Problems Relevant to Natural Gas Markets in Europe*, Barcelona, 23–29 October, 1970.**

arising not only as a function of the distance to be conquered by the pipelines, but also as functions of the lines' diameters and load factors. Thus, European-scale distances, the spatial intensity of energy demand and the existence of juxtaposed customers with contrasting temporal patterns of requirements for energy, have combined to produce a low transport cost component in the delivered price of gas from a locationally advantaged field like Groningen. Indeed, it can be argued that these factors combined to produce the world's most favourably placed natural gas resource and thus one which would provide keen and effective competition with existing energy supplies as, on a more localised scale, Po Valley, south-west French and Austrian gas had done earlier in Europe's post-war development.

b. The Groningen Monopoly

In such a situation, and given many producers as well as many consumers, one could have hypothesised a "scramble" for markets in which the output of the field was quickly bid up to its optimum rate and the price of the gas adjusted downwards until some customer could be found for the last unit of profitable production. As shown elsewhere, this did not happen as the development of Groningen was the sole responsibility of one joint enterprise, the three component elements of which, viz. Shell, Esso and the Dutch government, each calculated that their interests would be best served by a controlled production rate, coupled with a pricing structure which enabled super-normal profits – a monopoly rent – to be earned on the development. Thus, maximum planned production was set at the order of 50×10^9 cubic metres annually – compared with a possible rate of production of up to twice this level – whilst the well-head price was fixed at about 4.3 Dutch cents (= about 1.2 U.S. cents) per cubic metre, in a situation in which the field had low-cost producing conditions of a degree of excellence virtually unmatched elsewhere. And, in that these low-cost geological conditions were coupled with a rationalised producing operation of a scale and technical finesse which is possible only when a very large field is worked by a highly competent operator, then the overall costs of production, at the chosen level of output, must have been well below the average costs of gas production in the United States. Yet there, the average well-head price of gas is less than half that of Groningen.

In light of this evidence it seems a reasonable hypothesis to present the relationship between the supply of Groningen gas and the potential demand for it in energy intensive north-west Europe as one of an areal – or spatial – monopoly. Given this position, then the marketing policy to be followed was a matter of choice on the part of the suppliers: and their choice would naturally be one which maximised profits. The evaluation of profit maximising alternatives demanded close attention to the market possibilities (one can assume, with the range of production levels under consideration, that the marginal cost curve for the development of the field could be taken as being horizontal, thus enabling the supply function to be taken as given in the profit-maximising evaluations).

Marketing possibilities existed in a range of locations and in a range of outlets at each location so that the variables included not only delivered prices to consumers, but also transport costs which would, of course, be a joint function of distance, volumes along specific routes, and the load factors achievable given the temporal demand patterns of the potential customers. The result of this analysis of the Western European markets for Groningen gas – an analysis, incidentally, which was only possible because Shell and Esso had intimate knowledge of all the energy markets concerned – has, of course, never become public knowledge. However, from information that has been made available and from an examination of the nature and shape of the demand curves for natural gas within specifically located markets in Western Europe, it has been possible to construct an approximation to the marketing strategy of NAM Gas Export as it was evolved about 1966. This reconstruction is presented in Figure II-1.2 (p.139).

As already suggested, the strategy naturally aimed at profit maximisation in a situation in which the single supplier could choose its customers without competition from other gas sources of comparative significance. These were, however, several constraints which had to be taken into account. The most important of these were –

(a) the slow rate of depletion of the Groningen field as required by the Dutch government.

(b) the need to avoid frontier price discrimination in light of Dutch membership of the E.E.C. and the country's other international agreements on trade policy.

(c) the need to earn sufficient foreign exchange from export sales readily to satisfy the Dutch government that the new resource could play an effective role in strengthening the national economy.

(d) a belief that to push too much gas into West Germany would invite retaliation by the German government in the form of taxes on the product – designed to protect the indigenous coal industry.

The net result of the policy was a planned export (by the mid-1970s) of some 30 milliard cubic metres of Groningen gas per year. In the absence of competition, this gas could "cream off" the most profitable markets. This implied selling through distributors rather than directly to large consumers of energy and this, coupled with the chosen export price level, made its retailing possible only in premium outlets (plus some short-term availability of interruptible gas to bulk energy users, so as to achieve early high load factors on the main pipelines). In particular, the optimising solution precluded any efforts to serve – on a continuing basis – the large, low-price energy markets in Western Germany and other zones of heavy industry in Luxembourg and Eastern France.

Given this approach to its marketing, Groningen gas did not represent a major disturbance in the European energy situation. The way in which its minimal impact on the bulk energy market occurred is illustrated in Figure II-1.3 (p.140). Assume the gas is available at location OG with a selling price of SPG and with transport cost curves as shown (viz. by the upward sloping curves away from SPG, showing the price of the commodity increasing with distance away from OG). This gas now competes in the market area AB only between d. and g. Its only exclusive market lies between f. and g., whilst between d. and f. it shares the market with oil from supply point OP. The rest of the market AB is unaffected by gas from OG and so continues to be served without disturbance by oil from OP and OP_1 and coal from OC. We would hypothesize that this was the justification by the seller of Groningen gas to concentrate premium market sales, plus a very limited penetration of bulk energy markets (for example, between d. and g. – the equivalent of the markets of the Netherlands itself). The approach optimised profits so that there was no incentive to compete with coal and oil in other geographical areas. In order to substitute oil and coal over wider parts of the market AB, the price of the gas from OG would have had to be reduced. It is only when the price is considerably reduced rom SPG to SPG_1, that the gas can be sold over virtually the whole of the market.

Such behaviour would be sub-optimal for the monopoly gas supplier given that, with the much higher level of output involved, marginal costs were likely to rise above marginal revenues (which, of course, decline as market penetration is increased), thus reducing profits. Such behaviour will, therefore, only occur if there are forces which threaten the spatial monopoly that the single gas supplier enjoys in the premium market, so that his output is kept below the level at which he has calculated his profits are maximised.

In the absence of such external forces (viz. new high volume suppliers of gas), which would have threatened the real monopoly of NAM Gas Export, there was no reason why the development of the markets for Groningen gas should not proceed in line with the – theoretically justified – strategy described above. The strategy implied, of course, a severe limitation on the geographical and economic impact of gas on the West European energy economy. An expectation that the strategy was viable was not, however, restricted to the parties which stood to gain from it – Shell, Esso and the Dutch government. The imposition of limitations on gas' role in the market also became the generally accepted view of outside bodies which studied and analysed the situation. Thus in both official and private evaluations of the European energy situation in the late 1960s, natural gas was presented as continuing to play a relatively minor role in meeting the total European demand for primary energy.

This was a surprising conclusion to reach in light of the fundamentally much more favourable spatial relationships of gas supply and demand in Western Europe, compared with both the United States and the Soviet Union (See Figure II-1.1 (p.136)). In spite of these favourable relationships for gas in Western Europe, it was predicted that gas would become only 25 to 33% as important as it had already become in these two other regions of large-scale energy consumption. The prediction, therefore, can be justified only on the grounds that the spatial monopoly secured by Groningen gas is likely to be a stable and long-lived phenomenon. But these grounds are inherently weak in that the chosen strategy for marketing Groningen gas in itself encourages forces which would undermine it. Moreover, the familiarisation effect that this supply of gas has already had on Western European opinion, in general, and on potential customers, in particular, ensured that the latter would be actively seeking more of this preferred energy source at prices which made it attractive compared with other fuels. It can thus be argued that the development of European markets for natural gas as implied by the Groningen strategy is inherently unstable and that it will quickly come under pressure from consequential development – assuming they are outside the control of the three parties already involved.

It can indeed, be shown that such pressures have already so strongly emerged that any forecasts of future gas market developments in Western Europe based on an extrapolation of trends which emerge from the Groningen strategy are inherently unsound. Forecasts for the development of markets for the mid-1970s onwards must now be based on an anticipated availability of sufficiently large quantities of gas from a sufficiently large number of competing suppliers which will made the markets – including the bulk energy market – open to effective competition. Thus, natural gas will be more strongly substituted for both oil and coal in an increasing number of end uses and over rapidly expanding parts of Western Europe.

c. New Gas Supplies enter the Market

These forces – of new supplies and new suppliers – are both indigenous and external to Western Europe. The demonstration effect that Groningen has had on the search for gas within the region is already clear. It has already stimulated a higher level of exploration work. These efforts have not only produced significant additional reserves in existing producing areas (viz. S.W. France, Northern Italy, Northern Germany), but they have also led to the discovery of volumes of resources under the North Sea which already appear to be approaching the equivalent of about another Groningen. Such indigenous resources in themselves offer enough additional gas to present supplies from Groningen with effective competition.

Thus the U.K. has already discovered reserves enough to cover its gas market expansion in the 1970s and pressure will slowly build up – and more quickly in the event of new discoveries – for the exporting of some of this availability to the mainland. The wide price differential already established between British and Dutch gas is enough to make this possible. Gas discovered in Norwegian waters – both in gasfields – and as associated gas in the "giant" Ekofisk oil field – will most logically find its market openings in Southern Scandinavia and North Germany via a pipeline to the Danish coast. The supply price of the associated gas will, in particular, be low enough to enable Norwegian gas to undercut the present Groningen-based pricing structure in these areas. On land, recent new discoveries in S.W. France appear to justify the hesitancy reportedly shown by Gaz de France in signing larger contracts for the importation of Dutch gas as existing price levels. Similarly, with the discovery of gas in the Northern Adriatic as far as Italy's import policy is concerned. And finally, new fields have been discovered in the Netherlands itself and in Dutch North Sea waters. Already, in one case, gas has been contracted for sale abroad independently of the Groningen-based system and strategy. This gas, from the Bergen field in North Holland, has been sold *directly* to a bulk energy user at a price which is very little higher than the Dutch frontier price for Groningen gas: as such, it opens up a market in electric power generation in West Germany which has hitherto been considered as closed to natural gas and which was certainly impossible to tap whilst the gas moved through local distributors whose mark-up on the bulk delivered price put it out of the reach of low-price energy consumers. It now seems likely that many other similar kinds of energy users in West Germany will be seeking means of securing access to a source of energy which appears to offer the lowest cost way of making electricity and for producing industrial heat requirements on a large scale.

Notwithstanding the encouraging indigenous possibilities, if all potential gas customers had access only to these resources then it seems unlikely that all of them could be satisfied, given reasonable assumptions about reserves that will be available for development and the rates at which depletion will be permitted by the governments concerned. Whilst other governments seem unlikely to insist on depletion periods as long as the 35 years established for Groningen, any period of less than 20 years would probably be unacceptable because of continuing – if irrational – fears about the "future" of gas supplies. Fortunately for the potential customers, however, external supplies are being attracted to Western Europe by the opportunities offered by the markets for profitable sales. Contracts for the movement of such external gas have already been signed in respect of supplies from North Africa and from the Soviet Union. The reserves potentially explorable are, moreover, in terms of the likely level of European demand for them in the 1970s, essentially limitless as they are equal to some five or six "Groningens". Their supply availability depends only on the relatively short-term problems of the completion of the necessary infrastructure for their transportation to the markets.

Figure II-1.4 (p.143) puts these possibilities for marketing externally produced gas in the geographical perspective of the gas supply and demand centres within Western Europe. An indication is given at each such probable point of entry for external gas of the likely supply price at which it *could* be made available. This is the price at which sales would still be profitable to the supplier, rather than the price at which it is currently contracted or on offer. In the case of North African gas, this is because technological developments will bring prices down from their current levels and in the case of Soviet gas, lower prices of gas will be offered in order to get the volume of business which is now being sought

d. Competitive Impacts

Thus, in 1970, the natural gas industry in Western Europe stands poised between a situation in which consumption has been controlled by pricing policies based on the existence of real monopolies and one in which a plethora of new suppliers, gradually coming into competition with each other, promises a radically changed market patterns, both by region and by economic sector. As a result both new areas and new sectors of the West European energy economy will be opened up for this alternative to both coal and oil in a trend-breaking development for market expansion.

Part-cause and part-effect of this situation will be the emergence of a fundamental reappraisal of the strategy which has hitherto marked the sales policy for Groningen gas. Instead of a strategy based on export sales to premium markets over a relatively extensive part of the continent, and only in total up to the quantity whereby a monopoly profit can be earned, there will be a new strategy based on the maximisation of sales from a 20–25 year depletion of the presently declared reserves – for which there is an 0.95 probability of their even exceeding this volume. This will demand direct sales to bulk users in a geographically restricted area (to keep transport costs down and hence maintain the level of "netbacks" to the producers).

Groningen gas will thus become the "market leader" in the new competitive situation and will actively seek out markets in competition with oil and coal, as well as with natural gas from other supply points. Output will be expanded at least to the point at which the total demand curve intersects the long-run supply price curve so that the price will move down to eliminate the monopoly profit. Available gas per annum could thus become twice as great as planned by present strategy with the price falling to little more than half of the present border price (say to 2.26 Dutch centres per cubic metre = about 2.0 U.S. cents). Other supplies of natural gas will, in this changed situation, find their market openings in Western Europe around the periphery of the area of the continent within which the competitive position of Groningen gas will be unassailable.

In that this development is hypothetical, predictions as to the markets for natural gas which would arise from it must still remain speculative. But not to make such predictions at all – and to rely instead on the extrapolation of gas-use trends which have arisen from a set of conditions which are now under severe and increasing pressure – would mean the likelihood of an eventual unpreparedness for a fundamental change in the energy and natural gas economy of Western Europe. The signs of such a change are as apparent as those which indicate an approaching storm after a spell of calm and undisturbed weather. The meteorologist who failed to evaluate the implications of such approaching changing conditions would fail to in his professional duty.

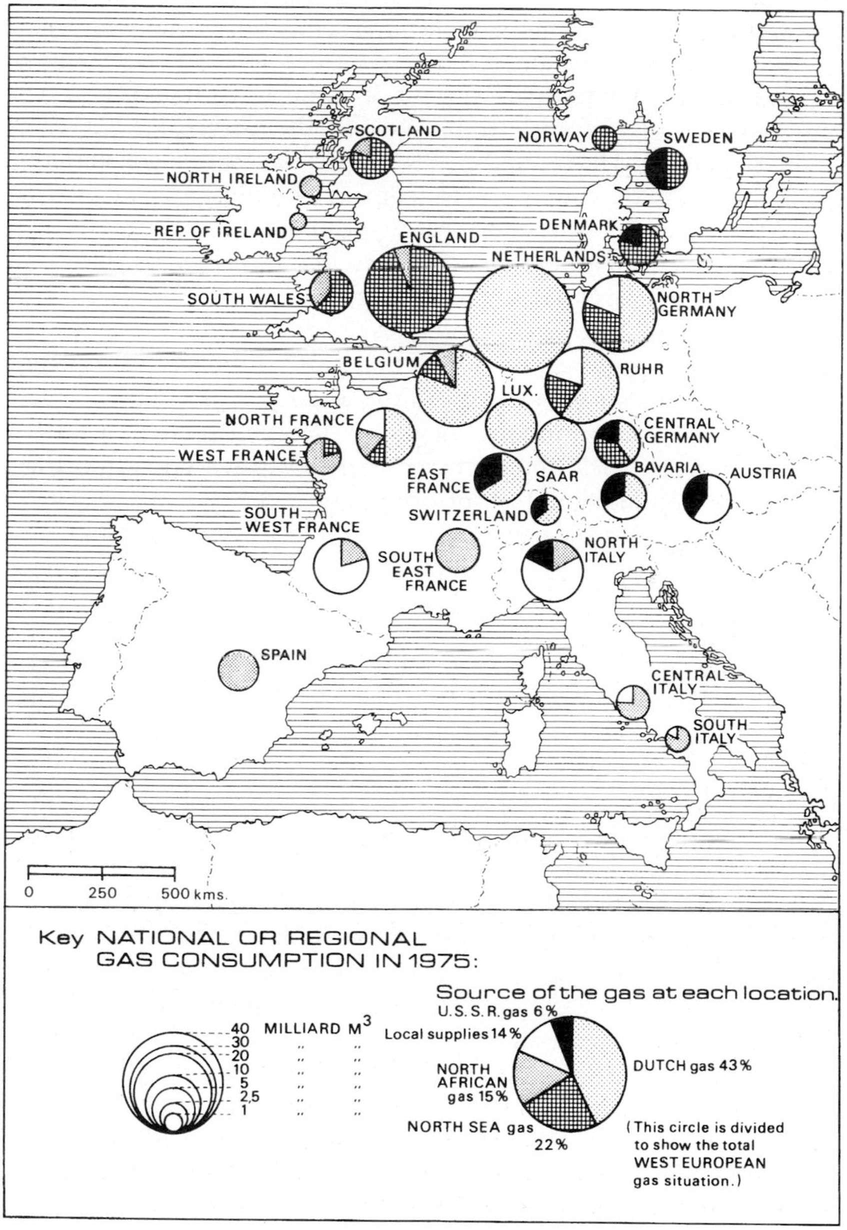

Figure III-1.2: Estimated western European natural gas consumption in 1975 – showing sources of gas supplies

The same would be true for those of us professionally concerned with the future of the energy economy, should we ignore the storm signs of approaching change from monopolistic to competitive conditions in the gas sector. This is the justification for speculation about the development of gas markets in Western Europe in a competitive framework. The results are set out in Figure III-1.2 showing the markets for gas which could have emerged by the mid-1970s. This shows that gas consumption within Western Europe could be some 275 milliard cubic metres; 50–100% greater that is, than that predicted in the 1969 OECD report as the possible 1975 consumption in Western Europe. The large difference can be explained partly in contrasting gas supply evaluations and partly in terms of contrasting gas demand considerations.

On the supply side, the OECD considered the idea of depleting the presently known reserves in as little as 20 years as unrealistic and presented only very cautious views on the possibilities of higher production rates which would emerge from new discoveries. In marked contrast, we consider it unrealistic not to notice the very high degree of probability that adequate new reserves will be discovered to justify a 20-year depletion curve for gas from existing known fields. It is worth noting in this respect that the U.S. has now lived quite happily for over a decade with a reserves-to-production ratio of less than 20 years and for well over two decades with a ratio of less than 25 years. Indeed, reserves to production ratios in excess of 15 to 20 years reflect either formidable continuing discoveries of new fields, or else monopolistic supply-side behaviour in a less than fully integrated geographical system of supply and demand. Given the competition for markets that we envisage, then higher rates of production than those that are currently scheduled from known resources would seem to be a matter of course.

On the demand side, our higher estimate of the future size of the market lies in the greater degree of optimism that we have in the market's short-term capacity to absorb much more gas than is currently considered possible by OECD and other organizations. Here again, the constraint that demand estimates have hitherto had to face arises from the acceptance that the rate of growth of gas markets will basically continue to be determined by the speed at which the so-called premium energy markets can be converted to using natural gas. Such a rate is necessarily low when one considers the amount of investment and the complex infrastructure that is required for such market developments, with its low average annual sales to the many millions of small consumers involved. If our alternative hypothesis is accepted viz. that Europe is on the brink of evolving into a fully competitive natural gas economy, then one has a situation in which suppliers actively seek low value outlets so as to build up sales volumes – and hence cash flows – as quickly as possible. Rates of growth in gas use will then consequently rise well above the levels anticipated by "conventional wisdom", as increasingly large numbers of large low-price energy users buy gas in bulk to substitute for increasingly expensive coal and oil products, the continued use of which also creates the atmospheric pollution problems decreasingly acceptable to majority opinion in Western Europe. If the rest of industrialized north-west Europe achieves the growth rate in natural gas use in industry similar to that now anticipated in the U.K. (and this should not be too difficult considering the faster rate of economic growth expected for most other countries in comparison with Britain), then a total West European gas consumption of the order of magnitude suggested in Figure III-1.2 is achievable.

The key market in this respect is Western Germany where the gas use suggested is well over double the amount currently considered likely by "conventional wisdom". But such conventional forecasts implicitly ignore opportunities such as that already taken by one major electricity authority in obtaining direct access to gas at a price well below that at which it could have bought the gas from a local distributing company; hence enabling it to use gas for electricity generation, rather than other fuels and nuclear power. Such opportunities are similarly being sought by North German industrialists based on cheap gas from the North Sea, and by bulk energy users elsewhere in the country using direct supplies from Russia whereby they can substitute gas for oil and coal. Such developments open up markets for natural gas to a level and at a speed which none of the "conventional wisdom" forecasts based on trend extrapolation can take into account. We would thus argue that the achievement of the use of cheap gas in bulk energy markets is only a matter of time, given competition between suppliers. When this is achieved, then the penetration of natural gas into West German markets at the level suggested here (over 70 milliard cubic metres per year) becomes a relatively short-term development. Such is the impact of a geographically integrated, competitive gas economy, as opposed to one based on a set of spatially separate monopolistic situations. The timing of the breakthrough from the one to the other in Western Europe must be the key question in any consideration of natural gas marketing possibilities in the 1970s.

Chapter III – 2

The Widening Gas Markets of the Early 1970s

A. An Alternative Marketing Strategy for British North Sea Gas: Exports to Europe*

a. The Existing Strategy

The existing strategy for marketing the 4000 million cubic feet daily (mcfd) of the North Sea gas that the British sector is expected to produce by 1975 and which has to be purchased by the Gas Council – as the sole purchasing authority on a "take or pay" basis – seems unlikely to be successful. Markets in the domestic, commercial and industrial sectors will be difficult to build up to this required level. This is, in part, because of consumers' difficulties of converting from other fuels, but in greater part it is the result of aggressive competition for energy markets from a large number of oil companies which are anxious to sell not only oil products, but also liquefied petroleum gases (LPG) and liquefied natural gas from overseas sources as a result of rapidly declining costs in moving LNG by ship.

b. Gas for Power Generation; at the Cost of Coal

Given such difficulties in building up its markets in these sectors at a sufficiently high rate, the Gas Council will have a very positive incentive to sell natural gas to the Central Electricity Generating Board for use in the conveniently located power stations of the East Midlands

*** An edited version of an article published in *The Times*, London, 13 August, 1971**

and South Yorkshire. These stations could be offered interruptible gas for less than the average price paid by the Gas Council to the gas producing companies, for both pipelines' load-factor reasons and to enable the Council to secure a cash flow from the volumes of gas that it has to pay for in any case under the terms of its contracts with the producers.

There thus seems a real possibility of an offer of natural gas by the Gas Council to the C.E.G.B., for delivery after 1971, at around 2.5d per therm delivered to these power stations. Protection for coal in the power station market, as extended by the last government, lasts effectively only until that year. An extension of such protection by the new government, given the availability of indigenous gas at such a low price, would be difficult to justify, particularly when the government will be under severe pressure from both the electricity and gas interests and when a denial of such outlets to the Gas Council will mean either that it has to pay the oil companies for gas it cannot use or that it has to shut-in its own discovered reserves in North Sea fields. In both cases the impact on the finances of the Gas Council would be very serious.

Given such incentives, the use of up to 1000 mcfd of gas in the power stations is not out of the question; sufficient, that is, to substitute up to 15 million tons of coal per annum, assuming a somewhat higher efficiency in the burning of gas compared with coal. With a steadily declining likelihood of the government preventing such a development, the only restraints on a Gas Council/C.E.G.B. agreement would appear to be those of a technical nature. None of these, however, would have other than a delaying effect on the switch to gas as the conversions of the boilers in the power stations concerned and the laying of the required branch pipelines from nearby points on the high pressure gas transmission system could easily be carried out during the course of a couple of summers.

Thus, within two years of an electricity/gas agreement a large slice of the market for East Midlands and South Yorkshire coal could disappear, suggesting that such a possible development in gas use could be one of the main factors occasioning the delay in the progress of the C.E.G.B/N.C.B. negotiations over long-term outlets for coal in electricity generation. The impact of such a development on the last chance of profitability for, and the maintenance of morale in, the coal industry does not need to be stressed. The potential impact of gas on coal is, indeed, potentially greater than the competition which coal faces from nuclear power developments for, quite apart from the long commissioning delays on nuclear power stations, their development mainly affects future coal production opportunities in the high cost coalfields. In contrast, the use of natural gas in the base-load power stations along the River Trent and in Yorkshire would eliminate markets for relatively low-cost coal, hitherto considered safe from competition.

It is partly on the assumption that high social costs – chargeable to public funds – would be generated by such a decline in the coal industry that one can argue for a limitation on the penetration of natural gas into the future power station market for coal. In greater part, however, such a limitation on Gas Council sales to the C.E.G.B. is justified by the existence of other markets for the gas which will, under the present strategy, go to the power stations at a price which is only marginally profitable for the gas industry.

Alternative markets could be created through state intervention in oil markets so that natural gas is substituted for oil products in all end uses where this is technically possible. Such a restraint on consumer choice was not, however, acceptable even to the last Labour government and the likelihood of such direct intervention under the present Conservative administration is probably zero. One must, therefore, turn to examine the marketing possibilities for upwards of 1000 mcfd of British North Sea gas in a Western European, rather than in a purely British, context. The rationale of a marketing strategy which looks partly to the possibilities of exporting gas to Europe is as follows.

The European Markets' Alternative

The Gas Council buys UK North Sea gas from the producing companies at the "beach". The general price level is just under 2.9d per therm delivered to the Council's transmission system. Allowing for transport costs in getting the oil from the off-shore field to the beach, we may thus assume a well-head price of about 2.2 to 2.5d per therm (depending on the location and size of different fields.) We may now assume that a line carrying British gas is laid under the North Sea to a landing point in West Germany. There would appear to be no technical barriers to the construction of such a 200 mile line. It might, however, best avoid the Dutch sector of the North Sea – for political reasons associated with gas transit through a third country – and thus take a somewhat longer route which would keep the line entirely in the British and West German sectors. Further assuming a 500–600 mcfd capacity line and a high load-factor, the costs of transportation would lead to a delivered supply price of the gas to large energy users in North Germany of about 3.5d per therm. A roughly similar delivered price could also be achieved for supplying British North Sea gas to Northern France and

Southern Belgium, assuming in this case that the gas would be taken through the high pressure UK transmission system (already completed or under construction) from Bacton in Norfolk to the south coast and then pipelined under the English Channel.

The physical movement of British gas to the mainland of Europe is thus certainly practical. Equally important is the fact that at the delivered cost as indicated above, British gas would be fully competitive with alternative supplies that are currently on offer to users in the regions concerned. For example, the bulk delivered price of Dutch gas to these markets – after allowing for the lower calorific value of gas supplied from the Groningen field – is approximately 4.0d per therm to northern Germany and Belgium and 4.5d to Northern France.

Moreover, in both Germany and Belgium the sales of Dutch gas are essentially to local distributing companies, the marketing policies of which increase significantly the effective prices at which the gas is available to consumers. Thus, British exports could specifically seek out particular large-scale energy users (chemical plants, industries with steam raising needs, power stations, etc) which could be served quite independently of the local distribution system and which would be delighted to have an opportunity of securing access to a new lower cost source of energy – in much the same way that one of the largest electricity utilities in the Ruhr has managed to obtain natural gas from a small field in the Netherlands independent of the distribution and pricing system for Groningen gas.

In political terms there can be no serious objections to the idea of British gas being offered to near European neighbours. Indeed, one could expect a highly positive reaction to such enterprise for neither Belgium or Germany are content with the relatively high prices which they have had to pay for their imports of Dutch gas in the context to date of the monopoly which it currently enjoys in these markets. Competition from alternative supplies would thus be welcome – as, indeed, it has already been in the case of Germany which, in spite of long-distance delivery problems and of political considerations, has agreed to buy gas from the Soviet Union in preference to more gas from the Netherlands. In France, all imports are the responsibility of the state entity, Gaz de France, but it too is also willing and anxious to diversify its sources of imports and is currently not only buying gas from Algeria as well as from the Netherlands, but is also negotiating for supplies from the Soviet Union.

In light of these favourable political, economic and technical factors there would appear to be adequate justification for examining export opportunities for British gas. A volume of 1000 mcfd by the mid-1970s – of which perhaps 600 mcfd would be sold in Germany and 400 in France and Belgium – seems well within the bounds of practicable and marketable possibilities.

New gas discoveries in British waters, leading to possibilities of an even high rate of annual production would enable these European markets to be developed to an even greater extent. Marketing of such additionally available British gas could also be done by the Gas Council or, as seems likely to be politically more attractive to the U.K. government, it could be undertaken by some of the North Sea producing companies, their having been designated to seek markets in Europe. AMOCO, for example, one of the largest producers, could well be interested in exporting some of its U.K. gas production that it has already agreed to sell gas to the Ruhr electricity company.

How UK gas exports are effected is a matter of political choice but, in any case, such sales would eliminate the serious and costly threat to the British coal industry as described above, and hence also eliminate the need for the government's requirement to meet the social costs of redeploying large numbers of miners and rehabilitating towns and villages in the coal mining areas affected – additional, that is, to the government commitments already made to undertake such financial responsibilities in other parts of the coal industry in which competition from oil has already led to a rapid rate of rundown in coal demand.

U.K. gas exports of 1000 mcfd would also produce gross foreign exchange earnings of at least £50 million per annum. This would be a significant contribution to Britain's balance of payments position in the long-term as the flow of foreign exchange would, of course, be virtually guaranteed for the lifetime of the normal 20 to 25 year gas supply contracts that would be signed with the European customer. Such earnings from exporting some of the country's gas resources would, incidentally, more than offset the foreign exchange costs that will shortly be incurred as Britain is forced to import coking coal for the steel industry – a type of coal that the N.C.B. cannot produce in adequate quantities at reasonable prices.

The combination of these advantages from this alternative marketing strategy for North Sea gas thus justifies serious consideration. Such consideration would also appear to be a matter of some urgency – for two reasons. First, to take advantage of the current great upsurge in demand for natural gas imports by the European countries concerned. A year's delay in British efforts may mean that this demand is assuaged by offers that are being made by other external suppliers – the Soviet

Union, Algeria, Libya, and even gas from the Norwegian sector of the North Sea – and by a new lower price marketing policy for Dutch gas, where proved reserves are already large enough to justify a major increase in exports.

And second, in the context of a requirement to plan the post-1972 future of the state-sector coal, gas and electricity industries more effectively. Whilst the possibility of large quantities of natural gas available for substitution in coal-fired power stations continues to exist, none of the state energy industries can adequately plan their future development – and hence their investment requirements which collectively constitute such a heavy drain on government resources. The elimination of this uncertainty through the implementation of the gas export-based strategy outlined would thus be advantageous to all the many parties concerned.

B. Dutch Gas Production Boost*

Groningen's Giant Gasfield

In 1972 the Groningen gasfield will become Western Europe's second largest source of indigenous energy. Its anticipated production this year of 58 milliard m^3 – will provide almost as much energy as West Germany's famous Ruhr coalfield (hitherto unchallenged as West Europe's most important energy source for nearly 100 years) and rather more than from the recently expanded East Midlands coalfield of the U.K.

Hopefully, this comparison of Groningen gas with the giant coalfields of western Europe will help to destroy once and for all the myth of the fields' relative unimportance in the European energy context – a myth which has been created in order to serve objectives such as profit maximization by the companies concerned and maximum possible foreign exchange earnings by the Netherlands: together with a desire that its development should not 'upset' existing energy markets – most specifically, the profitability of the Dutch refining industry! The significance of Groningen's energy, and one of the reasons for its high rate of growth in spite of earlier strategies designed to constrain it, lies in the ease with which it can be produced – most notably in terms of its low labour cost component. A handful of a few hundred employees are able

*** An edited version of an article published in *Petroleum Times* Vol 76, No 1931, February 11, 1972.**

to bring this tremendous amount of energy to the surface – in contrast with the more-than-100,000 miners required to produce roughly the same amount of energy in both the Ruhr and the East Midlands. It is in this great differential between the labour productivity in Europe's traditional energy supplying industry and that of the new source that one sees an example of the kind of development process which is turning west Europe into an affluent, mass consumption economy.

The development of Dutch gas has had less attention than might have been expected in a continent concerned with its increasing reliance on energy imported from parts of the world hardly noted for their political stability. Paradoxically, this has arisen partly because attention has been diverted to generally more newsworthy hydrocarbon activities in the North Sea; though the beginning of these was directly related to the evaluation of the Groningen gasfield as marking the southern end of a large and potentially rich hydro-carbon province. Successive discoveries of first, natural gas and, later, oil have focused attention on other parts of this geologically attractive region whose total promise now seems likely, within a decade, to release the western European energy economy from the iron grip of the OPEC countries and the international oil companies.

In the meantime, however, and for the first half of the 1970s it is the prolific, low cost and perfectly located Groningen gasfield that will not only play the most important role in reorientating the European energy economy several percentage points away from independence on imported oil, but will also continue to stimulate energy consumers located beyond its economic range to try to secure access to alternative supplies of natural gas.

Plans and Strategy

By 1967 the strategy for Groningen gas involved a maximum annual output of 45–50 milliard m^3 – roughly half of which was to be marketed nationally and the rest exported. Within the Netherlands prices were fixed to control the rate of substitution of gas for oil and coal in the interests of orderly marketing and a controlled rate of rundown of the Limburg coalfield, whilst the pricing of exports was such as to ensure that the natural gas was sold almost entirely in the so-called premium energy sectors. However, though the strategy was evolved through sophisticated techniques and was undoubtedly soundly-based on appropriate infrastructure developments – scheduled in the interest of the above mentioned aims of profit maximization – it has been overtaken by actual developments as a result of three main considerations.

First, it failed to take account of the rapid "demonstration effect" that increasing familiarity with natural gas has on potential users, such that it almost invariably becomes the preferred fuel (over coal and oil) so creating great pressures on suppliers to make more of it available.

Second, it ignored the response of other potential gas suppliers in and to Western Europe to seek a share in the highly profitable gas selling operation that the Dutch were creating – such that the monopolistically determined prices came under competitive pressures.

Third, the long-term contracts, though made at firm prices which were high in the context of Europe's low energy prices of the 1960s, made no allowance for the unexpectedly high general inflationary trend of the past two to three years, let alone for even bigger increases in oil prices since the beginning of 1970s.

The combined impact of these factors has thus gradually undermined the hitherto limited development strategy for Groningen gas and has pushed it into a current period of very rapid expansion indeed. The relationship between what was planned and what has happened is show in Table III-2.1

In the national market it was already becoming apparent by the early part of 1969 that the "demonstration factor" was hard at work in upsetting the planned market developments. The preference for natural gas in both the public distribution and the industrial sectors was becoming more marked, having been significantly stimulated in the previous year or so by the temporary price differential in favour of gas that had emerged as a result of the sharp upward increase in oil prices in the aftermath of the Suez War.

TABLE III-2.1
Dutch gas – planned & actual sales ($m^3 x 10^9$)

Planned Sales for →	1970			1972			1975		
	National	Export	Total	National	Export	Total	National	Export	Total
As planned in									
1967	13	8	21	18	12	30	25	25	50
1969	16.5	9.5	26	21	15	36	33	30	63
1971	–	–	–	30	22	52	40	35	75
1972	–	–	–	33	25	58	42	46	8
Actual or Likely Sales	20.1	11.3	31.4	35	27	62	50	55	105

This period of disequilibrium in the hitherto firmly controlled oil/gas prices relationship persuaded many existing oil customers to switch to gas and new energy customers to choose it automatically. This was technically possible within the framework of a transmission and distribution infrastructure which was already capable of handling even the highest possible seasonal peaks (as a result of the investment by 1969 of nearly 2,000 million Dutch Guilders – about £230 million. Neither was the investment capital required a constraint as Gasunie had no problem in raising long term loans, either privately or from the public, at relatively modest interest rates, averaging under 7 per cent. Thus, the 1967 forecast of national gas use of 25 milliard m^3 by 1975 soon began to look like a considerable underestimate of the real possibilities.

Change in Policy

The underestimate became an absolute certainty with a change in policy in 1968 which opened up most of the thermal electricity generating market to natural gas. The contracted interruptible supplies were, moreover, interruptible only when temperatures fell to a daily average of under 0°C – a relatively infrequent event in the maritime Netherlands. The change in policy came as a result of Gasunie's decision to seek additional markets and to improve its system load-factor through interruptible sales. It met with a ready response from the electricity authorities anxious both for stability in their fuel prices and for access to a fuel which reduced atmospheric pollution problems. Since 1971 both these factors have become much more significant given first, the uncertainty and confusion in the oil markets created by the collusion of the oil producing countries with the international oil companies and second, the rising concern for the environment.

Increasing Substitution

Thus, within the Netherlands there are now no good reasons why natural gas should not complete its substitution of oil (and the little remaining use of coal) in all those sectors of the Dutch economy where such substitution is technically possible and where there are no special reasons – such as institutional/organizational relationships and/or long-term contracts between oil/coal suppliers and their consumers – to inhibit a switch. By 1975 the Netherlands will have a gas-based energy economy (in much the same way as Texas or Louisiana where locally available gas has driven out the use of oil in most sectors), with at least 52 per cent of the country's total energy consumption being provided by the Groningen and other gas fields. The total annual natural gas use will be some 50 milliard m^3. Thereafter, of course, with substitution of other energy sources virtually complete, the rate of increase in the demand for gas will slow down to parallel much more closely the rate of increase in the overall demand for energy; unless and until natural gas becomes a viable alternative to gasoline and diesel fuel in the road transport sector.

When we turn from the Dutch to the European situation, we see from Table III-2.1 that gas exports have also moved ahead of the levels planned by the marketing strategy evolved by the late 1960s. Table III-2.2 specifies how reality has diverged from the plans in the case of each of the main external markets concerned.

The eventual degree of divergence between plans and reality is likely, however, to be even greater than indicated in Table III-2.2. Whereas the planned 1975 exports were thought of as virtually maximum possible annual deliveries, it now seems likely that the minimum level of 55 milliard m^3 could well be significantly exceeded, depending on whether the infrastructure required for delivering more gas is built up quickly enough, following revised contracts which call for faster deliveries of gas than have so far been promised.

Effect of Oil Price Inflation

There is, of course, no longer any doubt that markets exist for just as much gas as the Netherlands cares to offer to its neighbours, assuming that the base-load energy demand for steam raising for various purposes in the nearby highly-intensive energy-using industrial areas of West Germany, Belgium/Luxembourg and Northern France are

TABLE III-2.2
Dutch gas exports by 1975 (m^{3}x10^9)

	To Belgium/ Luxembourg	To France	To West Germany	To All Other countries	Totals
Planned by the late-60s Market Strategy	5.5	4.5	11.5	8.5	30
Currently Contracted (in July 1972)	9.6	10.5	20.6	6.0	47.6
Likely minimum with continued development	11.5	12.5	24.0	7.0	55

profitable and virtually insatiable markets just waiting to be won from oil.

Dutch gas export strategy originally discounted these markets as they were not thought likely to be profitable enough to justify the higher level of production required. However, between 1969 and early 1971 the strategy was already undermined to some degree by factors such as political pressure from Belgium which was anxious for access to an alternative industrial fuel; by AMOCO's decision to sell its Bergen field gas directly to German power stations, rather than through Shell/Esso supported and/or controlled German gas distributors whose high price mark-ups were containing sales; and by the "embarrassment" of being outbid by the Soviet Union for gas markets virtually on Groningen's doorstep at a time when official gas industry opinion in the Netherlands still asserted that Russian gas sales in Western Europe were politically, if not economically, impossible!

Since early 1971, however, the importance even of these factors has been overshadowed by a new and powerful influence, viz. that of escalating oil prices in western Europe in the aftermath of the oil companies' new agreements with the OPEC countries. The consequential strong upward movement in basic energy prices has made Dutch natural gas a "best buy" for many regions and consumers, even at the price levels which had previously been fixed to ensure its sale only in premium markets. This oil price inflation has, in effect, reduced the *real* price of Dutch gas by between 33 and 50 per cent and has thus stimulated a massive demand for it from industry, power stations and other energy consumers, all anxious to avoid the rising costs of oil.

Production/Reserves Ratio

Thus, with a combination of a 1975 national requirement for 50 milliard m^3 and an export potential for that date of at least as much gas again, the key question now becomes one of the adequacy of the resource base. Official statements about the inability of Dutch gas to sustain other than modest levels of output have dominated thinking about market development for so long that it becomes difficult to cast off the air of pessimism surrounding the subject.

The understatement of Groningen reserves throughout the development period can, it now seems, best be explained as being motivated by considerations of price maintenance. But with the price levels now firmly established on a long-term basis, making this low cost gas a high-profit energy source, the motivation for understating reserves and production potential has largely disappeared. Even if the understatement was never so motivated but was, instead, based on the most conservative possible calculations of the resource base and its production potential (viz. 95 per cent certainty over the existence of reserves and a 35 year depletion period, respectively) then today, the market opportunities which have been "forced" on Dutch gas producers are demanding a fundamental reappraisal of the commercial and national economic logic of this ultra-cautionary approach.

Additional Reserves

In 1968 reserves were declared at 1850 milliard m^3 of gas and they were scheduled to be depleted over 30 to 35 years. Since then about 100 milliard m^3 have been produced, but, in the meantime, reserves have climbed to almost 2500 milliard m^3 and no one talks seriously any more about depletion over eternity! Thus, an annual production of well over 100 milliard m^3 is now an achievable aim and is now all-but generally recognized as a highly desirable one. Even so, it will be well into the 1980s, even in the very unlikely event of no further gasfield discoveries anywhere else in the Netherlands or the Dutch sector of the North Sea, before concern will be necessary about the depletion of the 2500 milliard m^3 of reserves which are now assured.

By 1980, however, the continuing recovery of gas from the Groningen field, and the further knowledge this brings about the reservoir, seems likely to have further enhanced its ultimate production potential. The deeper drilling in Groningen (to 6000 metres) could well have located additional reserves, while recently discovered on-shore fields in the provinces of North Holland, Friesland and Drenthe seem likely to be evaluated upwards in the same way as with the Groningen field. And by then we shall have learned something of the size and nature of the so-far undeclared discoveries in off-shore areas. Overall, therefore, there appear to be very good grounds for much less pessimism about the future of Dutch gas developments in the course of the next decade.

There would, indeed, seem to be a good chance, even with production between now and 1980 averaging at least 120 milliard m^3 per year (implying a production in 1980 of about 150 milliard m^3), that Dutch gas reserves by then will still give a very comfortable production/reserves ratio of better than 20 years. By then Dutch natural gas production will not only be a more important energy source than both the Ruhr and the East Midlands coalfields – but more important than the total production of all the coalfields of western Europe by that time. Such has become the power of Europe's energy consumers' preference for natural gas.

C. Natural Gas' Potential as a European Energy Source*

Introduction

The arrival of Spring 1972 marks the end of the first of a series of dangerous winters for Western Europe in terms of its energy supplies. The dangers arise out of the upward pressures on oil prices from the combination of much higher taxes in the oil producing countries and the resolve of the oil companies to pass on these increases – and more besides – to their customers. In the meantime, customers face the threat of their supplies being cut off should the countries and companies not get what they want.

The inherent dangers failed to materialize in the winter of 1971–2 only because of a particularly favourable combination of circumstances. First, the lack of as much economic growth as had earlier been expected throughout western Europe, has led to a demand for energy well below that predicted. Second, the exceptionally mild winter throughout most of the continent has eliminated much of the normal seasonal demand for heating oils. And third, there has been an availability of Groningen gas as a substitute for oil products for large numbers of customers in the Netherlands, Belgium and Western Germany. This, moreover, at least in the Netherlands, has been at prices which were fixed well below the levels to which the oil companies had anticipated raising their oil product prices.

As a result of these factors, most of western Europe has emerged from the winter with oil prices – except those in the gasoline, kerosene and other non-competitive markets – well below the levels that the new tax reference crude oil prices would have indicated as likely – and with oil tanks throughout the continent full of products that customers have not needed to buy. Inadvertently, and at the cost of the oil companies, western Europe must already have achieved the 90 days oil stocks that the O.E.C.D. Oil Committee recommended should be gradually built up – to give temporary protection against threats to oil supplies.

Different in Britain

It is really only in Britain that the situation has been different. Here the oil companies achieved their profit bonanza as customers had to face fuel costs 50 per cent or more higher than a year earlier. A contrasting set

*** An edited version of an article published in the *Petroleum Times*, Vol 76, no. 1935, April 7, 1972**

of factors has been at work – with only the mild winter in common. First, the natural gas industry in the U.K. has had freedom to charge industrial consumers what the market would bear. It has thus escalated its prices along with those of oil, so enabling it to meet the rapidly increasing conversion costs for its domestic consumers from which inadequate sales revenues would otherwise not have made the conversion process a profitable operation.

Second, the threat and the ultimate reality of the national coal strike influenced coal consumers to turn to alternative energy supplies wherever possible —meaning oil in most cases, given the gas industry's apparent assumption that it is working with a very limited resource availability indeed, and one, moreover, that should be almost entirely committed to high price customers within the framework of a production rate related to a 20 years depletion of presently proven reserves.

Future development

There are three fundamental elements which will determine the main lines of development of the European energy economy over the next five years or so.

The first is the ability or otherwise of the coal industry – either through its own efforts or with the intervention of governments – to maintain its contribution to the total energy supply. This is a matter for controversy, with some people arguing that coal can and should maintain its position, given the existence of a more than adequate resource base from which to produce reasonably priced coal. Such a view, however, is difficult to sustain in the face of rapidly escalating labour costs which affect coal prices directly and immediately and in the face of difficulties in retaining an adequate mining labour force, given alternative employment opportunities. In the short-term there are probably over-riding strategic considerations – arising out of the second factor – which ought to persuade governments to bear the costs of maintaining coal output, particularly in Britain and Germany. In the longer term, however, a further downward trend in the production of coal must be expected – especially with the exploitation of indigenous reserves of gas and oil many times larger than the loss of coal. Table III-2.3 shows two alternatives for European coal production up to 1980. In neither case does the production indicated represent other than an underutilization of the potential offered by the resources. Most of these will remain for Europe's use a century or more hence, when the need for their development could well have produced different methods of exploiting them.

The second element is that of the steadily increasing pressure on oil prices and oil supplies exercised by the collective control exercised through OPEC over almost all oil that is sold outside the country of origin. OPEC, having enjoyed success in its first collusive ventures, is now poised to take advantage of its monopolistic position. In such a situation the declining power of the international oil companies must be basically oriented towards policies which enable them, through collusion with each other, to gather enough 'crumbs' of profit from the 'rich' oil exporting countries' tables to produce adequate returns on their investments – at, of course, the expense of the continuing disorganized oil consuming countries. Thus, as shown in Table III-2.3, the current conventional wisdom expectation of a rapidly rising demand for imported oil by Western Europe through the 1970s implies not only a

TABLE III-2.3
Western Europe's energy supply 1970–80 with contrasting assumptions on levels of indigenous oil and gas production

	1970		1975				1980			
	Actual		Existing Estimates		Author's Estimates		Existing Estimates		Author's Estimates	
		%		%		%		%		%
Total Supply	1475	100	1700	100	1750	100	2200	100	2300	100
Of which:										
(a) Indigenous Supplies										
Primary Electricity	40	3	75	4	55	3	270	12	150	7
Coal etc	385	26	315	19	250	14	260	12	200	9
Oil	25	2	40	2	100	6	50	2	500	22
Natural Gas	95	6	150	9	355	20	215	10	500	22
Total Indigenous	*545*	*37*	*580*	*34*	*760*	*43*	*795*	*36*	*1350*	*60*
(b) Imported Supplies										
Coal etc	45	3	55	3	50	3	80	4	50	2
Oil	880	60	1050	62	880	50	1285	58	775	33
Natural Gas	5	–	15	1	60	4	40	2	125	5
Total Imported	*930*	*63*	*1120*	*66*	*990*	*57*	*1405*	*64*	*950*	*40*

NB: All tonnages in millions of metric tons coal equivalent

costly, but also an insecure and uncertain way of providing the energy requirements for sustaining western Europe's economic development.

The third element involved is, in part, an alternative to the depressing prospects offered by the second. It is concerned with the possibilities of increasing the use of indigenous natural gas; also a matter of some controversy. Over certain aspects there is no doubt; viz. that the European energy consumer, like his counterparts elsewhere, prefers natural gas over any alternative energy sources for most end-uses: and that natural gas demanded can be produced and distributed in western Europe in a general cost/price framework which makes for reasonable levels of profitability in the industry.

European Gas

There is, nevertheless, a general, conventional-wisdom tendency to view European gas as an energy source of little more than marginal importance in quantitative terms, the availability of which will certainly make no difference to European energy prices. Thus, both official governmental and major-oil-company circles are still talking in terms of gas as contributing no more than 10 per cent of total energy demand by 1975 and thereafter only managing modestly to increase share of the energy market to 12% by 1980.

This is an outstandingly pessimistic view. Given consumers' preference for gas whenever and wherever it is available, and, even allowing something for institutional and political restraints on gas developments, it is predicted on the basis of an unreasonably static view of the size of the natural gas resource base in Western Europe. This is unreasonable because it implies a limitation on hydro-carbon discoveries which has not hitherto occurred in any other newly developing producing area of the world.

The view appears to emerge out of Austrian, Italian and French experience with gas discoveries in the early post-1945 period. In those cases relatively small and isolated gas field occurrences were "over-sold", with consequential later difficulties in maintaining the levels of production to which the producers had committed themselves.

Consequently, undue optimism has never been a failing in the evaluation of the resources and production potential of the Groningen field. Thus the total resources of the field have had to be repeatedly up-rated over the last decade so that its currently declared reserves of almost 2500 milliard m^3 have been barely touched by the production to date of only 200 milliard m^3. The only recently changed market strategy anticipated that production should peak at only 50 to 60 milliard m^3 per annum with a depletion period of up to 35 years.

Various pressures, previously described in part B of this chapter, eventually combined to undermine that strategy. Thus upwards of 80 milliard m^3 per year are already committed to customers in the Netherlands, Germany, Belgium, France and Italy from 1975 onwards. There is, moreover, now little remaining doubt, politics permitting, but that consumer pressure on the resources available will push the rate of off-take up to more than 100 milliard m^3 by 1975 – and still give a more-than-20 years' depletion period for the reserves. This alone almost meets the total European gas use expected for 1975 by conventional wisdom (see Table III-2.3.)

Elsewhere in Europe, already discovered and declared non-associated natural gas resources amount to roughly the equivalent of another Groningen – shared between the older producing nations already mentioned, the other smaller producing fields in the Netherlands, quite significant expansions in North German's gas fields in the last five years and, most important of all, the discoveries of large gas fields in the U.K. sector of the North Sea where, with recent re-evaluations of existing fields and the discoveries of some smaller new fields, the proven reserves now seem to be approaching 1000 milliard m^3. Taken together, these resources can *almost* sustain the author's estimate for indigenous natural gas' contribution to Western Europe's energy economy by 1975 – the equivalent of 355 million tons of coal per year (see Table III-2.3).

A 25 per cent or so expansion in the resource base, or a willingness to use existing reserves a little more quickly than the present 20–25 years depletion period, would enable this much higher contribution of gas to the Western European energy economy to be met. It is, moreover, inconceivable that exploration drilling in other parts of Western Europe's potentially petroliferous areas has not already produced knowledge of the existence of additional gas resources which, given appropriate political and other conditions for development, could make a contribution of this size to energy supplies by 1975. Two areas stand out in this respect, viz. Dutch off-shore developments and the discoveries in the middle North Sea basin between southern Norway and Eastern Scotland/N.E. England.

Dutch Offshore Potential

Political, legal and marketing strategy considerations have hitherto held up the exploitation of the Dutch sector of the North Sea, but recently changed circumstances have accelerated the tempo of exploration and a number of gas finds have already been made – with unofficial reports of at least five structures, one of which has estimated recoverable reserves of 500 milliard m^3; and another of 150 milliard. Assuming the others to be no bigger than the smaller of these two, then the offshore resources already add up to half-a-Groningen in fields which lie in shallow enough waters and near enough to the coast to be brought quickly and profitably into production on the basis of the existing well-head price for natural gas in the Netherlands. This price is some 50 per cent higher than the beach price paid for British sector gas.

North Sea Prospects to 1975

Further north in the North Sea a number of oil and condensate fields have been discovered. From each of these there are reported to be significant availabilities of gas, including enough in the Ekofisk group of fields in Norwegian waters to justify an *immediate* annual production rate of 12–15 milliard m^3 per year. This is on offer to the highest bidder from either the U.K. or West Germany – with the contracts awaiting Norwegian government permission to export the gas. Nearby oil fields on the British side of the median line also appear to offer roughly a similar gas potential, some of which could be landed on-shore U.K. by 1975, if the oil companies are offered a price high enough to justify investment in developing the facilities.

Unfortunately, the U.K. Gas Council's apparent willingness to rest content with the supply potential it has already achieved for 1975,together with the legal impossibility of any other body in the U.K. making an offer for this gas – the C.E.G.B., for example, would probably "jump" at the chance if it were not legally inhibited – may push development of this discovered potential back into the later 1970s. Even without this, however, the Dutch and Norwegian off-shore developments alone almost give the additional 25 per cent of resources required in Western Europe to make the total gas potential for 1975 the equivalent of the 355 million tons of coal shown in Table Table III-2.3.

Longer Term Potential

Beyond 1975 estimation of the rapidly developing situation becomes more difficult. However, given the build up of consumer demand as hypothesized above, we may reasonably assume that there will be sufficient incentives for producers to supply at least enough indigenous gas for it slightly to increase its share of the total energy market to 22% by 1980. As shown in Table III-2.4 this implies an annual production of about 500 million tons of coal equivalent viz. some 425 milliard m^3 of natural gas per annum, assuming that about 30% of the production will be of the lower calorific value Groningen-type gas. This implies an extension of the 1975 resource base by about 50 per cent, capable of sustaining an annual output of some 100 milliard m^3 per annum.

Though this may by the standards to date of natural gas developments in Western Europe appear to be a formidable task, it does, in fact, represent a relatively modest resource base extension by the standards of the world's other major gas producing provinces. Particularly in the light of the potential for hydrocarbons which has already been revealed by the drill or by seismic surveys in those parts of the continental shelf around north-western Europe additional to the areas already evaluated in terms of 1975 production possibilities. Over 40% of the necessary extension to the resource base could, as shown in Table III-2.4, come from the expected development of large gas reserves in the middle North Sea basin where large discoveries have already made, but not yet evaluated. Other reserves will emerge from discoveries to be confirmed in the northern North Sea basin and in other areas of western Europe's continental shelf.

Conclusion

Given the enthusiasm and willingness of several dozen oil companies involved to spend at least £200 million on exploration efforts alone in the new concession areas recently allocated around the shores of the U.K., and significant sums elsewhere in Europe's offshore areas, then the probability of *not* discovering additional gas reserves capable of sustaining an annual output of 500 milliard m^3 by 1980 must be very low indeed. Although the companies greater current incentive is to find and produce oil, this process automatically enhances the opportunities for finding gas. The latter, once found, will, within the framework of the attractive Western European energy markets, have no difficulties in securing sales outlets. Thus, post-1980 indigenous natural gas supply prospects for western Europe seem more likely to exceed the possibilities outlined in this article than to fall short of them and so push natural gas development potential beyond the level at which they meet 22% of the total demand for energy.

The net result will be a much healthier position for Europe as far as its relationships vis a vis the oil producing world is concerned. This seems to be such a desirable development as to justify a very considerable European effort to ensure that its natural gas resource potential is explored and exploited as quickly as technology and capital availability will permit, rather than at the significantly lower level implied by the

TABLE III-2.4
Western Europe: estimates of natural gas resources and production potential in 1975 and 1980

Location	1975 Reserves[1]	1975 Production Potential[1]	1975 Coal Equivalent[2]	1980 Reserves[1]	1980 Production Potential[1]	1980 Coal Equivalent[2]
On-Shore Netherlands	2500	120	135	2750	135	150
South N. Sea – British Sector	1000	50	65	1250	60	75
Austria, France, Italy etc	600	35	45	600	35	45
On-Shore West Germany	550	30	35	650	35	40
South North Sea – Denmark, Netherlands and West Germany	750	35	45	1000	40	50
Middle N. Sea–Britain/Norway[3]	?	25	30	1250	80	90
North N. Sea– Britain/Norway[3]	?	–	–	750	25	30
Rest of Continental Shelf[3]	?	–	–	250	15	20
Total	5450+	295	355	8500 minimum	425	500

1 *All figures in milliards of cubic metres (= $m^3 \times 10^9$).*

2 *Figures in millions of metric tons.*

3 *As these are presently mainly associated gas reserves, the 1975 production potential is a function of the oil off-take rate rather than the gas resource base. By 1980 non-associated reserves should be in excess of 2000 milliard m^3.*

present policies of the countries concerned and as currently anticipated by the European Economic Community, viz. for a maximum of 215Bcm of indigenous gas production by 1980, supplying only under 10% of Western Europe's total energy use.

Chapter III – 3

Constraints on Developments in the 1970s

A. The Dutch Gas "Shortage": Unanswered Questions and an Alternative Hypothesis*

Introduction

Recently announced changes in Dutch gas policy very effectively continue the "mystery" which has surrounded the development of the Groningen field almost since its discovery. From 1959 to 1969 a pretence was maintained to the world at large concerning the real size of the field. Even as late as the summer of 1969, in response to arguments by this author that the field's production was being held back by monopolistic pricing policies from its full annual production potential of 80–100 milliard cubic metres per year, Gasunie's marketing director said that an annual production in excess of 50 milliard m^3 was "unthinkable", given the nature and structure of the field.

Immediately after this, however, as consumer preference for gas over coal and oil really started to exercise its power (as users in the Netherlands, Germany, Belgium and elsewhere became familiar with natural gas for the first time), the "unthinkable" happened so that the expansion of sales moved rapidly ahead well beyond planned levels. The chronology of this development – resulting either from an under-evaluation of the market potential for natural gas or from a public

*** Published in Dutch under the title, "Het Nederlandse Aardgastekort: On beantwoorde Vragen en een Alternatieve Hypothese" in the *Economische Statistische Bulletin*, Vol. 58, No 2900, pp 422–5, 16 May 1973**

presentation of a set of figures which bore little relationship to the real aims and intentions of Shell, Esso and the Dutch government. As shown in Table III-2.2 publicly-declared plans and actual achievements diverged so that the declared planned production levels had to be rapidly adjusted upwards year by year – until finally, in early 1972, they forecast 1975 sales identical at a level identical with that hypothesised by the author some three years previously. This somewhat mysterious history of the development of the Groningen gas field has been previously described and examined (see "Dutch Gas Petroleum Boost" reproduced above as part B of Chapter III-2), but it has been necessary briefly to mention it here again as a background to more recent and even more mysterious views in government and industry circles on the prospects for gas.

Dutch gas presented as a scarce commodity

The prelude to this latest round of events was a speech in mid-April 1973 by the Head of the Mining Section of the Ministry of Economic Affairs, Mr. J. Jonkers. He announced that geological and resource base considerations made it imperative that the annual rate of production from Groningen be reduced from the planned 100 to 82.5 Bcm. This warning was, it seems, based on a report which expressed some concern for the gas recoverability rate from the field, as a consequence of some fracturing of the formation and the irregular advance of the gas/water interface. This report had, however, initially been submitted to the Minister in 1972 without it having had any effect on Gasunie's and NAM Gas Export's considerable sales promotion efforts in 1971/2 in a market which had become entirely a sellers' market; following the efforts by the oil companies to increase their oil product prices by between 30 and 50% in the aftermath of the Teheran and Tripoli agreements with the oil producing countries and the later collusion between the oil companies themselves. Thus the timing of Mr Jonker's speech was in itself mysterious, to say the least.

Whether intentional or not, this announcement brought into the open a debate which had apparently been raging within official circles for some time, with at least three main factions involved. First, the "conservationists" who wanted only a slow use of Groningen gas to ensure that it lasted forever! Second, one group of economic/policy advisers which saw the immediate marketing opportunities as giving the best chance there was ever likely to be for tying up outlets on long term contracts at highly profitable prices. And third, a second group of economists/policy advisers who saw the continued availability of large supplies of the, by now, relatively cheap Dutch gas (cheap, that is, within the framework of the higher price levels for oil products being sought by the oil industry) as a major element of weakness in the European energy pricing situation and who felt, therefore, that this should be eliminated in order to sustain the wider interests of Shell. Especially as Shell, in turn, was reminded by the other major oil companies of its obligations in this respect, given the benefits it had gained from the success of the U.S. oil companies' efforts in obtaining the permission of the U.S. administration to work together over oil taxes and prices, in spite of anti-trust legislation which theoretically forbade such collusion.

The Lack of Credibility for the Arguments

The first and the third groups combined to defeat the second group and decided to use the geological report on the potential production difficulties of the field as the reason to be given to the public for cutting back output and putting up prices.

The credibility of the policy change was, however, undermined by the figures attached to the statement. This said that Groningen could not sustain the expected 1975 output of 100–105 milliard cubic metres per year – so using a figure for a production potential that was nearly 20% higher than the January 1972 Gasunie figure of the maximum output achievable by 1975. In fact, the only previous use of the figure of 105 milliard m^3 as a 1975 possible production level was given by the author of this article in a previous evaluation in October 1971[1] – when it was received with the usual questioning as to the qualifications of the author to comment on such matters and questioning the validity of this forecast. Now, however, says the Ministerial report, this hitherto unofficially impossible annual production level of 105 milliard m^3 cannot be sustained as the field's output needs to be cut from an annual rate of depletion of 5% of the estimated reserves one of 4%. In absolute figures this implies an annual reduction of maximum output from 100/105 to 80/84 milliard m^3.

Could it be coincidence that this lower level of output is identical with this amount that Gasunie said in January 1972 that it had planned to produce: an amount, moreover that it was more than happy to sell on the basis of the then going prices. Now, in order to achieve this "reduced" level of output, the gas price to both big industrial users and to the local distribution companies will have to go up by an average of about 25%.

An Excuse for Price Increases

In brief, the whole saga of the "need" to reduce production savours of a device evolved to push through a price increase, required either to boost government revenues (in that the Dutch government takes over 70% of the total profits from gas sales in one way or another) and/or to eliminate the important element of weakness that Dutch gas prices have represented in European energy markets over the last 18 months or so. If it were just a result of the first factor then it would, of course, be mainly a matter for the Dutch people in as far as it represents an enhanced form of taxation. Nevertheless, it seems inappropriate to raise revenues in such an extra-parliamentary way, especially as it has been done on the basis of telling the people rather less than the whole truth: and also noting that the Dutch parliament has not had a full-scale opportunity to evaluate and determine gas policy since 1966 when the declared reserves of the Groningen field stood at only about one-fifth of the current level.

The co-owners of the field, Shell and Esso, were, moreover, unable or unwilling to provide any additional information on the nature of the production problems which had occasioned the need for a "reduction" in the planned rate of off-take. In response to a request from the author for such information, they wrote through NAM, "we see *no merit* in discussing the geological, petrophysical, engineering and environmental factors involved" (in the policy change). Only rarely, perhaps, does not get such an unequivocal statement of the secrecy with which the companies concerned like to shield their operations and the near contempt in which they hold the public's right to know the bases on which a country's natural reserves are being developed. It was this kind of attitude which had led to the nationalisation of the petroleum industry assets in countries such as Libya and Iraq. It is also the kind of attitude which necessitates the establishment of a powerful and expert countervailing force to evaluate and oversee production and pricing policies for Western Europe's newly developing oil and gas production.

The Impact on European Energy Consumers

Apart from the lesson to be learned for future European oil policy from the Dutch experience, one must also note that the Dutch government's effective decision to let natural gas prices for bulk energy users to float more or less freely upwards with increasing oil prices has important implications for energy consumers over a wider area of Western Europe. This is because the change in domestic pricing policy – involving higher prices – now enables NAM Gas Export to seek upward revisions in its contracted prices for Groningen gas supplied to West Germany, Belgium, France, Italy and Switzerland without offending with Common Market regulations. These do not permit price discrimination against consumers in member E.E.C. countries outside the country of origin of the commodity.

The importance of this lies not so much in the fact that foreign consumers of Dutch gas will have to pay an increased price for it (though this, of course, will have a significantly favourable effect on the profits of Shell and Esso and on Dutch foreign exchange earnings), but more in the fact that it will eliminate the remaining element in the energy pricing situation in Western Europe which remained potentially controllable and, moreover, outside the control of the oil companies. Their outline agreement with each other on oil marketing policies and infrastructure expansion plans controlled the situation as far as oil suppliers themselves were concerned.

The European coal industry has neither the capability nor the wish to challenge oil prices – as a result of its own problems of increasing costs, particularly labour costs which constitute over half of the industry's total – so that it generally welcomes the rise in energy prices as a means of reducing its financial losses. To date, available natural gas in Europe, apart from Dutch gas, has been too limited in quantity to establish market leadership as far as industrial fuels are concerned. Most notably in the United Kingdom, the gas industry, faced with heavy capital expenditure on extending its transmission and distribution system and on converting millions of residential consumers, has been more than happy to sell gas to industry at whatever higher price levels the oil companies cared to establish for their competing oil products.

Thus, with the upper price limit now removed from the Groningen gas, and with the bulk of the energy market controlled by the oil industry, Europe's only short term defence against a rise in oil prices is a continuation of the succession of mild winters (which over the last two years have had the effect of deadening the demand for heating oils) and a continuation of the current less-than-scintillating performance by the economy in which the relative stagnation has taken the expected growth out of oil industry sales so forcing companies – particularly those relative new to Europe – to seek required volumes of product sales by price cutting where necessary.

Nevertheless, these factors are both short-term influences only. The elimination of either one of them will cause an immediate hardening of energy prices, whilst the combination of a cold winter and a revival of economic activities would have a traumatic effect on European energy prices and, as oil industry profits soar to record heights, make self-evident the real motivations for the change in supply and pricing policies for Dutch gas.

B. The North Sea Gas Potential and Western European Gas Markets by the mid-1980s*

Introduction

The potential role of natural gas in the Western European energy market remains as perplexing now as it has been throughout the last decade. More than eight years ago I published a study, *Natural Gas in Western Europe* (see Section II chapter 1), which drew attention to the availability of a potential for gas production which far exceeded both the plans of operators and the expectations of governments in a situation in which there was no possible element of demand constraint on the potential to supply.

The essential element that could then be identified in the European gas market was the existence of a monopoly supplier seeking to achieve the highest possible monopoly rent out of the remarkable phenomenon of the largest non-associated gas field in the non-Communist world; viz. the Groningen field of the Netherlands in a concession allocated in its entirety to a joint enterprise of Shell, Esso and the Dutch government. The parties concerned were not only anxious to earn monopoly profits out of the exploitation, but were also jointly concerned with restraining the sale of gas such that their own interests in the maintenance of fuel oil and middle distillates markets would not be adversely affected.

I suggested in 1969 that this strategy would be undermined by competition from alternative supplies, notably by the development of gasfields in the southern part of the British sector of the North Sea (and possibly in other sectors too), and also by the attraction to the high-price gas market in Western Europe of gas supplies from the Soviet Union via pipelines from the massive West Siberia fields, as well as from other oil producing countries via the rapidly developing technology of liquefied natural gas transport. Based on these considerations an alternative forecast of the shape of the West European gas market by 1975 was suggested. This forecast may be compared with those at the same time by the energy planners in the O.E.C.D., the E.E.C. and various national entities. The sharp contrast between them, together with the situation as it developed by 1975, is shown in Table III-3.1.

The purpose of these introductory comments is not to allocate "marks" for the degree to which the forecasts were right or wrong, but simply to indicate the tendency by European "officialdom" to produce estimates of the future supply of natural gas based either on a static view

*** The edited version of an article published in *Geoforum*, Vol 8, pp 155–168, 1977.**

of the reserves situation or on an extrapolation of specific commercial and state policies which were in danger of being undermined by the forces of competition and/or change in the general energy and economic environment.

The Importance of Natural Gas in the West European Energy Economy

The above statement is as true today as it was a decade ago: indeed, probably even more so than it was then, because more governments and more companies have become involved, thus making the situation even more complex to analyse. This has had the effect of encouraging even more of the actors to pursue policies and to take decisions which, in essence, aim to limit the scale and speed of development of the resource.

TABLE III-3.1
Western Europe: a comparison of 1969 estimates of natural gas production in 1975 with the actual figures

Country/Region	1968 Production	Official Estimates for 1975	Author's Estimate for 1975	Actuals in 1975
		(in $m^3 \times 10^9$)		
The Netherlands	25.1	55	118	98
(of which Groningen)	(25.1)	(55)	(100)	(92)
West Germany	5.8	15	25	20
(of which on-land)	(5.8)	(15)	(20)	(20)
South North Sea (British/Danish Sectors)	2.3	30	38	37
Italy	10.4	12	20	15
France	8.7	6	10	11
Rest of Europe	<1	2	5	2
Total	*53*	*120*	*217*	*183*
(Estimates of % actual)		(65%)	(117%)	

Source: 1969 estimates from P.R. Odell, Natural Gas in Western Europe *(De Erven F. Bohn, Haarlem, 1969). 1968 and 1975 actuals from E.E.C. and national statistics*

In addition, there is another factor which now intervenes to isolate the evolution of the natural gas market from the supply and demand schedules as determined by competition in the energy market place. This is the existence of a more or less general belief that natural gas is an inherently scarce commodity, such that its discovery must be viewed in the context of saving it for the 21 century. By then, it is argued, it will be needed to provide a little residual light and warmth in a world otherwise devoid of readily available and usable sources of energy – in brief, the scarcity syndrome as originally set out in the first and ill-considered report of the Club of Rome and not un-related to the stories of gas shortages in present-day United States.

No one would advocate profligacy in the use of a fossil fuel, especially natural gas with its many advantages over the alternatives, but it is a far cry from such advocacy to a plea for a reasonable view on the development of the Western European natural gas market based on the following considerations:

(a) *The continent of Europe needs to reduce its dependence on oil imported from OPEC countries in the interests of security, both economic and political. Energy policies of individual countries, as well as of the E.E.C. and the I.E.A., are supposed to be changing economies and societies in such a direction. Yet the existence of known natural gas resources in and around Western Europe could, if they were developed effectively, cut an additional 50 million tonnes a year off the region's oil imports within the short space of time needed to develop the infrastructure to get the gas into the transmission systems.*

(b) *Europe stands at the beginning of its search for hydrocarbons and so far, in relation to the total potential, very little has been achieved. It is thus inappropriate, to say the least, for Western Europe to approach the question of reserves' decline and resource exhaustion in the same way as is currently necessary in the United States where the nation's resources have already been thoroughly explored and exploited for 50 years. The appropriate lesson to be drawn from the U.S., is to note how quickly reserves build up during the first twenty to twenty-five years of an effective exploration effort. On this basis, one can extrapolate a continuing rapid development of Western Europe's gas resources as the appropriate base on which to plan depletion rather than making plans merely related to the amount of gas already discovered.*

(c) *This can be done in light of the prospects already known for the gas resources of Western Europe and with an assumption on the anticipated exploration efforts. Such an approach shows that there is a high probability that the currently declared proven and probable reserves of gas are understated, and that available reserves will go on increasing for the foreseeable future, even after allowing for the gas which will be used in the context of what has now become a relatively slowly rising demand curve. This is illustrated in Table III-3.2 in which the potential annual supply of natural gas in Western Europe by the mid 1980s is also set out.*

TABLE III-3.2
Western Europe: its "proven" and possible natural gas resources and an estimate of their development by the mid-1980s

	Reserves declared 'Proven' plus 'Probable' in 1976 ($x10^9m^3$)	Reserves likely to be declared by the early 1980s ($x10^9m^3$)	Mid-1980s annual production potential ($x10^9m^3$)
On-shore Netherlands	2030	2100	105
South North Sea – British Sector	550	1050	50
South North Sea – Other Sectors	440	1250	55
On-Shore West Germany	310	450	25
Austria, France, Italy, etc.	420	600	35
Northern North Sea Basin – U.K./Norway	900	2500	115
Rest of European Continental Shelf (Ireland, Spain, etc)	50	350	20
Total	*4700*	*8300*	*405*

The Demand for Natural Gas

Within the context of OPEC oil at high prices, there is little likelihood that any natural gas produced in Western Europe would fail to find markets, except locally and in extreme circumstances of gas saturation of specific markets (eg, the residential sector market in the Netherlands). Otherwise, there are unlikely to be any near or medium-term future demand restraints through competition from other fuels. It is also reasonable to conclude that there need not be any significant political/environmental constraints on the use of gas, except in the case of its use for power generation for which restraints have been suggested in the belief that it is wasteful to use gas as the energy input into centralised electricity generating stations given their low 35 per cent conversion efficiency[2]. Finally, neither is there a high probability that the necessary infrastructure to transport and distribute gas cannot be built.

Overall, it is thus difficult to foresee much demand limitation on the use of natural gas in Western Europe for at least the next ten years. Its advantages as a source of energy are well-known and generally recognised. Its incorporation into the energy economy, at the maximum rate of development that the expansion of the production potential allows, still remains a function of the geography of its supply and the location of energy demand. This is illustrated in Figure II-I.1 (p 136), initially used in my 1969 study with the following note of explanation:

> *"In both the U.S. and the USSR the main energy consuming areas are remote from natural gas supplies. Despite these facts of geography gas has become a preferred fuel in both countries. Within Western Europe, on the other hand, the major supplies of natural gas are in the heart of the areas of heaviest energy consumption. Other things being equal, this situation should ensure the most rapid development and utilisation possible of Western European gas with consequent enhanced economic advantages for the continent's energy users"*[3]

The Supply of Natural Gas

The continuing basic validity of this earlier observation on gas in Western Europe is not in doubt: though there are now some constraints, as "other things have not been equal". For example, as already shown, there have been institutional (monopolistic) restraints on the production of Dutch natural gas. Now, almost a decade later, there is consequently a major difference between the production potential which the resource base development appears to indicate should be possible by the early 1980s (see Table III-3.2), and the latest O.E.C.D. figures for energy supply and demand in Western Europe to 1985[4]. These O.E.C.D. forecasts show only a range of 229–258 x $10^9 m^3$ of gas production, compared with the potential for a production of over 400 x $10^9 m^3$ which emerges from the analysis shown in Table III-3.2.

It is not that the O.E.C.D. figures for the future production of natural gas are so *very* pessimistic, when compared with the 1975 level of

$183x10^9m^3$. Within the context of an expected 47 per cent increase in energy use in Western Europe by 1985, the O.E.C.D. 'reference case' allows for a natural gas production increase of 45 per cent, and for one of over 63 per cent in its "accelerated policy" case. The latter contains, as one of its central elements, a stimulated level of natural gas production. Nevertheless, the difference between what is officially expected and what can be reasonably estimated as possible, is of the same order of magnitude as the difference which existed between the estimates made in 1969 for production levels by 1975 (see Table III-3.1).

In both cases the official view indicates an increase in indigenous production which is no more than 40 per cent of the increase which could be sustained by the full exploitation of the resources. It would thus seem that the same behavioural characteristics of official forecasters of the natural gas sector of the West European economy are again at work, viz. their inability, or unwillingness, to think in dynamic enough terms about the development of the gas resource base under the stimulus of an open-ended demand. Alternatively, however, one can argue that the O.E.C.D., as a multi-national energy planning agency, is influenced in its published forecasts by the reticence of member governments to commit their countries to the maximum possible production of natural gas. This is because such a commitment implies either too large exports to neighbouring countries so creating possible uncertainty over the likely availability of supplies of gas for domestic markets into the 21st century), or, even more simply, too high a level of production when set against other national energy supply decisions. Such decisions may simply seek to curb the rate of exploitation of hydro-carbons, or they may be related to pre-existing objectives for other sectors of the national energy economy (viz. coal and nuclear power) which are supported by powerful pressure groups that do not wish their interests to be under-mined by too much gas being brought to the market.

It is, indeed, only these factors that provide an adequate explanation for the continued under-estimation of the potential for indigenous natural gas in the Western European economy: given the absence of any really serious doubts over the size of the resource base which could be developed, or over the ability of the industry to produce and deliver the commodity, even from the adverse environments of the northern parts of the North Sea.

Table III-3.3 summarises, by country, the differences between the O.E.C.D. "Reference" and "Accelerated" cases, on the one hand, and the production potential estimates previously listed in Table III-3.2, on the other. The Netherlands, the U.K. and Norway are, in all cases, the dominant suppliers. The differences in the estimates for these countries are thus critical in the analysis of the overall situation.

Before dealing with these in detail, it is worth observing that in none of the other countries is there an official expectation that the 1985 production levels will partly reflect discoveries which have not yet been made, other than relatively small developments in reserves which will be necessary to maintain production levels in countries like France and Austria. This is truly remarkable, given that every country on the list plans an active exploration programme for natural gas in potentially petroliferous regions in which there must be some hope of success; otherwise no company or state entity would be willing to invest in exploration. Even so, the difference between the collective estimates for these minor producers together amounts to less than 30 x $10^9 m^3$. By contrast for the Netherlands, the U.K. and Norway the difference in each country's estimates is at least that great.

TABLE III-3.3
Western Europe: 1977 estimates of natural gas production by 1985

Country/Region	1975 Production	O.E.C.D. Estimates		Author's Estimates
		Reference Case	Accelerated Case	
		(in milliards – 10^9 – of m^3)		
Netherlands	96	92	111	146
United Kingdom	40	50	50	105
Norway	–	30	40	65
West Germany	20	20	20	20
Italy	15	17.5	17.5	20
France	10	8	8	12
Spain/Portugal	–	6.5	6.5	10
Denmark	–	2.5	2.5	5
Ireland	–	1.3	1.3	10
Austria	2	1.3	1.3	2
Total	*183*	*229*	*258*	*405*

Sources: O.E.C.D estimates from that Organisation's publication. The Outlook for Energy to 1985, *Paris, 1977. The author's estimates are based on production potential from gas reserves already discovered or likely to be available by 1980 so that there is time for the necessary production/transportation infrastructure to be built.*

Potential Supply from the Netherlands

The discrepancy in the case of the estimates is largely a function of contrasting expectations concerning the off-shore potential, though there is also a component relating to the author's expectation that the possible decline in production from the giant on-shore Groningen field will be more than made up by the exploitation of new on-shore resources which are now being found in the Netherlands as a result of an active exploration effort. The 'gap' in the interpretation of the offshore situation is, however, more important and emerges out of the information set out in Table III-3.4 and Figure III-3.1. This shows that the southern North Sea has many gas fields and discoveries. About 30 are shown in the Dutch sector on the map, whilst the more recently compiled information in Table III-3.4, indicates 51 discoveries. Thus, the official figure of a production of only 10 milliard m^3 a year by 1985 is almost ludicrously low, even given that it is based only on gas to be

TABLE III-3.4
Natural gas resources of the southern North Sea basin (excluding associated gas)

	Dutch Sector	British Sector	German/Danish Sectors
Total number of Gas Discoveries	51	31	3
Number of Discoveries Declared as Gas Fields	11	15	–
Governments' Declaration of Remaining Proven Gas Reserves (10^9m^3)	322*	552*	–
Number of Fields in Production	2	7	–
Other Fields with announced Production Plans	8	–	–
Current (1976) Annual Production (10^9m^3)	c.3	c.40	–
Estimate of 1980 Production from Fields currently on production or in development (10^9m^3)	c.10	c.42	–
Likely Remaining Reserves for all fields in each sector at summer 90% probability (10^9m^3)	1000+	1050+	50+
1980s Production Potential with Full Exploitation of the already discovered reserves (10^9m^3)	40+	50+	2–3

Notes: * *Arithmetic total "proven" reserves of declared fields. Based on all discoveries made and not just on declared fields. Based on 20–25 year depletion periods for the fields.*

produced from the small number of officially declared fields. Yet in 1977, exploration for new gas fields in the Dutch sector seems likely to reach a record level as companies at last respond to the opportunity the market for gas at prices which Dutch consumers now consider to be extraordinarily high (domestic gas prices in the Netherlands have more or less quadrupled over the last 3 years.)

Indeed, given information now available (regretfully little, given the high degree of secrecy maintained in respect of Dutch gas reserves) and what one can confidently extrapolate, a level of production of less than 40 x 10^9m^3 per annum from the Dutch off-shore reserves by 1985 would only be possible within the context of the unwillingness, on the part of the government, to allow reasonable developments to go ahead. Supply, in other words, will be constrained by an even more powerful institutional force than the N.A.M. monopoly over Groningen production in 1969.

Potential Supply from the U.K.

For the U.K. the situation is rather different and somewhat more complex. Three main factors are involved. First, developments in the British sector of the southern North Sea (see Figure III-3.1) are curtailed by the lack of incentives to the companies which have so far discovered a large number of fields in the area. All gas from these fields can only be sold "at the beach" to the British Gas Corporation and, as it is not prepared or not allowed, to pay a price which enables the companies to meet the opportunity costs on the investments required for the development of new fields, or even for the development of additional reserves from fields already in production, exploitation is not taking place and so reserves remain unproven and unused. This gas, together with that under the adjacent Dutch sector of the south North Sea, currently constitutes Western Europe's lowest-cost energy resource potential. Due to quirks of both Dutch and British policy, the benefits to be gained from its development are being foregone.

Second, the U.K.'s Gas Corporation tied itself, some years ago, to an open-ended commitment (both in quantity and price terms) to buy natural gas from the large Frigg field beneath Norwegian waters.[5] (See Figure III-3.2). Both the relatively high cost of this new gas (compared with the beach price of UK sector gas) and its annual availability, (at least 15 x 10^9m^3 per year, thus adding about 35 per cent to the level of supply), creates cause for concern over the ability of the Gas Corporation to market it successfully. This is because the Corporation is well aware that it is unlikely to be allowed to sell any gas for electricity generation as its use

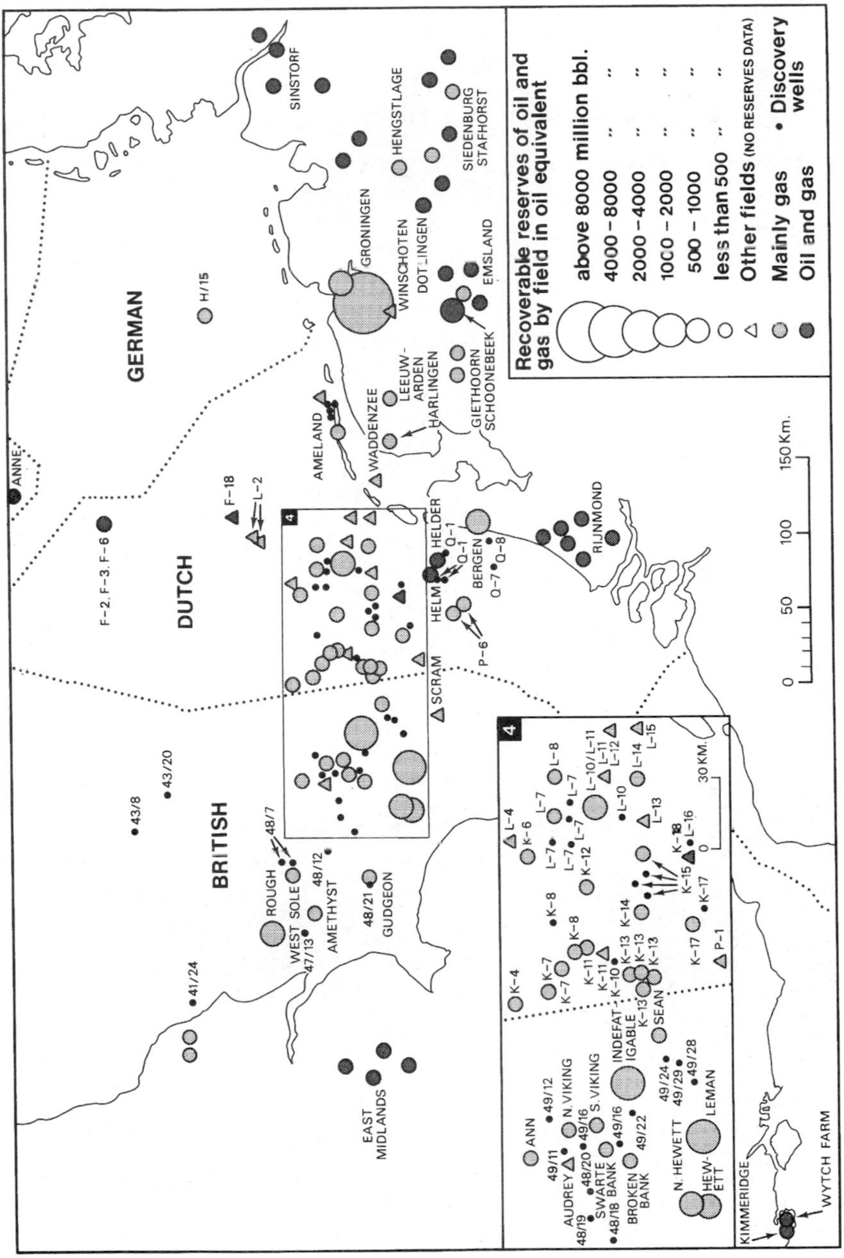

Figure III-3.1: Oil and gas fields and discoveries in the Southern Basin of the North Sea Oil province (see Figure III-3.2 for the key to the symbols on the map)

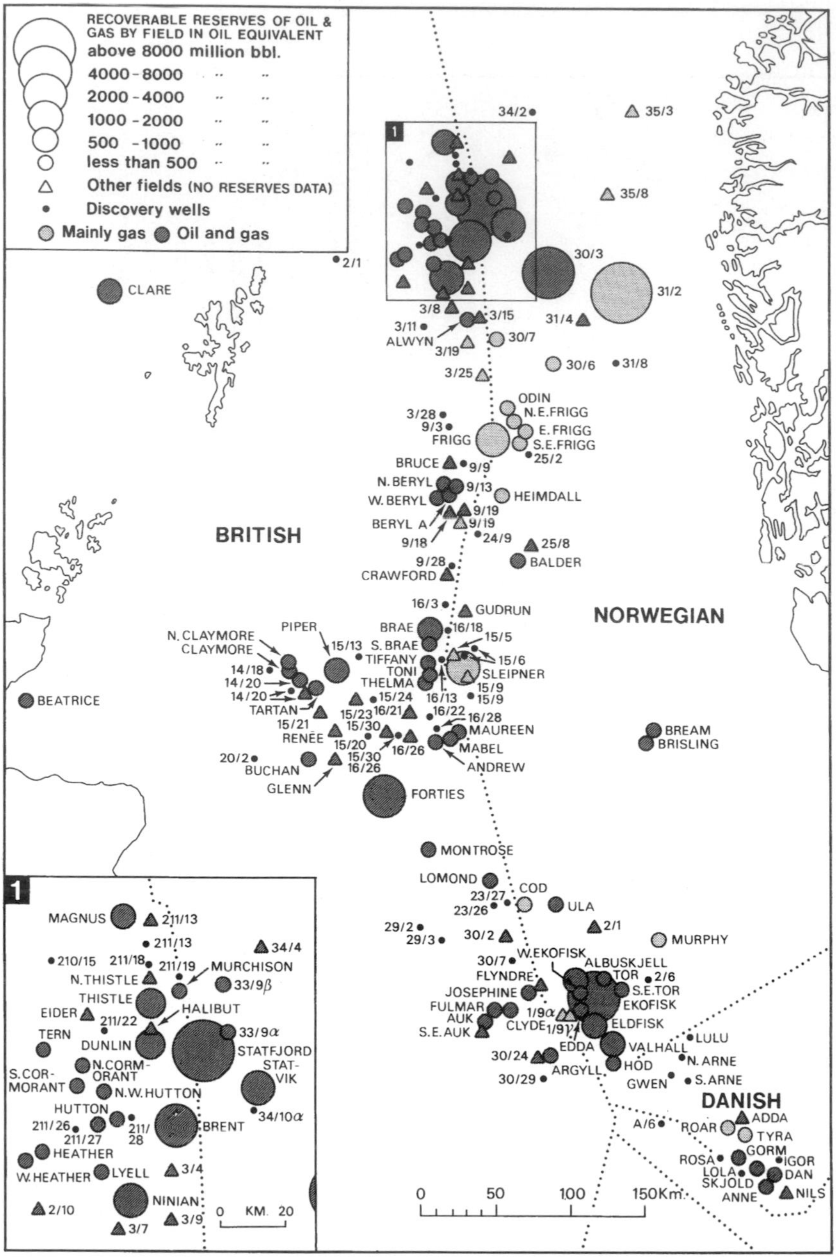

Figure III-3.2: Oil and gas fields etc, in the Northern Basin of the North Sea

would upset the one secure market for British coal. Thus, in concentrating on the marketing problem for Frigg gas (which, of course, imposes immediate and near-future cash-flow problems for the Corporation), it has to be less then fully enthusiastic about the rapid development of the even greater new production potential from the northern part of the British sector of the North Sea. Here many new oilfields with high gas-oil ratios have been discovered, so opening up the possibility of high annual rates of associated gas production by the early 1980s.

Third, there is the unwillingness by the U.K. to declare its production potential for 1985 at a higher figure than the 50 x $10^9 m^3$ designated in the O.E.C.D. study is because of the expensive question of the development of a multi-user pipeline system for collecting associated gas from the 30 or so north North Sea oilfields with collectable quantities of natural gas (see Figure III-3.3). This is partly a matter of expensive technology *per se*, but it is also a matter of politics, given the present Labour government's insistence that the operation should, in part, be a state enterprise so that the private sector companies involved cannot be certain that they will achieve an adequate return on the investment they are required to make. Detailed studies on the project are already under way and though urgency has been indicated, the required studies may be a way of buying time so that an appropriate policy can emerge. This is, however, not only time which will give the U.K. Gas Corporation a little breathing space for its new marketing efforts, but also time that will be expensively bought, either at the cost of oil production which has to be foregone, or of the associated gas that has to be flared.[6]

Indeed, so great are the difficulties involved so long as new supplies of British North Sea gas are viewed solely in a British context, that it seems self-evident that the preferred solution must lie in the U.K. linking its gas transmission system to that of the mainland of Western Europe, viz. across the Channel into France and/or Belgium. And yet this solution is, in itself, a problem because of the unwillingness of the U.K. to commit itself to long-term gas supply contracts for other member countries of the E.E.C. In 1975 the British government refused to allow British gas (from the Viking field in the southern part of the North Sea) to be sold to mainland European customers. Even now, in spite of E.E.C. legislation which formally forbids discrimination between E.E.C customers for any product from a member country, it seems likely that a similar decision would again be made. In the meantime, the potential for natural gas production from the U.K. will be restrained as in the case of the Netherlands largely for institutional, rather than sound economic or strategic reasons.

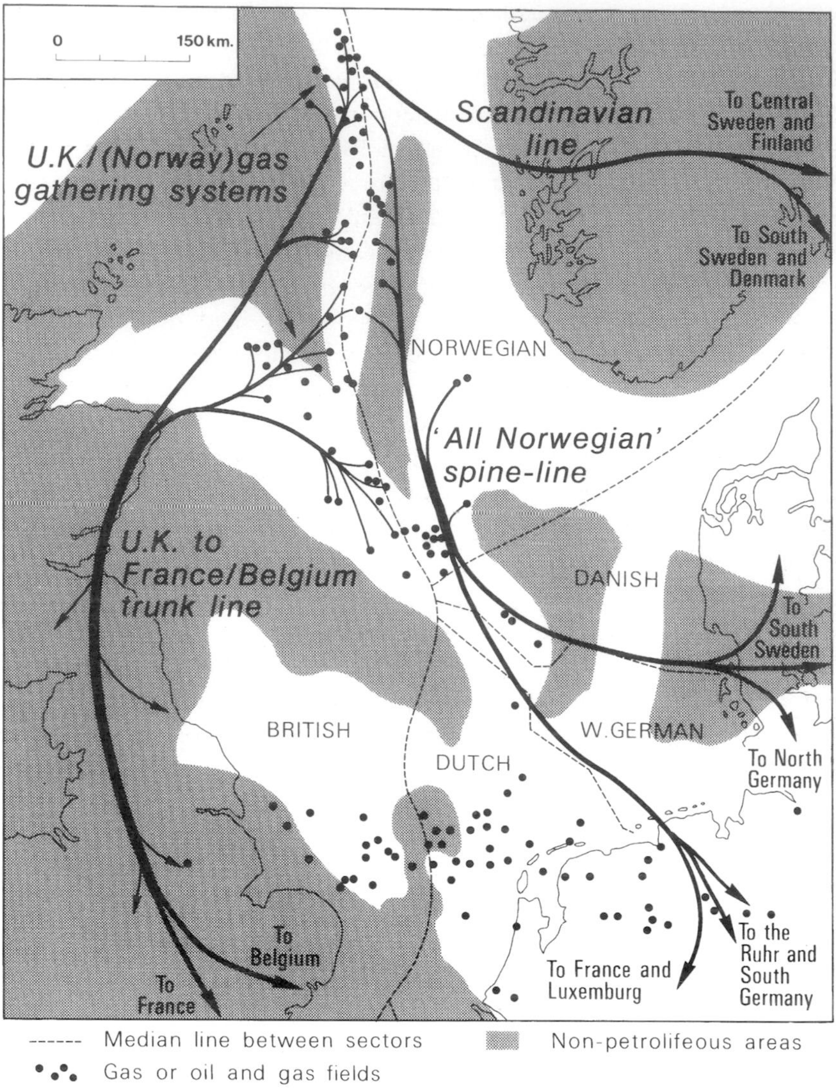

Figure III-3.3: The North Sea oil and gas province. Fields and discoveries etc, to December 1976 and alternative possibilities for natural gas collection and delivery systems from the Northern and Central Basins (N.B. The gas transport systems are hypothetical only: they do not represent definite plans in general or in detail as at March 31, 1977)

Potential Supply from Norway

This leaves consideration of Norway's gas producing potential. In respect of this, the O.E.C.D. forecasts specify the greatest proportional difference between 1985 "Reference" and the "Accelerated Policy" cases, with the latter indicating a 33 per cent higher availability than the former. Even so, the "Accelerated Policy" indicates so low a figure that it will take deliberately negative Norwegian gas production policies for it not to be exceeded. Unhappily it is such policies in respect of hydro-carbon production levels which epitomise Norwegian attitudes towards its North Sea hydro-carbon resources. If all considerations have to be evaluated in the light of accepted principle that the total production of hydro-carbons by Norway shall not exceed the equivalent of 90–100 million tons of oil per year then, given the reserves already found, some of which are currently under development, the production of natural gas *cannot* rise above the 30 x $10^9 m^3$ per year level. The rest of its stipulated maximum hydrocarbons' production "quota" will be fully taken up by oil from Ekofisk, Statfjord and other fields already proven.

In an earlier comment on this main element in Norwegian oil and gas policy, the issue of a limitation on Norway's gas production was put in the context of a Western Europe whose political and economic future is dependent on its success in reducing dependence on oil from the Middle East.[7] This suggests that the internal Norwegian reasons for formulating the policy decision in 1973 has been overtaken by events. This is confirmed by the fact that the assumption used in the calculations of the optimal level of Norwegian hydro-carbons production, viz. that all the country's resources would be fully employed in other economic activities, is no longer valid in the post-1973/4 oil price shock situation.

While the 1973 decision on oil and gas production limitations has not been formally reversed, there are recent indications that it is being seriously re-evaluated. It is reflected, for example, in Norway's licensing of a relatively large number of new blocks in the North Sea and in its decision, in early 1977 to allow exploration north of latitude 62° to begin in 1978.[8] The change thus opens up the possibility of re-appraising the level of natural gas production that could be achieved by Norway by the mid-1980s. Fields still to be evaluated, together with those that are still to be discovered over the next three to five years, contribute to the enhanced potential which may be conservatively put at the level of 65 x $10^9 m^3$ as shown in Table III-3.3.

As much of the potential relates to associated gas, from fields such as Statfjord, the production decision is often a joint one, involving oil as well as gas. Thus technical, rather than marketing, considerations can be paramount. If the gas "has" to be sold, rather than reinjected or flared, then transport considerations, particularly with respect to pipeline distances under water and the scale of the operation, may well be an overall determining factor for the volume which must be involved. Given the location of Stratfjord and other possible contributing fields to an enhanced supply of Norwegian gas (See Figure III-3.3), the quantities would seem likely to have to be larger rather than smaller.

If a "Norwegian" solution to the Norwegian North Sea gas potential is to be achieved, through the development of the north-south "spine-line" which could link a series of fields in Norwegian waters with Germany and/or Denmark (See Figure III-3.3), then a delivery system with a capacity for upwards of 20 x $10^9 m^3$ per annum seems likely to be required in order to keep the unit investment cost down to a level which ensures the marketability of the gas. This, when added to existing contacts, would take Norway's contribution of gas to the European market above even the 'Accelerated Policy' figure in the O.E.C.D. report. Such a system is attractive, not only because of the "Norwegian-ness" of the solution, but also because it ensures entry to the high-gas demand areas of West Germany and other mainland markets, where an oil related price would be an acceptable way of determining the gas price.

This, however, does not apply to the alternative solution, viz., the building of a system of collector pipelines from the Norwegian fields to tie into the proposed multi-user trunk gas lines running from fields in the British sector of the north North Sea to the U.K. mainland. In this case Norway could get away with a more limited scale of development, but only in exchange for difficult negotiations with the U.K. both over the charges to be made for the use of the lines to the U.K. and for using the U.K. as a transit state, if there were to be an attempt to get the Norwegian gas delivered to the mainland of Western Europe via the U.K.

Not that such a development would be without benefits to the U.K. Indeed, two sets of benefits would arise. First, there would be the added value to the British economy of this service function with, most specifically, the Norwegian throughput of gas in the British trunk lines more or less guaranteeing the short-term viability of the multi-user line project. Second, the U.K. would benefit in terms of the political opening that the transhipment of Norwegian gas to France or Belgium, would give to a British government in any attempt to "sell" the idea that some British gas could also move across the Channel, at both low cost and high value in terms of foreign exchange earning to the U.K. economy.

Such Anglo-Norwegian co-operation could thus enable Norway to restrict its supply of natural gas to a level lower than that necessary were Norway to go it alone in creating a separate transportation system required for the profitable exploitation of its own gas resources. A decision on this issue may well be the main element involved in determining whether Norway does nothing more than expected by the O.E.C.D. energy analysis, or whether it moves to a level of supply by 1985 which will take its total production of oil and gas beyond the previously determined upper-limit of the 1973 plan. This choice by Norway also depends on its own institutional response to the challenge and opportunities presented by the natural gas markets of Western Europe; the same sort of choice, that is, as has already been shown to apply in respect of the other two main suppliers of gas to serve Europe's needs and demands.

Conclusion

The Netherlands, Britain and Norway thus have much in common in their near-future relationships with the rest of Western Europe, particularly in terms of their ability and their responsibility to determine just how independent the region can become from O.P.E.C. oil in the 1980s. Let us assume that all three decide to pursue policies that maximise, rather than restrict, the production of their large natural gas reserves. Further, let us assume that all the additional natural gas will substitute the oil that the O.E.C.D. currently expects will have to be imported from the O.P.E.C countries. Within the context of the figures in the O.E.C.D. report on West Europe's total 1985 energy demand, and taking into account the total supplies of other sorts of energy available in Europe at that time, then the need for oil imports would fall from O.E.C.D.'s calculated 11.2 million b/d to only a little over one third of this amount (roughly 4 million b/d). It would thus, to all intents and purposes, make Western Europe independent of O.P.E.C. for its supplies of essential energy. This much, in respect of Europe's energy security, is at stake from the decisions on future levels of natural gas production that have to be taken by Norway, Britain and the Netherlands.

References

1 See P.R. Odell "Gas as a current source of energy in Europe" *Petroleum Times*, Vol 75, October, 1971, p. 18. See also Chapter III-1.3 for further details/argumentation for this high level of Groningen production by 1975.

2 This 'belief' is unjustified given the techno-economic advantages of producing electricity from gas in highly efficient combined heat and power plants and in combined cycle generating stations. See Chapter 6 in Section I of this book.

3 P.R. Odell, *Natural Gas in Western Europe: a case study in the economic geography of energy resources*, De Erven F. Bohn, Haarlem, 1969

4 O.E.C.D., *The Outlook for Energy to 1985*, Paris, 1977

5 The reason for the B.G.C. accepting such a commitment – in the light of lower cost gas availability from the south North Sea basin and a high expectation of gas reserves being discovered in British waters further north (as has now happened) – has never been satisfactorily explained: it seems to lie in the belief, by the former Chairman of the Corporation, that he was dealing with a scarce commodity and was concerned that he would not be able to cover his forward sales. This reasoning also explains why the B.G.C. stopped trying to sell gas to new customers at the same time. These mistakes have cost the B.G.C. dearly in respect of both its supply costs and its sales revenues.

6 The Dept. of Energy's announcement at the end of June 1977 that part of the Brent oilfield must be closed down until associated gas production can be collected nicely illustrates this point.

7 See P.R. ODELL, "Norway's Oil and Europe's Energy Needs; Bases for a mutually advantageous development policy", Financial Times Conference on *Scandinavia and the North Sea* Oslo, 29 April 1974.

8 This decision, however, has been 'put on ice' pending a thorough investigation into the Ekofisk field "blow out" in May 1977 and for further study of environmental questions.

Chapter III – 4

Gas' Expansion Prospects Re-emerge*

A Foreword

The introduction to the original version of this paper has been omitted as, in essence, it summarized the contents of the author's earlier paper, "The North Sea Gas and Western European Gas Markets by the mid-1980s" (published in this volume as part B of Chapter III-3.)

That earlier article emphasised the then-emerging constraints on Western Europe's gas production and use. This paper shows the great extent to which these constraints materialized. Production in 1985 was less than two-thirds of the potential that was previously shown to exist, whilst use was little more than 50% of that anticipated by the O.E.C.D. In particular, the constraints on supplies of gas from Europe's three largest producers (The Netherlands, the U.K. and Norway) proved even more emphatic than argued in my 1977 paper. Thus, the "major expansion" of natural gas in Western Europe, as predicted in this paper, is starting from a significantly lower base than alternative policies would have produced by 1985.

*** A paper presented to the 1985 North American International Association of Energy Economists Conference, *World Energy Markets: Stability or Continued Cycles*, Philadelphia, December 10–13, 1985 and subsequently published in the Proceedings of the Conference edited by W.F. Thompson and D. J. DeAngelo, Natural Resources and Energy Management, Western Press Special Studies, Boulder 1985, pp 286–330.**

1. The "Official" View of the Market – and its Implications

a. It is in the context set out above of a policy-generated scarcity of gas and of regulated markets (including an E.E.C. Directive which forbade the use of gas for power generation) that all official forecasts for indigenous natural gas' contribution to the Western European energy economy have been eminently pessimistic. This was clearly summed up in a 1982 special International Energy Agency study on gas in which the following conclusion was reached,

> *"Good prospects exist for increasing production ... but in practice this will depend not only on geology, technology and economics, but also on appropriate policy stances in the producing countries".*[1]

This was reflected in the I.E.A.'s energy supplies' forecasts at that time for 1990.[2] These indicated a decline in indigenous gas production from 160 million tons oil equivalent (m.t.o.e.) in 1980 to 157 m.t.o.e. in 1990. The I.E.A.'s most recent set of forecasts[3] is even more pessimistic showing a 1990 indigenous production of only 150 m.t.o.e. and a year 2000 level of production of only 95 m.t.o.e. In other words, the I.E.A.'s forecasts of future Western European gas production indicate acceptance of the current negative attitudes and policies on the exploration for, and the exploitation of, the continent's gas resources. These are expected to be maintained. In particular, the Netherlands, Britain and Norway will continue to fail to respond to the supply opportunities offered by their considerable reserves, and they will, moreover, be able to resist the now rapidly emerging much more competitive elements in the energy sector.

b. The history of regulation and control of the gas industry in Western Europe, as summarized above, certainly provides some justification for the I.E.A.'s extremely conservative view of the prospects for the future supply of indigenous natural gas. Since mid-1983, however, there have been developments in attitudes, policies and strategies in respect of the production of gas in Western Europe and its potential for meeting an increasing share of the continent's energy needs, viz.

i) There is growing embarrassment over the pre-emption of an increasing part of Western European gas markets by imports. Imports of gas have already almost doubled since 1980, as a result of take-or-pay contracts made with the Soviet Union and Algeria. Meanwhile, the production of indigenous gas has declined, in spite of the existence of an infrastructure capable of producing and distributing much larger volumes, a rapid build-up of proven reserves (see below), and the expectations by the governments of the producing countries in Western Europe that their production would have been greater. Commitments made by several Western European countries to take increased volumes of imports of natural gas in the coming decade now pose an even greater threat to indigenous production, unless the total market can be significantly expanded.

ii) This has lead to increasingly acrimonious relationships between Western European gas importing countries and the continent's main producers. The latter are anxious to secure additional outlets for their gas, in order both to achieve the increased employment of resources by the industries which supply and service gas production, and to generate the revenues which the governments need – and on which, indeed, they had already reckoned prior to the weakening of the gas market. They now find that their objectives in this respect are being increasingly thwarted by competition from imported gas supplies.

iii) There is increasing recognition that the policies of pricing gas at or above premium oil products' prices or even electricity prices' equivalent and of deliberately restricting the use of gas to high-value outlets is producing an uncompetitive energy pricing situation in Western Europe – compared with elsewhere in the world, most notably the U.S. where natural gas is viewed as a third-equal potential energy source, along with coal and oil, and is thus made available, as necessary and as appropriate, to low value energy markets, in competition with low-grade oil products, coal, and electricity.

c. These changes in attitudes towards natural gas markets are now producing moves towards the Western European equivalent of the U.S. recent deregulation of its gas industry. This development, with the consequent opening up of large markets in power generation and industrial steam-raising in competition with coal, fuel oil and nuclear power, provides the opportunity for the full exploitation of the large additional volumes of indigenous gas reserves which are already proven (see Table III-4.1), and for the exploration and discovery of additional volumes of reserves which, to date, remain undiscovered as a result of the hitherto restrictionist policy.[4] Belated recognition by policy makers of, *first*, that Western Europe is a gas-rich rather than a gas-poor region in terms of its reserves potential and, *second*, that it is economically desirable, while energy prices remain high, to find and produce these reserves quickly, rather than slowly combine to justify an hypothesis that the Western European gas industry is now at the beginning of a period of production expansion. This view which is markedly at variance with that of the I.E.A., as indicated above. A 26% increase in the contribution of indigenous gas to Western Europe's energy supply by 1990 and a 60% increase by 2000 seem possible. If so, then the prospective 240 m.t.o.e. of indigenous gas in the continent's energy supply in 2000 will be well ahead of the 160 m.t.o.e. which then seems likely to be contributed by indigenous coal.

2. Gas Reserves and Production Potential in Western Europe

a. The prospects for expansion of the production of gas in Western Europe are already widely recognized by the oil industry – with an increasing number of highly optimistic statements on the opportunities for additional gas production from the companies most directly concerned in the exploration and exploitation processes. British Petroleum, for example, has described the 1980s as "the decade for gas expansion in Western Europe". AMOCO has indicated outstanding prospects for gas discovery and production; while Shell, which discovered the giant Troll field in the Norwegian part of the North Sea in 1979, has expressed the view that this provides an opportunity to move total European production to a level higher than the best achieved to date. This view is shared by Mobil following a study which it made in depth of the North Sea's remaining potential[5]. Table III-4.1 shows the emerging situation on production, reserves, reserves development and production potential over the rest of the century in the countries of northwest Europe where most of the short and medium-term opportunities for an expanding indigenous gas industry are located. The relationship between the currently proven plus probable reserves and the 1984 level of production shows an R/P ratio of more than 40 years. This may be compared with a "necessary" minimum ratio of about 15 years and an R/P ratio of little more than 10 years in the United States, with a current annual level of natural gas production which is almost four times higher than in Western Europe. This low R/P ratio in the U.S. necessitates additions to gas reserves in each and every year which are always large enough to replace all the gas which is used annually and this has to be done, moreover, in a country which has been intensively explored for its gas reserves over many decades.

b. By comparison, Western Europe has an easy supply prospect. Even after allowing for the depletion of proven reservoirs sufficient to match the planned production of indigenous gas over the second half of the 1980s (as shown in Table III-4.1); and further assuming additional gas discoveries and field extensions which are modest relative to both recent experience and to the expenditure on exploration and field appraisal/development which is already committed, then the gas reserves which will be proven and probable in northwest Europe in 1990 will provide a reserves to production ratio which is higher than the current ratio of over 40 years. The ratio could, indeed, by then be in excess of 50 years and so make the region even more emphatically one of significantly under-utilized natural gas potential. The reserves situation by 1990 seems likely, in other words, to offer near certain prospects (viewed from the point of view of reserves availability) for major production expansion. The final column of Table III-4.1 shows that annual production from these northwest European countries could be at least $325x10^9m^3$. This is equal to almost 275 million tons oil equivalent: a production potential for natural gas from this part of Western Europe alone which far exceeds the I.E.A. forecast for a year 2000 production of indigenous gas of only 240 m.t.o.e. from the whole of the continent.

c. Even if one allows for a (highly unlikely) run-down of natural gas production elsewhere in Western Europe (where gas has been produced for some decades already, and where, indeed, as will be shown later, there are plans for maintaining or even expanding production), a situation as forecast by the I.E.A., of slow decline in the continent's gas output over the present decade, followed by a strong decline in the 1990s implies such a serious failure on the part of the gas industries and governments' energy policy makers to secure the production and utilization of the gas reserves of north-west Europe as to be an eminently unbelievable proposition. The forces that are already at work to ensure that the production of natural gas in Western Europe gets back on to a growth curve undermine the validity of such a pessimistic view. These forces have already been presented in general terms, but the following supply-side developments, in particular, should be highlighted in order to show how opportunities for the expansion of the supply of natural gas are already well under way – and why they are likely to become stronger

TABLE III-4.1

Natural gas in Northwest Europe: Production, reserves and potential ($m^3 \times 10^9$)

Country/Origin	Production in 1962	Production in 1973	Production in 1984 est.	Cumulative Prod to 1984	End-1984 Remaining Proven Probable Reserves (1)	Reserves/ Production Ratio (in years)	Planned Average Ann. Prod. 85–90 (2)	Likely Remaining Reserves 1990 (3)	R/P Ratio in 1990 (in years)	Potential Annual Production in 1990s (4)
Netherlands:										
Onshore	8	60	55	1015	1750	29	75	1550	27	102
Offshore	0	0	20	80	400			500		
W. Germany:										
Onshore	10	15	15	290	250	18	16	300	22	20
Offshore	0	0	0	0	25			50		
Denmark:										
Offshore	0	0	0.2	0.2	150	100+	2	200	100	10
Norway:										
North Sea	0	0	31	175	2500	80+	40	3000	100+	150+
North of 62°	0	0	0	0	substantial			1000		
United Kingdom:										
South North Sea	0	25	27	425	650			450		
Elsewhere on C.S	0	0	16	65	1050	40+	45	1450	45	45
Onshore	0	0.2	0.2	3	50			150		
Ireland:										
Offshore	0	0	2	7	60	30	2	75	37	4
TOTALS	*18*	*c.100*	*c.166*	*c.2060*	*7000+*	*over 40*	*c. 180*	*8725+*	*50+*	*325+*(5)

Notes: *(1) Based on Governments' and Companies' declarations of reserves*
(2) Based on announced production schedules by governments and companies
(3) After allowing for production 1985–89 and for new additions to reserves in the meantime
(4) Assuming no artificial restraints on production (eg. policy restrictions), but allowing for technical restraints
(5) This figure is less than individual countries' total because of temporal differences in peak production from country to country)

over the next year or so;

i) There will be an increase, rather than a decline, in production levels in the Netherlands as a consequence of the upward re-evaluation of existing reserves, the continuing discovery of new reserves, and the need by the government for the revenues it secures from gas production. It has, indeed, already been decided to maintain – and, if possible, to increase – the industrial use of gas; to sell more gas for power generation at least until 1987 or, in the absence of other markets, beyond that date; and to encourage the use of gas in combined heat and power installations, the development of which is being stimulated by government action. Meanwhile, some additional export markets for gas in neighbouring countries have already been negotiated and others are under examination.

ii) There will be a more rapid than currently expected increase in production from Norway after the later 1980s. This will be the result of changed attitudes on production levels and prices in response to the market opportunities which are developing in Western Europe, in the context of a growing appreciation that the real value of gas left in the ground is not increasing (on the contrary, it is already declining quite rapidly and will continue to do so). Moreover, Norway will also need to export more gas in order to achieve the level of tax revenues the country requires and the level of offshore activities which are necessary to keep the resources available to the Norwegian offshore industry fully employed. Politically, Norway will increase its efforts (so far unsuccessful) to persuade gas importing countries in Western Europe that Norwegian gas is preferable to Soviet gas. Indeed, a Norwegian/E.E.C. agreement on preferential treatment for additional gas from Norway seems highly likely by the early 1990s.

iii) There will be a significant expansion of British production as the result of the appreciation of the reserves of existing fields and the exploitation of new reserves, considerable volumes of which (over $1000x10^9m^3$) have already been found. Some production will by the end of the decade be marketed directly by the producing companies themselves under revised legislation which will make this alternative to selling the gas to the B.G.C. more attractive. This legislation also opens up the possibility of natural gas exports from the U.K. by one or more of the producing companies, following a challenge to the validity of the current government ban on exports of gas to other member countries of the E.E.C. The potential for gas production from the relatively mature areas of the UK continental shelf (which still retain substantial prospects for exploration and development) is estimated as rising to some $50x10^9m^3$ per year by 2000. In addition there are good prospects for greatly increased levels of on-shore gas production and for offshore gas supplies from so far undeveloped parts of the continental shelf – such as the West Shetland basin.

iv) There are also prospects for additional output from the other North Sea basin countries, viz. Denmark and West Germany, both of which are now belatedly pursuing intensified and more extensive gas exploration – and exploitation – policies. Expansion of their production will follow the expansion of the markets for gas.

v) Elsewhere in Western Europe there is a growing gas exploration effort in a wide variety of productive and potentially productive offshore basins (as illustrated in Figure II-6.2 (p. 242), and, increasingly, in reappraised onshore prospective regions (such as the Paris basin, the Pyrenees and in parts of the Alpine region). There is a near zero probability that all these efforts will be unsuccessful for if there were no chance of success then the exploration investments would not be made. With expected exploration successes, the currently generally low production levels in these countries (see Table III-4.2) have a high probability of being modestly increased over the period to the end of the century.

3. Contrasts between the 'Official' and the Alternative Indigenous Gas Supply Prospects

a. The contrasting estimates of country by country volumes of natural gas production for 1990 and 2000 are set out in Table III-4.2. The marked contrasts shown clearly demonstrate the unwillingness, or the inability, of the I.E.A. to take into account in its forecasts the opportunities for increased production offered by the rapidly evolving West European gas resource base. Except for the smallest producers (viz. Denmark, Ireland, Spain and Turkey, accounting in total for only about 5% of output), the I.E.A.'s forecasts of individual country's gas production in 2000 are not only lower than the levels of production for 1990, in spite of expected major additions to reserves in the meantime, but they are also lower than the production level achieved in 1984 – even though production in 1984 was constrained by demand conditions. This forecast has, moreover, been made in spite of a currently known availability of proven reserves which is more than adequate to sustain current levels of production well into the 21st century. And in the case of the three major producers (viz. the Netherlands, Norway and the UK), the I.E.A.'s forecasts of declining rates of production would, on extrapolation, bring production in those countries down to zero even before currently proven reserves are exhausted. Indeed, in the case of Norway production would be near-zero before 10% of its reserves are depleted.

b. Given the motivation of the countries concerned to continue to produce gas – in order to serve national economic interests, if for no other reason – and the forecast of a continuing strong demand for natural gas in Western Europe (as shown by the I.E.A.'s own expectation that gas imports will increase from 45 to 137 m.t.o.e. between 1990 and 2000), the I.E.A.'s forecast of an imminent or a very near future decline in Western Europe's natural gas production and the impending demise of the industry soon after 2000, is, quite simply, inexplicable and unbelievable. Unless, of course the Agency is forecasting that indigenous western European gas production will be unable to compete in the market place against imports. If this is the cause of the pessimism then it is worth noting the contrast with other views presented by the I.E.A., viz. that West European deep mined coal will be able to compete so effectively against imports that indigenous coal output will increase! The latter depends on the I.E.A.'s expectation that energy prices in the 1990s will rise in real terms to levels higher than those of the early 1980s so indicating an expectation of energy shortage. Thus the idea of market

TABLE III-4.2

Forecasts of natural gas production in individual Western European countries for 1990 and 2000 (in m.t.o.e.)

Country	1984 Production (estimated)	Forecasts for 1990		Forecasts for 2000	
		IEA	This Study	IEA	This Study
1. Major Producers					
Netherlands	55	50	65	24	70
Norway	34	26	30	4	55
United Kingdom	32	35	45	31	58
Sub Total	*121*	*111*	*140*	*59*	*183*
2. Minor Producers in 1984					
Austria	1.5	1	2	1	2
Denmark	>1	2	4	2	8
Germany (West)	15	14	16	13	17
Greece	>1	>1	>1	>1	1
Ireland	2	1	2	2	5
Italy	12	10	12	10	12
Spain	>1	1	3	2	4
Turkey	>1	>1	1	>1	2
Sub Total	*32*	*30*	*41*	*31*	*51*
3. Non IEA Member					
France	7	na	8	na	6
TOTAL for Western Europe					
	160	*149**	*189*	*96**	*240*

* *Includes 8 for France in 1990 and 6 in 2000*

Sources: *IEA Energy Policies and Programmes of IEA countries, 1983 Review, OECD, Paris, 1984.*

This Study: based on production potential from declared and expected proven/probable reserves, but with production limited by both supply side and demand (market) constraints.

constraints on the relatively much lower-cost Western European gas industry is also an unlikely, if not an impossible, explanation for the organization's gas production forecasts. To put it bluntly, the I.E.A.'s indigenous gas supply forecasts appear to be based on irrational argument, and an inconsistent set of supply/demand relationships.

c. By contrast, our alternative forecast for an expansion of Western Europe's indigenous gas supply is related to the essential parameters of adequate reserves' availability and the national economic and political interests of the producing countries. The alternative forecast indicates, however, a level of production which is significantly less than the potential to produce (as defined by reserves and their producibility over time). It thus takes into account both nationally-determined constraints on production levels – particularly by Norway – and also an expectation that there will continue to be gas demand constraints in the short to medium term. These constraints are the consequence of high prices, conservation in residential use through technological developments, the impact of the recession on industrial use, and competition from external suppliers of natural gas. The combination of these factors is, indeed, so powerful that the increase in the production of indigenous gas over the next three to four years is likely to be slow. Thereafter, the constraints will gradually become less effective so that eventually the level of production will build up rapidly. This will apply particularly in the mid-to-late 1990s when the integrated Western European gas supply and transmission system will have been extended to all parts of the continent – including Scandinavia, the British Isles and the Iberian peninsula. (See Figure II-7.1 – on page 263 – for the present network). Lower real transmission costs and enhanced security of supply as results of the fully integrated network then serve to stimulate gas use – and hence production from a still growing reserves' base. Natural seems likely to become Western Europe's most important source of indigenous energy very shortly after the turn of the century.

4. Competition from Imports of Natural Gas

a. In the meantime, however, the indigenous natural gas supply industry has to fight off increasingly virulent competition from imports. The I.E.A. clearly believes that the industry will not be successful in this respect. It forecasts that imports will increase almost four-fold from their present level by the year 2000, whilst indigenous production first stagnates and then declines. By 2000, according to the I.E.A., imports will contribute almost 60% of Western Europe's total gas supply. With our more realistic alternative view on the supply prospects for indigenous natural gas a much more modest increase in the volume of imports may be predicated. Over the period to 2000 they seem likely to increase from just under 40 m.t.o.e. to 85 m.t.o.e. per annum. This implies a modest increase only in the share of imports in the total supply of Western Europe's gas, viz. from just over 20% in 1984 to about 26% in 2000.

b. The world is, of course, rich in discovered, but so far unexploited, gas and western Europe has become something of a "jampot" for potential supplying "wasps" from outside the continent for well over a decade. This condition has arisen, as indicated above, basically because the regulated high price of gas in Western Europe since the early 1970s provided a very profitable market for external suppliers. Two other factors have also been of importance; first, the size of Western Europe's energy market, and second, the geographical intensity of that demand. These factors create market conditions which are eminently suited to its being served by pipelined natural gas. In this respect the Western European energy market is a more interesting and suitable prospect for domination by natural gas than the energy markets of either the United States or the Soviet Union. The wide open spaces of both of those countries diminish the economic viability of natural gas in the generally distant markets. Nevertheless, in spite of these relatively adverse geographical conditions in both the U.S. and the Soviet Union, gas use has expanded to fill about 30% of the energy markets in the two countries. By contrast, natural gas in Western Europe has, to date, only achieved a little over half this degree of penetration of the energy market; a rough measure perhaps of the extent to which the gas industry in Western Europe has been restrained.

c. As demonstrated in Chapter III-3 in a discussion of individual country's policies, factors of nationalism have inhibited the speed of development of Western Europe's indigenous gas resources, compared with the level of development that would have been achieved under a continental-wide, unified politico-economic framework, akin to those of the U.S. and the U.S.S.R. In spite of the attractions of the market even imports of natural gas to Western Europe have been inhibited from some potential suppliers. The following factors have been important in this respect:

i) The large-scale contracts for imports that are required, to enable economies of scale in transportation to be achieved, necessitate the joint agreement of two or more of the potential importing countries – as, for example, with the plans, now on ice, to import additional supplies of L.N.G. from Algeria and Nigeria. This factor has also retarded the expansion of pipelined imports from the U.S.S.R. in, for example, the unfulfilled plan to pipe Soviet gas via Finland to Sweden and on into other northern European countries.

ii) Large-scale imports require a commitment to a high level of capital costs, the investment of which creates an inflexible delivery system involving possible political risks – as, for example, with the proposed natural gas line to Western Europe from Iran via Turkey and South East Europe.

iii) A commitment to a major scheme for the delivery of gas from a single supplier creates conditions in which the supplier can then exercise both economic and political pressures for changing the terms of agreements – as, for example, in the attempt by Algeria in 1982/3 to secure a higher price than that originally agreed for the gas it supplies to Italy through the Trans-Med pipeline.

iv) The high transport costs involved, both via long distance pipe-lines and, even more so, by L.N.G. tankers, imply a netback-value of the gas to the suppliers below that which they prefer (viz., the energy equivalent value of crude oil exports). This means that agreements have been difficult to achieve, and/or that the gas, on delivery, has not been available at a cost which enabled it to be priced at a level which permits the expansion of the natural gas market. This is particularly true, as in the recent past, under conditions of stagnant or declining energy demand, and a plentiful supply of other sources if energy; notably, of competing petroleum products which have become available at prices discounted from those against which the prices for gas have been determined.

d. The combination of the impact of these factors seems likely to continue to constrain the growth of natural gas imports to a rate which is well below that forecast by the I.E.A. over the rest of the century. This likelihood is enhanced in the context of the ability of the Western

European gas producing countries to produce – competitively – much larger quantities of gas than the I.E.A.'s forecasts indicate. In this respect it is worthy of note that the I.E.A.'s 1983/84 forecast for gas imports in 1990 is 28.5% lower than its 1981/82 forecast[6].

e. Nevertheless, some new contracts for imports, to provide gas additional to that agreed under earlier contracts, have been made over the last few years. These include enhanced L.N.G. supplies from Algeria to France and Belgium (at prices now well above today's market prices so that the importers are negotiating reductions in the quantities involved); gas through the now completed Trans-Med pipeline from Algeria, via Tunisia, to Italy; and, most important of all, additional supplies (of up to 20 m.t.o.e.) of Soviet gas through the spare capacity in the pipelines to Western Europe which are already in operation (or in which additional capacity can be created through the installation of new compressor stations). These agreements provide near certain additions to imports over the rest of the 1980s, as the contracts have generally been concluded on a take-or-pay basis. They will lead to a total of about 55m.t.o.e. of imported gas by 1990.

5. The Impact of Competition from Imports

a. The ability and willingness of the potential exporters of more gas to Western Europe to sell their gas, if necessary, at a lower price than that which they originally thought they might get is an option open to them, given the relatively low costs of gas production in the countries concerned. The phenomenon has already led to an expansion of the market for imports at the expense of some indigenous production – particularly exports from the Netherlands which are now running well below their anticipated levels[7]. Additional imports of gas by many Western European countries in the later part of the decade seem likely to intensify this difficulty for the Netherlands' planned level of exports. They also adversely affect Norway's production prospects as a result of that country's sluggish response to date to the challenge of the weakening energy market since 1981, and the potential over-supply of gas in Western Europe throughout the next ten years at least.

b. There is, moreover, an increasingly likely prospect that the countries exporting gas to Western Europe will, within the next few years, in order to make their exports more attractive, offer gas under conditions of price and deliverability which will make it competitive with imported coal, and thus well below the ex-refinery fuel oil equivalent price – to which gas for industrial use in Western Europe has generally been priced to date. Indeed, the first move in this direction has already been taken with the inclusion of a price-of-imported-coal component in the indexing calculations for pricing imports of Soviet gas to West Germany. Should this development escalate into a full-blooded attempt by gas sellers to compete with imported coal in Western Europe, then the expansion of the gas market could be even faster than the alternative forecasts above suggest. Such a development would have the effect of increasing indigenous gas supplies to some extent (as some Western European producing countries responded to the challenge), and imported supplies to a much greater extent. This would produce, overall, a higher contribution of natural gas to Western European energy supplies than the 245 m.t.o.e. forecast above for 1990, and a total market for natural gas which is more than the forecast of 325 m.t.o.e. by 2000.

c. Though such a development would have some positive impact on the size of the total demand for energy, the greater effects would be on the markets for coal and oil products. Further substitution of oil and coal by gas would take place and the share of gas in the total energy market would rise to more than 26% forecast above for 2000. In this event, the natural gas market in Western Europe would, by the end of the century, be moving closer to the characteristics of the gas markets in North America and the Soviet Union – with a considerable use of gas in bulk energy uses, in spite of the current E.E.C. and national directives to minimise gas use in power generation and industrial steam-raising. It is necessary to emphasise that this possible development in Western Europe is *not* resource-constrained. A combination of the full exploitation of indigenous resources (to provide an annual supply in excess of 240 m.t.o.e.) and a continental-wide integrated transmission system, plus the construction of expanded delivery systems for gas from outside the continent (whereby an annual deliverability of more than 150 m.t.o.e. would be possible) would place no strain on the ultimate availability of resources, nor on the funds necessary to enable the exploration, exploitation and transport investments to be made. This upside potential for natural gas over the next 15 years does, indeed, provide the element of greatest uncertainty in the future of Western Europe's energy supply possibilities.

d. Although oil prices are likely to remain the main determinant of the outlook for Western Europe's energy prices over the short and the medium-term, the continuing relative decline in the role of oil in total energy supplies will open up the possibility of "price leadership" moving elsewhere in the energy sector by the early to mid-1990s. Price competition between the suppliers (both indigenous and foreign) of the rapidly expanding volumes of natural gas, on the one hand, and the suppliers of imported coal, on the other hand, could by then have become acute – and so provide the "leadership" thereafter in the further evolution of the energy pricing situation. Oil prices would then, of course, have to become responsive to natural gas/imported coal prices competition.

e. Meanwhile, indigenous and imported supplies of natural gas, the prices of which are still closely related to oil prices, will be increasingly effectively marketed to out-compete oil products – both fuel and gas oils. The present OPEC oil-related price for gas in Western Europe gives, however, a general price level which is so far above the long-run supply price of natural gas for the volumes which are likely to be demanded that the declining real price for gas, which will emerge as a result of the falling oil price, seems unlikely to have an adverse impact on the evolution of the gas supply. Bulk gas for large users will become available relatively much more cheaply in order to keep markets moving ahead in the way required by the expanding supply. The large economic rents generated by low-cost gas production (relative to market values) will be squeezed, but this process will mainly reduce unit government revenues from gas production; a development which could well lead to an enhanced motivation for the governments concerned to seek to expand production levels in order to sustain revenues from their natural gas industries.

6. Conclusions

a. The changes now under way in the gas industry in Western Europe are thus taking it away from the constrained, regulated and high price system of the 1970–83 period. Instead, competitive forces are likely increasingly to dominate the prospects for the industry over the rest of the century. This change of structure is in the context of a growing appreciation that there are more than adequate reserves – both indigenous and imported – to sustain the long-term growth of gas markets at the expense of other sources of energy.

b. This predicated evolution of a competitive, expanding and declining real price gas market in Western Europe undermines the bases on which gas policies in Western Europe have been determined in recent years. The relevant issues in this respect are as follows:

i) It arouses scepticism over the validity of the central parameter in much European gas planning, viz. that there are relatively near future dangers (before 1995) over the adequacy of available resources to sustain existing, let alone increasing, levels of demands. The widely accepted – and much implemented – idea of "gas depletion policy", with its implication of restraints on the level of gas production which might otherwise be achieved in the short-term, is related to this parameter.

ii) It raises doubt over the related implicit (or even the explicit) assumption lying behind the idea of gas production controls; viz. that gas saved for use over the long term has a value which exceeds its value in current and near-future additional use. This would be true only with unrealistic assumptions of rising real price for the commodity and a constrained resource base which creates significant "user costs" in the short term. Neither of these has more than a low probability of being correct so that policies which emerge from them are unlikely to be appropriate.

c. On the supply side, Western Europe is well-blessed with gas reserves and production potential (as shown in Tables III-4.1 and 2) and it is self-evidently economically disadvantageous for the countries concerned to pursue policies which diminish the levels of production of its resources. One economic reason for restricting supply is related to the possible danger of a country becoming too dependent on the production of gas. This is not the case when the gas potential is viewed in the Western European context. The industry, even if doubled or tripled in size, would remain of limited importance relative to the region's economy as a whole. There is even less justification for the fear, often expressed, that gas used today, or in the near future, means less gas will be available for the mid-to long-term future – so exposing our grandchildren to the danger of a scarcity of resources. The development of Western Europe's gas potential is, even after 20 years of production , still in its early stages. The geographical size and the emerging petro-geology complexity of the continent's potential gas-bearing regions – both onshore and offshore – indicate that many more decades of activity will be required before the totality of the resource base is properly known and evaluated. Equally lengthy periods of exploration/exploitation have been necessary in all other parts of the world where gas has become important – most notably, in the United States. The continuity of the search for, and the development of, gas resources depend, however, on incentives to the exploring and producing companies. The most effective incentive is a high positive cash-flow from the depletion of reservoirs that have been discovered and this, in turn, depends on the expansion of profitable markets for the commodity. In as far as markets are constrained and/or profits to the producers are otherwise too limited, then the exploration/exploitation efforts will atrophy – so that resources which exist will not be discovered and thus cannot be made available for use, either for the short or the longer-term future. Depletion policies do not save gas for the future; they ensure that gas which could be used to meet future needs is never found and, in the meantime, they also diminish the positive effects on the economy that higher gas production rates could secure. This truth is slowly becoming known in Western Europe.

d. On the demand side, gas policies in Western Europe have largely evolved in the belief that gas is a premium fuel to be used only in so-called premium markets. This is valid only in part; there are clearly uses for gas in which it commands a premium price. But these markets are limited. In the most important gas-using countries – the United States and the Soviet Union – much gas goes into industrial steam raising and power generation. All other countries outside Western Europe with developing gas production potential use their gas in these non-premium markets. These markets are now becoming increasingly important in Western Europe[8] with the recognition that gas has to be flexibly priced to allow it to serve a range of potential markets in Western Europe. Thus flexibility in gas pricing is now on the way to becoming the norm in Western Europe, while less and less is heard of the previously presented view that natural gas ought not to be burned "wastefully" in industrial boilers or in power stations. On the contrary, there is instead the beginning of a very positive attitude to such gas use – not only for economic, but also for environmental reasons.

e. The latter development emerges from the recognition that the use of natural gas in power stations and for industrial steam raising provides a solution to the "acid rain" problems: a problem which has become one of rapidly increasing concern across the continent and in respect of which action is already required by many national governments and will soon be required by EEC regulations. The use of natural gas promises to be a lower-cost alternative for solving the problem than either the process of SO_2 removal from coal-fired power station stacks or the expansion of nuclear power. This is particularly the case when natural gas use for power generation opens up attractive economic potential for dispersed (rather than centralized) production of electricity, with waste-heat recovery achieved as a matter of course for industrial and/or residential and commercial use. Such combined heat and power production based on natural gas is, indeed, an increasingly important element in energy policy developments in many Western European countries. In this context maximising the use of natural gas is the preferred option by the increasingly important "Green" lobby: its support adds to the likelihood of expanding natural gas use in Western Europe.

f. Events since 1983 and the potential developments discussed above in the evolution of the natural gas markets in Western Europe over the coming years will make the future role of the gas industry very different from its role over the last 20 years. Controls on gas production and the regulation of its use and price are on their way out under the pressures of rapidly changing circumstances so that, as already in the United States and the Soviet Union, natural gas in Western Europe will eventually achieve rough parity, along with oil and coal, in providing the energy required for the region's continued development.

References

1 I.E.A. *The Prospects for Gas to 2000*, O.E.C.D., Paris, 1982, p.19

2 I.E.A., *Energy Policies and Programmes of the I.E.A. Countries, 1981 Review*, O.E.C.D. Paris, 1982

3 I.E.A., *Energy Policies and Programmes of the I.E.A. Countries, 1983 Review*, O.E.C.D. Paris 1984

4 As, for example, in the case of the U.K.'s potential reserves, as demonstrated in the recently published study by the UK Offshore Operators Association (U.K.O.O.A.), *Potential Oil and Gas Production from the UK Offshore to the Year 2000 (London, September 1984)*. In this study, which deliberately chose to work with conservative development potential parameters, the UK's gas production for the year 2000 is nevertheless shown to be significantly higher than the highest production level achieved to date.

5 M. Frisch (Mobil North Sea Ltd), "The Supply Outlook for Gas from the North Sea" *Petroleum Review* Vol 39, No 462, July 1985 p 36–43

6 See the I.E.A.'s 1981 and 1983 Annual Reviews; *op. cit*

7 Recently published West German trade statistics for 1984 show that its imports of Dutch gas decreased by 7.3% compared with 1983. By

contrast, West Germany's 1984 imports of Soviet gas were up by no less than 24% to provide about 35% of the country's total gas imports. The trade statistics indicate that the average unit price of imports from the Soviet Union was only a little more than two-thirds of the unit price of Dutch gas.

8 In 1984, for example, the Netherlands, Denmark and Italy decided to use (more) gas in power stations.

Chapter III – 5

The Unexploited Gas Market Opportunities of the 1980s*

a. Developments to date: the Key Parameters

First, the industry on anything other than a local scale (as in south west France, north Germany and the Po Valley of Italy) is recent.[1] The first major discovery, now known as the Groningen field in the northern part of the Netherlands and currently estimated as having had almost $2500x10^9m^3$ of initially recoverable reserves, was not made until 1959 and it did not begin to produce commercially until 1965. The important British south North Sea gas fields date from the late-1960s, and the Norwegian gasfields now in production from the mid-1970s. The super-giant Norwegian offshore Troll field (with a currently estimated $1500x10^9m^3$ of recoverable reserves) was discovered in 1980 and is not scheduled to begin production until 1993. (See Figure II-7.4 on p. 263 for the geography of discoveries/potential supply in Western Europe). Because the major discoveries and production are recent and the history of natural gas use is short (See Table III-5.1) and because the potentially gas-rich areas are very extensive especially offshore around much of the long coastline of Western Europe, (see Figure II-7.3), the natural gas industry in Western Europe is still in its youth – if not its childhood. Even the Groningen field is little more than 40% depleted (on currently declared reserves figures) after 23 years of production at annual rates

*** Based on an article, "The Western European Gas Market: the Current Position and Alternative Prospects published in *Energy Policy*, Vol 16, No 5, October 1988. Note that a German translation was published some months previously in the *Zeitschrift fur Energiewirtschaft*, Vol 12, No 2, June 1988.**

which reached more then $90x10^9m^3$. The annual production potential of gas in Western Europe is not yet near its peak – or plateau – level, and there are decades of exploration yet to come.[2]

Second, the economic geography of Western Europe, with its short distances between points of gas production and areas of use (See Figure II-1.1 on p. 136) coupled with the spatial intensity of energy demand in

TABLE III-5.1
Natural gas use in Western Europe 1961–1986 (in m^3x10^9)

Country	1961	1966	1971	1976	1981	1986
Austria	1.2	1.9	3.3	4.5	4.3	5.0
Belgium/Luxembourg	–	0.1	6.4	11.3	10.5	8.4
Denmark	–	–	–	–	–	1.2
Finland	–	–	–	0.9	0.7	1.1
France	4.4	5.9	12.2	20.9	27.0	25.7
Greece	–	–	–	–	–	0.1
Iceland	–	–	–	–	–	–
Ireland	–	–	–	–	1.0	1.5
Italy	6.5	9.0	13.8	24.2	25.1	31.0
Netherlands	2.3	3.3	27.1	36.3	33.0	35.5
Norway	–	–	–	–	–	–
Portugal	–	–	–	–	–	–
Spain	–	–	0.5	1.6	2.3	2.6
Sweden	–	–	–	–	–	0.2
Switzerland	–	–	–	0.6	0.9	1.0
United Kingdom	–	0.9	19.9	38.1	44.7	54.0
West Germany	1.6	4.0	22.8	39.9	45.5	44.2
Totals	*16.0*	*25.4*	*105.8*	*178.3*	*196.9*	*211.5*
Countries (out of 18) using Natural Gas	5	7	8	10	11	15
Natural Gas as a Percentage of Total Energy Use	1.8%	3.3%	9.7%	13.4%	14.7%	15.2%

Source: *Derived from B.P.'s 1987* Annual Review of World Energy *and earlier years' issues of B.P.'s* Statistical Review of the World Oil Industry

the industrial, commercial and residential sectors, provided a highly attractive market within which gas could compete against other energy sources. Furthermore, the increasingly effective high-profile environmentalists' lobby against present levels of atmospheric pollution – including the so called greenhouse effect – has now given the gas industry an increasingly important advantage over oil and coal. Additionally the anti-nuclear power lobby has embraced the idea of increased natural gas use in super-efficient combined cycle power stations as the most effective alternative to the expansion of nuclear generated electricity.

Third, the transmission and marketing of natural gas over much of Western Europe has been entrusted to state owned, or state controlled, entities which generally emerged from the 19th century town-gas industry orientated to limited and local markets. By the 1960s the town-gas industry was already in severe economic difficulties as a result of the rapidly rising real cost of suitable coal for gas making, and competition from oil and electricity for traditional markets. The large scale availability of natural gas from the mid-1960s in those parts of Western Europe where most town gas production was concentrated rescued the gas industry from an early demise. Initially, however, it also kept it in the hands of managements which generally had limited horizons – both geographically and in the range-of-markets' terms. This produced a perception of gas as an energy source with limited market prospects so that there was a general tendency to confine the sale of natural gas, viewed as a substitute to town gas, to the highest value market outlets only.

Fourth, this limited and high-value of gas markets was, moreover, accompanied by an even more widely held perception of inherently very limited supply of natural gas in Western Europe so that an adequate physical availability of supplies to meet future demands was questioned. It was, indeed, generally assumed by Western Europe's energy policy makers in the 1970s that the early exploration for gas had not only given an instantaneous 100% accurate figure of the sizes of discovered fields, but had also indicated, at a high level of probability, the totality of the continent's gas resource base. Policy makers at national and the European levels thus argued that the rational and orderly planning of markets required that no commitments were accepted for marketing gas other than that for which reserves had already been proven.

The concept of "undiscovered but discoverable gas" (on the basis of which markets for natural gas have been expanded elsewhere in the world) was ignored or rejected as a relevant input to future gas market developments in Western Europe.[3] Instead, the prime objective and concern of gas policy makers and planners became the "need" they perceived to close the so-called future "supply gap": that is, the situation in some 20 to 25 years in the future when there would necessarily be a gap between the level of demand (calculated on the basis of an assumption of a slowly rising demand for gas in these markets which the planners designated for it use) and the continuing availability of producible supplies from already proven reserves. No consideration was given in the calculations to the prospective further development of reserves from already known fields and to the additions to reserves as a result of the continuing exploration process. Supply potential in Western Europe was thus ignored in the evaluation of future markets. In other words, a self-justifying limitation was imposed on the prospects for gas markets in Western Europe. This was, moreover, done in the context of institutional/organizational arrangements which served to encourage such constraints; in a system, that is, as will be shown later, which virtually eliminated a competitive approach to markets.

Fifth, most Western European governments have allowed – or, indeed, even required – gas prices for the consumers to be set at levels which are higher than the prices of alternative energies in different end uses. Thus, gas prices for steam-raising in industry were set higher than fuel oil or indigenous coal equivalents (and hence well above the lower price of imported coal), while for commercial and residential markets gas prices were fixed above gas oil prices and, sometimes, even above the price of electricity in countries in which this was sold for space and water heating and for cooking. Additionally, as if to make doubly sure that natural gas was inhibited from competing in markets in which there were thought to be national energy interests and policies to protect (notably in respect of the maintenance of indigenous coal production and to ensure success for the policies of nuclear power expansion), many national governments – and later the European Commission – "prohibited" the expansion of natural gas use for power generation[4] – not only in relatively thermally inefficient condensing stations, but also in much more efficient combined cycle and combined heat and power modes.

b. The Consequences of Irrationality

The very modest expansion of gas' share of the Western European energy market from under 10% in 1971 to just over 15% in 1986 (see Table III-5.1) is conventionally presented as a success story for a 'new' source of energy, particularly in the reports and the public relations exercises by the companies and other institutions involved. In reality it is no such thing. The increase merely represents a minimal achievement by consumers in securing a little more of their generally preferred source of energy against a combination of restraining commercial and political interests on both the supply and the demand sides.

Given, *first* a plethora of indigenous gas resources (as shown in Table III-4.1 on p.376) and a large import potential for additional gas from a number of external sources (see Figure II-7.4 on p.263); and *second*, the great uncertainties over supply and price which have beset the international oil market since 1972 (when Western Europe depended on external oil for over two-thirds of its total energy needs); and *third*, the clear inability of either the indigenous coal or the nuclear industry in Western Europe to expand their contributions to total energy supply to anything like the degree written into the policy makers and planners' expectations, the mere five-point increase in gas' percentage contribution to the energy market over the past decade and a half represents a failure by the gas industry and government energy policy makers to accept the opportunities offered by natural gas for changing Western Europe's energy system. The failure clearly reflects the anti-natural gas expansion policies and strategies based on misconceived notions of the scale of the indigenous gas resource base, the decisions not to exploit relatively low-cost imports of gas, and the protection of a wide range of energy markets against natural gas use.

Thus, in the mid-1980s West European gas sector one finds the double irony of an under-exploited supply potential co-existing with an under-developed market. The misconceptions over gas supply and gas markets have, moreover, not simply been allowed to persist by the powerful club of gas transmission and distribution companies/institutions (some state and some private); they have been deliberately encouraged by them. In their still largely unconstrained exercise of power in the gas markets the entities continue to persist with inflexible long-term (20 to 25 year) strategies designed only to ensure secure access to volumes of gas which just match the deliberately limited sales in restricted high value markets. Their managements' principal objective, in other words, appears to have been to find guaranteed long-term supplies just adequate to meet their pre-determined calculations of market penetration, chosen in such a way that they do not have to worry much at all about competition from alternative energy sources. The strategy overall reflects a "satisfying" approach by management which, though anxious to be seen to be doing a technically excellent job, has no stomach to accept the challenge of gas as the new energy source or to respond to the opportunities for a competitive approach to Western Europe's energy markets.

Thus, on the one hand, is the primary concern for the decades'-long paper-contracts to "avoid" the long-term so-called "supply gaps" and an associated concern for the provision of technically efficient, though probably over-engineered and too high cost[5], transmission and storage systems (See Fig. II-7.4 on p.263); while on the other, there has been inadequate attention to research which aims to expand the market opportunities for gas. Thus, the gas industry's research facilities have been little concerned with work to improve the thermal efficiency of gas-fuelled combined cycle power plants; or with work on gas-fuelled motor vehicles or gas refrigeration or gas-driven heat pumps for small-scale applications.[6] By contrast, the European electricity industry has devoted major R. and D. efforts to the opportunities for developing new markets – under the stimulus of a generally competitive and expansionist attitude by the management of the electricity industry towards its product and its potential for expanding its markets, even in the context of a relatively slowly growing Western European energy economy.

In the light of the attitudes and policies adopted by the European gas industry since the early 1970s, it is hardly surprising that inter-governmental organisations in Western Europe – notably the E.E.C. and the International Energy Agency – have in their forecasts of the prospects for the Western European gas market predicated[7] a declining supply of indigenous gas and slowly rising volumes of imports to meet the limited degree of market penetration that gas is supposed – or allowed – to achieve. Whilst it would be unfair to describe the managers of West Europe's gas industry and the region's energy/gas policy makers as having had a collective death wish, it would be even more unjustified to conclude that they are now desperately anxious to take appropriate steps to see that the industry grows up. The present state of the gas industry in Western Europe can, as a result, perhaps best be described as one of "premature middle age" – if not one of "early senility"!

c. Consequential Prospects for West European Gas

This inertia partly reflects the interests of the main actors in the system, viz. the club of large transmission/distributing companies, together with many of the largest suppliers, including NAM in the Netherlands, Statoil and British Gas; and partly the absence of an effectively organised pressure group which aims to achieve fundamental change in the structure of the gas market. Consumer watchdog bodies have been convinced that gas is such a scarce commodity that their central concern has been for the very long-term availability of supply to premium end users – and hence they oppose gas' expansion into so-called non-traditional markets. Most European countries' governments continue to have favourite uneconomic energy investments – such as high cost coal production in Britain and West Germany, and heavily capital intensive nuclear electricity in France and Belgium – which they continue to protect against increased gas use. This plethora of forces would be unhappy with an upheaval in the restrained gas market which they have helped to create; while some of them would even now be against taking action which sought to eliminate the possibility of such development.

Moreover, no significant political grouping – of either left of right – is positively and unequivocally for a major expansion of gas production and/or use. Even the environmental lobby – which recognizes the utility of increased gas use as a pollution-reducing strategy – is inhibited in its support by its equally fundamentalist conviction that gas, as a fossil fuel, is inherently scarce in spite of the evidence to the contrary, as shown in Table III-4.1 on p.376 – or as demonstrated in independent geological and economic investigations of the most likely size of the natural gas potential in both offshore and onshore fields.[8] Finally, the European Economic Community, which has striven hard in other sectors to ensure that markets are competitive and non-discriminatory, has failed to intervene in the oligopolistically organised European gas industry to produce a less regulated and constrained system. On the contrary, as indicated above, it has intervened to restrict gas markets' development by directing against its use for power generation.[9]

In spite of this highly regulated gas supply situation in Western Europe there is, nevertheless, an element of rivalry between the gas suppliers within Western Europe, albeit not yet strong enough to have been converted into an overt declaration of their willingness to act competitively. This is a result of the now grossly excessive availability of gas relative to demand at current price levels. Indeed, if Northern America can be described as having a "gas bubble" of supply potential,[10] then Western Europe might be said to have a proverbial gas-filled "barrage balloon! Except for the U.K., whose gas industry remains in stable equilibrium, gas producers elsewhere in Western Europe are all suffering from lower than expected sales over the last five years, resulting in less than fully used capacity in both production and transport facilities. Restrained demand, coupled with the increasing availability of gas from new suppliers (including suppliers from outside Western Europe), has become the essential determinant of recent and current – as well as expected – levels of production . This is in sharp contrast with the widely propagated view until the early 1980s that physical limits on production would determine the size of the industry in Western Europe. This view of the future of gas was portrayed not only by national and inter-governmental policy makers, but also by some of the producing companies themselves.[11]

In some countries gas from other European suppliers was contracted (usually on a take-or-pay basis) in the expectation of scarcity. This imported gas has since had to be used preferentially over indigenous gas as markets have failed to expand to the degree expected. For example, gas bought from Norway under the contracts for Ekofisk area and Heimdal gas has replaced the use of some indigenous German and Dutch gas. In Italy and France imports both from other European countries, as well as production from national fields, have been substituted by imports from elsewhere. Even some discovered UK gas has remained unexploited because the British Gas Council chose instead to buy supplies from the Frigg field in the Norwegian part of the North Sea.

Exporters of gas to Western Europe were, for a period, attracted like wasps to the proverbial open pot of honey. The sales they negotiated were not only a result of their expectations for future markets, given the "supply gaps" to which European gas policy makers so frequently referred, but also a result of the very high prices which the West European gas distributors were prepared to pay,[12] "safe" in the knowledge that the gas they bought could be sold on to captive customers at above oil equivalent prices; and in their confident – though mistaken – expectations that the very high oil prices of the early 1980s would go on going up.[13] More recently, however, as a consequence of the constrained market objectives of the west European gas industry and of falling energy prices in the aftermath of the 1986 oil price collapse, exporters of gas to Western Europe have discovered not only that they have built gas export capacity they cannot fully employ (for example, the Soviet and Algerian gas export lines, and the Libyan and Algerian L.N.G. facilities), but also that they have been obliged to renegotiate contracts in respect of both volume and price.[14]

Thus, the gas exporters to the Western European market have been forced to reduce – or even abandon – their expectations. This has been exacerbated in the aftermath of the carefully orchestrated deal made for the first stage development of Norwegian Sleipner and Troll reserves and their transport to mainland European markets (See Figure III-5.1 on p.399) between Norway (through Statoil), the other entity mainly concerned with producing the field (viz. Shell), and the 'Club' of

European gas purchasing entities (led by Ruhrgas).[15] The deal has generally been presented as a major breakthrough" for gas in Europe – because of the volumes of reserves involved (at least 1500 milliard m^3),

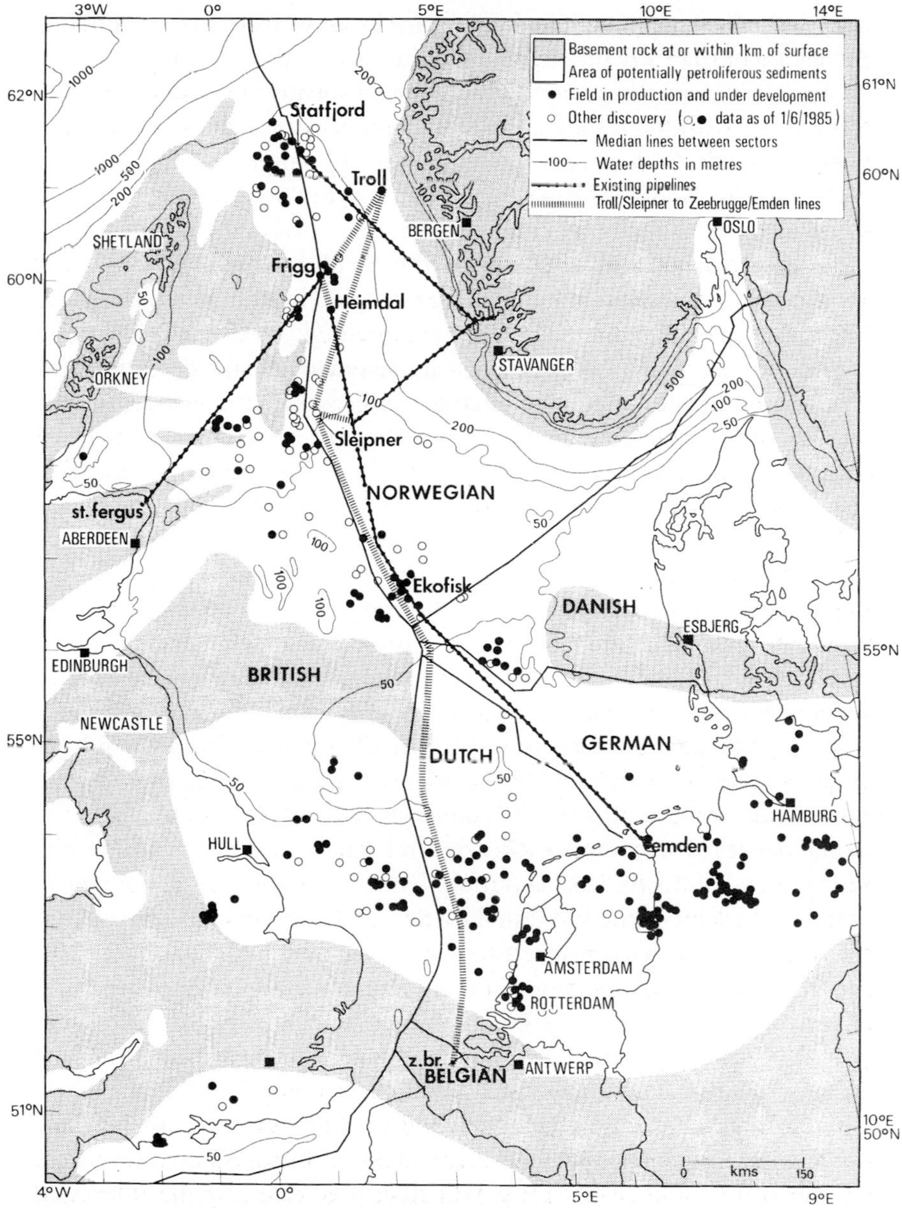

Figure III-5.1: The North Sea basin showing Existing and Proposed Norwegian Gas Pipelines to Markets in 1985

the new technology required for the production system, and the "needs" to have guaranteed large scale and long term contracts for the gas before the work can start.[16]

In a technical and a political sense it is a breakthrough: technically, because it ensures the exploitation of large reserves from a more difficult environment than any other previously tackled in the North Sea; and in political terms, because it not only brings Norway in a more positive relationship with the rest of Europe, but also because it demonstrates the European-wide influence of Shell, with its interests in both the production and the transmission/distribution functions, as a key actor in the region's gas strategy-making strategy.

In economic terms, however, the Sleipner/Troll deal can most satisfactorily be interpreted as the means whereby the *status quo* in the European gas market is to be maintained. Indeed, given the comprehensive – and the exclusive – involvement of all the parties with a vested interest in a basic "no-change" gas market situation (in terms, that is, of allocating relatively high price gas to a limited range of customers with high value end-uses), it is not at all surprising that this "new" Norwegian gas is presented essentially as the development which will fill the "supply gap" predicated by the parties concerned from the mid-1990s onwards. The interested parties' supply-side hypothesis in this respect is related essentially to their expectations of the "need" to replace "depleting" Dutch Groningen gas supplies,[17] while their demand side forecasts hypothesise only a very slowly expanding gas market in Western Europe to 2000; and, thereafter, stagnation – or even decline.[18]

In other words, the interest in the Sleipner/Troll deal – and the shape and form which it has been given by the signatories to the agreement – indicate that the development is orientated essentially to the maintenance of the strait-jacket within the confines of which the European gas industry has been moulded and curtailed for almost thirty years. In the light of the interested parties' success in extending well into the future the gas market structure which they have hitherto imposed on the region, the prospect for an alternative more competitive and expansionist future for the gas market in Western Europe this side of the year 2000 has been greatly reduced – especially given the continued absence of any E.E.C. effort to impose a more competitive structure on the industry. And this is in spite of the now widely accepted much higher potential production possibilities from Western Europe's own resources (as shown in Table III-4.1 on p.376) than predicated by the interested parties and the increasingly strong consumer preference for the expansion of gas use as an alternative to alternative energy sources.

d. An Alternative Prospect for the West European Gas Market

There is only one remaining element for breaking the European gas market out of its contemporary strait-jacket. This relates to a possible high-level, high-stakes geo-political play which involves much more than natural gas *per se*. It is, indeed, a function of the recent momentous changes in both external and domestic policies of the Soviet Union. First, there is the continuing potential for agreements on major multilateral reductions in arms and in tension between the Soviet Union and the West; and second, the attempt by the Soviet Union to introduce new economic policies which are orientated to securing the enhanced production of goods and services in order to meet the demands of the population in the U.S.S.R. itself. This combination opens up the possibility of a more rapid exploitation of the Soviet Union's vast natural gas resources, accompanied by an option for a massively increased flow of Soviet gas exports to Western Europe. Indeed, it is possible to argue that, without such an increased production and export of Soviet gas, the plans of Gorbachev have a much lower probability of being realizable.

Elsewhere[19], I have discussed the range of issues involved for the Soviet Union in this increased potential for gas production and exports and for the economic and political objectives which could be derived from the development. In this paper it is more appropriate to concentrate on the gas component *per se* and on the implications of greatly enhanced Soviet gas exports to the West European gas market from the early 1990s.

First, there is no supply-side reason why the current 35 milliard cubic metres of Soviet gas exports to Western Europe should not – in the fullness of time – be increased threefold, fourfold, or even more.[20] The 'fullness of time' is itself a variable, with a heightening prospect that it could well be a shorter, rather than a longer, period in the context of a significant degree of détente between east and west, whereby the Soviet Union is able to devote sufficient resources to exploiting its gas reserves.

Second, in the aftermath of the Troll deal, whereby the limited market horizons of Western Europe's gas industry and its policy makers have been satisfied for the next generation (given the very high probability that Troll I will be followed by Troll II and even Troll III), such additional Soviet gas exports will have to be made available at a price which enables Western European gas markets to be significantly expanded – notably into the so far little exploited large scale steam-raising and power generation sectors. This implies Soviet gas being made available at a price which is related to, albeit not as low as, the price of the lowest-cost alternative, viz. imported coal. This is currently available at less than half of the fuel oil and one third of the gas oil equivalent prices.

Consumer prices of most gas sold in Western Europe have hitherto been related to these relatively expensive oil products.

Third, in that the Soviet union will – for its own economic reasons – insist that any enhanced flow of its gas to Western Europe is based on counter-trade arrangements and so, in effect, offer guaranteed markets in the Soviet Union for a much expanded volume and range of European goods and services (in an international situation of increasing difficulties elsewhere in the world for Europe's relatively high-cost exports), there will be many interest groups in Western Europe – ranging from political parties, through organised labour and to the supplying industries themselves – which would welcome the creation of such an opportunity for additional exports. Their combined influence could easily exceed that of the Western European gas and other energy industries which can be expected to oppose the Soviet offer of much more gas on the grounds that it will be a serious energy market destabilizing force: especially since more Soviet gas would, through its indirect effects on all sectors of the energy market, bring lower prices all-round for Western Europe's energy consumers.

If there were to be a breakthrough in the currently constrained Western European gas supply situation engendered by a new Soviet policy, then, because it would emerge as an essential element in Soviet economic restructuring and expansion, it is likely to be a long-term phenomenon. Thus, other gas suppliers to Western European markets – both external (such as Algeria) and internal (notably Norway), but not excluding the Netherlands and the U.K., the two other European countries with large gas reserves as shown in Table III-4.1 on p.376) would be obliged to compete. The resultant supply-side competition for rapidly expanding gas markets could mean the end of the present regime of inertia and regulation in the European gas market; or at very least, lead towards its severe and significant modification.

e. Natural Gas' Expanded Role: Western Europe's Most Important Energy Source?

Given these developments, then natural gas' approximately 17.5% share of the total Western European energy market in 1987 would become but an interim staging post en route to a much higher achievable percentage contribution – of up to 30% in the final analysis.[21] Gas would not only continue to replace oil products in a widening range of end-use markets, including transportation, but it would also begin to offer more effective competition in the electricity sector. Gas would increasingly compete in end-use markets hitherto more or less exclusively available to the latter (such as compression and refrigeration) and it would also become a rapidly growing fuel for alternative forms of power generation.[22] In this latter market, gas-fuelled, high-efficient combined-cycle electricity production would become a generally preferred option over coal and nuclear-fuelled conventional less efficient condensing power stations, while gas-fuelled combined heat and power (CHP) projects would be relatively much more attractive as the basis for the energy systems for many more industrial and commercial users.

As indicated earlier in this paper, Western Europe has an energy market structure which is inherently much more favourable to high levels of gas use than in either the United States (and Canada) or the Soviet Union. This is the result of the much higher geographical intensity of energy demand in a much smaller and more densely populated part of the world, with the advantages these phenomena give to natural gas' use – basically because of the inherently lower unit costs of transmission and distribution in such circumstances. Inertia in the gas industry itself and its regulation by government as policy makers have hitherto inhibited – and, indeed, continue to inhibit – the expansion of gas use to a level well below its "rightful" share of the Western European energy market.[23]

A change in the fundamentals of the gas market in Western Europe, sparked by the potential availability of much increased volumes of lower-cost gas from the Soviet Union in the context of progress towards east-west détente, now seems to be a serious possibility for undermining the opportunity which the gas suppliers, distributors and governments in Western Europe have hitherto had to earn high profits and large revenues, respectively, by making relatively low-cost gas available to customers at high prices. The potential benefits which are achievable from a major expansion of the gas market seem likely to be quickly recognized as much too important to be denied to Western Europe's energy consumers. In these circumstances, pressures and actions to end the restrictive practices of existing gas market institutions and the inappropriate gas production and use policies of governments would not be long delayed.

This alternative prospect for the major expansion of gas markets in Western Europe, would, of course, mean the continent becoming a relatively much more important gas-using part of the world from the mid-1990s on. An eventual more-than-doubled gas use and a near-doubled share of gas in the total energy supply (in a relatively slowly growing energy market), will make the position of natural gas in the Western European energy system markedly different from its present situation – and from the prospects, as described above, as emerging from a continuation of present policies and strategies. It is, indeed, not impossible, with a post-1995 development of compressed natural gas as serious an alternative to petroleum products for automotive use, that natural gas could, in the early part of the 21st century, become the single most important component in Western Europe's total supply of energy.

The contrast between the currently-expected year 2000 energy supply situation (with an assumption of no essential change in the

TABLE III-5.2
Alternative potential Western European energy supply patterns in 2000 under
A. Continued constraints on natural gas markets
B. The development of a competitive gas market
(quantities in millions of tons of oil equivalent)

Energy Sources	Estimated for 1987			Alternative Patterns in 2000					
	Energy Supply Pattern			A			B		
	Total	Indig.	Imports	Total	Indig.	Imports	Total	Indig.	Imports
Natural Gas	205	165	40	230	125	105	350	205	145
%	18.3			18.9			27.8		
Oil	595	210	385	550	120	430	560	220	340
%	53.1			45.1			44.4		
Coal etc	230	175	55	305	165	140	230	130	100
%	20.5			25.0			18.3		
Primary Elec.[1]	90	90	–	135	135	–	120	120	–
%	8.0			11.0			9.5		
Total Energy Supply	*1120*	*640*	*480*	*1220*	*535*	*685*	*1260*	*675*	*585*
%	*100*	*57.1*	*42.9*	*100*	*43.9*	*56.1*	*100*	*53.6*	*46.4*

1 Primary electricity (including nuclear power) converted to m.t.o.e. on the basis of the heat value of the electricity produced

constrained and regulated gas markets)[24] and the situation as it could emerge from an expansion of gas supply and gas markets is presented in Table III-5.2. The alternative prospect would serve to ensure that Western Europe not only has energy costs no higher than other industrialised regions of the world, but also an energy supply system with inherently greater thermal efficiency conversion prospects. This, coupled with the significantly reduced use of both coal and oil products would ensure that the energy system overall becomes relatively much more environmentally-friendly.[25] In addition, the major potential for growth in market-located ('on-site'), gas-fuelled combined cycle and combined heat and power generating capacity, together with the substitution of gas for electricity in a number of end-uses for which electricity has hitherto been dominant (and which uses were, moreover, expected to provide significant growth markets for electricity), will serve to slow down – and eventually even reverse – the trend towards the so-called "electrification of society": for sound economic reasons. Such a change would undermine the sole remaining argument for any further expansion of nuclear power in Western Europe, and could even open up the option for a concerted move towards a no-nuclear system, thus meeting the preferred energy sector option of most of the population of Western Europe – in the aftermath of Chernobyl.

References

1 See P.R. Odell, *Natural gas in Western Europe: a Case Study in the Economic Geography of Resources*, De Erven F. Bohn, Haarlem, 1969 for an early analysis of the then initially emerging gas market prospects.

2 See, for example, recent independent studies on the prospectivity for continuing and expanded hydrocarbons' discoveries and exploitation both onshore and offshore Western Europe, by Midland Valley Exploration Ltd., Glasgow, 1986 and 1987. See note 8 below for details.

3 This approach has provided the basis for the Dutch Gasunie's annual 'gas afzet plan'. Each year Gasunie specified its long-term plans for marketing the gas available to it. Availability was defined as 4% of the declared volume of proven reserves from the Dutch gas fields, plus the volumes of imported gas for which contracts had been signed. On average, however, over each succeeding year rather more gas has been added to reserves than has been used. For example, in 1965 the marketing plan was related to reserves of about $1575x10^9m^3$. Since then approx. $1470x10^9m^3$ has been used, but nevertheless, at the end of 1987, some $1800x10^9m^3$ of proven reserves still remain to be used.

4 This 'prohibition' was implemented by EEC Council Directive (no. 75/404/EEC) concerning the use of natural gas in public power stations.

5 This phenomenon is similar in character to the "gold-plated" and high cost approach to North Sea oil and gas production facilities, in the context of high oil and gas prices and 100%+ write-off allowances against high tax rates. Since the fall in the oil (and gas) prices and reductions in tax rates since 1986 North Sea developments costs have fallen by up to 40%. The motivation for cost control has not yet diffused through to the gas transmission sector as the entities concerned operate generally in circumstances of monopolistic pricing in national or regional markets.

6 These are potential new markets for gas which have been defined and analysed in the United States. See American Gas Association, *The Outlook for Gas Demand in New Markets, 1986–2010*, Arlington (Virginia), August 1986.

7 See, for example, successive issues until 1985 of the I.E.A.'s *Annual Review of Energy Policies and Programmes of I.E.A. Countries (O.E.C.D. Paris)* and the E.E.C.'s *Review of Member States' Energy Policies*, Brussels. More recently, both organisations have been obliged to change their views in the light of the overwhelming evidence of a massive upside potential for increased indigenous gas production (relative to potential for increased indigenous gas production) and unlimited (relative to forecast demand) potential for imports from elsewhere. The impact that these Organisations' earlier mistaken analyses and predictions had on the prospects for the gas industry in Western Europe has, however, been formidable. Thus, their earlier views will continue to be an important restraint on the expansion of gas production and gas used for a long time yet.

8 Midland Valley Exploration Ltd., *1. Future Oil and Gas Developments in the UK North Sea: Reserves of the Undeveloped Oil and Gas Discoveries. 2. Hydrocarbon Prospectivity of Onshore basis in France. 3. The Potential for Further Major Hydrocarbons Discoveries Onshore UK*, 14 Park Circus, Glasgow G3 6AX, 1987.

9 See Note 4 above for details of E.E.C. Directive controlling gas in power station use. Recently (in 1986 and 1987) the Energy Directorate of the E.E.C. has made efforts to have this Directive rescinded, but this has not yet been achieved. Meanwhile, faced with surpluses of natural gas availability a number of EEC countries have built or converted power stations to burn gas. No action has been taken by the EEC in respect of these breaches of the Directive.

10 The U.S. Dept. of Energy/Energy Information Agency *Annual Energy Outlook*, 1986. (Washington 1987) comments as follows, 'The relatively low rate of decline in gas reserves, together with a variety of regulatory roadblocks ranged against the marketing of gas, accounts for a build-up in surplus production capacity known as the "bubble". The study estimated the surplus in 1986 at 3.0–4.6 tcf (=85–130 x $10^9 m^3$). This, however, is a surplus measure against a reserves to production ratio of only 10 years. Applying the same criteria to the Western European

reserves position (as shown in Table III-6.1 on p.412 with end-1986 remaining reserves of $7900x10^9m^3$) then, compared with actual production in 1986, Europe's "bubble" of potential additional production exceeds $6000x10^9m^3$. Even measured against a 25 year reserves/production ratio the "bubble" of non-exploited gas for Western Europe is more than an order of magnitude larger than the surplus in the United States.

11 See the I.E.A. and E.E.C. reviews previously footnoted (Note 7). The limited potential to produce which they reported reflected the nature of the information being supplied by national governments and/or state owned/controlled gas industry companies. For example, the Dutch *Gasunie* persisted in the emphasis it gave to gas "scarcity" in its *Annual Reports* until the middle 1980s.

12 This phenomenon applied generally across the range of the West European monopolistic gas transmission/distribution companies. Most notably, however, Distrigaz of Belgium broke ranks with the rest when it agreed in 1982 to a "take or pay" contract for L.N.G. supplies from Algeria at a base price which included a component indexing the gas price to Algeria's crude oil export prices. This had the effect of creating an imported gas price above the norm of fuel oil equivalence as the latter always sells for less then crude oil price.

13 For an analysis of the expectations in the early 1980s of future oil pricing developments see P.R. Odell, "The Deregulation of Oil: the International Oil Market Framework" in the *Proceedings of the 18de Vlaams Wetenschappelijk Economisch Congres*, Brussel, May 1987 (Published by the Belgian Vereniging voor Economie, Brussels, 1987).

14 Algeria has had to renegotiate its price and other conditions for L.N.G. and pipeline gas exports to a number of Western European countries since 1985 under the threat of the importing countries' wide choice for alternative future supplies. France's demands for more favourable terms have, however, been particularly tough (and the French market is especially important for Algeria) so that the negotiations have not yet been successful. Formal arbitration proceedings now seem likely to be invoked by France to resolve the issue.

15 For a description of the Troll deal see J.P. Stern, 'Norwegian Troll gas: the Consequences for Britain, Continental Europe and Energy Security'. *The World Today*, Vol 43, No 1, January 1987 pp1–4.

16 *Ibid.*

17 In this respect too, the expectations of the policy makers are at variance with the facts. As indicated above in Note 3, depletion of Dutch reserves is not a consequence of production. Additional production has, on the contrary, generated continuing reappraisal-of-reserves and of the exploration processes so that reserves volumes have, over a 22 year period, been more than maintained. The Groningen field, in particular, has had its recoverable reserves up-graded many times since production

commenced so that its remaining proven plus probable reserves today are little different from the proven and probable reserves of the field as calculated when production began in 1965 (at about $1500x10^9m^3$). There does appear to be a possibility – derived from the technical data available in increasing volume as a result of production experience – that additional gas from the Carboniferous source rock is migrating into the Permian reservoir as production of the gas originally in place in the reservoir proceeds.

18 The International Energy Agency (in its publication *Natural Gas Prospects*, OECD Paris 1986) forecasts gas demand in OECD Europe to increase to $248–280x10^9m^3$ by 2000 (1987 use will be about $210x10^9m^3$) and to 258–305 by 2010. The EEC in its most recent evaluation of gas markets comments that "based on forecasts from member states… there will only be a limited growth in demand. This is particularly striking in the 1990s when a fairly flat picture is presented with demand leveling off at around 200 mtoe per annum". See *Energy in Europe*, No 6, December 1986, p28.

19 See P.R. Odell, "Gorbachev's new Economic Strategy: the Role of Gas Exports to Western Europe", *The World Today*, Vol 43, No 7, July 1987, pp 123–5

20 The ability of the Soviet Union to increase its exports to this extent emerges from the continuing rapid expansion of its annual production which is planned to rise from $712x10^9m^3$ in 1987 to $1050–1150x10^9m^3$ by 2000. When new export lines planned or under construction are completed by 1994 and added to the export capacity in existing lines, $110x10^9m^3$ of gas will be capable of being exported – and there are no reasons why additional export capacity cannot be developed thereafter. The International Institute for Applied Systems Analysis *International Gas Study* – involving a major contribution from Soviet scientists – provides the background to these prospects. See H-H Rogner, *A Multi-Regional Study of European Energy Markets*, IIASA, Luxenburg, 1987

21 Even without the development of new markets for natural gas, a 30% natural gas share of the total energy market does not come close to market saturation. In the early 1970s gas' contribution to the Dutch energy supply exceeded 50% and even now, following the deliberate policy of reducing gas use for power generation its share is about 43%.

22 See Note 6 for details of expectations in the United States for the development of new markets for natural gas.

23 There are no formal 'rights' or 'wrongs' on the shares of different sources of energy in total national or regional markets. By 'rightful' in this context one means a share of a market which is not restricted by monopolistic behaviour on the supply side, or the exclusion of markets by governmental edict on the demand side. Without such constraints the role of natural gas in the Western European energy economy, based on supply and price and on competition with other energy sources, would be greater than it has so far been allowed to become.

24 See Note 18 for references to the forecasts for the future contribution of natural gas which reflect the continuation of present policies and strategies.

25 For example, the use of compressed natural gas instead of gasoline for cars and light trucks/vans would eliminate (or almost eliminate) the emissions (from the point of fuel extraction/production to the point of consumption) of carbon monoxide, SO_x and particulates; and reduce NO_x emissions by almost 40%. See N.E.Hay (Ed) *Guide to New Natural Gas Utilization Technologies*, Fairmont Press, Atlanta, 1985. Emission reductions by the use of gas in combined cycle power plants in place of coal in conventional condensing stations would be even greater.

Chapter III – 6

A Fundamentally Different Framework in the 1990s*

Introduction

The period from the mid-1970s to the late 1980s was, as shown in the two preceding chapters, one of relative stagnation in the Western European gas market. Annual indigenous production barely changed and gas use increased only modestly with the lack of dynamism arising from highly effective institutional constraints on both the supply and demand sides. Since 1989 there have, however, been a series of developments – economic, political and technical – which are serving collectively to establish a fundamentally different framework for Europe's gas market. These hold out the promise of massive expansion over the next 10 to 20 years. The developments are described and evaluated in the first part of this paper and the results are then used to predicate prospective supply and demand relationships to 2010. Finally, some geo-political aspects of the developing gas markets in Europe as a whole will be considered.

Western Europe's Indigenous Gas Reserves

From the data presented in Table III-6.1 it can be seen that the process of gas reserves' development in Western Europe has remained a highly dynamic one. Over each decade since 1956 gross additions to proven and probable reserves have exceeded the cumulative production

*** Based on an article, "Prospects for Natural Gas in Western Europe" in *The Energy Journal*, Volume 13, Number 3, 1992, pp.1-19.**

of gas by at least a factor of two, to produce a steadily rising "total original recoverable reserves" figure and an always generous reserves to production ratio. In spite of this, a widely-held conventional view emerged which held that European gas reserves were scarce as a result of which limits were imposed on both gas exploitation and gas use at the level of the mid-1970s[1].

This view was particularly strongly held in The Netherlands (at the time by far and away the leading West European gas producer, accounting for over 60% of the region's total production), where it led to decisions to reduce industrial and power generation demand at home and to freeze exports to the amounts already contracted.[2] As a result, Dutch gas output peaked in 1976 and quickly fell back to two-thirds of its earlier levels. As shown in Figure III-6.1, this was quite unnecessary: the appreciation of the reserves of the giant Groningen gasfield – responsible for about 65% of total Dutch gas production since 1976 – has kept up with cumulating volume of gas extracted. Remaining reserves

TABLE III-6.1
The evolution of Western Europe's natural gas production and reserves from 1956–1986 and a forecast for 1996 ($m^3 \times 10^9$)

Year	Cum. Prod. to date	Cum. Prod. in the decade	Year-end proven & probable reserves	R/P ratio (in yrs)	Total original recoverable reserves	Gross additions to reserves in decade	Net additions to reserves in decade
1956	50	35	500	40	550	–	–
1966	225	175	1,900	76	2,125	1,575	1,400
1976	1,150	925	4,350	27	5,500	3,375	2,450
1986	2,700	1,550	6,900	38	9,600	4,100	2,550
Forecast							
1996	4,450*	1,750*	9,250**	46	13,700	4,100	2,350

* *These production figures assume the continuation of present policies.*

** *On the basis of field discovered by end-1986. Since then more fields have already been found and there is, of course, a near-zero probability that no more fields will be found, given the continuation of an expensive exploration effort in many West European countries and their off-shore areas.*

are thus still at the level at which they were declared when production from the field first started in 1965.

The situation is similar for the other gas-rich countries in Western Europe. In the UK, the British Gas Council, having rapidly enhanced both its own production and encouraged, through its purchasing policies, other producing companies to do the same, thereafter seemed satisfied with a plateau level of gas availability of about 40 Bcm. Its policy thus inhibited exploration and exploitation activities, in spite of which reserves continued to build through fields' appreciation and some new discoveries. On the Danish offshore, the sole concession holder and

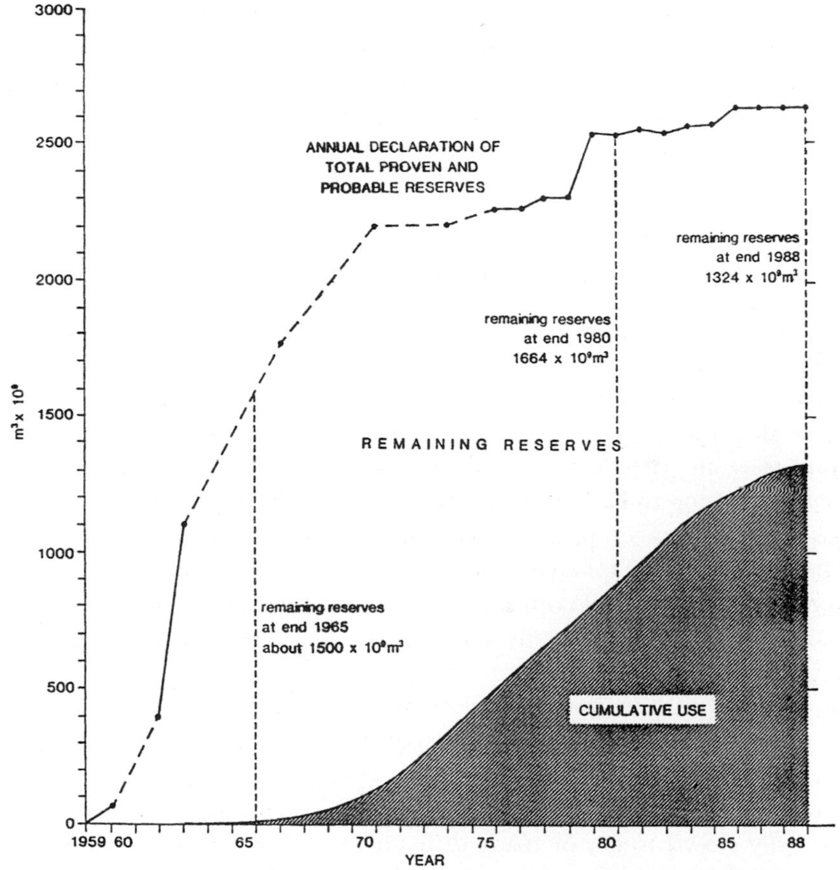

Figure III-6.1: The Groningen Gasfield: Cumulative Production (1965 to 1988) and the Evolution of Proven and Probable Reserves (from discovery in 1959 to 1988)

operating company, the DUC, was unable or unwilling to extend and intensify its exploitation efforts so that production remained related to its initial discoveries. Meanwhile, new reserves were being found. Norway developed gas production to over 25 Bcm per year within five years in the 1970s, but then held output at approximately the same level throughout the 1980s in spite of continuing new discoveries and the annual addition of large volumes of new reserves.[3]

For all supplying countries the principal motivation for limiting production was the conservation of discovered reserves. This, it was argued, was partly to "ensure" supplies for future generations and partly because, in the context of a general expectation at the time of permanently rising real oil prices, it was thought that gas left in the ground had an inherent worth greater than the extra net revenues which could be achieved from the short-term sale of more gas. The collapse of the oil price in 1986 and its subsequent later lower real value undermined this second argument in favour of limited gas production. This can be defined as having marked the beginning of new West European attitudes to the exploitation of its expanding gas reserves.

The Threat to the Institutional Limitation on Gas Markets Posed by the EC's Single Europe Act

The negotiations on the establishment of a Single European Market (SEM) started coincidentally with the collapse of the oil price and the greater realism this generated over gas policies. By 1988 it was already clear that the energy sector, in general, and the gas and electricity industries, in particular, with their essentially national characteristics, were not going to be left out of the economic integration process.[4] For gas, the Commission proposed a set of measures designed to achieve the liberalisation of the gas transmission systems (certainly third party access and, possibly, imposed common carriage conditions) and an end to the monopolistic market structures which existed in all member countries to a greater or a lesser extent.[5] The steps under consideration clearly implied the introduction of some competition extending across national frontiers and thus presented the prospect for an expansion of gas demand under conditions of competitive supply, particularly in the industrial and power generation markets which in most EC countries had either been formally closed to gas or made difficult to exploit.[6] Indeed, gas use in these markets in 1988 was no greater than it had been in 1978.

Recognition of the threat posed by the SEM to the protected and/or monopolistic base on which the gas industry in Western Europe had developed led to strong pressures both by the gas entities themselves and by their home countries' governments to "water down" the liberalisation and competitive elements in the package.[7] The issues have not yet been resolved, but the opposition has already obliged the EC to postpone proposed Third Party Access (TPA) rights to gas transmission systems for a small number of very large users in each country and to cross-border movements where spare capacity can be shown to exist for at least three years beyond the original January 1, 1993 deadline. Moreover, no agreement has been achieved for terminating the monopolies. As already seemed likely in 1990,[8] the due date of January 1, 1993 for the completion of the SEM for the energy sector will now see no more than the beginning of a process of change in the Euro-gas market. Many opponents of change are still prepared to have the liberalization and de-monopolisation proposals tested in the European Court so that the process seems certain to stretch out over the rest of the 1990s – well beyond the new date of January 1, 1996 when the "single internal market" should be completed.[9]

Nevertheless, the modest proposals which have been agreed for implementation from 1993 will open up gas supply and price negotiations to some 500 large users. In the context of readily available additional gas these negotiations will lead to an enhanced contribution by natural gas to the power generation and industrial markets. Moreover, the EC's more radical proposals have begun to change perceptions of the opportunities open to natural gas. Market expansion on the scale perceived necessitates additional "players" to ensure access to the capital required, to accelerate the rate of investment and to create alternative negotiating possibilities with potential new gas suppliers.

As a result of recent legislation in the UK and the work of the gas regulatory regime which that legislation established, this is already happening. New sellers and transporters of gas are already in the market in competition with British Gas and making progress, albeit slowly.[10] In Germany, Wintershall (a subsidiary of the giant BASF chemical group) is committed to building a new trans-national transmission system independent of that created over the past 25 years by Ruhrgas and its sister companies. In Italy, the government is considering the establishment of supply and marketing alternatives to those provided by the state-owned natural gas corporation, in recognition of the fact that Italy's preference for more gas in the country's energy system necessitates a broader institutional structure. Even in The Netherlands, where Gasunie had hitherto had a *de facto* monopoly over gas purchasing and transport, the SEP (the Association of Electricity Producers) was able to buy gas directly from Norwegian suppliers for fuelling two new generating stations: to be located, ironically, in the province of Groningen directly above the Groningen gasfield!

Overall, there are thus already significant breaks in the tightly-knit gas industry's organizational and operational structure which has hitherto completely dominated the gas industry in Western Europe. Movement towards more liberalisation and competition within the industry now seems irreversible and unstoppable, but there is still far to go before the fully integrated gas market within the Community is in place.[11]

Environmental Concerns will Enhance the Competitive Position of Natural Gas

Environmentally more friendly energy policies are now a requirement of EC members. Indeed, the need for a high level of environmental protection was specifically included in the Single Europe Act.[12] Apart from other energy related environment issues, the EC proposals for measures to secure the stabilization and eventually the reduction of CO_2 emissions from 1990 levels imply significant changes in energy use trends and patterns.[13] The encouragement of the use of natural gas is widely seen as the lowest cost route toward the achievement of the CO_2 emissions objective. Generally, therefore, compared with other sources of energy, gas enjoys a relatively easy passage against the strengthening environmentalists' anti-energy use pressures. Thus, on the one hand, governments' (and the EC's) energy policy makers are now reacting positively to expanded gas use, while, on the other, the gas industry itself is coming to perceive the additional profitable opportunities for investment which has been opened up by the environmental issue.

The Impact of Changes in Eastern Europe and the (Former) Soviet Union

The speed and the intensity of political and economic changes in Eastern Europe and the (former) Soviet Union add both important supply and demand side components to the European gas market. On the supply side, earlier strategic fears for too great a degree of dependence on gas from the Soviet Union, in the context of the East/West conflict,[14] no longer have much remaining validity. Though serious doubts remain on the short-term politico-economic outlook for the USSR's successor states, positive mid- to longer-term development prospects in a number of the new republics can be correlated strongly with the enhanced and efficient exploitation of their large gas reserves. This implies the necessary involvement – for financial, managerial and technical reasons – of western oil and gas companies: conceivably leading eventually to a situation in which the prospective hydrocarbon resources potential of these new countries become highly-ranked in the exploration and production budgets of the companies concerned.

Such a supply-side response would, however, be muted were it not for the now clearly defined market opportunities for the prospective much increased volumes of gas from such investments. Western Europe has, of course, been a significant market for Soviet gas for almost two decades. In 1991 such imports exceeded 60 Bcm and there are contractual commitments already for somewhat larger annual volumes during the rest of the 1990s. Additional supplies from Russia – and other former Soviet republics – do not pose an acceptability problem, providing the supplies emerge from the involvement and investments of western companies – but they will, of course, have to be delivered competitively to the distant Western European markets.

In terms of both relative importance in total supply and of market competitiveness, additional gas supplies from the former Soviet republics seem likely to be even more significant for the countries of Central and Eastern Europe, including the former East Germany. Though these countries have received Soviet gas through the lines running from the Soviet Union to Western Europe, the amount involved has generally remained limited, viz. to only about 10% of their total energy demand. This limitation reflected the East European countries' needs to maximize indigenous energy production – mostly coal and lignite – for both ideological and balance of payments reasons.[15] Energy use patterns thus remain orientated to solid fuels, generally burned inefficiently and without environmental protection devices. This has long been a matter of concern in neighbouring Western European countries, but changing the adverse environmental situation caused by inefficient energy use has now also become a priority for the new Eastern European administrations.

Though indigenous gas production on a larger scale than hitherto is a medium-to-longer-term prospect for some East European countries (especially Poland), much increased gas imports are the only realistic energy/environment solution for the 1990s. For reasons of both location and the existence of transmission facilities and other infrastructure, additional supplies of gas from Russia and other former Soviet republics will provide for most of the increased demand. Alternative supplies from the Netherlands, Norway and North Africa will make no more than a limited contribution in this period.

Nevertheless, the incorporation of much more gas into the energy systems of the east and central European countries, so bringing their energy use patterns more into line with those in Western Europe and, as a result, offering the opportunity for a more closely integrated and intensified transmission system between exporters and importers across the continent, will be an important element in linking the two hitherto strongly separated parts of the continent.

The Influence of Recent Political Developments Outside Europe

The traumatic events in the Gulf in 1990/91 have once again provided a reminder of the inherent political difficulties which influence the supply and price of international oil – even though, in the final event, the actual supply problems arising from the loss of both Kuwaiti and Iraqi oil from the market were constrained by the unexpectedly high "surplus" production capacity which proved to be available elsewhere in the international oil supply system. Nevertheless, there was a six month period of trauma over oil prices as the average internationally traded price of oil between August 1990 and January 1991 exceeded the average price of the first seven months of 1990 by 75%.

This sharp reminder of the uncertainties of the international oil system led both the EC and individual European countries to renew their resolve to strengthen their "away from oil" policies: in order to inhibit potentially adverse political and economic disruptions in the future, given a perception of both continuing political and even military problems in the Middle East and an expectation that the Gulf producers will inevitably re-dominate the world oil market – and the price of oil – before the end of the century. In this context the perception of the attractiveness of natural gas as an alternative to oil has been further enhanced in European energy policy making circles. Indeed, after decades of relative indifference to gas as a source of base-load energy – alongside oil, coal and nuclear power in various sectors of the energy economy: and a decade and a half since 1976, when an EC energy policy directive specifically excluded the use of gas for power generation, the Community has moved to the other extreme – and has elevated gas to the central role in its energy supply strategy for the next 20 years.[16]

Demand Expansion Prospects – and the Supply Response

Demand Prospects

The combined impact of the recent stimulants to gas use are already showing through. In 1988 gas use in Western Europe was 221 Bcm, an increase of only 24% in gas use over the 12 years since 1976. Preliminary figures indicate that gas use in 1991 was about 255 Bcm, an increase of 15% in only three years. Moreover, whereas in the 1980s the share of gas in the energy market had stuck at about 15%, it is now approaching 20%. The substitution of other energy sources by natural gas, the use of which has, hitherto been mainly restricted to the residential/commercial sector, is now being extended to the industrial and power generation sectors of the economy. Gas use in these latter sectors actually declined for a decade after 1976 as a result of pricing and policy conditions imposed by a combination of structural and political factors. Since the late 1980s such constraints have been gradually lifted in most European countries, culminating in the withdrawal in 1991 of the 1976 EC Directive against using gas for power generation.[17]

After 1993, as the EC moves towards the establishment of a single energy market, the opportunities for additional natural gas will become generalized across the continent: particularly among the 500 or so large industrial and power station users who will, as shown above, secure the right to buy their gas from suppliers other than their local gas company. An EC-imposed right-of-access to unused capacity in trans-national lines for gas will be an integral part of this process of opening-up under-developed markets.[18]

Major developments along these lines will necessarily have to await the construction of new and dedicated lines for such gas transmission, but such expansion is already underway. This, in part, is at the geographical peripheries of the market, viz. in Spain, Ireland and Southern Scandinavia; and, in part, in a number of countries, notably the UK and Italy, where government policies explicitly or implicitly support such developments. There are, on the other hand, some countries where policies and/or competition from other energy sources continue to hold up the expansion of the gas market. France is one of these. There, the dominant Electricité de France, with a surfeit of nuclear generating capacity, needs to retain its high share of the space heating and industrial markets. Even more important in this respect is the former West Germany – the largest energy market in West Europe. Gas use here declined in the first half of the 1980s and even by 1989 use was only marginally above that of 1980. This overall situation was, however, the net result of sharply contrasting trends in the country's gas markets. On the one hand, residential use rose 50% over the decade, but this was entirely offset by the declining use in other sectors: notably a 40% fall in power generation – in spite of a 15% expansion in electricity output. Moreover, oil products – in the context of an efficient, aggressive and competitive oil industry – generally continued to compete effectively against gas in many markets. Finally, policy objectives in the 1980s (relating to successive governments' preferences for coal and nuclear power) were, in essence, negative for gas.

Officially – at both the Federal and Länder levels – and in institutional circles (such as Ruhrgas and the VDEW – the German Electricity Association), little significant expansion of the role of gas in the country's energy economy has been anticipated, let alone encouraged. In particular, there remains a strong prejudice against gas for power generation. Thus, even though Germany is committed to an overall 20% plus reduction in CO_2 emissions by 2000, the official view of the likely contribution of gas to this objective is modest. Even the planned reduction in indigenous coal production is to be compensated by increased coal imports and additional nuclear power.

A recent independent study of the West German energy economy[19] has, however, cast doubt on the rationality of the country's official energy policy. It argues that even more high-cost indigenous coal production will have to be abandoned and that neither coal imports nor nuclear power can effectively substitute it – for a combination of economic and environmental reasons. It has thus indicated the necessity of expanded gas use, particularly in power generation: to produce at least a 25% increase in total gas demand by 2005.

Actual gas demand developments by 1991 are, in fact, already running well ahead of these independent expectations, in the context of the planned major expansion of the country's capacity to import and transport gas through a transmission and distribution infrastructure to be built independently of the hitherto monopolistic operation directed by Ruhrgas. Given the prospect of a near future breakthrough by the West German gas market into a more competitive mode and a recent much more positive view of gas' potential contribution to the energy economy by the Federal Economics Ministry,[20] major expansion of gas use in Germany now seems more likely than not. Its market for gas will, it seems, keep up with the expansion already confidently expected by the UK and Italy. Overall, forecasts of the West European gas market prospects for the period to 2010 show 50 to 100% increases in total demand over 1990 levels.[21]

As already indicated above, the countries of Eastern Europe have an even stronger motivation to expand gas use. Here up to a threefold increase in gas use seems likely arising mainly from the substitution of coal and lignite (and of some nuclear electricity from unsafe stations) for both economic and environmental reasons.

The Supply Response: from Indigenous Resources

As shown above, there were until very recently demand limitation constraints on indigenous supply. One monopsonistic constraint has, however, already been broken in the case of the UK – Western Europe's second largest gas producer. Here exploration and producing companies are no longer dependent on a single gas purchaser's decision as to whether or not to buy the available gas. There are now alternative buyers with access to markets. Note, however, that even so a decision to develop a field still requires formal Department of Energy acceptance of the development plan proposed for the field.[22] This requirement not only leads to delays while official evaluation takes place, but it also means that the ability of government to control the amount and the timing of production is retained.

Not even the creation of the Single European Market by the planned date of 1993 will make any direct difference to the national regulation of gas' exploitation in Western Europe. The principle of national sovereignty over resources development appears never to have been questioned by the Community and will thus remain intact. Indirectly, the Single Market's requirement for non-discrimination between Community-based companies in respect of the allocation of concessions[23] could mean the introduction of greater competition in the bidding process for gas exploration and exploitation rights, and thus possibly enhance the level of activities required from the successful companies. Nevertheless, this cannot guarantee the development of discovered reserves, as national governments could still exercise their right to deny or delay exploitation on grounds of national strategic or economic considerations. The Dutch government, for example, has recently reiterated its view that decisions on the exploitation of the country's gas reserves are of strategic importance and should be reserved through the monopoly rights of Gasunie (50% state owned) in respect of purchases of offshore gas and gas transmission across the Netherlands.[24] And Norway, in any case, remains, for the time being at least, outside the Community.

Thus, the prospects for a wholly de-regulated and a de-nationalised indigenous gas supply within West Europe remain largely excluded for the 1990s. Nevertheless, this negative prospect does not exclude the evolution of a significant supply-side response within the confines of the existing systems. Compared with the earlier perception of impending scarcity, there is now a general recognition of the expanded supply potential from the discoveries already made. Additionally, there is enthusiasm for the likelihood of continuing exploration success for the range of gas-prospective geological plays which have been established over the past 30 years. These are capable of being further explored and exploited for significant volumes of gas at costs which make profitable developments possible at current market prices. Higher net-back prices to producers – which may be expected in the 1990s – will enhance the potential.

Gas market expansion in the UK – arising from organizational changes in both the gas and the electricity industries and from the critical importance of additional gas use in enabling the country to meet its EC environmental obligations (in respect of both SO_2 and CO_2 emissions) – have already stimulated production from just over 40 Bcm in 1988 to an estimated 51 Bcm in 1991. Given the continuity of the demand factors indicated above and a supply-side response to the competition for additional industrial and power generation sales, an increase in UK gas producion to at least 60 Bcm per year by the late-1990s is highly probable: with the likelihood of at least a long-duration production plateau or of yet further increases in annual output thereafter, if the extensive areas of the country's continental shelf and slope which to date remain wholly or largely unexplored turn out to be gas rich.

The Netherlands has already been Western Europe's biggest gas producer for almost 30 years. After rising to a peak annual production in 1976 of 101 Bcm, annual output fell back (to an eventual low in 1988 of 57 Bcm) as a consequence of a major policy shift. At that time there was a declared official expectation that not even this low level of output would be maintained through the 1980s after which there would be a yet steeper decline as a prelude to the near exhaustion of reserves soon after 2000.[25] Repeated arguments against the validity of a policy which ignored the dynamics of the processes of gas reserves' discovery and appreciation had no impact for well over a decade.[26] Finally, however, in 1989, a combination of continuing reserves growth and the growing competition from Norway and the USSR in the traditional markets for Dutch gas led to a change in Dutch policy. New contracts were signed with foreign customers, industrial gas users at home were assured of available gas for the long-term and the electricity generation market was re-opened to natural gas. Production responded and in 1991 it was already up to 69 Bcm. Even so, annual additions to reserves have still exceeded production by a significant amount. Thus, ability to produce up to 80 Bcm per annum through the 1990s now seems to be a near certainty, followed by only a modest decline to a somewhat lower plateau level of production; and with the "final" depletion of the country's gas reserves "postponed" indefinitely. Behind this revised prospect lies an expectation of continuing gas discoveries form an intensified exploration effort on new – and reallocated – concessions.

Good though the gas supply prospects now are for the UK and The Netherlands, they are still relatively modest compared with Norway's potential. Its remaining proven reserves alone are of the order of 4,000 Bcm, representing well over 100 years supply at the estimated 1991 rate of production. The agreed initial development of the Troll field, (Europe's second super-giant field after Groningen), together with that of several other large fields, are scheduled to boost annual output to at least 35 Bcm by the mid-1990s. Thereafter, further expansion is technically and economically possible, but how much more will depend on government policies and parliamentary decisions – rather than on reserves availability and/or developability.[27]

Politically, Norway's closer association with the EC in the short-term through the 1991 EEC-EFTA agreement[28] – and its possible EC membership after 1996 along with Finland and Sweden – together with its economic need to have more of its gas resources exploited, should serve to engender sufficient new field-development decisions quickly enough to enable the upward curve in annual production to go beyond the presently planed 35Bcm. An amount on the order of 45 Bcm per annum by 2000 is highly probable, with yet more significant increases, thereafter, as Norwegian gas is sold to new markets in the UK and the rest of the EC; as well as in the countries of East Europe, where diversification away from complete dependence on Russian gas will be a strategic objective of the countries concerned.

Overall, after also taking into account the potential for additional production from other (smaller) West European suppliers (viz. Italy, Germany, Denmark, Spain, Ireland, France and Austria, in descending order of importance), the estimated indigenous 1992 production level of about 190 Bcm could be increased by around 50% by 2010. This indigenous supply expansion is, however, at the bottom end of the range of estimates of annual gas use by 2010. Thus, there will be major opportunities over this period for enhanced flows of gas to Western Europe from external suppliers.

The Supply Response from External Sources

Existing flows of gas from external suppliers to Western Europe have reached about 90 Bcm per year; this comes mostly from the former Soviet Union and Algeria in the ratio of approximately 2:1. Contracts to expand the flow to 120+ Bcm per year by 2000 have already been signed (80 Bcm from the former Soviet Union and 40 Bcm from Algeria). The security of these expected flows has, however, been called into question by recent political events, viz. the break-up of the former into a set of independent sovereign states facing huge politico-eocnomic problems; and, for the latter, the possibility of a fundamentalist Islamic government achieving power. Moreover, the expansion of the West European market arising from the recent changes described above, less the indicated enhanced indigenous supply, suggests that more than 120 Bcm per year will have to be imported from external sources after 2000.

Neither the gas-rich successor states to the Soviet Union nor Algeria have gas reserves availability problems in meeting Western (and Eastern) Europe's expected future demands. Nor is there an insurmountable transport problem in getting the gas to market, given the transmission lines that are either already in operation, under construction or projected. These will be capable of handling the volumes which seem likely to be involved – at least until 2000 – and, in the meantime, additional lines could be built. Moreover, the political and economic problems of Russia and its sister republics and of Algeria are a powerful motivation for these countries to increase their gas exports to Western Europe. Additional gas exports are a better and more likely means than any other alternative for these countries securing the revenues which will assist them to develop.

But the inflow of investments for gas developments that Russia and Algeria need to achieve production and transport capacities which are capable of serving the European gas market potential open to them for the long-term, now awaits clear signs of suitable politico-economic environments in which international oil and gas companies would feel it safe to invest. On the assumption that such investment starts to flow within the next couple of years or so – and on the further assumption that there is no subsequent relapse in confidence on the part of the investors – then Russia and Algeria between them should not only be able to meet their existing and planned commitments to supply upwards of 120 Bcm of gas to Western Europe and 60+ Bcm to Eastern Europe by 2000, but also be able to go on meeting the near-totality of Europe's growing import needs into the early decades of the 21st century.

If not, then alternative more distant suppliers of gas (Nigeria, Venezuela, Iran, Qatar etc.) – which will otherwise find only niche markets in Western Europe inside the next 20 years – may be able to justify large investments in significant exports facilities for producing and shipping liquefied natural gas to Europe: as Indonesia, Brunei and Australia have already been able to do for the Japanese market. Such more distant alternative sources of gas might, however, run foul of the consequences of an energy-sector politico-economic accord which embraces Europe from the Atlantic to the Urals – and beyond; as envisaged by the European Energy Charter when originally proposed by The Netherlands in 1990. The East-West cooperation intended by the Charter implies that West Siberia, the Barents Sea and parts of the former Soviet "southlands" (from Azerbeijan to Kazakhstan) will become to the European gas market what Alaska plus Canada could be to that of the United States, viz. distant gas suppliers of significant dimensions from resources of large proportions capable of meeting the needs of the markets concerned for decades into the future.

Algeria and the other gas-rich countries of North Africa lie, however, much closer than the former Soviet republics to the main European growth markets for energy on the north side of the Mediterranean basin. This geographical propinquity opens up the prospect of a long-term economic relationship across the Mediterranean based on the potential availability of large reserves of developable gas. This component in the evolution of the European gas market depends, however, on changes in what currently appears to be a low-prospective compatibility between the political systems on the two sides of the Mediterranean basin. Whilst it is increasingly clear that the East/West dimensional rapprochement is not only hoped for, but also being actively sought by all parties concerned, the same cannot presently be said of the North/South relationship. The economic sense that it would make continues to be subordinated to the political difficulties involved.

Thus, to a greater or lesser degree, for both future gas imports from the gas-rich successor states of the former Soviet Union and in respect of a similar potential from the countries of North Africa, the central issues facing energy policy markers and gas industry investors and operators are more in the domain of international politico-economic relationships, than of simple and straightforward market orientated efforts to match supply with demand. Neither the short-nor the long-run availability of more than enough gas to meet the highest conceivable level of annual demand are in doubt. Nor is the ability of the industry to engineer and successfully to operate a geographically much extended and a more intensively structured transmission system; one, that is, that operates to common standards over a region stretching from the Atlantic to the Urals and from the Arctic to the Sahara and thus forming a "continental" gas market of geographic dimensions roughly similar to those of North America including Mexico.

Indeed, including the gas market of the relevant European part of the former Soviet Union, the total amount of gas produced and used in

Europe already significantly exceeds production and use in the North American market. Yet a potential for doubling the demand for gas in West and East Europe has been indicated in this paper. Even when one sets against this expansion the prospect of a potential decline in gas use in Russia and other European republics of the former USSR – as the efficiency of production, transmission and use is improved from its existing abysmally low levels, the gas market in the wider Europe, as defined above, seems likely to continue to move still further ahead of that in North America over the coming 20 years.

Moreover, the existing differential in terms of the two markets' values – consequent not only on the size difference, but also on the very much lower average unit price which gas commands in the North American energy market – will also continue to widen. While real price increases in North America are generally forecast to be relatively modest,[29] those in Europe, on the contrary, are generally seen as likely to rise much more strongly. The current average unit value of gas in Europe (on entering national transmission/distribution systems) is already only a little under $3 per million Btu's. This could rise (in $1990) to $4 by the end of the century and towards $5 a decade later.

The current high – and expected even higher real – price for gas reflects the general principle applied through most of Western Europe since the 1970s, and now being applied in Eastern Europe and soon to be applied in Russia, viz. the principle whereby gas is priced at an appropriate oil-related value (generally a mélange of gas oil and fuel oil prices to give, in round terms, a "heavyish" crude oil equivalent value). This principle may be modified in the more liberalized market anticipated through the application of the Single Europe Act, but only in respect of the manner of the gas price calculation for large users, as a consequence of competition between gas and (imported) coal in the industrial and power generation sectors. Even in these markets, however, the higher efficiency at which gas can be burned in plant, which is also less capital intensive than that required for burning coal, will enable gas to be sold at a significant premium to coal. This may, indeed, turn out to be not very different from the present 30% differential between coal and fuel oil prices. Thus, the imported coal-determined gas price for these large users need not decline much – if at all – from present levels. And even if there were significant reductions in average gas prices to large users across Europe, then the adverse impact on the overall average value of gas seems likely to be largely offset by increased unit charges to the smaller users in the other market sectors: as the element of the cross-subsidisation of prices between different classes of consumers is brought to an end in the new market situation.

Finally, one should note that gas prices in Europe would necessarily be increased by the implementation of the EC's proposed progressive CO_2 energy tax over the period 1993–2000.[30] However, in as far as the CO_2 component in the tax (50%), as proposed by the Commission, would differentiate the tax-rate imposed in favour of gas and against oil and, even more so, coal, it would serve to allow the gas price excluding the tax to rise in real terms, relative to the pre-tax prices of coal and oil. After including the differentiated tax payment, the tax-paid price of gas would suffer little, or even no, net disadvantage against the other fuels: and thus the price increase would have little or no impact on the preference for gas over the other energy sources.

The potential 20% rise in the net-back price to gas producers arising from the proposed CO_2 tax would be a powerful supply-side incentive. It would enable more intensive reserves development as well as enhanced production and transmission facilities to be financed, so ensuring the long-run availability of both indigenous and imported supplies to Europe's rapidly expanding natural gas market over the coming decades.

Notes and References

1 See Peter R. Odell, "The West European Gas Market: the Current Position and Alternative Prospects, *Energy Policy*, Vol.15, No.5, October 1988, pp.480–493. Included as Chapter III-5 in this volume.

2 See Ministerie van Economische Zaken, *Energienota*, Den Haag, 1974.

3 These issues were discussed in greater depth in P.R. Odell, "Constraints on the Development of Natural Gas Resources with Special Reference to Western Europe" in C. Campbell Watkins and L. Waverman (Eds) *Adapting to Changing Energy Prices*, Oelgaschlager, Gunn and Hain, Cambridge, Mass. 1984. pp. 55–70. Included in this volume as Chapter III-4.

4 Commissions of the European Communities, *The Internal Energy Market*, COM (88) 238 final, Brussels, June 1988. Agreed by the Council of Energy Ministers, 24 September 1988. See *Energy in Europe*, No.12, March 1989, pp.12–13.

5 Commission of the European Communities, *Energy in Europe: Special Issue on The Internal Energy Market*, Brussels 1988.

6 See E.E.C. Council Directive, No. 75/404/EEC.

7 Royal Institute of International Affairs, London and Science Policy Research Unit, University of Sussex, *A Single European Market in Energy, R.I.I.A., London, 1989: and "Eurogas attacks EC Gas directives", International Gas Report*, Financial Times Business Information Ltd, London, Issue 197, February 21, 1992 p.11.

8 P.R. Odell, "The Completion of the Internal Energy Market: On the Need to Distinguish the Hype from the Reality", Proceedings of the *Forum Europe Conference*, Brussels, January 1990.

9 See "The Subsidiarity Principle", *Energy Economist*, Financial Times Business Information Ltd, London, No.124, February 1992, pp.6–8.

10 "British Gas/Office of Fair Trading declare Peace – for how long?", *International Gas Report*, Financial Times Business Information Ltd, London, Issue 199, March 20, 1992, pp.3–7.

11 Royal Institute of International Affairs, *op. cit.*

12 *The Single European Act*, December 1985, Article 100A(3).

13 Commission of the European Communities, *A Community Strategy to limit Carbon Dioxide Emissions and to improve Energy Efficiency*. A communication from the Commission to the Council, 1991.

14 See J. Estrada et al, *Natural Gas in Europe: Markets, Organisation and Politics*, Pinter Publishers, London, 1988, pp.179–182.

15 George W. Hoffman, *The European Energy Challenge: East and West*, Duke Press Policy Series, Durham, N.C., 1985.

16 Commission of the European Communities, "Energy for a New Century: the European Perspective", Special Issue, *Energy in Europe*, July 1990.

17 See above, footnote no.6

18 Commission of the European Communities, Directive on the Liberalisation of the EC Gas Industry, 17 January 1992.

19 Prognos AG (Herausgeber), *Energieprognose bis 2010. Die Energiewirtschaftliche entwicklung in der Bundersrepublik Deutschland*, Edition mi-Poller, Landsberg/Lech, 1990.

20 "German Energy Policy to favour Gas", *International Gas Report*, Financial Times Business Information Ltd., Issue 194, January 10, 1992, pp.7–8.

21 Commission of the European Communities, *Energy for a New Century*, Energy Balances to 2010, Brussels, 1990; DRI, *European Natural Gas*, 1987–2010 reported in Financial Times *International Gas Report*, Issue 197, 21 February 1992, pp.12–13; P.R.Odell, "European Gas Prospects" in the Proceedings of the *Gas Daily* 4th Annual Natural Gas Conference, Vienna, May 1991: International Energy Agency *Natural Gas Prospects and Policies in OECD Countries,* OECD, Paris, 1991; Netherlands Energy Research Foundation, *European Economic Integration and the Prospects for Natural Gas*, Petten, 1990.

22 Department of Energy: *Development of Oil and Gas Resources of the United Kingdom*, HMSO, London, 1991, p.24 viz. "Under the terms of petroleum production licences, development work and the production of petroleum may only be carried out with the consent of the Secretary

of State, or under a development and production programme approved by him." See D.M. Marks, "European Community Public Procurement Proposals; Implications for the Oil and Gas Industry", *Oil and Gas Law and Taxation Review*, Vol.7, No.3, 1988/9, pp.89–91.

23 Proposals by the Commission of the European Communities in respect of this requirement are currently in the early stages of discussion with member countries.

24 Algemene Energieraad, *Ontwikkelingen in het Gas Beleid: Advies van de Minister van Economische Zaken, Den Haag, November 1991.*

25 Ministerie Van Economische Zaken, *Energienota 1974*, (Den Haag, 1974) and *Nota Energiebeleid 1979* (Den Haag, 1979).

26 P.R. Odell, "Constraints on the Development and Use of Natural Gas Resources with Special Reference to Western Europe", *op. cit.*

27 Uniquely, planned field developments in Norwegian offshore waters have to be approved by the Storting, the country's parliament. See the Norwegian Ministry of Petroleum and Energy's annual report on the Norwegian Continental Shelf for details of approvals given each year: both for original development plans and for later amendments.

28 The EEC-EFTA agreement, in essence, extends the application of non-discriminatory economic relations to member countries of both organisations. This implies Norway's acceptance of the principle of equal opportunities of EC member countries' investments in its oil and gas industry: to date Norwegian interests have had preferential treatment.

29 Gas Research Institute: *1992 Edition of the GRI Baseline Projection of US Energy Supply and Demand to 2010*, GRI Strategic Planning and Analysis Division, Washington, DC, 1992.

30 See above footnote no.13.

Chapter III – 7

Gas Use and Imports to Increase with Adequate Low Cost Supplies*

1. Antecedents

Over the near 40 year history of the European natural gas industry to date, long run supply costs have hardly mattered. To the east of the former 'Iron Curtain', East European countries' gas was imported in the context of COMECON trade agreements within which, of course, prices were unrelated to western economic concepts (Dienes and Shabad, 1979: Hoffman, 1985; and Park, 1979). Within the rest of Europe,[1] the effectively state-owned, controlled and/or regulated gas industry has generally evolved under the influence of planned developments, rather than market forces. (Odell, 1969: Estrada, 1988). Supply side opportunities were controlled and constrained as, indeed, were market developments. In the absence of competition, market values have been generally unrelated to long-run supply prices. Economic rents, collected partly by producing and distributing companies, and partly by the governments of the gas-rich countries, have constituted a formidable percentage of total revenues generated by the industry since

1 Designated in this paper as Western Europe, comprising all European member countries of the OECD – except Iceland, which is irrelevant to issues of gas, and Turkey, which provides a bridge for natural gas imports from the Middle East and Central Asia and is thus an external actor in the emerging European natural gas market.

* **Published in *Energy Exploration and Exploitation*, Vol.15, No.1, 1997. It is an updated version of a paper originally presented at the SNC Energy Day in Stockholm in October 1996 under the title "The cost of longer-run gas supply to Europe".**

1965; the year in which the exploitation of the super-giant Groningen gasfield, with extremely low costs of production, began. (Odell, P.R., 1969 and 1988)

2. The Link between Oil and Gas

The deliberate and definitive link between international oil prices and the prices paid to indigenous gas producers – and for most gas subsequently imported into western Europe – was established in the early 1960s through the strategy jointly evolved between the Dutch government and the Shell/Esso consortium responsible for the discovery and development of the Groningen field (Odell, 1969 and 1988), This created a barrier to the expansion of Europe's gas markets in the period of low and falling oil prices through to the early 1970s. Thereafter, with the onset of OPEC's control over international oil prices, gas prices in Europe were enabled to rise to levels of more than an order of magnitude above the costs of supply.

This, in itself, curbed demand, but, in addition, for reasons related to a perception of gas scarcity, policies designed to limit natural gas use were imposed.[1] Moreover, the Netherlands, Britain and Norway, also imposed restraints on potentially highly profitable gas production so artificially equilibrating the market at a scale which was well below that which would have been achieved under competitive market conditions. (Odell, 1988 and 1992)

The subsequent two-thirds fall in the oil price (from its peak level in 1981/82) has brought gas prices tumbling by approximately the same percentage. Given that gas has gradually come to be recognised as a preferred source of energy – for environmental as well as for strategic/security reasons – and given that such considerations are often reflected in favourable tax conditions for gas *vis à vis* other fuels, then gas demand has been stimulated (from 200 Bcm in 1981 to over 300 Bcm this year). On the other hand, the fall in the price of gas appears to have had virtually no impact on the availability of supply. At most, the price fall led to the premature termination of production from a few small fields in France, Germany and elsewhere as they became uneconomic to continue in operation. Meanwhile, as far as production potential in Europe is concerned, this has, as shown in Table III-7.1, continued to strengthen, related essentially to the continuity of investments which

1 National policy steps to achieve this objective were subsumed in a European Economic Community decision in 1975 to restrict the use of gas in power stations, viz.. *EEC Directive* no. 75/404. (See Odell, 1988).

have ensured reserves additions in excess of year by year depletion. This is expressed in the continuing build-up of proven and probable reserves over the four decades since 1956.

3. Indigenous Reserves Development

For the 40 years of energy sector traumas to 1985 – arising from the volatility of the international oil market – the data show that demand limitations and production constraints led to reserves building up very strongly. Reserves additions were a multiple of the volumes of gas produced in each of the successive decades. Even in the decade since the price collapse of 1986 gross additions to proven and probable reserves amount to over 3000 Bcm, compared with the 10 year cumulative production of under 2000 Bcm. Of the total indicated original recoverable reserves at the end of 1995 of 12,345 Bcm from the fields discovered, some 62% remain to be used. Even in the context of $17 per barrel oil (see Table III-7.2), Western Europe's potential to produce natural gas remains formidable: with an approximate 35 years reserves to production ratio at the 1995 rate of production. Supply prospects for indigenous gas will be examined later in this paper, but it is first necessary to set the framework of prices and demand within which those prospects need to be evaluated. The bases and assumptions from which these are derived are set out in Table III.7.2.

TABLE III-7.1
Evolution of Western Europe's natural gas production and reserves 1956–1995 (in $m^3 \times 10^9$)

Period	Cum. Prod. to date	Cum. Prod. in the decade	Period-end remaining recoverable reserves	R/P ratio (yrs) at end of period	Indicated total original recoverable reserves	Gross additions to reserves in decade	Net additions to reserves in decade
Pre-1956	50	35	c.500	c.40	550	not known	not known
1956–1965	225	175	c.1,900	c.76	2,125	1,575	2,450
1966–1975	1,150	925	c.4,350	c.27	5,500	3,375	2,450
1976–1985	2,700	1,550	c.6,500	c.36	9,200	3,700	2,150
1986–1995	4,645	1,960	c.7,700	c.35	c.12,345	c.3,160	c.1,200

Sources: The author's compilation from various sources.

Contemporary Prices for Gas in Europe

Aspects of the contemporary pricing of gas in Western Europe's six major consuming countries (responsible for about 93% of total West European gas use in 1995) are set out in Tables III-7.3, III-7.4 and III-

TABLE III-7.2
Oil price assumptions: and related gas price equivalents 1995–2025 in 1995 $

	1995	2005	2015	2025
a. Forecasts for the Average Price of Internationally Traded Crude ($/bbl) f.o.b. [1]	17.05	17.40	19.60	25.65
b. Nominal Value of Gas at f.o.b. Crude Oil prices + $0.50/bbl freight costs. ($/MMBtu) [2]	3.22	3.29	3.69	4.80
c. Average Gas Price Equivalent at the Frontier/Beach ($/MMBtu) [3]	2.67	2.85	3.45	4.75
d. Forecasts for Ex-refinery Oil Products Prices ($/ton) [4]				
i. Gas Oil	151	160	190	255
ii. Low Sulphur Fuel Oil	104	110	125	165
e. Equivalent Gas Price, based on 50% Gas Oil and 50% Fuel Oil ($/MMBtu) [5]	3.21	3.39	3.95	5.28
f. Weighted Average Gas Price to End Users ($/MMBtu) [6]	5.68	5.45	5.95	7.35

Notes

1 *The 1995 value is the US Dept. of Energy calculation of the average price of the 26 most-traded crudes for the period October 1994 to September 1995. See P.R. Odell,* Global Economic/Energy Prospects, *1994–2020 prepared for the IIASA/Stanford International Energy Workshop, 1995 for the derivation of the forecasts.*

2 *Based on calorific equivalent of 5.45 million Btu per barrel.*

3 *See Table III-7.3 for the source of the 1995 values. Assumptions for other years are related to the author's oil price forecasts.*

4 *The author's own forecasts for annual average prices for cargoes ex-Rotterdam and Mediterranean refineries. From unpublished work for private clients.*

5 *Based on gasoil with 40.55 million Btu/ton and 7.5 bbls/ton and low sulphur fuel oil with 39.23 million Btu/ton and 6.7 bbls/ton.*

6 *See Table III-7.3 for the derivation of the 1995 price. Assumptions for the other years are based on a continuing exchange rate of DM 1.46 to 1 US dollar, and a 10% reduction per decade in the relationship between the average gas price to end users and the average gas price at the frontier/beach.*

7.5.[1] In Table III-7.3 note, first, the relatively small variation from country to country (except for Belgium)[2] in the average border or beach/well-head price of gas; second, that the border price for gas in 1995, at an average of about $2.67/MMBtu, was 20% below the equivalent of the landed internationally traded crude oil price of $17.05/bbl; and, third, that there are major variations in end-user gas prices (excluding all taxes) both between types of users and, for the same category of users, between the six countries. A calculation of the average end-user price per country (weighted by the percentage of use in each of the sectors designated in Table III-7.3 line c.i) shows almost as wide variation (of up to 22%) between countries as the variation in border prices. If, however, one relates the former to the latter (shown in line c. ii of Table III-7.3) then, compared with Belgium, where user prices are least marked-up from border prices, gas prices to consumers in France and Germany are 29.1% and 26.3%, respectively, more expensive. Nevertheless, even for Belgium, the costs of inland transmission, distribution, sales and profit margins amount to 75% of the cost of the gas as delivered at the border by the suppliers.

These data appear to indicate the existence of not inconsiderable "fat" in many gas end-user prices in much of Western Europe, suggesting that the contemporary limited competitiveness of gas may well be as much related to the structure and organisation of internal national markets as to possible over-pricing by suppliers. Note, in passing, that these data on prices do not indicate that market liberalisation in the UK had, by 1995, produced significantly lower overall pre-tax consumer prices: Italian and Dutch prices remained very close to those of the UK. The weighted average prices for the six countries taken as a whole are shown in Table III-7.4. On average, gas consumers pay more than twice, and domestic users over three times, the border price: even without taking end-user taxes into account.

Table III-7.5, on the other hand, contrasts fully tax-paid end-user gas prices with the equivalents for competing sources of energy. Major differences in the incidence of taxes on energy between countries pick out the UK as the country with lowest prices: but in as far as this low-tax situation applies generally to energy, other energy sources in the UK

1 Based on research by the author as part of the contribution by Energy Advice Ltd (London) to the study, "European Gas Markets; Prospects to 2020". This was published by Prognos AG, of Basel on *Die Energiemärkte Deutschlands in Zusammenwachsenden Europa – Perspektiven bis zum Jahr 2000* for the German Federal Ministry of Economics, 1995.

2 Belgium's higher price reflected its continuing exposure to an earlier long-term contract to purchase Algerian LNG at a crude oil equivalent price.

TABLE III-7.3
End-user prices in the six main gas using countries indexed to border (beach) prices of gas, 1995

	Germany	U.K.	NL	France	Italy	Belgium
a. Border Prices						
i. In $/MMBtu	2.72	2.60	2.72	2.78	2.71	3.25
ii. In Pf/kWh	1.35	1.27	1.35	1.37	1.35	1.59
Index: Lowest = 100	*106.3*	*100*	*106.3*	*107.9*	*106.3*	*125.2*
b. End-User Prices in Pf/kWh excluding VAT and all taxes. (Figs in brackets show % of total use in each sector)						
i. Domestic	4.69 (30)	3.67 (45)	3.79 (35)	4.71 (37)	4.07 (30)	4.57 (32)
Index: Border Price – 100	*347*	*289*	*281*	*344*	*301*	*287*
ii. Commercial	3.30 (15)	2.15 (13)	3.53 (13)	3.43 (16)	3.23 (10)	3.43 (10)
Index: Border Price = 100	*244*	*169*	*261*	*250*	*239*	*216*
iii. Medium Industry – Firm	2.87 (8)	1.81 (5)	2.12 (6)	2.03 (10)	2.15 (9)	2.24 (8)
Index: Border Price – 100	*213*	*143*	*157*	*148*	*159*	*141*
iv. Large Industry – Firm	1.77 (23)	1.71 (6)	1.74 (14)	1.67 (14)	1.86 (15)	1.89 (13)
Index: Border Price = 100	*131*	*135*	*129*	*122*	*138*	*119*
v. Large Industry–Interruptible	1.69 (6)	1.27 (9)	–	1.61 (14)	1.74 (15)	1.56 (13)
Index: Border Price = 100	*125*	*100*	–	*118*	*129*	*98*
vi. Power Stations	1.93 (16)	1.46 (19)	1.59 (25)	1.60 (2)	1.69 (19)	1.53 (17)
Index: Border Price = 100	*143*	*115*	*118*	*117*	*125*	*96*
vii. Feedstocks	1.58 (2)	1.00 (3)	1.36 (7)	1.52 (7)	1.61 (2)	1.43(7)
Index: Border Price = 100	*117*	*85*	*101*	*111*	*119*	*98*
c. Weighted Average of sector prices in b.						
i. Price (Pf/kWh)	*2.98*	*2.54*	*2.65*	*3.09*	*2.63*	*2.79*
ii. as % of Border Price	*221*	*200*	*196*	*226*	*195*	*175*
d. Index to Lowest Value in c.						
i. by weighted Absolute Prices	*117.3*	*100.0*	*104.3*	*121.7*	*103.5*	*109.8*
ii. by % over lowest mark-up from Border Price (Belgium = 100)	*126.3*	*114.4*	*112.0*	*129.1*	*111.4*	*100.0*

Sources: Border prices are derived from the official data for Belgium and the UK and, for other countries, from estimates published by European Gas Markets. End-user prices are from European Energy Pricewatch and Energy Advice's research.

are still generally lower priced than gas (on a calorific basis, without taking contrasting efficiencies in use into account). But the price differentials working against gas relative to other fuels are generally even worse in the other countries, except for the Netherlands. There, only power station coal is significantly lower priced than gas. Significantly, the 42.6% share of natural gas in total energy use in the Netherlands stands at well over twice the European average of 18.8% (B.P., 1995), indicating how far gas' substitution for other sources of energy could go in the rest of Europe given gas pricing policies which are more competitive with alternative energy sources.

In Belgium, other sources of energy are generally very competitively priced: gas thus has no significant price advantage in any sector (except over low sulphur fuel oil in power generation), so it is hardly surprising that gas use is restricted to 18% of total energy use. The same is true for Germany (where gas also accounts for 18% of energy use), except in respect of coal as a consequence of the continued use for political reasons of much higher cost indigenous coal production. In

TABLE III-7.4
Weighted average gas prices[1] in West Europe's six main gas using countries, 1995

a. Average Border Price in	$/MMBtu	$2.72
	Pf/kWh	1.37 Pf
b. End User Prices in Pf/kWh (share of sector in Total Market)		
	Domestic (35.6)	4.14
	Commercial (13.3)	3.15
	Medium Industry (8.1)	2.08
	Large Industry – firm (14.4)	1.68
	Large Industry – interruptible (9.0)	1.58
	Power Stations (16.9)	1.65
	Feedstocks (3.8)	1.41
c. Weighted Average[1] Overall 2.77 Pf		
– as % of Average Border Price		202%

Source: Derived from Table III-7.3

1 *Weighted averages based on total gas use per country shown in Table III-7.6 and percentage use per sector shown in Table III-7.3.*

France (with only 12% gas use), cheaper alternatives to gas are available in most sectors. In other words, over much of Europe gas pricing policies (sometimes including energy taxation policies) inhibit the expansion of

TABLE III-7.5
Comparative prices of gas and competing fuels by sector in the six main gas-using countries, 1995
(gas prices in pf/kwh including vat where applicable and all taxes)

Sector	Germany	UK	NL	France	Italy	Belgium
a. Domestic: Gas	5.8	4.0	4.8	5.6	8.4	5.8
Indexed Gas Oil (Gas = 100)	<u>*65*</u>	*74*	*104*	*93*	*127*	<u>*53*</u>
b. Commercial: Gas	3.7	2.2	3.8	3.4	6.4	3.4
Indexed Gas Oil (Gas = 100)	<u>*78*</u>	*118*	*97*	*116*	*134*	<u>*67*</u>
c. Medium Industry: Firm Gas	3.2	1.9	2.4	2.2	2.4	2.3
Indexed Gas Oil (Gas = 100)	<u>*88*</u>	*129*	*146*	*160*	–	<u>*99*</u>
Indexed L.S.F.O. (Gas = 100)	<u>*54*</u>	*92*	*100*	*89*	*81*	<u>*68*</u>
Indexed Coal (Gas = 100)	*119*	*88*	–	*115*	<u>*60*</u>	<u>*48*</u>
d. Large Industry: Firm Gas	2.1	1.7	1.9	1.9	2.1	1.9
Indexed L.S.F.O. (Gas = 100)	<u>*81*</u>	*100**	–	*104*	*82*	<u>*81*</u>
Indexed Coal (Gas = 100)	*181*	<u>*64*</u>	–	*85*	*68*	<u>*58*</u>
e. Large Industry: Interruptible	2.1	1.3	–	1.8	2.0	1.6
Indexed L.S.F.O. (Gas = 100)	*84*	*135*	–	*108*	<u>*98*</u>	<u>*101*</u>
Indexed Coal (Gas = 100)	*188*	*87*	–	*86*	<u>*72*</u>	<u>*70*</u>
f. Power Stations	2.3	1.5	1.7	1.8	1.7	1.5
Indexed L.S.F.O. (Gas = 100)	<u>*76*</u>	*117**	–	*108*	<u>*95*</u>	*99*
Indexed Coal (Gas = 100)	*92***	*70*	*86*	*86***	<u>*68*</u>	<u>*78*</u>

N.B. *The lower the indexed value for alternative fuels, the greater the price competition for gas. The two lowest values for each alternative fuel are underlined. Belgium and Germany appear to have the least price-competitive gas in most sectors.*

★ *Fuel oil in large industry and power stations in the U.K. is 2% sulphur.*

★★ *Coal price to German and French power stations is calculated after allowance for subsidies to indigenous coal.*

Sources: See Table III-7.3.

gas demand. This limits the economies of scale which could otherwise be achieved and so adds yet another factor making the border supply price of gas less significant to the prospects for the evolution of the European gas market than is often portrayed.

5. Demand Expansion: and Implications for Supply

Notwithstanding the continuing constraints on gas use, the European market for gas is, and will remain, in an expansionist mode (Pauwels, 1994; Prior, 1994; Stern, 1990).

The potential for such expansion is considered in depth in other contributions. Here, as a background only to considerations of the prospects for the supply price, we predicate an expansion of gas use in Europe over the 30 years to 2025 as shown in Table III-7.6.[1] Overall, this involves a near 100% increase in gas use in Western Europe and a two-and-a-half times increase in its use in Eastern Europe from estimated 1995 levels. These increases imply a cumulative use of gas over the 30 year period of more than 16,000 Bcm: compared with the cumulated use of under a third of this amount over the period of 40 years, 1956–1995. Meeting this potential demand clearly involves both the discovery of many new fields and the recovery of large additional reserves from existing fields within Europe, together with ready access to increasing volumes of imported gas. Thus, for the first time in the history of the European gas industry, demand increases will directly influence upstream activities and, in due course, reduce the contemporary highly favourable gas reserves-to-production ratios to more modest levels.

Under these demand circumstances the cost of longer-run supplies could thus become a critical variable in the further expansion of the industry. We shall argue below, however, that such a development is likely to be of only modest proportions during the time period under consideration in this paper. In order to test this it is necessary to look separately at the indigenous supply potential, on the one hand; and, on the other, at the imported gas contribution. Undue dependence on the latter may well be viewed as a serious enough problem to justify demand side constraints, irrespective of considerations of relative prices (Pauwels, 1994).

1 This is derived from research by the author for his contribution to Prognos, 1994. It is compatible with other recent forecasts of European gas demand (as, for example, by Williams, 1993; Pauwels, 1994), and as reported in Prior, 1994. Though note that these studies were generally confined to the period to 2010. For consideration of the long run supply price issues this is, from a 1995 base, too short a period. Hence the need in this paper to extend the period by another 15 years.

TABLE III-7.6
The evolution of European gas use by country: 1995–2025 (in Bcm)

Country	Actual Use 1995	Potential Use in: 2005	2015	2025
a). Western Europe				
Germany	74	85	102	118
United Kingdom	73	95	105	116
Italy	48	72	90	104
The Netherlands	37	40	43	46
France	33	43	52	62
Belgium	11	16	20	25
Spain	8	18	25	34
Austria	7	10	11	13
Finland	3	7	9	10
Denmark	3	6	8	10
Ireland	3	5	6	7
Switzerland	2	4	5	6
Sweden	1	3	6	8
Luxembourg	1	1	2	2
Greece	negl.	2	5	8
Norway	negl.	2	3	5
Portugal	–	5	7	12
Sub-Total	*304*	*414*	*497*	*582*
b). Eastern Europe				
Romania	24	27	29	32
Poland	10	16	23	29
Hungary	10	14	16	22
Czech Republic	7	13	16	23
Slovakia	5	8	10	14
Bulgaria	5	6	8	11
Former Yugoslavia	4	6	8	11
Albania	negl.	2	3	4
Sub-Total	*65*	*92*	*113*	*146*
c). Overall Total	*369*	*506*	*610*	*728*

TABLE III-7.7
Natural gas in Western Europe, 1961–1995

Year	Indigenous Production (m^3x10^9)	Imports (m^3x10^9)	Total Use (m^3x10^9)	% Dependence Imports	No. of Countries Using Gas	Gas Use as % of Energy Use*
1961	16	0	16	0	5	1.8
1966	24	1	25	3.8	7	3.3
1971	104.5	1.5	106	1.4	8	9.7
1976	164	19	183	10.4	10	13.4
1981	176	24	200	12.0	11	14.7
1986	179	39	218	17.8	15	15.2
1991**	198	72	270	26.7	15	19.4
1995**	221	83	304	27.3	16	22.0

* *Primary and nuclear electricity calculated on the heat value of the output.*

** *Includes data for the former East Germany.*

Source: Compiled by the Author from various sources.

6. Indigenous Supply Prospects and Prices

This historic importance for Western Europe of its indigenous gas resources is shown in Table III-7.7. Such indigenous resources have always provided most of the gas used and continue to do so to the extent of almost 70% of the gas supplied. Given the indicated expansion of demand over the next three decades, the role of indigenous production will remain important to the secure evolution of the market. This presupposes, of course, that known indigenous gas reserves are not only profitable to produce, but also that continued investment in the exploration for, and the exploitation of, additional reserves remain economically viable. Given the prolific nature of the occurrence and the potential massive availability of external resources, enhanced gas-on-gas competition might well drive gas prices down, so that the economic producibility of the European reserves become an issue of importance and concern.

Of the total remaining proven and probable reserves shown in Table III-7.1, over 60% fall into the former category; and, by definition (B.P., 1995b), are thus recoverable at present costs, prices and technology.

Given the relative maturity of the North Sea oil and gas province within which most of these reserves are located and given the production and transportation systems which are already in place, an expanded production can be anticipated at modest costs. There is a known availability of gas of at least 4500 Bcm at costs which require no or small (real) price rises to make the operations profitable. Beyond this, there are the probable reserves (which given their occurrence in a large number of individual fields have an overall high probability of being available) of at least another 2000 Bcm which can also be brought profitably to market within the constraint of present prices. This already favourable situation for larger-scale future production from known reserves is, moreover, made yet more attractive by the continuing prospect of real-costs' reduction as a result of further progress in exploitation technology. (Handley, 1995; Williams, 1992)

Even more important these reserves' estimates relate only to the gas in fields already discovered and declared commercial by 1993, with a modest 5% allowance for reserves' appreciation by the end of 1995. This is a highly conservative view of these fields, ultimate potential given, first, the certainty that, on average, the statistical population of discovered fields will continue to exhibit the phenomenon of reserves' appreciation. Even the 36 year old Groningen field – for which the estimates of its initially recoverable reserves have been up-graded so many times that they are now declared at twice the level of 1965 when production began – still has some potential for further upward reappraisal. (Odell, P.R., 1992) There is a similar prospect for the largest and long-since discovered offshore gas-fields in the southern basin of the North Sea; while, for the to-date little exploited fields of the central basin, the process of effective reserves' evaluation – based, as it must be, on production experience – has barely begun. (Kemp, 1994; Stoppart, 1994)

In the Norwegian sector of the North Sea a recent upward revision of the estimates of reserves (Norwegian Petroleum Directorate, 1993) has not only greatly enhanced the length of period over which plateau rates of production can be maintained, but has also pushed the country's potential peak production above 100 Bcm per year. Except for the Ekofisk and Frigg groups of fields, Norway's gas exploitation history is, indeed, still in its early stages. Given, however, the massive scale of the investment that has already been made in both new production facilities (as, for example, in the Troll fields) and in pipelines to transport the gas to the mainland of Europe (Royal Norwegian Ministry of Industry and Energy, 1995), coupled with the modest marginal costs which will be incurred in due course in the additional exploitation of the fields, the depletion of whatever volumes of gas are eventually defined as producible is not in doubt at netback values close to today's beach/border prices (see Table III-7.2), measured in real terms.

In addition, there is a second major reason for noting the conservatism of current indigenous reserves' data for Europe, viz. the certainty of a continuing process of discovery of new fields in western Europe in the context of high exploration expenditures in both known and prospective gas rich provinces. Outside the North Sea, the mid-and north-Norwegian and the west of Shetlands basins have already been proved hydrocarbons rich. (Norwegian Petroleum Directorate, 1993: Department of Trade and Industry, 1995) The former, in particular, have already been shown to be gas prone so that proposals for the initial developments of the already discovered fields are already under way. Other areas around the coast of Europe have been demonstrated as having more modest resources, but these could eventually be developable in appropriate locations, viz. in relation to emerging gas demand patterns, given the ability of advancing technology to lower field development costs. These, however, will only become 'necessary' to sustain increasing European gas production in the longer term.

Meanwhile, on the assumption that indigenous gas will only be able to secure about 60% of future west European gas markets, in the face of competition from external suppliers, then the reserves which are known, probable and possible (within the limits set by expected market values expressed in real terms) from the exploitation of existing and planned fields' developments will not only meet the total cumulative requirements to 2025 (some 8000 Bcm), but will also provide at least the minimum carry-over of reserves of about 4000 Bcm for an adequate R/P ratio of 10 to 12 years at that date. This implies additions to reserves over the next 30 years which are approximately equal to the gross additions to reserves over the last 20 years.

The exploration and exploitation history of Europe's upstream offshore gas industry is still relatively short so that the learning process for effective production continues. Thus, the inevitability of rising costs through the reserves' depletion process seems likely to continue to be offset in large part by the downward pressures on costs engendered by technological advances. (Adelman, 1990: Handley, 1995) And as this is starting – in 1995 – from a situation in which prices are well above the long-run supply price (so that sizeable rents, taken mainly in the form of taxation and, less so, as super-normal profits, are still being achieved), then, under increasingly competitive conditions, a downward, rather than an upward, trend in the border/beach price of indigenous gas over the rest of the century is the more likely prospect. Note that this development is, indeed, already under way in the UK where current beach prices for gas are only about half those of earlier contracts. On the other hand, expected rising real international oil prices post-2000 (see Table III-7.2) will eliminate the validity of that lower gas supply-price as the most appropriate base from which to price European gas. As has been the case throughout most of the past 40 years, the profitability of indigenous gas production seems likely to remain assured by the expected development of the price of international oil. On the basis of this range of considerations the potential most likely evolution of European gas production is as shown in Table III-7.8. Growth continues throughout the period – albeit at a much reduced rate after 2005 – to give an overall increase of over 60% over the next three decades.

7. Europe's Gas Imports

Concern for higher priced gas imports – sometimes expressed as "the inevitability of higher prices" (Williams, 1993) – relate in part to exaggerated expectations for the volumes of gas which Europe will have to import. These expectations emerge from the forecasts of much more limited prospects for indigenous production than those indicated above (Pauwels, 1994: Prior, 1994). Conventional wisdom has long been pessimistic in this respect. For example, views presented through to the late-1980s indicated that the level of indigenous production would have passed its peak by 1990 (International Energy Agency, 1986) so that after only another decade, even within the context of a slow growth in gas use, Western Europe would depend principally on gas imports. Most forecasts on indigenous gas prospects remain relatively pessimistic; differing only in their views on the rate of decline in production from the late 1990s (Prior, 1994). As shown above, we do not share this view – indeed, have never shared the view – because of the existence of the very extensive potentially petroliferous areas and of the ready availability in Europe of financial and technological resources whereby long continuing exploration efforts can be sustained (Odell, 1969 and 1988). The success of those efforts in finding and adding to reserves thus enables output to keep moving ahead. Table III-7.7 above shows that this view has to date been fully justified – in spite of deliberately constrained production policies from time to time by the main producing countries.

The process will, as argued above, continue and, as shown in Table III-7.9, restrain the rate of increase in import needs. Thus, from an estimated 119 Bcm of imports in 1995 (almost three quarters from Russia and most of the rest from Algeria, there will be relatively modest increases to 167 Bcm in 2005; to 231 Bcm by 2015 and to 316 Bcm by 2025. The 30 year cumulative volume of gas involved will be no more than 6,500 Bcm. Assuming the continuation of Russia and Algeria as the only suppliers and of the present 75:25 exports, split between them, then the call on the former's currently proven reserves of 48,000 Bcm would be under 10%: but would be 45% of the 3,600 Bcm of Algeria's presently declared reserves. The latter will have to be expanded if Algeria is to maintain its present contribution to Europe's imports.

7.1 The Supply and Price of European Imports until 2004

To date, both Russia's and Algeria's access to European markets has been made easy; partly by the political constraints on production in Europe and partly because Germany and Italy, respectively, have, in return for gas imports, achieved important economic advantages in respect of the markets they have secured for goods and services required by the gas industries of the two countries (Estrada, 1988). Germany and Italy, in turn, ensured success in the construction of the capital intensive transportation infrastructures which were required to get Russian and Algerian gas to Europe. For the two gas supplying countries there were equally important gains. Its export contracts with European customers enabled Algeria to produce gas with significant economies of scale and thus to secure low-cost gas for use in its own economy. For the former Soviet Union – and, more recently, Russia – the role of gas exports in sustaining the country's hard-pressed economy has been crucial. (Dienes, Dobozi and Radetski, 1994)

These considerations – for both Germany and Italy, on the one hand, and for Algeria and Russia, on the other – continue to apply: though now, of course, in the context of enhanced political problems (Estrada, 1995). Ignoring the latter, the call on the resources of the two exporters – in terms not only of their gas reserves, but also of their resources of management and finance whereby they can achieve the potential for this higher level of exports to Europe as specified above – remains modest, compared with the benefits which the exports will generate.

The two countries will thus be highly motivated to secure the largest share possible of the European gas markets open to imports. As shown above, at least through the two decades to 2015, the markets for imports will remain quantitatively restricted from a combination of relatively modest increases in demand and by the expansion of Europe's own output; most notably from the North Sea and adjacent areas, but not excluding proportionately larger, though volumetrically limited, increases in gas production in a range of both west and east European countries. This is consequent upon the enhanced exploration and development efforts which will occur, especially in Eastern Europe where opening the upstream gas industry to international expertise, management and investment will, first, stop and, then, reverse the decline in output which has been under way since 1988.

The outlook for prospective exports of gas to Europe will be particularly restricted for the first decade – with, as shown in Table III-7.9, a forecast requirement for only an additional 48 Bcm in 2005 compared with 1995. Contracts are already in place sufficient to meet most of this modestly increased requirement for imports – mainly on the basis of oil-related prices as used in the existing system. In the current situation of energy over-supply it seems unlikely that the external

TABLE III-7.8
European gas production potential; 1995 to 2025
($m^3 \times 10^9$)

Country	1995 Actuals	Potential in: 2005	2015	2025
a. Western Europe				
United Kingdom	72	98	114	124
The Netherlands	66	70	68	62
Norway	31	69	90	105
Italy	18	23	25	25
Germany	16	16	13	10
Denmark	5	9	10	11
France	3	2	2	2
Ireland	2	3	4	4
Others	2	2	2	5
Sub-Total	*221*	*292*	*328*	*348*
b. Eastern Europe				
Romania	17	25	28	32
Hungary	5	7	8	10
Poland	4	6	7	7
Former Yugoslavia	2	2	2	4
Others	1	4	5	9
Sub-Total	*29*	*44*	*50*	*62*
c. *Overall Total*	*250*	*336*	*378*	*410*

TABLE III-7.9
Gas production, use and trade in the West and East European gas markets, 1995–2025 (in Bcm)

Country	Actuals		Forecasts for:	
	1995	2005	2015	2025
a). Western Europe				
Gas Use	304	414	497	582
Gas Production	221	292	328	348
of which:				
– used in the Countries of Production	*149*	*186*	*205*	*215*
– exported to other West European Countries	*72*	*95*	*108*	*115*
– exported to East European Countries	*0*	*11*	*15*	*20*
Net Imports from External Supply Sources	83	130	183	252
Import Dependence	(27.3%)	(29.8%)	(37.1%)	(43.3%)
b). Eastern Europe				
Gas Use	65	92	113	146
Gas Production	29	44	50	62
of which:				
– used in the Countries of Production	*29*	*42*	*47*	*58*
– exported to other East European Countries	*0*	*2*	*3*	*4*
Imports from West Europe	0	11	15	20
Net Imports from External Supply Sources	36	37	48	64
Import Dependence	(55.4%)	(43.5%)	(43.4%)	(45.2%)
c). Total Europe				
Gas Use	369	506	610	728
Indigenous Production	250	336	378	410
Imports from External Sources	119	167	231	316
Import Dependence	32.3%	33.0%	37.9%	43.4%

Sources: See Tables III-7.6 & 8 for gas use and production by country, respectively. Forecasts of intra-European trade and gas imports to Europe are derived as indicated in footnote 1 on p.435.

suppliers will be able to improve on this method of pricing. Indeed, excess supply relative to demand, coupled with the sellers' strong motivation to try to sell more gas, seems more likely to lead to gas prices which move independently of oil on the down-side, though still reflecting oil equivalents on the up-side (see Table III-7.2) in the event of the international oil market strengthening over the coming decade. Given their already existing gas production and transport infrastructure, either in place or under construction, whereby most of the expected 2005 level of exports can be handled, neither Algeria nor Russia seem likely to face any problems of economic viability in fulfilling their obligations and opportunities to supply gas at prices which are little, if any, higher, in real terms, than those of today.

For Russia, in particular, the continuing national decline in gas use will free up both the gas itself and the infrastructure capacity required to make it available for export (Stern, 1995). Thus, the profitable export of all the Russian gas for which markets are likely to be found in Europe by 2005 is hardly in doubt: except, of course, in the context of internal political problems which could both disrupt the supply/transportation system and significantly increase costs.

For Algeria, neither the economics of, nor the capacity for, enhanced exports up to the contracted level for 2000 of 60 Bcm per year, compared with the approximately 30 Bcm which is being exported to Europe in 1995, appear to pose any problems. (Estrada, 1995) It is once again only political problems leading to the partial break-down of the country's gas production and transmission systems, which could pose a physical restraint on gas availabilities and/or the imposition of higher costs.

7.2 The Supply and Price of Imports, 2005–2014

For this second decade, uncertainty over Europe's gas imports increases given that it is a period when, as can be calculated from the data in Table III-7.8, the annual rate of increase in indigenous production falls to about 1.2%: well below the 2.8% annual rate of expansion foreseen until 2005. Additionally, though all but one of Algeria's export contracts extend beyond 2005, this is not the case with Russia whose contracts are mostly valid only through the early years of the 21st century. Of the indicated demand for imports of about 200 Bcm by the middle of this period (extrapolated from Table III-7.9), less than one half is already contracted under contemporary supply and pricing criteria.

Algeria, as already suggested above, may well be able to increase its approximate present 25% share of Europe's total imports to as much as

35% by 2005, but its ability to maintain its enhanced position thereafter depends on the necessary expansion of its proven reserves base and on the level of costs at which larger volumes of gas can be exploited and moved to markets via new or expanded pipelines to southern Europe. However, within the context of an oil price rising from today's level of \$17.5 per barrel through to \$19.60 per barrel (in 1995 \$) by 2015 (see Table III.7.2) there will be scope for real increases in the delivered gas price (up to a maximum of approximately \$3.45 per mmBTU). Such a price would provide a powerful economic motivation for further investment in Algerian gas exploration, production and transportation and thus provide for the expansion of supplies to Europe to the ± 90 Bcm which Algeria might wish to try to secure in competition with supplies from Russia (Pauwels, 1994). However, alternative North African supplies – which could, in part, make use of the same trans-Mediterranean transmission lines – could also offer a significant potential addition to the flow of gas to southern Europe by that time. Libyan and Western Desert Egyptian gas provide these possible competing alternatives. This gas lacks effective alternative external markets so that there will be no competing demands for it and thus no upward pricing pressures on the supplies.

Russia, however, seems likely to aim to provide up to two-thirds of external gas supplies to Europe in the 2005–14 period as a result of both economic considerations and from the viewpoint of the policy interests of both supplier and customers. In the context of the plenitude of Russia's gas reserves, the basic technical competence and the organisational strength of its gas industry (measured by Russian standards) and the interest of international oil and gas companies in the exploitation of Russia's gas resources, it is impossible to forecast just how development will proceed and what specific projects will be completed so as to be able to deliver gas to Europe in the decade beginning in 2005 (Stern, 1995). There are a number of gas resource development opportunities which could be exploited to provide more than sufficient gas to meet the prospective European demand for Russian gas (rising from about 120 Bcm in 2005 to at least 145 Bcm in 2014). It does, however, seem more likely than not that the continuity of the rehabilitation process in respect of both the pre-existing production and transmission systems will continue to be central to the profitable export of additional gas to Europe after 2005. This will be in the likely context of an internal demand for gas which even then remains below historically higher levels (as a result of the achievement of a price-motivated increased efficiency in gas use in Russia itself by that time), and of a renewed pipeline system in which transportation losses will have been much reduced from today's high levels by the application of western technology, so offering greater throughput capacity. (Dienes, 1994: Pauwels, 1994; and Stern, 1995).

Such gas will be competitive relative to the border price (of up to $3.45/mmBTU) which the European gas market will have to bear in the context of the slowly rising international oil price after 2005. The availability of enough Russian gas sufficient to equilibrate the European market through such supplies, seems highly probable early in that decade when about 120 Bcm a year will be required. By that time, however, it is conceivable that initial developments in major new gas regions – notably Yamal and the Barents Sea – will be poised to make a contribution to Russian gas supplies. Even if these new supplies are initially costed for delivery at more than $3.45/mmBTU, and may thus have to be "eased" into the market on the basis of a degree of cross-subsidisation from the profits on the production of gas from already established areas, this could still be in Russia's interests to do, in order to be ready for the opportunities post-2010. By then, economies of scale in the new areas and through the new lines will become effective enough to sustain the profitable exploitation of the new gas in its own right.

North Africa plus Russia will thus be able, politics permitting, profitably to supply Europe's import requirements through the decade from 2005 to 2014 at border prices which reflect internationally traded oil equivalent values. Over the decade, as shown in Table III-7.2, these seem likely to rise from $2.85 to about $3.45 per mmBTU. This oil equivalent value does, of course, only set the upper limit to gas prices imported into Europe, so as to ensure that gas markets can expand in competition with oil and coal. Under competitive gas market conditions, as seems not unlikely given the plethora of potential supplies, gas prices could well fall below this upside restraint. How far below depends on the degree of competition which emerges between the main external players, viz. Algeria and Russia (both, by that time, with a possible range of individual supplying companies); and between these external suppliers and the competing indigenous producers.

Other potential battles for market shares could, of course, produce very different contributions to supplies from those shown in Tables III-7.8 and 9. Central Asian gas suppliers, viz. Turkmenistan and Uzbekistan, would also like and, indeed, expect to be additional competitors for a share of European gas markets in the 2005–2014 period. Given, however, both the geo-politics and the economics of the Central Asian to Europe transmission lines, investments in this gas for

European markets, even at the oil equivalent prices indicated above, seem likely to be marginally viable at best. Even with the utilisation of best international gas industry practices, the most modern technologies, the required political will and only modest claims for transit fees from Turkey and other intervening countries, getting a gas production and transportation system in place and working effectively so to enable Central Asian gas to be economically deliverable to Europe so early in the 21st century currently seems unlikely. Except in the event of a cooperative arrangement whereby Russia transits the gas to the border with Europe. This would be a lower cost option, but it implies Russian control over, and hence limitations on, the deliveries of Central Asian gas in a situation in which Russia would see the gas as competing for its own export opportunities.

7.3 The Supply and Price, 2015–2024

In speculating on the third future decade uncertainties multiply; but so do the number of possibilities for serving the then still growing Euro-gas market. Accessible resources by 2015 will have multiplied as a result of continuing exploration and development investment in the meantime in response to market challenges. Cumulative use in the decade will most likely exceed 6,600 Bcm of which less than 60% are likely to come from indigenous resources, leaving almost 2,700 Bcm to be provided by imports; rising from some 230 Bcm in 2015 to about 320 Bcm by the last year of the decade.

Such a larger market will clearly have space for Libyan and Egyptian, as well as Algerian, gas from the North Africa. The main expansion of Russian supplies from 2015 will have to be derived from the newly developed reserves of the Yamal Peninsula, the Barents Sea and other regions of the country; depending on where Russian and foreign companies decide to make their investments over the next twenty years. Furthermore, the initial development of gas in Turkmenistan and Uzbekistan for export to European markets will at last be stimulated by the much larger-scale market opportunities and thus be able to secure significant economies of scale in the operations. And, finally, even some low-cost-to-produce Middle East gas – from Iran and Qatar particularly – could well have been developed for export by pipeline to Europe. Thus, Europe's expanding gas markets will become served by a much expanded – and lengthened – transport network implying, of course, higher average transmission costs compared with the earlier decades, albeit in the context of low cost gas at the well-head, compared with the likely supply costs of indigenous gas by that time.

The ability of such longer-distance gas profitably to enter the European market will be related, in part, to its low supply price at the production locations and, in part, to the higher real prices at which international oil will then be traded – as a consequence of tighter supply/demand relationships emerging in the international oil market by that time from the high and still growing demand for oil products in South Asia, the western Pacific Rim and other parts of what is now the developing world. Over the ten year period from 2015 we forecast the oil price to rise from $19.60 to $25.65 per barrel (in 1995 $). In this oil price context the competitive entry price for gas at the borders of Europe will increase from $3.45 to $4.75 per mmBTU (also in 1995 $).

Though the increase in the size of the Euro-gas market for imports will have become considerable by this third decade in the future, the volume of gas involved will remain, nevertheless, relatively modest in relation to the volume and diversity of gas reserves' developments which, as described above, will be stimulated by the opportunities offered by European outlets. Thus, investments in the meantime by national and international interests in many regions capable of supplying Europe after 2015 in the context of an anticipated higher price for gas as a result of oil market developments, could conceivably lead to a situation in which too much infrastructure capacity is created and thus to a potential over-supply of gas. The continuation in such circumstances of a squeeze on prices arising from gas on gas competition would, of course, lead to an average border price for the 230 to 320 Bcm of gas required per annum which is below the oil equivalent prices indicated above.

Large though an import of up to 320 Bcm of natural gas to Europe may seem, compared with the present situation (of only 130 Bcm), it is worth noting that in energy terms the volume of imported gas demanded pales into relative insignificance in the perspective of Europe's relatively recent history as an energy importer. Over 20 years ago – in 1973 – Europe imported well over 800 million tons of oil: the energy equivalent of almost three times that which could be provided by 320 Bcm of natural gas imports some 30 years hence.

8. Conclusions

The issues and arguments which underpin the thrust of the paper can be summarised as follows:

- The assumption of a growth market for gas in Europe predicated on the basis of a fundamental shift in the perception of gas availabilities and on gas' role in the energy market; and the argument that this shift will undermine the structure of the industry as it has evolved to date, whereby both supply and demand have been curtailed by monopoly and monopsony to volumes below those which would have developed in a competitive market.

- The argument that Europe itself is gas-rich beyond present general perceptions and that the economics of developing indigenous supply over the next 30 years to a level 60% higher than in 1995 are soundly based.

- The view that oil-equivalent prices will continue to set the upper limit to well-head/beach/border prices for gas and that this expectation creates the prospects for an "easy" gas supply situation throughout the 30 year period under review.

- The claim that Europe is favoured by the existence of a set of external gas-rich regions with a total potential for supplying gas well beyond the highest possible market prospect: and that it is Europe, rather than any other market or markets, which provides the best opportunities for those suppliers to achieve returns from the exploitation of their gas.

- The argument that in the context of a highly competitive supply side prospect, actual well-head/beach/border prices for gas in Europe could be under the upper limit set by oil-price equivalence.

- The idea that, in this situation, the supply price for gas may not be the main pricing problem facing European gas users. It is the non-competitive structure and behaviour of gas sellers which have kept prices high and restricted market growth. The competitive gas supply development predicated in this paper will need to be matched by a degree of market liberalisation whereby the 'fat' created by the protected sellers' behaviour is eliminated.

References

Adelman, M.A. (1990), 'Mineral Depletion with Special Reference to Petroleum,' *Review of Economics and Statistics*, Vol.72 pp.1–10.

Ausems, A. (1994), 'Prospects for North Sea Oil and Gas: The Netherlands,' *The Study Group for International Commercial Contacts, Energy Week Conference*, (London).

B.P. (1996), *Statistical Review of World Energy*, (London).

Department of Trade and Industry (1995), *The Energy Report; Oil and Gas Reserves of the United Kingdom*, (London: H.M.S.O.).

Dienes, L. and Shabad, T. (1979), *The Soviet Energy System* (Washington D.C.: Winston).

Dienes, L., Dobozi, I. and Radetzki, M (1994), *Energy and Economic Reform in the Former Soviet Union*, (Basingstoke: MacMillan).

Estrada, J., Bergesen, H., Moe, A. and Sydnes, A. (1988), *Natural Gas in Europe, Markets, Organisation and Politics*, (London: Pinter Publishers)

Estrada, J., Bergesen, H., Moe, A. and Martinsen, K.D. (1995), *The Development of European Gas Markets*, (Chichester: J. Wiley and Sons).

European Commission (1994), *Energy in Europe, 1993 Annual Energy Review*, (Brussels).

Handley, R.L. (1995), 'Upstream Opportunities created by the new Gas Markets,' *Petroleum Review*, Vol.48, No.574, pp.500–503.

Hoffman, G.W. (1985), *The European Energy Challenge: East and West*, (Durham N.C.: Duke Press).

International Energy Agency (1986), *Natural Gas Prospects*, (Paris: O.E.C.D.)

International Energy Agency (1991), *Natural Gas Prospects and Policies*, (Paris: O.E.C.D.).

Kemp, A.G. et al. (1994), 'Low Oil Prices, Prospective Activity Levels of the UKCS and the Effects of Cost Savings,' *The Study Group for International Commercial Contacts, Energy Week Conference*, (London).

Ministerie von Economische Zaken (1995), *Olie en Gas in Nederland*, 1994 (Den Haag).

Norwegian Petroleum Directorate (1993), *Improved Oil Recovery – Norwegian Continental Shelf*, (Oslo)

Odell, P.R. (1969), *Natural Gas in Western Europe; A Case Study in the Economic Geography of Energy Resources*. (Haarlem, De Erven F. Bohn N.V.)

Odell, P.R. (1988), 'The West-European Gas Market: the Current Position and Alternative Prospects,' *Energy Policy*, Vol.16, No.5, pp.480–493.

Odell, P.R. (1992), 'Prospects for Natural Gas in Western Europe,' *The Energy Journal*, Vol 13, No.3.

Odell, P.R. (1995), 'Global Economic/Energy Prospects, 1994–2020,' International Institute for Applied Systems Analysis, *International Energy Workshop* (Laxenburg).

OPAL/Energy Advice (1995), *European Energy Pricewatch* (Walton-on-Thames).

Park, D. (1979), *Oil and Gas in Comecon Countries* (London: Kogan Page).

Pauwels, J-P., Possemiers, F., Swartenbroekx, L. and Lievens, M. (1994), *Géopolitique de L'Approvisionnement Énergétique de L'Union Européene au XXIe Siecle, Vol 1* (Bruxelles: Bruylant).

Prior, M. (1994), 'The Supply of Gas to Europe,' *Energy Policy*, Vol.22, No.6, pp.447–454.

Prognos, A.G. (1994), *Die Energiemärkte Deutschlands im Zusammenwachsenden Europa: Perspektiven bis zum Jahr 2020*, (Basel).

Radetski, M. (1994), 'World Demand for Natural Gas: History and Prospects,' *The Energy Journal, Special Issue on the Changing World Petroleum Market*, pp.219–236.

Royal Ministry of Industry and Energy (1995), *Fact Sheet, Norwegian Petroleum Activity* (Oslo).

Stern, J. (1990), *European Gas Markets, Challenge and Opportunities in the 1990s*, (Aldershot, Dartmouth).

Stern, J. (1995), *The Russian Natural Gas 'Bubble: Consequences for European Gas Markets* (London, Royal Institute of International Affairs).

Stoppard, M. (1994), *The Resurgence of U.K. Gas Production*, (Oxford: Oxford Institute for Energy Studies).

Williams, J.R. (1992), *Natural Gas Reserves and Production – the European Picture*, (London, Shell International Gas).

Williams, J.R. (1993), *Europe: Competing for Sources of Natural Gas* (London, Shell International Gas).

Chapter III – 8

European Gas 2001

A. Retrospect and Prospects*

Introduction

The current ideologically-based European Union policy of gas liberalisation is well-nigh invariably presented in a highly favourable light, especially as a means for securing lower prices for final consumers through effective competition between suppliers. It also poses, however, serious problems for, and great potential dangers to, Europe's welfare. In this prospect it is thus threatening to repeat the negative consequences of equally ideologically-based gas sector policy decisions in earlier decades. An awareness of those mistakes of the 1970s and the 1980s is much more than simply knowing something of interesting elements in the history of European gas.[1] Such a background is, on the contrary, an essential component in an evaluation of the ability of the EU to act wisely in respect of its Gas Directives.

Retrospect 1 – Groningen: its Ordered Market and its Sequels

The discovery and exploitation of the Dutch Groningen gasfield constituted a mega-shock for the coal and oil based western European energy economy of the 1960s. Gas in unbelievable – and downplayed – volumes, producible at costs as low as those of Saudi Arabian oil production, was developed under monopolistic conditions

*** The Keynote Speech at the Petersburg Forum, Stadtwerke 2002 in the Gästehaus der Bundesrepublic Deutschland, Bonn, 25 February 2002.**

under the jointly-created system determined and owned by the Dutch state, Royal Dutch/Shell and Esso (Exxon). As shown above, an ordered market covering half of Europe (see Fig II-1.2, p.139) was deliberately created and imposed.[2] Between 1961 and 1971 indigenous gas production and use in Western Europe increased from 16 to 106m^{3}x10^9, with gas' share of total energy use rising from under 2% to almost 10% (see Table III-5.1,p.399)

The significance of Groningen lay, however, not only in its own gas-supply potential, but also in its demonstration effects, both geological and politico-economic. It quickly stimulated the search for additional gas reserves in the south North Sea geological province, as shown in Figure I-5.1 (p.107). New, albeit smaller, fields were found both on-shore and off-shore the Netherlands by the Shell/Esso company, the Nederland Aardolie Maatschappij (NAM). This company enjoyed exclusive rights to Dutch gas (and oil) exploration and exploitation rights until 1969 and thus had a decade after its discovery of Groningen to consolidate its supply control over the commodity.

To the west of the offshore median line in the South North Sea, exploration for gas began in 1963 and quickly proved significant reserves. Though Shell/Esso had concessions and achieved exploration successes very quickly, the consortium was not able to incorporate its UK gas into its European supply and distribution system (Figure II-1.2, p.139). On the contrary, its planned pipeline from Groningen to deliver gas to England was abandoned. Instead, the UK government stipulated that all offshore gas must be landed in the UK and sold to the national Gas Council which was accorded monopsonistic rights. This enabled the UK to exploit and market its natural gas as a national good, so initiating the country's subsequent 30-year divorce from the Eurogas system.

Meanwhile, the mainland western European monopolistically developed and controlled market based on Groningen gas sold at high prices (relative to other energy sources)[3] became an attractive objective for gas from external suppliers – notably pipelined gas from the USSR and liquefied natural gas, plus pipelined gas some years later, from Algeria. This development is shown in Figure II-1.4 (p.143). Later in the decade, gas discovered in the Norwegian sector of the middle and north North Sea was offered to the Euro market through the first of the pipelines built to north Germany: in the context of high oil-indexed gas prices at which the mainland gas transmission companies were prepared to purchase supplies. As with Russian and Algerian gas, these supplies were also incorporated into the pre-ordered system within the framework of which overall supplies and use increased by almost 60%

over the short five year period from 1971–1976. The ordered Groningen system essentially remained in place, though on a much larger scale and over a wider area, through a complex set of inter-relationships between a handful of suppliers and transmission companies.

Retrospect II – Gas-Unfriendly Perceptions and Policies

Coincidentally, the initiation and expansion of the natural gas industry in Europe took place over the same time period as that of a rising concern for the adequacy of energy resources for sustaining global and regional economic growth.[4] Though the claims for energy scarcity were largely specious, they were used by many of the entities involved in Europe's emerging gas industry as a justification for high prices and limitations on the rate of expansion. These entities were thus not motivated to challenge the perceptions of scarcity propagated by ill-informed institutions which lacked any understanding of the dynamics of gas exploitation. Prospects for development were therefore generally expressed pessimistically in terms of how few years there were of potential supply in the context of reserves already proven: rather than in the correct context of proven reserves as the "shelf-stock" of the industry which are replenished as and when required by further exploration and exploitation investments. The "institutionalisation" of this error was most noteworthy and significant in the case of Gasunie's use of a 25-year proven reserves requirement as the up-side limitation on its willingness to market gas.[5]

Needless to say, other vested interests – notably those seeking to protect coal and nuclear power against natural gas – used these arguments to secure state subsidies and other advantages for maintaining or expanding their market shares in many European countries' energy sectors.

More generally, in terms of energy policy decisions, formal restraints on gas supply were imposed by governments. Thus, each of the three important gas producing countries in Europe – the Netherlands, the UK and Norway deliberately constrained gas production levels.[6] Additionally, moreover, the European Commission issued a Directive in 1976 formally prohibiting the use of gas in so-called "non-premium" markets and especially in power generation.[7] This remained in force for over 20 years over which period the EC's Energy Directorate failed to recognise the scale and scope of natural gas' potential contribution to Europe's energy supply in the long-term.

The consequence of these perceptions and policies was near-stagnation in Western Europe's indigenous gas production. It grew from

$164m^3x10^9$ in 1976 (at 10 times the level of 1961) to only $198m^3x10^9$ by 1991, an increase of a mere 19% over 15 years! Gas use was also constrained, albeit less dramatically in the context of the gas supplied by exporters to western Europe, most importantly, the Soviet Union and Algeria. They achieved considerable market growth (from 11.6% in 1976 to 36.3% in 1991) at the expense of indigenous suppliers. Overall, the increase in the share of gas in the total energy market from 1976 to 1991 was a mere 1.8% (13.4 to 15.2%).

By the late 1980s the pessimistic view of gas had become so imbued that official forecasts for the industry's prospects indicated a near-future cessation of its growth overall (so that use in 2000 was forecast to be no higher than 1989) and a decline in the indigenous production of gas. The alternative view we took at that time, viz. for a more than 50% growth in use by 2000 and a 30% growth in indigenous output was somewhat contemptuously dismissed as academic speculation. The contrasts between the forecasts are set out in Table III-8.1 and are compared with the actuals.

Retrospect III – a Change of Gear

In the contexts of falling international oil prices (to which the border prices of gas throughout most of Europe have been indexed) in the 1990s and a much enhanced deliverability of gas via an enhanced transmission system between suppliers and markets – so leading to price competition between the suppliers – gas use in western Europe has been much intensified over the last decade.

TABLE III-8.1

Contrasts between the 1989 forecasts for gas in the Western European energy economy by 2000: and the actuals for that year

(in m^3x10^9)

	Official IEA/EU Forecasts	P.R. Odell's Forecasts	Actuals in 2000
Indigenous Production	155	228	262
Imported Gas	100	143	122
Total Gas Use	255	371	384
% Share of Gas in the Total Energy Economy	17.9%	26.7%	26.3%

This supply-side pressure on previously established prices reduced border gas prices by 47% between 1991 and 1999. It has, moreover, been accompanied by closer government examination of, and concern for, the monopolistic behaviour of the continent's small number of transmission companies (reflecting the situation put in place by Shell and Exxon during the period of Groningen gas dominance). Margins earned by these companies and by local distributors have thus been subject to more and more government intervention – so also leading to real price reductions: in reaction to, or in consideration of, the EU's intensified competition policy in general and to the Gas Directive for liberalised markets, in particular.

Finally, belated recognition of Europe's indigenous gas reserves/resources potential has finally emerged. In spite of enhanced depletion of known reservoirs – particularly by the UK where production increased almost two and a half times over the decade – reserves remaining have continued to increase. This process has now gone on for almost 50 years – as shown in Table III-8.2 – and continues even during the current decade, in spite of the rapid increase in gas production and use.

TABLE III-8.2
Evolution of Western Europe's natural gas production and recoverable reserves, 1956–2005 (in $m^3 x 10^9$)

Period	Cumulative Production to Date (by decade)	Period-End Remaining Recoverable Reserves	R/P Ratio (years) at end of Period	Gross Additions to Reserves (by decade)	Indicated Total Original Recoverable Reserves
Pre-1956	50 (35)	500	40	Not known	550
1956–1965	225 (175)	1,900	76	1,575	2,125
1966–1975	1,150 (925)	4,350	27	3,375	5,500
1976–1985	2,700 (1550)	6,500	36	3,700	9,200
1986–1995	4,645 (1960)	7,700	35	3,160	12,345
Forecast for:					
1996–2005	7,395 (2750)	8,195	28	3,425	15,590

Source: Author's calculations

In 2001 western Europe's gas use was of the order of $400x10^9m^3$ and its share of the energy market is now above 26%, with the most noteworthy developments in the power generation and industrial markets – at the expense of both coal and oil. Every country is now linked into the gas transmission and delivery systems, so making Europe a near fully-fledged three fuel economy. Indeed, gas is now poised to become the region's single most important energy source – having overtaken coal in 1993 and now closing the gap with oil in many countries. Exceptionally, gas overtook oil in the UK in 1999: and is likewise ahead of all other energy sources in some countries of the former Eastern Europe.

Prospects I – the Inconsistencies of Liberalisation

The analysis of natural gas' evolution in Europe shows that the earlier constraints on its expansion effectively disappeared in the 1990s. This, in part, was a result of liberalisation measures, but the development owed much more to the range of other factors, as specified above.

Under the continuing importance of these factors, plus the rapidly increasing influence of gas' environmental advantages over coal and oil (most notably its much lower contribution to CO_2 emissions), growth in its use will persist irrespective of the organisational structure of the industry. Unhappily, the EU's ideological commitment to liberalisation seems likely to threaten, rather than encourage, the prospects for growth, given the inconsistencies inherent to the process as set out in the Liberalisation Directive and the consequential uncertainties which serve to undermine confidence across the industry from exploitation to final consumption:

First, the inevitable uneven application of the Directive will give rise to at least a decade of unpredictability for both companies' and users' fortunes and lead to the creation of an unacceptably wide range of levels of benefits, dependent on the location of the interested parties with respect to the contrasting speeds and forms of the liberalisation procedures.

Second, liberalisation in itself produces conflict between environmental objectives and economic considerations, notably over the taxation of gas use (leading to higher prices and lower demand), so threatening netback prices and, hence, revenues achievable by suppliers.

Third, the elimination of long-term, take-or-pay contracts may be expected to reduce prices in the short-term, but it could well increase them over the longer term as their absence enhances upstream investment risks for suppliers.

Overall, these factors arising as a consequence of strategies based on a free-market ideology carry the seeds of supply problems post-2010, given the expected market developments shown in Table III-8.3.

TABLE III-8.3
The evolution of the overall European gas market, 2000–2020

	Actuals	Current Expectations for	
	2000	2010	2020
Gas Use	428	580–650	675–900
Indigenous Gas Production	282	320–350	350–425
Imports from External Suppliers	146	260–300	325–475
Import Dependence	34.1%	44.8 to 46.2%	48.1 to 52.8%

Source: Stern, J.[8]

Prospects II – Liberalisation at what Price?

The currently expected modest international oil prices to 2020 (at \$20±\$3 per barrel in 2000 \$) will sustain an indexed border price for gas in Europe of only \$3+ per mmBtu. This could restrain both indigenous gas production and gas import availabilities, given an expected long-run gas supply price curve rising to \$4 per mmBtu.[9]

In these circumstances of oil-indexed gas pricing – whereby gas prices have, generally, to date been kept well above its long-run supply price – the impact of an ideologically imposed liberalisation could mean that future gas prices will rise above their oil equivalents (as in the United States in the recent past). This would not only be intensely ironic (as liberalisation is promoted as a structural change which will reduce prices), it would also be a market de-stabilising developmentt.

Prospects III – Liberalisation and Supply Security

The liberal ideology does, of course, presume a competitive gas supply system, but this leads to the exposure of the producers in respect of the viability of their upstream investments, given both the long lead-time for the development of production capabilities and the high volume of investments required for securing the required economies of scale in the associated transmission systems.

Under such conditions arising as a necessary consequence of the free-market structure demanded by the EU Gas Directive, there will be a powerful motivation for suppliers' collusion. This seems likely to lead to a formal or informal Organisation of Gas Exporting Countries to

Europe (OGECE), possibly with an OPEC-like production quota system for its members (Russia, Turkmenistan, Iran, Qatar, Libya and Algeria), none of which need be "bound" to the requirements of free trade in gas as an inherent element in the Liberalisation Directive.

Norway, which while remaining a non-member of the EU has given commitments to the Union to pursue non-restrictionist policies, would under the circumstances of an OGECE nevertheless have an option to become a member (at the cost of counter-measures which would be taken by the EU) and so enhance the price levels achievable: or it could become a "free-loader" by its ability to ensure that the price of its gas to Europe reflects the going-price generated by the cartel of suppliers external to the European system.

Such possible developments would, of course, not only be adverse for the anticipated Eurogas market (see Table III-8.3), but it would also be more generally disadvantageous for Europe's economy. Thus, the increasing dependence on gas supplies from external sources (from only one-third in 2000 to +50% in 2020 as shown in Table III-8.3) arising as a consequence of liberalisation, requires much more attention than that which it has been given to date.

Conclusion: Gas is NOT "just another commodity"

With energy customers increasingly committed to gas supplied through expensive and dedicated pipelines, flows have to be uninterrupted. "Alternative supply sources", the usual backstop argument by liberalisation proponents, are a theoretical concept only – except in the long-term (by when the customers are dead) as they imply costly investments in production and transmission facilities to ensure their instant availability.

In these circumstances long-term, take-or-pay contracts – as sought by suppliers – are "of the essence". As is the requirement for the continuing involvements of governments in both the negotiations for, and in the monitoring of, the agreements. The EU will outlaw such contracts at its peril – both political and economic.

As with international oil (now "re-ordered" as the United States came to recognise the inherent perils of a "free" market), so an ordered Eurogas market – at least in respect of relationships between EU gas importing countries and external suppliers – is not only the safer option, but also the most likely way collectively to maximise the benefits to the parties concerned. In this context, "back to Groningen" could well become the Eurogas industry's rallying cry – if the first decade of liberalisation proves to be as tumultuous as predicted.

B. The UK Gas Industry and the Liberalisation of European Markets: an Alternative Analysis*

a. Introduction

i. This note seeks to clarify some of the issues in the Report relating to the future supply of natural gas, in general, and to the impact of the liberalisation of European gas markets, in particular, on gas availability. The Report argues that "there is considerable uncertainty about gas imports in the long-term" (para. 91) and that the UK's security of energy supply would best be achieved if left to market forces and that these "would not operate effectively without greater liberalisation of continental European markets". (para. 92)

ii. As the gas reserves and resources potentially available for use in Europe (including the UK) constitute about 70% of the global total, while Europe is responsible for only 18% of current world use – with this share more likely to decrease than to increase over the next 20 years – there can be no uncertainty over supply *per se*. Potential suppliers are, indeed, gathered around Europe like proverbial "wasps around the jam jar", with the jar in this case containing, first, high value gas markets and, second, a set of consumers ready willing and able to pay for the commodity in steadily increasing volumes.

iii. Presumably, therefore, the "uncertainties" to which the Report refers must be related specifically to the questions of how soon and how much the UK will need to import – and from which specific sources of supply such requirements can be most securely and cheaply obtained?

b. The UK's future gas import needs will be limited

i. The Minister of State for Energy has suggested that the UK is likely to be up to 50% dependent on gas imports by 2010 and some 90% dependent by 2020. These forecasts emerge,

* **Published as an Appendix to the House of Commons Trade and Industry Committee's Report on *Security of Energy Supply*, HMSO, London, January 2002, pp. 72–76.**

however, from inappropriate values attached, first, to gas demand developments; and second, to UK gas production potential.

ii. On the demand side, the rate of expansion of UK energy use has been predicated at about 2% per annum; implying that the 226 million tons oil equivalent used in 2000 will rise to 275 million tons in 2010 and reach 335 million tons in 2020. In reality, this has a very low probability of being achieved. Even with no more than a continuation of the very modest energy conservation and efficiency efforts to date, it is highly unlikely that incremental energy demand over the next 20 years will be almost five times that of the last 20 years, (viz. 110 mtoe compared with 23 mtoe). Given energy policy objectives, which are now orientated much more strongly to demand constraints, the probability of such future levels of energy use is close to zero.

iii. It is much more realistic to assume that future demand growth will continue to expand at no more than the 0.5% rate of the past 20 years; that is, to about 238 mtoe in 2010 and just over 250 mtoe in 2020. Within this framework, already certain or planned supplies of oil, coal, nuclear and renewables will deliver some 138 mtoe by 2010 and 145 by 2020, thus indicating a residual demand for gas of about 110 Bcm in 2010; and about 115 Bcm by 2020. Gas demand in 2000 was already 96 Bcm, with indigenous production significantly higher at 108 Bcm, indicating a net export from the UK of about 12 Bcm.

iv. Even if indigenous gas production were to fall dramatically from its present level – to some 82 Bcm in 2010 and to 58 Bcm by 2020 (that is, by roughly the decline expected by the UK Offshore Oil Operators' Association from the depletion of presently discovered reserves), then net import requirements for gas would be only 28Bcm in 2010 and 58Bcm in 2020 or, approximately, 25% of gas demand in 2010 and 50% in 2020. Additional recoverable reserves of UK continental shelf gas are, in the meantime, however, virtually certain to emerge both from re-evaluated fields and from new discoveries. This is a consequence of the industry's intensified efforts to achieve such an objective – based on the Department of Trade and Industry's current mid-range estimate of 2100 Bcm of remaining recoverable gas.[10] In the absence of adverse regulatory constraints on the industry's propensity to invest, production will be some 25 Bcm per year greater than the 82 Bcm and 58 Bcm in 2010 and 2020, respectively (as indicated above). Thus, the UK's net gas import needs could fall to near zero in 2010 and to no more than 33 Bcm in 2020.

c. An Accord with Norway will solve the supply "problem"

i. The security of UK gas supply for the next 20 years is thus at worst an issue of relatively minor dimensions: and, at best, it will be non-existent. In any case, net imports sufficient to meet the reasoned estimated shortfall can be virtually guaranteed from the exploitation of the massive additional gas reserves of Norway. In 2000, Norway exported 49 Bcm (all to mainland European countries). An additional 20 Bcm per year are already contracted, largely to the same set of purchasers, for delivery by 2004/5. Thereafter, however, another 50 Bcm per year are predicated for development, but with most of this gas not yet contracted for sale: though note that BP has already negotiated modest purchases of Norwegian gas for delivery to the UK from 2004 for a 25 year period under take-or-pay conditions.

ii. Technical discussions between the UK and Norwegian authorities on possible joint developments have already been scheduled for the very near future (possible by March this year). These preliminary moves need to be expanded into a major Anglo-Norwegian politico-economic agreement, whereby the best interests of both countries can be achieved. Such an agreement would seek to ensure a maximised net-back value at the well-head for Norway's large potential additional production, on the one hand; while for the UK, it would not only guarantee the volumes of required imports of new Norwegian gas, but also generate financial benefits from a joint venture for the transit of the remaining gas via the UK to the mainland of Europe. Ironically, the principal barrier to such a mutually beneficial agreement for the two countries could well be the incompatibility of OFGEM's stance on "competitive" pricing for gas delivered to the UK, with Norway's preference for long-term take-or-pay contracts, whereby the high levels of investments for the long-term, large-scale gas production in its Northern waters can be justified.

d. Other "guaranteed" source of gas

i. In the context of such an Anglo-Norwegian accord, the UK's needs for gas imports from other sources would thus be minimised – or even eliminated, except for seasonal requirements for high winter demand. Such relatively minor import volumes will be readily available – albeit at higher prices, given the seasonality of demand – from alternative sources. These could be *either* as pipelined imports from the Netherlands, where significant seasonable flexibility of supply is built into its gas production, transport and storage systems, and for which gas imports more than adequate capacity for moving it to the UK through the inter-connector from Zeebrugge (currently being expanded to handle 24 Bcm of gas per year) will already be in place; *or* as LNG imports, to terminals which can be built with relatively short lead-times for development, from an increasing number of possible suppliers within relatively short sea distances of the UK, viz. Algeria, Egypt, Nigeria, Trinidad, Venezuela and even Norway's Barents Sea Snohvit field, currently under development as an LNG export facility.

e. The UK and the non-liberalised mainland European gas industry

i. In the demand/supply prospects as set out above, the UK seems unlikely to be required to depend much, if at all, prior to 2020 on any other external suppliers of gas to Europe, viz. Russia and Algeria as existing suppliers; or Libya, Turkmenistan and Iran as potential suppliers within this period. Under such conditions, the concerns expressed in the Report for the slow process of gas liberalisation in most other EU member countries are misplaced. The UK's demands on the use of the pan-European transmission system already in place will be minuscule, even in the context of the system's already existing capacity to move some 300 Bcm of gas per year from the exporting to the importing countries. The UK's suppliers' lack of guaranteed third-party access rights to the system and the absence of price transparency would, at worst, thus be a minor inconvenience.

ii. Meanwhile, the mainland European gas transmission system is not only being increasingly reticulated and geographically extended, it is also being opened up to regulated or negotiated third-party access – with published tariffs – in ways which suit the needs of both exporting and importing countries.

iii. The antecedents to the evolution of the European mainland gas transmission network and the trading mechanisms for the gas itself have been very different from those that have emerged in the UK (following the UK's decision in 1972 not to allow gas exports to the rest of Europe, thus preventing integration). While supply-side competition for gas in the UK did not effectively emerge until the early 1990s, competition between the suppliers of gas to mainland European markets started as early as 1971 (with the entry of Russian gas to the market). By 1990 this competition had become so intense that prices dipped well below crude oil equivalents and enabled European consumers to enjoy much lower (pre-tax) prices – in all sectors except residential in which UK prices were held to an artificially low level through government intervention. This situation persisted until as recently as 1999, when European pre-tax prices eventually rose above those in the UK as a consequence of the strong upward price movements of crude oil – to which Euro-gas prices are indexed. But UK consumers' new-found price advantage from that situation has already proved very temporary, as the differentials disappeared with the falling oil price in 2001.

iv. Meanwhile, moreover, the generally high downstream profits in gas distribution in most mainland European countries – arising from monopolistic conditions – are now being brought under severe downward pressure through governments'-induced interventions which require structural changes in the national systems. Such changes are not, however, necessarily being effected in the competitive, but highly regulated, manner chosen to secure price reductions in the UK. This UK approach is very widely non-acceptable in many other European countries in the context of a continuing belief in the public service nature of the gas (and electricity) supply industries, often under municipal control; with gas and electricity revenues used to subsidise other public sector services. This constraint on UK-style liberalisation measures being implemented elsewhere remains highly pertinent.

v. There is yet more significant constraint on the ability and willingness of other EU countries to switch their gas purchasing arrangements from long-term, take-or-pay contracts to short-term trading. This reflects these countries' status already as major gas importers – and as even more import-dependent gas users in the future. Only Denmark and the Netherlands are self-sufficient. Many countries' 'dash-to-gas' has hardly begun, so that gas still accounts for only 19% of total energy used, compared with 38% in the UK. By 2010 gas dependence in mainland Europe is expected to increase to about 24% and by 2020 to 30% and will involve 315 Bcm of gas imports in 2010 and 390 Bcm in 2020. Given these expectations, guaranteed long-term arrangements for imported supplies are of the essence and it may be assumed that such arrangements will continue, irrespective of the requirements of the EU's Competition Directorate for a switch to short-term competitive markets. Such requirements are widely considered to be irresponsible.

f. The views and requirements of the countries exporting gas to Europe

i. The gas importing countries' motivations for the continuation of long-term contracts for gas imports are thus already powerful enough to ensure the sustenance of the 30 year old system, except for minor changes relating to joint sales (as recently made by the Norwegian state-directed selling agency), to destination clauses and to re-sale procedures. The importers' concerns are, however, greatly strengthened by the views and demands of the external suppliers – most notably Russia and Algeria, but also involving both Norway and also potential new suppliers. All of them have participated in recent gas suppliers' summit-meetings. At these meetings exporters expressed their 'indignation' at the EU's attempts to undermine the gas supply system on grounds of 'restraints on trade' – even to the extent of declaring existing contracts as unacceptable under EU competition law.

ii. The exporters argue that ending long-term take-or-pay contracts will mean that both volumetric and pricing risks will be moved to the producer, compared with the current sharing of those risks between exporters and importers (with the latter guaranteeing volumes). Liberalisation, the exporters say, does not take account of their needs and, moreover, they have not been consulted on the changes. Gazprom summarised the evolving situation as one in which it would be unable to conclude new contracts. Both Gaz de France (given France's 90% dependence on imports now and over 95% by 2010) and Ruhrgas (given Germany's current 80% import dependence and also rising) are giving strong support to the exporters' arguments.

iii. The French, German and other countries' concern for the adverse impact of liberalisation on prospects for gas supply and price is taken even further by their fears that the alienation and provocation of the external suppliers of gas to Europe will lead to the creation of a formal oligopoly of suppliers, viz. an Organisation of Gas Exporting Countries to Europe – OGECE, through which supply constraints and consequential higher prices will be the end result – to the disadvantage of all gas users in Europe. It is worth noting that neither Russia nor Algeria – nor Libya, Egypt and Iran – have signed and/or ratified the European Energy Charter which requires a free-trading regime.

iv. In light of such formidable opposition to gas trade liberalisation it is not very surprising that a mid-2001 ruling by the EU that a 'prioritised' infrastructure-connector for electricity between Germany and Poland involves such a large investment that those that have invested need 25 years of *exclusive* rights to use the line (in order fully to cover their costs); and that this requirement should 'supersede competition issues'. This appears to establish a precedent for treating infrastructural connectors for gas movements between states in a like manner.

v. Coincidentally, an E*U Report on Energy Infrastructure*[11] has recently been published and has designated a set of five prioritised requirements for the gas industry. These include the connection of networks between the UK, Netherlands, Germany and Russia; likewise, between Algeria, Spain and France; underground gas storage construction in Spain, Portugal and Germany; and the development of pipeline systems between the EU and both the Middle East and the Caspian basin (plus LNG facilities in France, Spain, Portugal and Italy) with the collective object of guaranteeing the availability of at least 20% of *peak* daily demand for all parts of the EU, (so implying a 100% guaranteed supply for most of the year). Should such EU-prioritised gas infrastructure developments be given exemption from competition rules – through the protection of investments made in the facilities against a 'required' third party access to the infrastructure for 25 years, then gas market liberalisation objectives would be well and truly thwarted for the whole of the period to 2020.

g. Optimal Policy for the UK

i. Given the prospect outlined above, then the development of the UK's gas industry would seem likely to be best effected by a combination of maximum efforts to maintain the volumes of indigenous production (as suggested above in para. b.iv); by an accord with Norway to make-up most of any shortfall in the context of a broad, mutually beneficial agreement (paras. c.i and c.ii above); and by the development of LNG peak shaving facilities (para. d). The use of the interconnector with the European mainland would remain as an optional extra for deliveries at peak periods.

ii. None of these proposals should create any great concern as to their practicality: and neither in respect of the level of costs involved. In the context, that is, of realistic forecasts of future levels of energy demand, in general, and of gas demand, in particular (as argued in paras. b.ii and b.iii).

iii. The UK's security of gas supply for the next two decades is certainly *not* a problem of accessing large volumes of gas from distant sources, separated from the UK by many intervening complexities. On the contrary, supply security is, in essence, a matter of good and mutually beneficial relations with Norway as a complement to the establishment of a number of nationally achievable aims. These are, first, much higher efficiencies in gas use, especially in the domestic sector (responsible for one-third of total consumption) through measures including fiscal ones; and, second, favourable policies for maximising indigenous gas exploration and exploitation through the abolition of regulatory practices which undermine investments in gas production, new terminals and extended transmission systems. The latter should pay particular, but not exclusive, attention to the UK as a transit country for large volumes of Norwegian gas en route to mainland Europe.

References

1 See above, Chapters III- 5, 6 and 7 for contemporary studies and commentaries on inappropriate European gas policies and actions over that period.

2 See Chapter II-1.

3 *ibid.*

4 As shown and specified in chapters I-3, I-7 and II-7 in Volume 1 of this book, *Oil and Gas, Crises and Controversies 1961–2000: International Issues*, Multi-Science Publishing Co., Brentwood, 2001

5 See above, Chapter III-3.

6 See Chapter III-3, for specific measures taken by these countries to restrain supplies.

7 EC Directive on gas use, no.75/404/EEC.

8 Stern, J.P., *Traditionalists versus the new economy: competing agendas for European gas markets to 2000*, Royal Institute of International Affairs, Briefing Paper, New Series no.28, London, 2001

9 See Chapter III-8 for elucidation on this development.

10 Department of Trade and Industry, *Oil and Gas Resources of the United Kingdom*, H.M.S.O., London, 2001.

11 European Commission, *Report on Energy Infrastructure*, Brussels, 2001.

Section IV

On Oil and Gas Politics, Policies and Structures

Chapter IV – 1

Oil and Western European Security*

In 1912 Sir Winston Churchill concluded that the British government needed to be involved in the business of exploring for and producing oil. A decision was thus taken to invest government funds in the Anglo-Persian oil company, which was at the time short of working capital for its exploration efforts in Russia, in order to safeguard supplies of oil for the Royal Navy. This British government intervention in the oil industry turned out to be an isolated case of state action, principally because repetition of it proved to be unnecessary for almost another half-century. Apart from the needs of navies converted from coal to oil-burning ships and a slow growth in the demand for gasoline arising from the development of motor transportation after 1918, Britain – as with most of the rest of Western Europe – remained, until after the second World War, almost entirely orientated towards the use of coal in their energy economies, and most especially in respect of the continued use of coal for power generation and as the fuel for industry of all types.

Even as recently as 1950, Europe's energy markets, which were technically open to competition between the two fuels, still remained some 90 per cent dependent on coal. This meant that any interruption in oil supplies would have had only a marginal effect on the continent's level of economic activities. Even in the field of overland surface transportation, railways remained the dominant mode for both passengers and freight, and they, of course, still utilized largely coal (or

* **Published in *Brassey's Annual – Defence and the Armed Forces*, W. Clowes and Sons Ltd, London,1972, pp. 64–75**

indigenously produced hydro-electricity) as their source of motive power. As a result of these considerations, Western Europe's essential economic requirements for oil remained limited. In strategic terms the continuing role of the United States as the source of much of Europe's requirements for gasoline and diesel fuels and the growing importance of exports from Venezuela, resulting from the rapidly expanding activities there of European and US oil companies, provided guarantees of the continued availability of necessary oil products in all circumstances —except that involving loss of control over the sea routes from the oil supply points in the Western Hemisphere.

This strategic consideration apart, the foreign oil policies of the UK and the Netherlands, and of France to a lesser degree, also had to take into account the international operations of their own oil companies, especially in the Middle East whose great potential for oil development had been substantially revealed in the late 1930s and the early post-second world war period. In that oil-producing region, however, European powers had had gradually to accept the greater power of the United States in securing opportunities for its 'international' oil companies to participate in Middle West oil development. Though this development was certainly not entirely welcome from many points of view, it did,nevertheless, add to, rather than detract from, the security-of-supply position *vis-à-vis* foreign oil. In as far as Europe's oil had to come increasingly out of the Middle East, then the presence there of American oil companies, automatically backed by the US government and, in the final analysis, with US military power, gave a greater chance that Europe would enjoy maximum possible security in the availability of oil.

Coal to Oil

Since 1950 oil requirements in Western Europe have grown not only for use in the conventional outlets in motor transportation, aircraft and so on, but also – and much more significantly from the point of view of the total quantity required and in the increasing range of products demanded – in respect of oil's substitution of the use of coal in rail traction, in industrial, commercial, and domestic use and in the production of electricity. A slow rate of substitution in the early and middle 1950s – consequent upon difficulties facing the expansion of the coal industry – was followed, after 1957, with a much higher rate of change from coal to oil. This reflected a changing price relationship between the two fuels, as the price of coal increased under the influence of wage inflation in the European coal-producing countries, while the price of oil declined under the impact of a rapidly increasing supply from very prolific and low-cost new fields in Venezuela, the Middle East and elsewhere; and as a large number of new oil companies moved into the business of supplying oil to Western Europe for the first time and, in so doing, undercut the earlier gentlemen's agreements that had kept up the prices of the traditional supplying companies.

With this fundamental change in the price-relationship between coal and oil, the non-coal-producing countries in Western Europe and, later, the coal-producing countries themselves, turned their backs on the indigenous product and adopted energy policies designed to give them access to lower cost imported energy. In the first group, there were countries like the Netherlands and Italy which had formerly been disadvantaged as far as their potential for industrial development was concerned by virtue of their lack of coal. Now they turned to oil as their basic energy source and, for the first time in the industrial period, secured energy-cost advantages over coal-producing countries like Britain, Belgium and West Germany, where, for a variety of reasons, coal had to be protected by means of taxes on competing oil products, so raising their cost above that in the countries without coal resources.

Even the protection of coal, however, only slowed up the process of change, so that one after another European countries became orientated to oil for their basic energy needs. In 1971, even Britain, whose post-war coal industry was more efficient and competitive than that of other European countries, finally became more dependent on oil, rather than coal, for its energy needs. As a result of these developments – coupled, of course, with an overall growth in energy use as a result of its expanding economy – Western Europe's total demand for oil exceeded 600 million tons in 1971, compared with under 50 million tons only twenty years earlier. Moreover, almost all of the total requirement had to be imported.

The Security of Supply Risk

During the thirteen years of most rapid growth in oil demand, between 1958 and 1971, the question of security of supply was pushed into the background (in favour of a greater concern for cheap energy). Nevertheless, in spite of greater attention to other considerations, the issue of supply security was never entirely lost sight of, particularly in light of the fact that the United States became an oil-importing rather than an oil-exporting nation. Moreover, there was also a relative decline in the importance of oil imports from other parts of the Western Hemisphere – notably Venezuela – supplies from which could be considered to be inherently more secure than those of the Middle

East. Security of supply arguments were certainly used by local coal-producing interests as part of the case for the favourable treatment within Britain, Belgium and Germany; and were also used in the advocacy by the main coal-producing countries of a coal-based European energy policy; compared with countries like Italy and the Netherlands which were strongly in favour of changing over to 'cheap' oil as quickly as possible. As a result, both in the relevant coal-producing nations and in the European Common Market as a whole, there was some success in sustaining levels of coal output above those which would have been possible in a situation in which the oil industry had been quite free to compete with coal. In the UK, for example, it was estimated that the imposition of a tax on fuel oil protected indigenous coal production totalling about 18 million tons per year – approximately 10 per cent of total production.

By the late 1960s, however, an increasingly wide differential between the price of oil and the prices which had to be paid to cover the rising costs of indigenous coal production led to an increasingly wide acceptance of the idea that European coal output and use should be seriously curtailed. In the UK it was calculated that only 80 million tons of coal production per year was economically viable. Over the rest of Western Europe only an equal quantity of competitive coal output seemed possible. Given these figures, it appeared likely that no more than 10–15 per cent of Western Europe's energy would come from indigenous coal by the second half of the 1970s. As hydro-electricity and nuclear power were together unlikely by then to supply more than another 15 per cent or so of the total energy requirement, Western Europe's energy economy would become dominated to the extent of some 70–75 per cent by imported oil.

Given this rapidly increasing dependence on oil in Western Europe and the unwillingness or inability of governments to curb it to anything other than a very limited degree, and with an expectation that oil demand would grow from its 600 million tons in 1971 to over 1,000 million tons by 1980 – see Table IV-3.1 on p.504, then strategic questions, related to questions of the sources of the required oil imports and the ways in which control could be exercised over the supplying countries' production levels, became important. As far as the latter was concerned the need to ensure some European interest in oil produced elsewhere in the world was recognized; through Shell and BP – Anglo/Dutch and British companies, respectively – as well as through smaller French and Italian oil companies. This was seen to require the ability of European governments effectively to communicate their viewpoints on strategic matters to the companies, particularly in the context of the various Middle East crises, during which oil politics become vitally important.

Notwithstanding the existence of such European-based oil companies, one must note that most of the oil used in Europe was – and, indeed, still is – produced, transported, refined and distributed for foreign – mainly American – companies, over which, of course, in the final analysis, political control rests elsewhere. In conditions of crisis they could theoretically 'retire' to the other side of the Atlantic. Realistically, however, given the existence of NATO and the OECD both with membership drawn from both sides of the Atlantic, one could reasonably suppose that any necessary action required in Europe from such companies to ensure the continuity of oil flows would be assured through pressure exerted by the US government. Nevertheless, in spite of this re-assuring view, the very high degree of responsibility exercised by non-European companies over Europe's oil supplies remains a serious issue, particularly with the contemporary development whereby the oil companies concerned are being forced to accept the participation in their operations of the state-owned companies of the countries where oil production is concentrated.

Middle East Oil and The Arabs

This, however, remains a less serious issue and a lesser potential problem than the second one indicated above; viz. the question of the sources of Europe's oil. In previous periods, as previously shown, it came mainly from the US and the Caribbean and was at danger only in respect of its shipment across the North Atlantic. After 1950, however, the United States itself became a net importer of oil. Even so, the strategic significance of the loss of this secure source of oil for Western Europe was offset by a major increase in the availability of oil from Venezuela – which at that time was also considered to be a secure oil supply source, with production in the hands of the major international oil companies. It met the major part of the growth in Western Europe's demand for a time, but post-1958 Europe's expanding needs could be met only by rapidly rising supplies out of the Middle East.

By this time, as a result of the pre-war oil discoveries which were quickly evaluated after 1945 and of even more important post-war discoveries of new fields, the Middle East had already been shown to have the world's largest deposits. So much so, in fact, that by the mid-1960s roughly two-thirds of all known oil reserves outside the Communist countries was estimated to lie in the region. Moreover, this oil was, in general, very low cost to produce to that both quantity and profit maximization considerations combined to persuade the oil companies to serve Western Europe from the oilfields of the Middle East. The required transportation system for this oil was orientated, in part, to the use of the Suez Canal for tankers plying from the Persian Gulf directly to Western European import terminals; and, in part, to pipelines from the Gulf to the Mediterranean coast and thence by tankers to Europe. In both cases, of course, the transportation routes lay through other countries, thus adding a further potential danger of interruption to supplies in the event of political or other difficulties in the region.

It is neither appropriate nor necessary in this article to consider the already well-known post-war political and other changes in the Middle East. Most of these have served to create a situation of instability with, indeed, periodic outbreaks of hostilities. Moussadeq's nationalization of Persian oil in 1951 was an early manifestation of the type of difficulties which could be expected as a result of oil industry development by foreign companies. This was an important, though certainly not the only, factor which persuaded the international companies to switch the bulk of their oil-refining activities from locations in the producing countries (of which the Abadan refinery in Persia was the most important example), to locations in the oil consuming countries, particularly of Western Europe. Thus, in the post-1950 period. Western Europe's oil was increasingly imported as crude oil for local refining here, rather than in the form of oil products ready for immediate consumption from refineries in oil-producing countries.

This shift in refinery location did, of course, represent a strategic safeguard in that European-located refineries could, within rather broad limits set by technical considerations, switch their sources of crude oil in the event of supply difficulties. This, however, gave relatively limited flexibility in a situation in which approximately 80 per cent of the crude came from a group of countries, all members of which were more or less subject to the same political and other problems. Indeed, in both the most serious Middle East crises – in 1956–57 and again in 1967, when the presence of Israel in the region caused hostilities – Western Europe suffered difficulties over its oil supplies. In the first crisis, only the continuing ability of the US to turn on an increased flow of supplies at short notice from its shut-in total production capacity saved Western Europe from a serious economic and, possibly, political crisis. In 1967, the supply position for Europe was saved only by the fact that the Arab effort to impose a boycott on their Western European oil markets was less than wholehearted.

This 1967 division of opinion within Arab countries as to whether or not Western Europe should be held to ransom through their being denied oil supplies was a first real indication that a fundamental reappraisal was under way in the oil-producing countries of the Middle East as to the role that their oil should play in determining their relations with the rest of the world. From an earlier unanimous declaration that Arab oil should be used directly as a weapon to secure policy ends (that is, by threats to cut off supplies to countries considered unfriendly to Arab causes), there was now an increasingly important faction which argued that as much oil as possible should be sold with as high taxes per barrel as could be won from the oil companies. The increased revenues which would thereby become available to the governments could then be used both for strengthening their economies and for bolstering the political and military positions of the Arab world.

In a real sense this revised appreciation by the Middle East oil-producing nations of the significance of their oil reserves, and of the way to use them in pursuing their national interests, represented a diminution of the short-term strategic threat to Western Europe arising from its continuing dependence on imported oil originating mainly from this part of the world – in spite of some efforts to diversify their sources by the companies responsible for supplying the oil. Moreover, the most significant geographical diversification of supply sources – that into North Africa – only reduced the strategic danger arising from the transportation component: the oil still came from countries with cultural and other links with the Middle East, and thus made it subject to interruption for the same reasons as most Persian Gulf oil. Diversification elsewhere – such as that given by rapidly increasing production from Nigeria and some extra output from Venezuela and other parts of South America and Africa – did not even provide enough additional oil to meet Western Europe's annual incremental demand. Thus, by 1970, Western Europe's dependence of Arab oil remained almost as high as ever.

The Period of Cheap Oil is Over

However, within the framework of a system of oil-producing nations which no longer sought to use their oil as a direct weapon in their battle with the west, but rather as a means of achieving economic and political strength through maximizing revenues and income from oil activities, one might argue that the failure to diversify sources of supply was of lesser importance than had hitherto been believed; and that the main problem associated with European dependence on oil had become an economic, rather than a strategic, one. In this respect, therefore, one must note the serious economic implications for Western Europe of collusion between the important oil-producing and exporting countries generated by the formation of the Organisation of Petroleum Exporting Countries (OPEC). In December 1970 – after a decade of efforts – OPEC took a collective decision to increase the taxes they levied on each barrel of oil and gave notice of their intention to escalate the tax levels each year until 1975.

The international companies, also working together, had no alternative but to accept the revised conditions on which they could work their concessions; but they, in their turn, certainly hoped jointly to create a situation in which the additional revenues they achieved from higher prices charged to their Western European customers far outweighed the additional taxes they had to pay in the producing countries. Thus, in a situation in which the net impact of the additional taxes on oil company's costs could be calculated at a maximum of $0.60 per barrel, the oil companies talked of price increases some 200 to 300 per cent higher and, of course, of continued escalation in the prices of petroleum products following the successive annual tax increases which were threatened through to 1975.

These changes were, by 1971, enough to indicate that the long period of falling oil prices in Western Europe was over – with, of course, no possibility by this time of any effective competition for oil from the continent's now much diminished and seriously weakened coal industry, the costs of which were, under conditions of continuing wage inflation, rising even faster than oil prices.

The tax changes, and the consequential price increases, are not, however, the only elements in the new situation now governing Western Europe's availability of oil. Additionally, the international oil companies, which have hitherto not only worked, but also controlled, the oil fields in the producing countries, are now under an obligation to hand over part of their equity interest in each and every oil field and in all other oil-industry activities (such as refining and transportation facilities) in the producing countries to the governments of the countries themselves. No final decisions on the scale of the participation of the countries have yet been made, but negotiations revolve around producing countries' claims for equity participation ranging from 20–51 per cent with their bids for shares in the ownership of the companies at prices which are related to the nominal book value of the enterprises, rather than to any other way of evaluating the worth of the investments and of the oil reserves known to exist. Whatever the final result, one may be certain that it will in any case be one which also implies higher prices for oil products both in the short run and the longer term, as the companies are obliged to give up part of the profits they had hitherto anticipated receiving; to which development they must react by trying to obtain an enhanced cash flow from each of the fewer barrels of oil which they will now be responsible for selling.

Thus, economically, Western Europe would appear to face a period of generally increased price levels for its essential oil requirements. The alternatives would be either a set of unprofitable oil companies, which is not likely to happen given their gentlemen's agreement to work together to prevent themselves from being unduly squeezed between producing and consuming countries: or a partial cessation of the flow of oil, following the failure of the oil companies to agree to the demands of the oil producers. This latter possibility would, of course, bring us back to the strategic aspects of the whole situation – for Western Europe can no longer be considered as a viable entity in any situation which denies it access to an inflow of much less than 600 million tons of oil per annum. Even this assumes that the demand for oil could be cut by up to 20 per cent through a system of curbs on inessential uses.

A Bleak Prospect

As indicated above, however, the threat to Europe's viability now comes not so much from the dangers of unilateral, political action on the part of the oil-producing countries (acting either in what they appear to believe to be their own interests or under the influence of other outside powers) simply to cut off oil supplies for long periods; as it does from the producing countries' newly found collective strength successfully to secure control over the supply and price of oil so that they receive the bulk of the financial benefits flowing from the demand for oil; or, worse still, from the possibility that an unwillingness to pay the price demanded may lead to a refusal to sell. The fact that the oil-producing nations' motivations for refusing to continue to send oil to Western Europe has changed does not alter the nature of the danger facing Europe as a continent whose total economies – and not just its transportation sectors and its military machines – now depend upon a continuation of the oil flow. The only exception to this would be in respect of very short-period interruptions in supplies, given that shortfalls of this kind can be covered by means of the not inconsiderable emergency supplies which governments have required shall be available within Western Europe itself.

Thus, the outlook is bleak – and it is, moreover, still generally considered to be getting worse as a result of the expectation of a continually rising demand for oil and from its still slowly increasing share of the total energy supply. The way this is currently expected to develop from 1970–80 is shown in Table IV-1.1 with, as indicated, a demand for over 1,000 million tons of mainly imported oil in 1980, when it will account for 70 per cent of Western Europe's total energy supply. Over this period, moreover, recent significant finds of oil in places like Australia and Canada – whose resources could be considered more secure than most others – are unlikely to be developed sufficiently to make much difference to the general level of dependence on traditional oil-producing countries.

The North Sea Discoveries

This sombre picture is, nevertheless, relieved by a development within Western Europe, the significance of which has to date been played down and under-rated – most of all perhaps because of Western Europe's unfamiliarity with the oil problem, in general, and with possible indigenous developments of oil and gas, in particular. Tables IV-1.1 and 2 show alternative set of predictions for the 1975 and 1980 Western European energy supply position. The revised forecasts do, indeed, show, respectively, a slowly increasing dependence on imported supplies of oil and, alternatively, a falling dependence; with the latter indicating a really fundamental change from the present conventional wisdom on the future situation.

TABLE IV-1.1
Western Europe's energy supply 1970–80, existing and revised (author's) estimates
(In millions of tons of oil equivalent)

Energy source	1970	1975 Estimates		1980 Estimates	
	Actual	Existing	Revised	Existing	Revised
Total	*980*	*1,100*	*1,150*	*1,500*	*1,550*
Of which:					
Hydro, nuclear power	25	50	35	120–160	100–140
Coal	290	245	200	150–230	130–180
Oil	605	720	660	940–1060	810–900
Natural gas	60	110	255	170	420

There are several essential points about the revised sets of estimates. First, they do not imply any changed relationship between the use of oil and the use of coal. On the other hand, they do indicate both the likelihood of a very significantly expanded production and use of natural gas – resulting in a reduced demand for oil – and a potentially high output of indigenous Western Europe crude oil from the North Sea and other off-shore areas.

In that natural gas is almost always a preferred fuel over oil, given its advantages in handling and in cleanliness and given the ability of the Western European economy to develop the sophisticated infrastructure required for its distribution, then the only limitations on the development of natural gas use lie in a too-small resource base and in price considerations. Europe's natural gas resources were until the late 1960s thought to be very limited indeed. Since then the proving of the Groningen field in the Netherlands as one of the largest in the world, the discovery of significant resources in the southern part of the British sector of the North Sea and the further discovery of additional large gasfields – and reserves of gas associated with oil fields – in the middle and northern North Sea oil and gas basins, have fundamentally changed this position so that a production potential of the order of magnitude indicated in Table IV-1.1 already exists – irrespective of any further discoveries which may be made in the next decade as a result of continuing exploration. This potential can, of course, only be realised by expenditure on the required infrastructure and by companies and/or governments setting pricing policies which encourage both its production and its use in the energy economy. Given the diverse nature

TABLE IV-1.2
Western Europe's oil supply 1970–80
Existing and revised estimates of imported and indigenous oil supplies
(In millions of tons)

Oil supplies	1970	1975		1980	
		Estimates		Estimates	
	Actual	Existing	Revised	Existing	Revised
Indigenous	15	25	70	35–40	250–350
Imported	590	695	590	900–1025	460–650
Total	*605*	*720*	*660*	*940–1060*	*810–900*

of the ownership of the resources – both between companies and between countries – these developments are by no means axiomatic. However, as the use of European natural gas is so much in the economic and strategic interests of Western Europe, any narrow nationalistic or company pressures designed to inhibit its rapid development must somehow be effectively constrained.

The same is true of policies towards the recently discovered oil resources of the North Sea and elsewhere. Here, again, there are potentials for disagreements between company and company over the sharing of the returns from fields which straddle concession lines; between company and government over the sharing of the profits to be made from production; and again, between the governments of European countries in the joint efforts that are required for the successful production, transportation and marketing of oil. All such disagreements could delay the build up of production capacity to the potential level indicated in Table IV-1.2 for 1980.

This would, unhappily, be in a situation in which, first, the limited amount of exploration to date has discovered enough reserves to justify an annual production of at least 150 million tons; and second, in which the continuation of exploration at a still increasing rate in areas shown already by seismic survey to have oil potential is considered 99 per cent certain to reveal the existence of further giant fields.

In brief, Western Europe's newly discovered potential for large annual outputs of both natural gas and oil, even in relation to the energy demands of an increasingly affluent continent, could put quite a different perspective on the significance of imported oil in the Western European economy. There is a good chance, given the effective exploitation of the oil and gas reserves following government agreements and suitable arrangements between companies and governments, that Western Europe's demand for imported oil could reach a peak by the mid-1970s. Even by then, as shown in the Table IV-3.1, it will represent a lower percentage of total energy demand. Thereafter a relatively static, or even a falling requirement for imported oil – to a 1980 figure well below to only slightly higher than the 1970 figure – will, of course, mean that imported oil gradually supplies a smaller part of total energy demand (and possibly no more than 30 per cent by 1980).

This hypothesis on the development of Europe's energy markets over the next decade or so may, in 1972, seem unreasonably optimistic and perhaps incapable of achievement. However, it should be noted that the dynamism of the oil industry has already produced two post-war decades – the 1950s and the 1960s – in which the quantitative and qualitative changes were greater than those that are now suggested for the 1970s. Should the 1970s, in fact, see the hat-trick as far as the third decade of fundamental change in oil industry affairs in Europe is concerned, then both Western Europe's economic and strategic position is going to be much more secure than seemed likely only a few years ago, when the only energy policy alternatives for the future were excessive build-up of oil stocks, or abandonment of cheap energy policies in favour of considerable support for high and increasing costly indigenous coal. Now, even a further run-down of the coal industry and a slower-than-hitherto expected expansion of nuclear power can be viewed with something approaching equanimity – for at least the second half of the decade.

Danger Period

Nevertheless, the three to four years to what could be achieved by 1975–76 (viz. a rapid build-up of indigenous oil and gas production), constitute a real danger period as far as Europe's security of energy supplies are concerned. In this period Western Europe has no alternative to continued dependence on imported oil for the energy supplies that it requires for its continued economic well-being and expansion and for the maintenance of its energy security. Perhaps the best one can hope for during this time is a continuation of the series of warm winters which in 1970–71 and 1971–72 so reduced the expected demand for oil that shortfalls in availability did not materialize. Instead, all available tanks were filled with reserve stocks, refineries had unused capacity and prices fell away from the higher levels which the oil companies had hoped to sustain in the aftermath of their tax agreements with the producing companies. These unusually warm winter weather conditions coupled with the slight recession in economic activities in Western Europe, plus the rapid increase in the availability and use of Dutch natural gas have, so far, in this current danger period combined to minimize the impact of significantly changed relationships in the oil world – both in terms of control over resources and over the international division of economic benefits from its production. But this still leaves three critical years before Western Europe can begin to breathe more easily concerning its essential requirements of energy – and only then in the event of both individual European governments and European organizations evolving policies adequate to ensure the rapid expansion of the continent's indigenous hydro-carbon resources.

Bibliography

(a) For background reading on the world oil industry see:

P.R. Odell *Oil and World Power; a geographical interpretation* (Penguin Books, Harmondsworth, 2nd Edition, 1972).

E.T. Penrose *The Large International Firm in Developing Countries* (Allen and Unwin, London, 1968)

(b) For more specific reading on the oil industry in Western Europe see:

W.G. Jensen *Energy in Europe*, 1945–80, (Foulis Books, London 1967).

H. Lubell, *Middle East Oil Crises and Western Europe's Energy Supplies*, (John Hopkins Press, Baltimore, 1963).

S.H. Schurr et al., *Middle Eastern Oil and the Western World*, (John Hopkins Press for Resources for the Future, Washington, 1971).

(c) Recent very significant developments concerning oil and Western Europe are necessarily still available only as articles. The following additional articles by the author of this paper will help to fill out some of the points discussed briefly here:

'Europe and the International Oil and Gas Industries in the 1970's' *The Petroleum Times*, vol 76 nos. 1929 and 1930, January 1972. (See Chapter I-4 in this volume).

'Natural Gas as a European Energy Source', *The Petroleum Times*, vol. 76, no. 1935, April 1972.

'Supply and Transportation Patterns of European Oil to 1980, *The Petroleum Times*, vol 76, no. 1937, May 1972.

'Europe's Oil Dependence,' *National Westminster Bank Quarterly Review*, August 1972. (See Chapter I-3 in this volume).

Chapter IV – 2

Bases for Western Europe's Strategic Response to the Oil Crisis – Demand Constraints and Indigenous Production*

Introduction

The energy crisis in Western Europe – as in most of the rest of the world – emerges out of the quite fundamental shift in power which has occurred in the oil world since 1971[1]. The shift in power has occurred as the hitherto competitive position between oil companies and oil producing countries has been eliminated by collective agreements to constrain the supply and hence create the opportunity for earning monopoly profits: an opportunity of which full advantage has been taken in the specially favourable conditions emerging out of the Arab nations' reaction to the war with Israel. An institutionalised shortage of oil has thus been created – and whilst the institution of the O.P.E.C. cartel remains in control of the situation, we must expect the situation to continue to get worse with a traumatic effect on the oil prices that industrial and other consumers will have to pay. The impact of the shift in power on long established levels of energy prices in Western Europe is shown in Figure IV-2.1 in which the relationships between posted prices for crude oil, realised prices for crude oil, landed prices of crude oil in the U.K. and the price paid by industry for its fuel oil are demonstrated for the period since 1955[2].

*** An edited version of a paper presented to the International Institue for Management Technology's Forum on *The Energy Crisis in Europe – Problems and Possible Solutions*, Milan, Italy, February 5/6, 1974.**

Whilst the institutionalised shortage of oil is maintained, posted prices may be assumed to continue to trend upwards – how steeply is anyone's guess, but here we have assumed a 10% per annum rate from 1974 to 1980. Realised prices – which, as the diagram shows, diverged from posted prices in the 1960s under the stimulus of competition for markets between the producing companies – moved back after 1970 to the posted price levels as the companies found they had no further need to compete for sales and they will remain, we may anticipate, at or near that level. At first, this will be by choice on the companies' part (in order to enable them to maximize their profits) but, later, of necessity, as the producing countries 'participate' (nationalise) the oil companies and then only make oil available at the officially posted prices – as, indeed, is already happening within the framework of the dangerous bilateral oil supply arrangements being made between governments. These are dangerous in that they build an element of stability into what would otherwise be an inherently unstable situation, as is the case with all oligopolies.

Thus, the average landed price of crude oil to the U.K. will move up in response. In the future the hitherto significant variations in prices introduced by varying transport costs etc, will become relatively insignificant in influencing this average landed price. From this one can move to the price which really counts, viz. the price of the oil delivered to the final consumer. This, as shown in the diagram, was relatively static for the whole period from 1955 to 1970 – at about £10 per ton. The already significant 33% increase between 1970 and 1972 is, however, likely to be modest in relation to the price levels for the period up to 1980. Extrapolation from known movements in crude oil prices will, even if the oil companies do nothing more than escalate their historic margins on their refining and distributing activities by 10% per annum in order to meet inflation trends and to improve the hitherto modest rates of return which they have earned in these sectors, lead to fuel oil prices of £55 per ton by 1975 and to almost £90 by 1980. They could, however, rise even more, as shown by the higher trend line on the diagram to £75 and £120 per ton in the two years, respectively. These price increases would roughly maintain the proportional relationship between landed crude oil prices and fuel oil prices in the market place.

Though this implies much increased profits in these sectors for the oil companies, one has to recognize that these may be essential in order to compensate the companies for their loss of profitability in the oil producing countries. Finally, it must also be borne in mind that fuel oil production has traditionally provided only a relatively small part of the fixed costs of refining oil, as these have generally been orientated to other less price conscious products. Given the level of prices for oil products now in prospect for Europe, the opportunity for – and the wisdom of – allocating downstream costs away from the industrial and commercial products will be very much smaller.

Energy Conservation

In the face of price developments of this magnitude it seems highly likely that increased efficiency in energy use and energy conservation in Europe will become important managerial aims, requiring serious steps which can and must be taken. These, moreover, must not only be in the field of technical improvements in energy systems, but also in adjusted production schedules, new product mixes and new marketing and distribution systems etc.[3]. Such a demand response to the new supply and pricing situation will, of course, apply mainly to price conscious sectors of

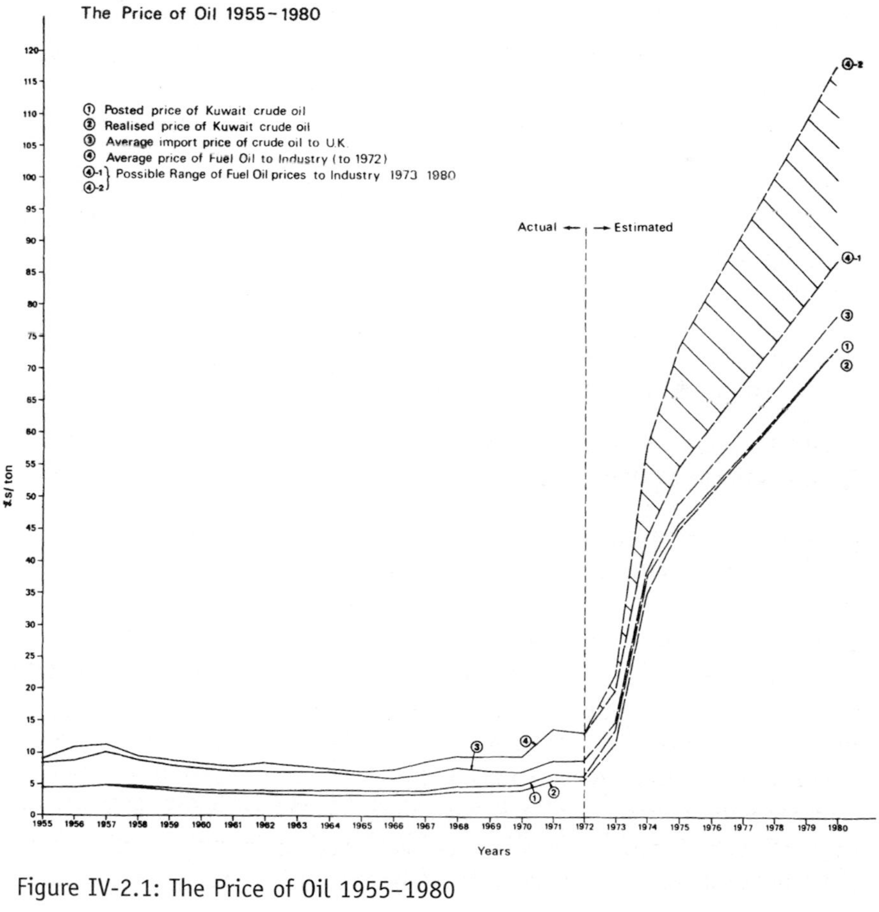

Figure IV-2.1: The Price of Oil 1955–1980

the economy – whilst, elsewhere, demand inelasticity and/or an ability to pass on price increases to consumers or tax payers will limit the impact of the higher oil prices on levels of consumption.

It is because of this that it seems essential that mechanisms other than price be used to constrain demand in European economies – with the objective of eliminating all growth in oil use and so holding consumption at overall levels no higher than those of 1972. This, of course, implies a significant and, indeed, a quite deliberate restraint on freedom of consumer choice including, for example, the ownership and use of particular kinds of cars to be determined by factors other than the ability of an owner – or his employer – to pay for it and its fuel! And in the encouragement of less energy intensive rail and water transport modes for the movement of freight, rather than much more energy intensive trucks and aircraft! And in the household sector where assessed needs, rather than personal choice, could determine how much energy is used. And, finally, a requirement that industry and commerce get used to the idea of producing and justifying an annual energy budget whereby it demonstrates its energy efficiency.

The diminished freedom of choice implicit to such developments will generally be considered to be offensive in itself. The development also involves an extension of the bureaucracy and it implies significant changes in the values which have been created by the mass-consumption societies over the last generation. It may thus be triply unacceptable, but the alternative of continued dependence on high cost and uncertain supplies of oil from overseas – with a possibility always present that factors and forces external to our European way of life can create chaos and confusion at any and every moment in time – could be even more unacceptable?

A Longer-Term Strategy – Component 1: Restraints on Energy Use

Over the last twenty years we have accepted the validity of an open European energy economy in which cheaper, readily available supplies of foreign oil achieved the dominant role – at the expense of known indigenous resources. This is especially true, of course, for European coal, which, though remaining potentially available in very large quantities, became unwanted and unused with the consequent need to close hundreds of mines. There developed, in fact, an overriding degree of irresponsibility towards the use of indigenous resources: a development which was encouraged by the notable powers of persuasion of the international oil companies, with their interests in maximising sales of foreign-produced oil to Europe.

We thus became convinced that we could not and should not pursue any kind of energy policy savouring of an autarkic approach – in marked contrast with the United States and the Soviet Union in both of which the minimisation of energy imports has generally been a declared objective of national policies. Even when a quite modest exploration effort in British and other north-west European off-shore waters clearly demonstrated a potential availability of very large resources of both oil and gas, it was still argued by some (notably the oil companies) that these should not make any essential difference to the required openness of our energy economy. Hence, we have remained largely unresponsive to the recent challenge to our established views on the structure of the energy supply – and have continued to act and to plan for increasing oil imports for the whole of the foreseeable future – as shown in the most recent official estimates of energy use in Western Europe for 1980 and 1985 in Table IV-2.1. Such complacency has now been rudely shattered by the fundamental changes in the international oil system – changes which

TABLE IV-2.1
Official and alternative views on Europe's energy supply
(in million of tons of coal equivalent)

	1973 Approximate actual use		1980 Estimates				1985 Estimates			
			Official		Alternative		Official		Alternative	
	Mmtce	%	Mmtce	%	Mmtce	%	Mmtce	%	Mmtce	%
Total energy	*1550*	*100*	*2250*	*100*	*1900*	*100*	*2850*	*100*	*2350*	*100*
Oil-Total	970	63	1500	66	835	43	1725	63	840	36
Indigenous	30	2	50	2	500	26	195	7	640	27
Imported	940	61	1450	64	335	17	1530	56	200	9
Gas – Total	135	9	265	12	575	30	385	14	790	34
Indigenous	125	8	215	10	500	26	300	11	640	27
Imported	10	1	50	2	75	4	85	3	150	6
Coal – Total	400	26	280	12	310	16	310	11	350	15
Indigenous	360	23	205	9	230	12	220	8	250	11
Imported	40	3	75	3	80	4	90	3	100	4
Primary Electricity	45	3	210	10	180	9	330	12	370	15
Total Indigenous	560	36	680	31	1410	74	1045	38	1900	81
Total Imported	990	64	1570	69	490	26	1705	62	450	19

eliminate the last excuse we have for not investigating the option of an autarkic energy policy.

On the other hand, the motivation for seeking self-sufficiency with, say, a Project Energy Independence 1985, (similar to the United States' Project Energy Independence 1980, already accepted by all significant interest groups there) lies, of course, in the tremendous burden of cost and, perhaps even more important, the uncertainty which continued reliance on imported oil places on our economy.

The opportunities for even being able to consider the idea of self-sufficiency in energy emerge out of two developments. First, out of the deliberately constrained demand policy – as outlined above. Table IV-2.1 shows the massive difference that would arise in the total use of energy in the period up to 1985 from such a policy. Officially, there is an expectation of a continued 5% per annum growth rate in energy use over this period – assuming continued economic growth at about 3–4% per annum and the persistence of trends towards increased use of energy intensive transportation (cars, trucks and aeroplanes) and the electrification of society. In contrast, we would suggest that demand restraint – engendered in part by price increases and in part by a modest curtailment of the freedom of consumer choice – could reduce the energy growth rate to 3% per annum between now and 1980 and to 4% p.a. thereafter. The result of this is a 15% reduction in the expected rate of energy use by 1980 (1900 instead of 2250 m.m.t.c.e. in Western Europe) and a 17.5% reduction by 1985 (2350 instead of 2850 m.m.t.c.e.). The evidence already available on the possibilities of utilising energy more efficiently – by improved methods of energy use, and by some modest changes in life styles – suggests that such reductions in total energy use are in essence a question of appropriate policies – both public and private – being followed, rather than a question of our having to accept the idea of sub-normal or even zero economic growth[4].

A Longer-term Strategy – Component 2: Self-sufficiency in Energy Production

Constraint over demand – as indicated above – is not only the essential element in the strategy required over the next seven years or so to overcome the problems caused by an adverse energy supply situation and a greatly increased price for oil. It is also a pre-requisite for the achievement of a longer-term aim for self-sufficiency in this vital sector of the European economy. Within the framework of a constrained demand, as shown above, there is a potentially achievable level of 74% independence by 1980, and 81% by 1985. Within this European framework, moreover, particular countries, most notably Britain and Norway, will achieve complete or near complete self-sufficiency with import requirements restricted to special products such as crude oil for blending and coal for coking purposes etc.

As also shown in Table IV-2.1, the potential change in the relationship between indigenous and imported energy does not emerge out of a massive revival of the coal industry or from a much more rapid than expected growth in the availability of primary energy (viz. hydro-electricity and nuclear power) – nor yet from the development of any exotic types of energy production. Indigenous coal and other solid fuels production is not even expected to be maintained at its 1973 level, though the effect of revised government attitudes towards the coal and brown coal industries, in Britain and Germany in particular, will enable indigenous coal's contribution to be a few percentage points higher than hitherto currently expected in both 1980 and 1985. Primary electricity, on the other hand, still seems unlikely even to make the contribution officially expected of it by 1980 – because of delays in commissioning nuclear power stations for both technical and environmental reasons. Post-1980, however, assuming by then that such technical and environmental difficulties have been overcome, and further assuming positive results of various governments' present efforts to stimulate faster nuclear development, primary electricity's contribution to the total energy economy may also be about 3% higher than officially expected. Even so, indigenous coal plus primary electricity could, by 1985, contribute only just over a quarter of Europe's total energy supply (rather less than it did in 1973).

The really important – and hitherto largely unexpected and unexplored – indigenous energy contribution comes from European oil and natural gas[5]. In small part this will be from long-known producing areas – in, for example, the Po Valley of Italy, south west France and northern Germany – but, in much the greater part, it will be the result of the massive exploration and development of the North Sea oil and gas province, the end-December 1973 status of which in terms of size and location of discovered oil and gas fields is shown in Figure II-2.2 (p.173). Although exploration in the province is still in its early build-up stage, no fewer than twenty 'giant' fields, each with at least 500 million barrels oil equivalent of recoverable reserves, have already been discovered and initially evaluated, together with at least 60 other significant finds of oil and gas. On the basis of the exploration history of the province and what is known about opportunities which remain to be explored, together with the known commitments to further exploration efforts in all sections of the basin (including those which have been under exploration for some time already, but where potential new oil and/or gas bearing structures remain to be tested), a model has been developed to simulate the development of the province over the period 1969–2019.

The parameters built into this model include annual rates of discovery; successive appreciations of recoverable reserves; the rate of development of the resources; and the depletion rates of the fields etc. Constrained random choices were made by the computer for the individual values of the variables. The ranges of values were established in the light of the knowledge about the behaviour of the province and the general behaviour of major oil and gas provinces. A series of simulations have been run on the computer and from these overall production curves for the whole period were established. A high/low range of production possibilities for all years over the period up to 2019 was determined[6].

The dynamics of the oil potential from the basin are shown to be a continuing upward curve of production until near the end of the century. The rate of increase in output beyond 1985 still runs ahead of the rate of increase of demand for energy, thus indicating the ability of indigenous oil to continue to play an increasingly important role in the total energy requirement, even without the extension of the oil search to other favourable areas around the coast of Western Europe (see below). Because of the nature of the occurrence of natural gas, viz. usually in combination with oil, with the later normally being the object of the search and its recovery the main determinant of the way in which a field is exploited – we have not yet found it possible to develop a model to stimulate the future long-term development of natural gas production. Work continues on this but, in the meantime, one can, nevertheless, suggest with confidence a Western European gas production potential by 1980 which, in energy terms, is just about the equal of the expected indigenous oil availability by then. This is based on the currently known reserves of marketable gas and an estimate of the potential rates of production which can be expected from these reserves by the early 1980s. Thereafter – given the continued discoveries of both new gasfields and of oilfields with high gas/oil ratios – we may reasonably assume the production of as much additional gas as oil by 1985[7].

These estimates of the future availability of indigenous energy have been presented at a Western European level – which, given the opportunities presented by, and the requirements of, the E.E.C. in terms of non-discriminatory policies on the flows of goods between member states, would appear to be the most relevant region for this sort of development. It would, however, be foolish to ignore the nationalistic claims for priority in the use of oil and gas and/or exclusive national rights to take decisions on production levels in particular areas – as already seen in a dispute over the export of Dutch off-shore gas to Western Germany[8] or in indications of the British government's attitude towards British oil[9]. Two points need to be made to put these nationalistic views in perspective. One is that with the quantities of indigenous production indicated as being likely to become available over the next decade or so, all individual producing countries – including Britain – will actually need to have export markets in order to enable development to go ahead at the indicated rate. Thus, the "Britain + the Netherlands + Norway" versus "the rest of Europe" argument will be rendered irrelevant. And secondly, in response to the argument that each individual country's resources should be saved for the demands of that country in the distant future, instead of being used for meeting nearer future demand over a wider area, the potential future availability of oil and gas from very much larger regions of off-shore Europe than are currently under exploration and development should be noted. The number and extent of the potential areas are shown in Figure I-5.2 (p.109) and clearly indicate that the development of the North Sea basin potential will eventually be joined by even more extensive opportunities for finding resources in other European offshore areas.

This could eventually lead to a reversal of the now expected north to south flow of energy within the framework of an integrated European-wide system, the establishment of which should be the prime aim of European energy policy over the next few years. It is also within the framework of such an integrated system that a European pricing system for energy – related to the cost of producing energy in Europe – and not to the internationally traded price of oil – could be established. This is an idea which was tentatively explored elsewhere[10]. It has now become even more fundamental to the well-being of the European economy, given the further massive increases in intentional oil prices since December 1973. The consequently much increased discrepancy between the energy prices which European energy consumers now seem likely to have to pay (see Figure IV-2.1) and those which will apply in the United States, where the domestic economy has already been protected from the international escalation of oil prices, calls into question Europe's continued ability to compete with the United States on world markets and indicates the desirability of similar action on energy prices being taken here. The opportunity for such action can emerge only from the two essential elements of Europe's future energy policy as developed in this paper – viz. first, immediate constraints on demand and second, a fundamental re-evaluation of the opportunities for the production of indigenous energy in Europe on the basis of which a high degree of autarky in the sector can be achieved.

References

1 See P.R. Odell, 'Europe and the International Oil and Gas Industries in the 1970s *Petroleum Times*, Vol. 76, Nos. 1929 and 1930, January 1972

2 See *The Energy Crisis*, a monograph written by P.R. Odell and published by Energy Advice Ltd. and Industrial Market Research Ltd, and available from I.M.R., 17 Buckingham Gate, London S.W.1.

3 See *The Energy Crisis*, op.cit. for a full discussion of these possibilities.

4 A recent study in Sweden – which is already just about twice as energy efficient as the U.S. – has indicated that a 25% saving in energy use could be made even there, if appropriate policies are followed

5 See P.R. Odell, "Indigenous Oil and Gas Developments and Western Europe's Energy Policy Options" *Energy Policy* Vol.1, No.1, June 1973 for a full discussion of this issue.

6 Details of this modelling procedure and the results from it have since been published as a monograph, viz. P.R. Odell and K.E. Rosing, *The North Sea Oil Province; an Attempt to Simulate its Exploration and Exploitation, 1969–2020*, Kogan Page, London 1975. See Chapter II-3 of this volume.

7 See Chapter II-1, II-2 and III-1 in this volume for a full presentation on data and prospects for European gas developments.

8 See Chapter III-5 of this volume.

9 See Chapter II-2 of this volume.

10 See P.R. Odell, "Using Europe's off-shore Energy Supplies in the Interests of the Consumer" a paper presented to the Financial Times Conference on *The North and Celtic Seas*, London, December 5, 1973.

Chapter IV – 3

The EEC Energy Market: Structure and Integration*

Introduction

Western Europe's traditional energy supply pattern – based on the production and use of indigenous coal and hydro-electricity – produced a long period of stability in the continent's energy market in which traumatic changes were the exception rather than the rule. For example, expansion of the old-established areas of coal production into adjacent areas as mining at increasing depths became possible was more important in meeting the growing demand for energy in the late nineteenth and early twentieth centuries than the coal obtained from the opening up of entirely new regions of production. Thus, geographical changes in the supply patterns of basic energy were essentially local.

Meanwhile, the secondary energy industries (that is, thermal electricity and town gas production) gradually orientated their operations and organisation to the steadily rising demands and to the rather slowly changing technologies of production and distribution systems – within the framework of unchanging fuel input patterns built around coal supplies (Manners, 1972; Chisholm and Manners, 1971).

In terms of economic development, the coal-based national economies of north-west Europe flourished. But around their peripheries countries like Italy and Denmark lay at a disadvantage, as the near symbiotic relationship between coal and industrialisation

* **Originally published in P. Lee and P.E. Ogden (Eds.), *Economy and Society in the E.E.C*, Saxon House, Farnborough, 1976, p 63–8.**

established some of the main elements in western Europe's geography of economic development (Odell, 1974a). This is, however, not simply a matter of historical interest. The spatial patterns of industry and population produced by coal-based economic growth remain a major determinant of the geography of development in the oil and gas sectors of the modern western European energy economy.

The European Coal and Steel Community

The importance of the energy sector in the early postwar moves toward European integration was given institutional recognition by the creation of the European Coal and Steel Community. Coal and steel were still the commanding heights of the postwar economy and they continued to generate rivalries and fears among the nation states of Europe over questions such as discriminatory pricing and monopoly behaviour; and more fundamental issues such as the 'nationality' of the coal resources of the Saar and the iron ore deposits of Lorraine. The ECSC was, moreover, intended not only to eliminate the causes of such disputes, but also to stimulate the expansion of coal and steel production in the aftermath of wartime destruction at a time when it seemed that Europe's future must depend on the success of such efforts (Jensen, 1967).

Thus, as far as the mainland of western Europe was concerned (following the failure of the UK to 'join' Europe), the ECSC might well have produced an integrated energy sector out of the hitherto nationally orientated industries. However, the progress it made in this direction was undermined by the inability of the indigenous coal industry to expand sufficiently to sustain the rapid rate of economic growth generated in the aftermath of postwar reconstruction. The consequential problem of energy supply was met by governmental encouragement of the use of imported oil. Such encouragement however turned out to be hardly necessary, for there was a 'natural' rapid growth in the import of oil, reflecting its use both as a supplement to coal and as a substitute for it in an ever-widening range of uses.

The impact of this external competition, with its adverse effects on the indigenous coal industry, was to create a geographical realignment of economic interests in western Europe. On the one hand, countries such as Belgium and France were left with labour-intensive coal industries to protect and, on the other, countries like Italy and the Netherlands were concerned only with securing access to energy at the lowest possible cost. West Germany's decision to opt for an open energy economy, at the expense of the viability of its own considerable coal industry, emphasised the division within Europe over the possible approaches to the energy sector. This division resulted in the virtual exclusion of the energy sector from the continuing inter-governmental moves towards economic integration. So, from being a major element in the early sectoral thrust towards integration, energy policy became of minor importance, concerned only with technical matters (for example, standards and nomenclature) and tariff equalization on trade in energy products. Such steps turned out, however, to be of little importance compared with the non-governmental sponsored integration in the European energy market achieved between late 1950s and 1970 by the activities of the international oil companies as they met the growing energy demands of the continent's rapidly expanding economy.

The West European Oil Market

The immediate postwar pattern of the oil industry reflected its relatively small-scale nature. Import terminals for oil products were supplied from export refineries in the USA, the Caribbean via the existing transportation infrastructures. The European refining industry, which had been built up before the war had been largely destroyed in 1944 and 1945, and its reconstruction was a slow process. In such conditions, there was little need for any 'Europeanisation' of the oil industry's organisation – except in its use of the waterway system based on the rivers Rhine and Meuse. But here the essential element was the transport artery itself, rather than the organisation and structure of the oil industry.

The national structure proved, however, to be very short-lived as the international oil companies set out to exploit the potential for a mass market for their products in western Europe. In this development, both the international oil companies themselves and the potential for the large-scale use of oil were of importance in determining the changes in the structure of the oil industry across the continent – and hence in the continent's energy sector which came to be dominated by oil.

The development of a European mass market in oil is a well-known phenomenon (Odell, 1975, ch. 3), the essential elements of which are demonstrated in Table IV-3.1. The growth in the demand for oil for the period from 1951 to 1971 stands out clearly. In meeting this rapidly rising demand the large international oil companies simply treated western Europe as a unified market which could be developed on the most rational basis possible. This, in turn, involved the creation of an international organisation and infrastructure integrated on a European scale.

TABLE IV-3.1
Western Europe: the use of energy and oil 1951–71
(in millions of tons coal equivalent)

	1951			1961			1971		
	Energy	Oil	%oil	Energy	Oil	%oil	Energy	Oil	%oil
Total western Europe	*696*	*79*	*11*	*854*	*273*	*32*	*1388*	*779*	*57*
of which									
UK	238	22	9	260	66	25	307	133	43
W. Germany	141	5	3	205	47	23	320	166	52
France	100	14	14	116	37	32	201	128	64
Italy	38	8	21	61	35	57	144	108	75
Denmark	9	2	27	14	8	60	26	25	92

In their organizational structure for western Europe, moreover, the oil companies treated the region as the complement of the Middle East. They ensured that western Europe became the principal market area for the rapidly increasing production potential of the prolific oilfields of the Persian Gulf and other parts of the Middle East and, somewhat later, of north and west Africa. Thus, western Europe became an essential part of the companies' so-called eastern hemisphere operations (in contrast to their western hemisphere operations in the Americas). Most of them chose to run these from London which thus emerged as the capital city of the European oil system. Only two other cities offered any competition to London in this respect; viz. Paris, which was the administrative and technical centre of the French multinational Compagnie Française du Petrole and of the influential Institut Français du Pétrole which developed significant international responsibilities within French oil policy; and the Hague in the Netherlands, where Shell had located the headquarters of its European operations. With these minor exceptions the oil industry in western Europe has been organised, directed and controlled from London. The latter thus came to provide the centre in which the structure and the size of the European energy sector was effectively determined from the middle 1950s through to the early 1970s, without respect for the boundaries of individual countries or of the existence of the EEC.

Meanwhile, the oil companies were organizing an extensive and comprehensive infrastructure for the transport, refining and distribution needs of the industry in developing the bulk market for oil. At one extremity this system involved the establishment of an increasingly intensive distribution network, with a repeated spatial patterning of facilities designed to meet local needs for oil products. Each facility was, nevertheless, orientated specifically to the particular demand characteristics of the region which it served.

The flow of oil products through the integrated distribution and marketing networks depended on supplies from the refineries. These, in turn, required a crude oil supply system dependent on tanker transport from overseas and, for the inland refineries in western Europe, on crude oil pipelines from import terminals at, for example, Rotterdam and Genoa. The overall development produced by these needs is shown in Figures IV-3.1 and 2. These demonstrate the intensity of the refinery expansion and pipeline network demanded by western Europe's use of oil. The trans-national crude oil pipelines shown on the map epitomize the geographically integrated nature of the oil market. Even France provided a point of entry – Marseille – for one of the crude oil systems, despite the application by successive French governments of nationalistic oil policies limiting the ability of the oil companies to integrate their operations on a European scale. Until the onset of the recent crisis over oil supplies and prices, France also planned a second system originating from the new mammoth tanker terminal at Le Havre.

Only the UK, for obvious reasons, remains isolated from this system of crude oil pipelines, but in respect of the large-scale flow of oil products from refineries – a flow largely dependent on water transport – its position was little different from that of other countries in Europe.

For this component in the physical integration of the oil system, the main influence has been the geography of individual companies' infrastructures. These have been built up over time by decisions on the location of import terminals and refineries and by participation and co-operation with other companies in the use of pipeline systems. The Shell system is shown in Figure IV-3.3. This demonstrates how its international infrastructure complements its organizational structure in a successful private enterprise response to the challenge and opportunities of the European energy market.

Natural Gas in Western Europe

Until the mid-1960s natural gas production in western Europe was small-scale, developed and organised nationally and of only local importance, notably in south-west France and northern Italy. By the mid 1970s, however, it had changed its scale dramatically, but it still remained an essentially national phenomenon, because of the high degree of national control exercised over the production of the resource. This

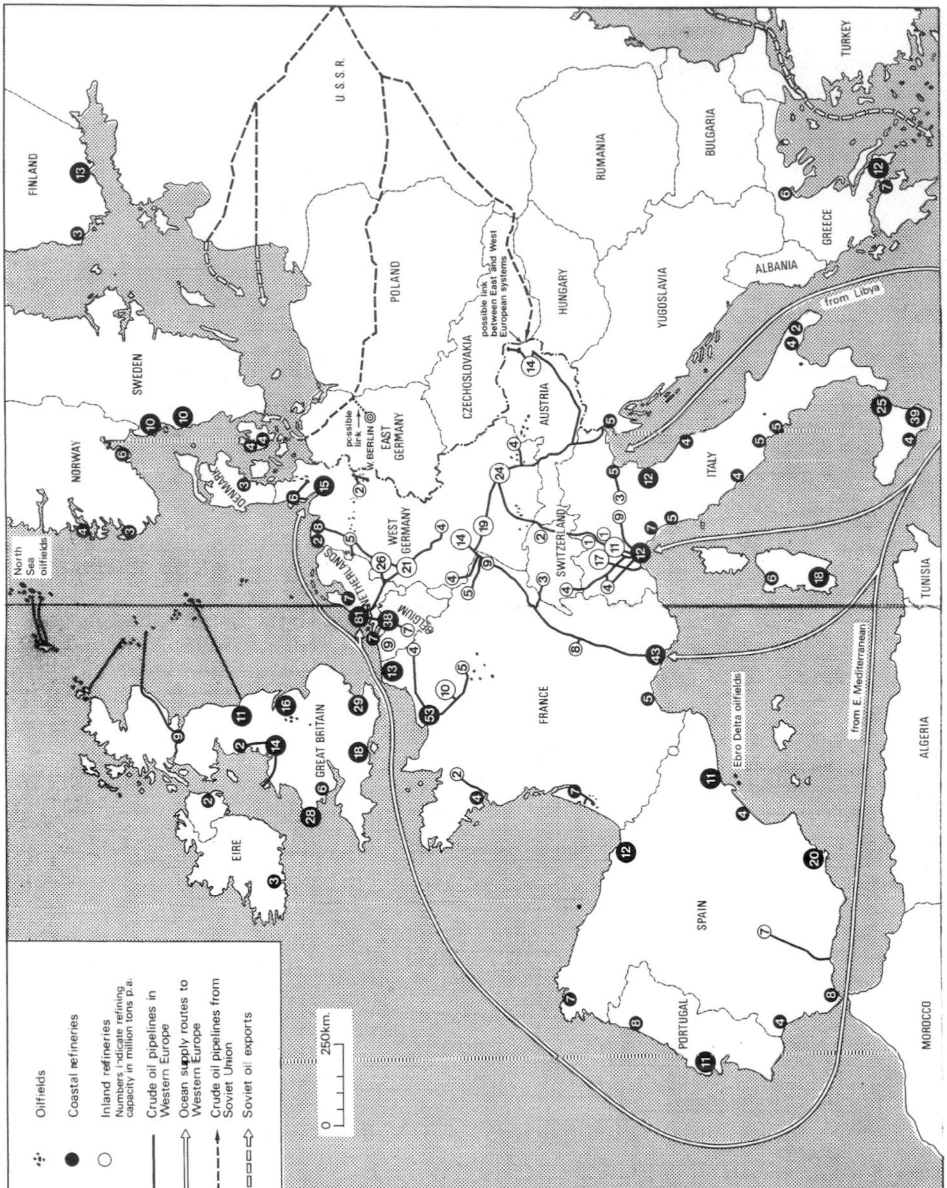

Figure IV-3.1: The Pattern of Oil Refining in Western Europe, 1975

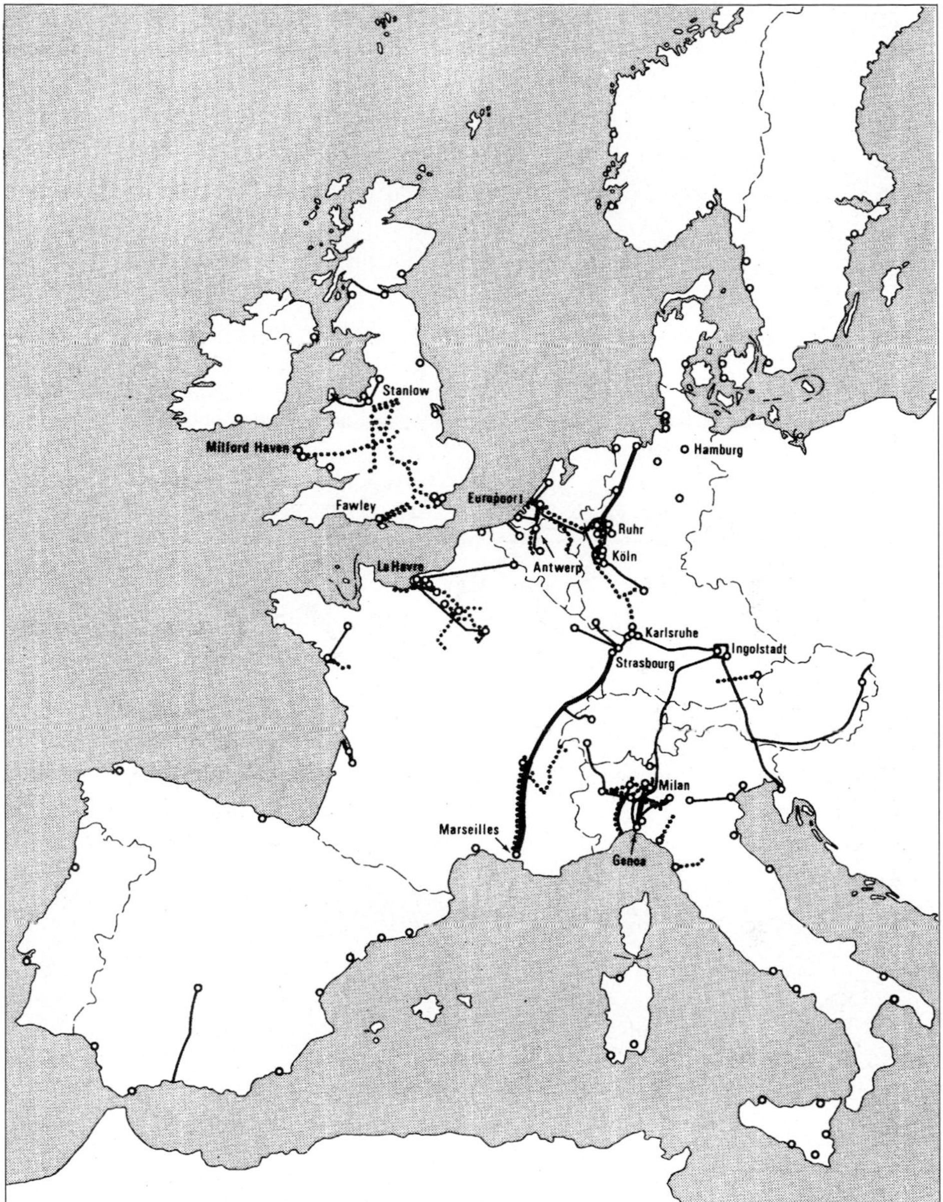

Pipelines

3 }
2 } Crude oil
1 }

o Oil Refineries

········ Oil products

Figure IV-3.2: Oil pipelines in Western Europe, 1975

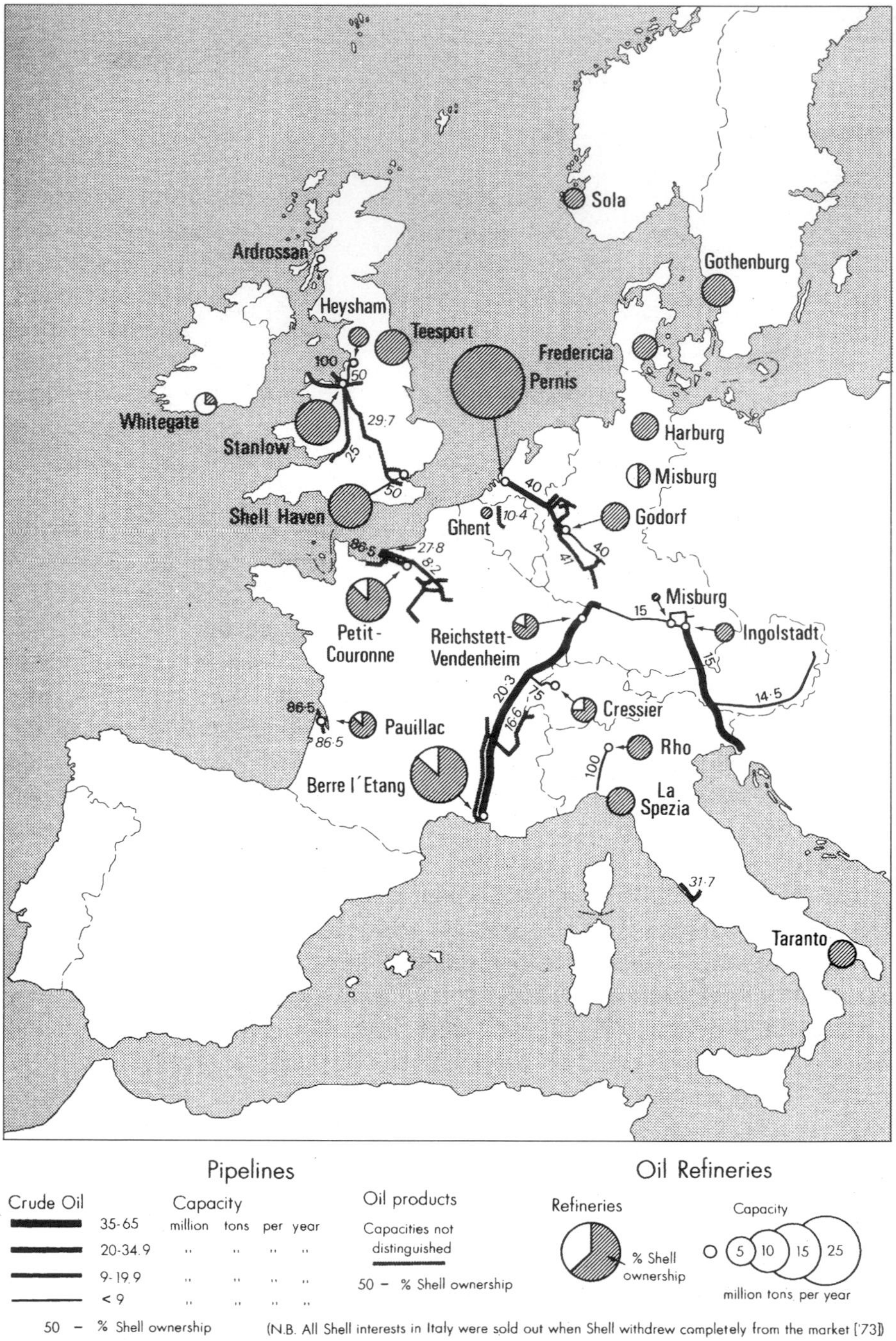

Figure IV-3.3: The Shell System in Western Europe

control has been maintained and extended in spite of the interests in natural gas production by the very same oil companies which, as shown above, successfully organised the oil market at a European level with little interference in their activities from the various national governments.

The significance of this nationalistic control over the gas industry became apparent in respect of the large quantities of natural gas discovered under the UK sector of the southern part of the North Sea. This is a production location which presented alternative marketing strategies. UK government control over this gas was, however, absolute – at least in terms of where the gas was to be sold and the price at which it would be marketed, even if not over the question of rates of production and of development of the fields[1].

Thus there was no question of UK North Sea gas being sold outside the UK. Indeed, in 1971 government permission was refused for UK gas to be sold in West Germany where the companies responsible for exploiting the Viking field, (viz. Conoco, an American oil company and its partner, the British National Coal Board) had secured markets. The UK attitude towards its gas resources was then emulated by the Netherlands in respect of its offshore resources. In 1973 the Dutch government refused to agree to a gas sales agreement, made between the potential producer (Placid Oil) and a consortium of West German customers, on the grounds that the gas would be needed in the Netherlands itself.

This Dutch government intervention in the gas market was surprising in two respects. First, the Netherlands was a member of the EEC (unlike the UK at the time of its decision) and was thus under an obligation not to discriminate between national customers and customers in other EEC countries. Second, the Netherlands had until then pursued a European marketing policy for its massive reserves from the Groningen field, in an effort to maximize returns from the exploitation of the commodity. In fact, to the end of the 1960s, Dutch national gas enjoyed a monopolistic position as a premium fuel in the enormous energy market of the mainland of north-west Europe (Odell, 1969).

An earlier European-orientated view of the potential spatial structure of the new natural gas industry emerged, not surprisingly, out of the interests of Shell and Esso. They saw the opportunity to develop this new energy source in a similar geographical framework to that used

1. But note that even decisions in these matters depend on price levels. Government control over prices implies governmental influence over other sorts of decisions as well.

for imported oil. Their marketing strategy, which was approved at the time by the Dutch government, required the development of sales and of a transport distribution infrastructure as indicated in Figure II-1.2 (p.139). The European scale of the operation became self-evident as the markets developed more or less as planned. The only major exception was with the proposed sales to the UK, where the market opening planned for Dutch gas was thwarted by the early discovery and development of gas in the UK sector of the North Sea.

A further factor which encouraged an integrated gas market within the EEC was the need for an international approach to the imports of natural gas. There have been two aspects to this. First, a shared infrastructure was required for importing gas from the Soviet Union. Soviet interest in the west European market for exports of its gas (and for gas from Iran for which the Soviet Union could act as an intermediary), arose initially out of the monopoly prices charged in the gas markets of the EEC by Shell and Esso. Even at distances of more than 5,000 km from the source of local production, Soviet gas could be delivered profitably at the price established by Dutch gas (Figure II-1.4, p.143). Thus Soviet offers were of great interest to Italy, West Germany and France and a series of bilateral arrangements became necessary to rationalize the delivery system for the imported gas.

Similarly, large-scale imports of liquefied natural gas (LNG), by very expensive tankers which are able to unload only in specially equipped terminals, also necessitated an international approach to the negotiations with the potential exporting countries. This need also produced *ad hoc* European arrangements with a joint West German, Dutch and Belgian project for LNG from Algeria and Nigeria. Additionally, Franco-Italian co-operation was initiated for developing trans-Mediterranean supplies.

However, the most significant and the most recent element in the creation of a European gas market has been the emergence of a potential major supplier of natural gas in western Europe itself. This is Norway, whose sector of the North Sea has quickly proved to have a high potential for gas production. Although the Norwegian government has tried to limit the rate of development of production as a matter of policy, the rate of build-up of its supply potential to other parts of Europe has been rapid. Figure IV-3.4 shows the development of the Norwegian gas system in the North Sea context, and the possibilities for its expansion as the resource is more fully exploited. This will lead to a very close relationship in the energy sector between Norway and the rest of Europe, in spite of the former's decision in 1973 to stay outside the EEC.

This relationship between a potential European supplier and the large energy consuming and importing nations of western Europe is made even more important by the potential availability of very large quantities of oil as well as of gas from the Norwegian sector. This is considered in detail in a later section of this chapter as it constitutes such a radical change in the geography of Europe and so could provide a new motivation for future efforts to integrate Norway into the EEC.

The EEC Energy Market at the Time of the 1973 Energy Crisis

It is clear from the preceding discussion that there were important common components within the energy markets of the member countries of the EEC at the time of the energy crisis. Most significant of all was the dominance of imported oil. There was, in addition, a common residual production and/or use of coal for reasons of social or regional policy and for specific energy uses, most notably in steel and electricity production. Furthermore, the production and/or use of natural gas in some countries, notably the Netherlands, had quickly assumed the main role in determining the share of the energy market which was to remain open to imported oil. During the late 1960s and early 1970s the expectation that the impact of natural gas production and supply on energy markets would be significant led to artificial supply constraints on its development in an effort to protect investments already made in the oil sector[2].

This concern for the impact of low-cost gas has, however, proved to be the final element in the twenty-year period since the mid-1950s of energy abundance in western Europe. Under such circumstances there had been little motivation to plan for the energy sector. Thus, by 1973, the structure of the energy market in Europe owed most – by far – to the results of competition between the international oil companies. In addition, and in part as a response, the structure also owed something to the nationalistic approaches to the coal and natural gas industries and to the initial steps taken in the field of nuclear power development by countries such as the UK and France. Though the level of international integration achieved in the energy sector of the European economy probably exceeded that in any other sector apart from agriculture, on which so high a percentage of official effort had been expended, it is

2. All EEC/OECD forecasts in the 1960s of the rate of development of the natural gas supply in western Europe and of its rate of incorporation into the energy economy underestimated the potential for gas. This was because the forecasts failed to appreciate the dynamic nature of the development of the resource base and worked instead with the very misleading 'proven reserves' concept.

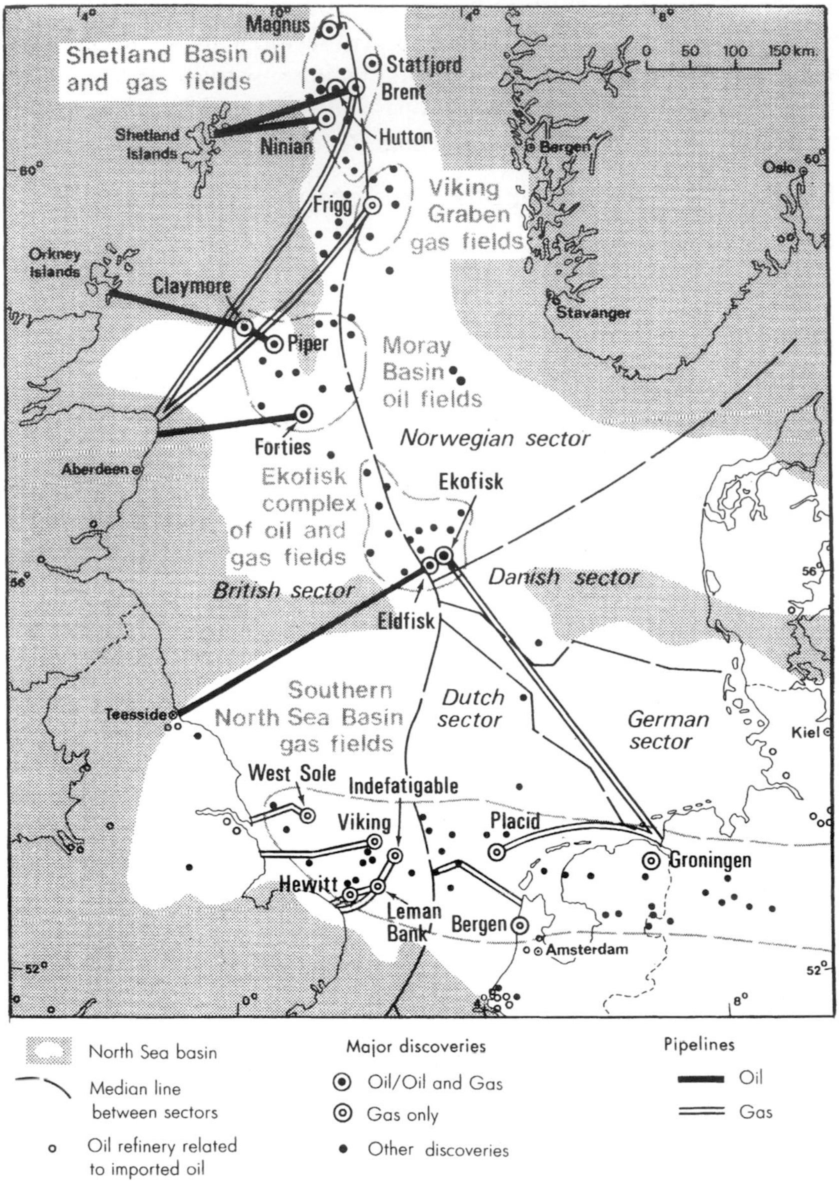

Figure IV-3.4: North Sea oil and gas fields and the pipeline network by 1975

paradoxical that almost none of this integration was a result of governmental efforts. National governments – and the EEC administration – had remained content for more than a decade to opt out of positive involvement in a sector which created no apparent problems for the progress of the west European economy. On the contrary, the falling real price of energy over the period, (see Figure IV-2.1 on p.493), arising out of intensive competition between an increasing number of oil companies, had been one of the region's most important economic advantages. This was especially so as it coincided with a period in which the USA moved into an era of higher-cost energy as a result of its government's efforts to protect indigenous oil against competition from the oil available from international sources. The consequence of this development was to make even more oil available in the open western European market, so intensifying competition between the companies to supply its expanding energy demands.

The Impact of the International Oil Crisis

The international oil system changed fundamentally in 1970 (Odell, 1971 and 1975) but the significance of the change remained largely unrecognized until 1973. The constraints on production imposed by OPEC and its decisions to quadruple the price of oil constituted a traumatic shock to the European economic and political system. In 1976 it remains an open question whether this shock will lead to a greater or a lesser degree of integration in the European energy sector, as the shock has set conflicting processes in train.

On the one hand, each country has intensified its national approach to energy in order to reduce the impact of the crisis. Thus, individual countries are searching for alternative sources of indigenous energy with a view to preserving them for national use and are also seeking a greater degree of security of supply through special arrangements with energy exporting nations. On the other hand, the stimulus of crisis conditions gave the EEC the incentive to start work on an energy policy which would structure the market towards the use of European resources and a revised (downward) estimate of the evolution of the demand for energy.

EEC expectations on the supply side, however, related essentially to the possibility, and the perceived desirability, of a crash programme for nuclear power. This limited view of the options open reflects three considerations: first, the influence of the only effective EEC institution in the energy field, Euratom; second, the remoteness of Brussels from the widespread public opposition to the development of nuclear energy on grounds of safety; and third, the lack of knowledge in the EEC of Europe's indigenous oil and gas potential. The latter reflected the institutional weakness in respect of the oil and gas sectors which, as shown above, had been organised after 1955 by others than officials of the European Commission.

Thus, to date, no post-crisis agreement on the restructuring of the energy market has been achieved at the EEC level, in spite of a pressing need for such a policy in the oil sector. This need arises out of stagnation in the market and from the desirability of reducing the uncertainties of supply through appropriate stock-holding policies. Thus, the reduction in the size of the oil market between 1973 and 1975 (instead of the expected 12–15 per cent increase) and the new forecasts of no or little growth over the next decade has led to a crisis of overcapacity in the oil industry's refining and transportation infrastructure.

In such a situation national interests produce economically irrational decisions when viewed from the European level. This is exemplified by the pressure exerted by the West German government for German refineries to be run at higher throughputs than the more efficient refineries in Rotterdam. Furthermore, strictly national regulations on the stockpiling of oil also lead to economic irrationality, as unnecessary investments are made in new facilities. Such country-by-country calculations of stockpiling needs ignore the possibility of an integrated attempt to use the surplus storage capacity which is available in underutilized tankers and refineries; while new investments are being made in additional storage facilities at a time when the continuing discovery of indigenous reserves of oil and the development of production capacity is rapidly eliminating the need for emergency stock, except for the period up to 1977/78. This particular failure to take the significance of North Sea oil and gas into account is, however, part of the continuing lack of appreciation of the significance of these energy resources for the whole structure of the west European energy market. It is with this aspect that the last section of this chapter is concerned.

A Return to an Indigenous Energy Base for the EEC

The first policy requirement of a return to an indigenous energy base is a decision by the EEC to pursue as high a degree as possible of conventional energy self-sufficiency as a basic strategy for action. The

3. Compare this attitude in western Europe with the attitudes in the USA and the USSR where there is a fundamental and continuing concern for the domestic availability of energy resources.

willingness to take such a decision depends, however, on a knowledge of the resource base. This, however, is not available because of the lack of research into the field of study – the economic geography of resources – which western Europe came to perceive as wholly unimportant, given the belief in the advantages of international trade in energy[3].

Assuming, however, that the option of an autarkic energy policy becomes acceptable for consideration, then in practical terms such a policy can be pursued only at an integrated EEC level. The essential interdependencies in the flows of energy, especially oil and gas, demand such a scale of approach. In addition, the relationships between the EEC and Norway, with its small population, low demand for energy and the largest European potential for oil and gas production in the 1980s, will assume a position of prime importance in the outlook for the structure of the European energy market.

Such an approach to western Europe's future energy economy is obviously also highly interventionist. Moreover, if maximum self-sufficiency is the policy aim, the intervention cannot simply be restricted to the supply side of the energy market. It must also be extended to the demand side, involving all regions and sectors of the economy of the EEC, in order to control the overall use of energy and to ensure that patterns of use reflect the availabilities of alternative sources of energy (Coppock and Sewell, 1976).

Thereafter, the details of the new structure of the energy market can be worked out. One essential element would be the establishment of a 'floor' price for indigenous energy production. Another would be the creation of a 'quota' regime for imported energy, with external supplies regulated either by long term agreements on quantity and price with friendly suppliers, or by an auction-based purchase by external suppliers of rights to sell energy to the EEC. The institutional elements would then be followed by the development of an economically rational infrastructure for the production, processing and distribution facilities necessary to meet the region's energy needs.

Beyond this a longer term aim could be the establishment of systems of energy use, such as 'Coalplexes', in which the combination of coal, oil, natural gas and nuclear power facilities could ensure the optimal use of the inherent energy content of the input fuels, together with integrated electricity/heat systems which eliminate the waste of the heat generated in the production of electricity in conventional power stations (Odell, 1974b). In all such moves towards a rational structure of both the supply of and the demand for energy, the spatial element is an essential component. Attention to these aspects would seem likely to produce the probability of fundamental changes in the geography of the European energy market in the period from the mid-1970s through to the end of the century.

References

M. Chisholm and G. Manners (eds.) *Spatial policy problems of the British economy*, Cambridge University Press, Cambridge, 1971.

J.T. Coppock and W.R.D. Sewell (eds.) *Spatial aspects of public policy*, Pergamon 1976

W.G. Jensen, *Energy in Europe 1945–80*, Foulis, London 1967.

G. Manners, *Geography of Energy*, Hutchinson, London, 1972.

P.R. Odell, *Natural Gas in Western Europe: a case-study in the economic geography of energy resources*, E.F. Bohn, Haarlem, 1969.

P.R. Odell, 'Against an oil cartel', *New Society*, no 437, 11 February 1971, pp.230–2.

P.R. Odell 'Energy and regional development' in I. Bencze and B. Gyala (eds.), *Regional studies, methods and analysis; selected papers of the IGU Conference*, Hungarian Academy of Sciences, Budapest, 1974a.

P.R. Odell, *Energy needs and resources*, Macmillan, Basingstoke, 1974b.

P.R. Odell *Oil and world power*, 4th ed., Penguin, Harmondsworth, 1975.

Chapter IV – 4

British Oil Policy: a Radical Alternative*

A. Introduction

This report seeks to highlight what appear to be the most important issues involved in the continuing development of UK offshore oil. It also makes recommendations for action in respect of strategic and tactical policy questions.

The development of the supply of offshore oil (and of gas), the cost at which it (they) can be produced and the price at which it (they) can be marketed, clearly constitute the main elements which will determine the shape of the energy sector of the UK economy in the 1980s and the 1990s. Oil and gas in general, and British produced oil and gas in particular, are the energy sources preferred by most British energy consumers. Alternative sources of energy such as coal and nuclear power are really only relevant in the context of the need and ability to supplement oil and gas. This is a clearly defined consumer preference. I would suggest that it should not be thwarted by policies which seek to limit the contribution of British oil and gas to the energy economy. This is of immediate importance in light of the very much higher resource costs which are involved directly in developing the supply of alternative energies. It is even more important to bear in mind that there would be considerable additional costs related to required infrastructural and other changes in society which would enable it to absorb very much more

*** Extracts from a report under this title, based on a study made by the author when he was a Special Adviser to the U.K. Secretary of State for Energy between 1977 and 1979. The report was subsequently published by Kogan Page, London in 1980**

energy in forms other than oil and gas. By 1979, all main energy users in the socio-economic system (with the single exception of the electricity generating industry) have become largely oil and gas dependent[1].

Such a firm view of the way in which the British energy economy should evolve simplifies, to some degree, the study of energy policy issues, most notably because it defines where the main effort should be directed. However, it does not, by any means, eliminate all the difficulties facing the country's energy planners. In analysing and defining the role of oil and gas there is a set of geological, technological, economic and political influences at work which serve to create problems of comprehension, let alone of evaluation. This report seeks to expose these problems in a systematic way.

B. External Politico-economic Influences: the Necessary Responses

Current international economic and political influences on oil developments are particularly difficult to analyse and evaluate in a world which is still going through a revolution in the mechanisms for determining the supply and price of oil. Consequently, the opportunities for most effectively developing the oil resources of the UK Continental Shelf have, to date, been judged only in the context of what has become a rather simplistic view of the future of oil, viz, that the commodity will inevitably become scarce and increasingly expensive over the time horizon at which it is necessary to look for planning purposes. For example, in the Department of Energy's *Energy Policy Green Paper (Energy Policy: A consultative document*, HMSO, 1979), we read:

> *'Even the most optimistic forecasts see oil supplies levelling off and then falling before the turn of the century and the most pessimistic sees demand overtaking supply as early as 1983. Perhaps the most likely outcome is that world oil supplies will begin to level off in the late 1980s, reach a peak in the early 1990s and decline thereafter' (pp. 11–12).*

The Green Paper further assumes 'for the purposes of the forecasts, that oil prices will rise gradually to around \$20–25 per barrel

1 This comment is not intended to imply that UK energy sources other than oil and gas should not be developed or encouraged. The output and development of the coal industry, in particular, ought, in my view, to be maintained. This would necessitate a continuing programme of new production ventures and of research and development on the future use of coal. This would, at least, be an insurance policy for the post-1995 period just in case resource-based limitations then make the country's oil and gas-based economy unsustainable.

(in 1977 dollar terms)'. It should be noted, however, that this implies an annual rate of increase in the real price of oil of only three per cent. Under almost any circumstances this is a lower percentage figure than one would have to use to discount the value of any oil reserves which are deliberately kept unproduced for use in the distant future. Thus the assumption does not provide an effective argument for a restrictive approach to the depletion of the UK's offshore oil resources. Nevertheless, it is being used to justify a conservationist approach to oil resources. Don't produce too much too quickly, it is argued, it is better to save oil for an uncertain future – or rather for a future in which oil will inevitably be in short supply[2].

Thus pressures have built up for restricting the rate of output of individual fields. In addition, it is argued that plans for the rate of exploration and development of the rest of the UK Continental Shelf should be cut back. Such unjustified pressures constitute a handicap to the effective exploitation of offshore oil. First, through the adverse effect that the results of such pressures have on the overall attractiveness to the oil companies of expenditure in the UK's oil provinces, relative to their opportunities in many other geological provinces around the world. Second, through the effects that such pressures have on the economics of producing an individual field. They not only have an unfavourable influence on the relationship between the discounted value of the oil and the levels of investment needed to produce it, but they also increase the investment risks. This is because the companies will be unable to use the production and transport installations which they have built in the harsh environmental conditions of the North Sea for longer-term production.

The adverse consequences of a restrictive approach to the exploitation of Britain's offshore oil will not, moreover, be limited to the exploiting oil companies. They will also affect the national interest. This is because the approach can lead to the failure to maximize the present value of revenues from oil production. It can also lead to a limitation on the favourable balance of trade arising from offshore developments. Neither the documents to which I have had access, nor the discussions that I have had within the Department of Energy have convinced me that effective evaluations to test for the impact of these important considerations have yet been made.

2 If oil is going to be in 'short supply', however, then the price over the next 20 years would increase to very much more than the maximum of $25 per barrel specified in the planning assumption. There appears, in other words, to be a serious inconsistency in the government's *Energy Policy* presentation.

This represents only one element of doubt about the validity of the present policy strategy for the development of the country's offshore oil resources. And it may, indeed, be the minor element. An even more important aspect is the relatively high probability (over 50 per cent in my view) that the official interpretation of future developments in the international oil supply and price situation (*viz*, steadily increasing real prices as a result of an inherent scarcity of the commodity) is incorrect.[3] Given a belief in scarcity, then the prospects for UK oil development can be assumed to be independent of what is happening elsewhere in the world. Developments in the international situation will never impinge on the ability of all British oil to find a market at continuously increasing real prices. This conclusion seems to be accepted without question as a basis for policy making. But such an approach is not justified, given the following considerations:

(a) UK oil is already and will continue to be fully integrated into the supply/refining and distribution networks of the major international oil companies. These may not want, or indeed be unable, to absorb all the UK oil they could produce. This depends upon the results of alternative supply, refining and marketing options open to them and on the basis of which they will seek to optimise the returns from their operations at the international level, to the detriment of some national interests.

(b) Sometime before 1985 the major oil exporting countries seem likely to have to take quota or even pricing action to

3 It would make this report too long and take it away from its main theme if I were to advance my alternative views in detail here. I have recently done this in the final chapter of the revised 5th Edition of *Oil and World Power* (Penguin Books, Harmondsworth, 1979). The essence of my argument is that the world's potential oil (and gas) resources are at least twice and possibly up to four times larger than the quantities which are assumed in the 'inherent scarcity' hypothesis. Moreoever, as the demand for oil is now growing much more slowly (at less than 1 per cent per annum instead of at 7–8 per cent), even a less optimistic view of future oil resources would enable the industry to continue to expand well into the second quarter of the 21st century. Current constraints on oil resources developments are thus nothing to do with the ultimate size of the resource base. Rather they are technico-economic (in respect of the difficulties associated with working in new habitats for oil) and institutional (in respect of the unwillingness and/or inability of the oil companies to operate in most parts of the world where potential resources remain undeveloped). However, both sets of constraints are capable of being removed, thereby opening up the prospects for plentiful supplies of oil over the whole period during which the world's economic systems are likely to want to use oil in increasing quantities: a period which one may assume will be terminated sometime in the first half of the 21st century when alternative energies become available on a large scale and make oil a less economic source of energy.

ensure that their exports secure preference in markets which are weakening, because of the over-supply of oil in a severely constrained demand situation. The Western European oil market seems most likely to be approached in this way by member countries of OPEC. Given that the demand for oil in Western Europe now seems unlikely to grow very much, it is the area in which competition from British and other non-OPEC oil is likely to be felt most strongly and in which many OPEC countries are building up a high degree of inter-dependence with Western European countries (in trade, commerce, finance investments etc). This inter-dependence is being developed so that the OPEC countries will be able to use those relationships to ensure that they can achieve their oil export plans. European countries will thus be susceptible to pressures to take specified quantities of oil from the OPEC countries on which significant sectors of European industry and commerce now depend.

(c) The UK appears, so far, to have made no efforts to ensure guaranteed long-term outlets for its own rapidly increasing oil production. The oil companies do not seem to be required to do this and BNOC appears not to want to do it.

(d) Although it is not the most likely development, there is, nevertheless, a chance of an international oil market collapse in the later 1980s – the period with which UK oil policy is now most concerned. This arises from the prospects for important new sources of supply (including the North Sea, Mexico, China and North America) in a situation of extremely slow growth in the demand for oil – partly as a consequence of deliberate conservation measures and partly from low rates of economic development. As a result an OPEC member may become unwilling or be unable to stick with the organisation's minimum oil price decisions. If this happens it will place at risk a part of Britain's existing capacity to produce oil. Some UK oil would not be worth producing compared with alternative supplies available. This would reduce the value of the oil that is produced, and would have both balance of payments and government revenue implications. It would also undermine most of the potential for continuing exploration and development activities on the so far undeveloped and/or unexplored parts of the UK Continental Shelf. Such exploration and development would be seen as too high cost and/or have too high a risk potential to make the necessary investment worthwhile.

There thus seems to be the need for a study of the relationships between various possibilities for the future international oil situation and the prospects for the continued, successful exploitation of the oil resources of the UK Continental Shelf. The objections of such a study would include:

a) evaluating the alternative options open to the UK in its offshore oil development in the light of contrasting international developments in the world oil market;

b) indicating ways of protecting British oil wealth *against* a possible deterioration in the world oil price in general, and/or attempts by other oil-producing countries to 'corner' part of the markets which are important for UK oil;

c) suggesting mechanisms to ensure that future decisions on how much British oil to produce are not made by the producing companies solely to optimise the overall returns on their operations at the international level: how, and how far, in other words can decisions on British oil development be divorced from internationally orientated corporate decisions?

d) indicating how the *use* of UK oil production (in the international market) can optimise British, rather than individual company, interests and/or the interests of the countries in which the companies have their headquarters.

There is, in other words, a requirement for positive UK strategic oil planning, so that the country is not simply obliged to react, let alone to bow, to external pressures. Instead, UK oil policy should aim to be influential in helping to determine the global oil situation for the next critical 10 to 15 years. Current energy policy is based on the belief that there is little the UK can do to affect the world price of oil, or the development of the international oil situation. This is not only defeatist:

it is also untrue. The UK needs a policy which ensures its effective control over some of the world's most significant oil reserves, in terms of their location in the centre of the world's most energy-intensive using region and in terms of their occurrence in a region which is politically stable. An appropriate strategy could be of crucial importance in helping to stabilise the international oil situation. Moreover, in using its oil power in this way the UK could also help to ensure its own well-being in an uncertain world.

C. The National Politico-economic Environment and the Exploitation of Offshore Oil Resources

The results of the traditional concession system

(a) Over the last 10 years new and expanded legislation and an increasing number of regulations have served to 'tighten up' the freedoms originally enjoyed by the concessionary companies in respect of exploration, field development and, of course, the tax obligations of the successful companies. I do not propose to define what was 'wrong' with the system to start with, or even to analyse the effects that individual changes in the concession and the tax arrangements have had on the system over the years. Such exercises have been undertaken and there is a high degree of awareness of the results.

(b) An examination of the evolution of the system and of its present state persuades me, however, that the government still lacks the essential means to initiate and to guide developments once concessions for exploration have been granted to individual companies or consortia. In essence, we appear to have arrived at the point where the government can – and, indeed, sometimes does – stop things happening, but it is still largely unable to cause things to happen. We thus have an essentially negative government approach to the development of oil on the continental shelf.[4] This is the case in respect of several different aspects of the concession system as it has evolved to date. These include:

4 For the moment I am discounting the significance of the British National Oil Corporation (and the British Gas Corporation) in this respect. Their significance – and the limitations on their importance – are dealt with later in the report.

i) *The work programme and the speed of its implementation on the concessions granted*
In theory, the government, on granting a licence, requires a specified work programme to be undertaken. In practice, this cannot be enforced as a company unwilling to complete a programme may simply hand back a block. Thus, 38 'obligatory' wells from the Fourth Round of Licences have not been drilled.

A failure to drill a well sometimes appears to be accepted by the government because 'new geological evidence' shows that further exploration would not, according to the company, be worthwhile. But such decisions run counter to the requirements of the licensing procedures. A company secured a particular block partly because it promised a specific work programme. It thus ought not to be permitted to withdraw without, at least, paying a penalty equal to the investment it has not made; nor should it be allowed to retain an interest in any part of a license for which the originally required work programme has not been fully completed. The British North Sea and the rest of the country's continental shelf is, like the rest of the oil-producing world, full of surprises, and a well not drilled is a penalty to the nation. It has been deprived of information from the drilling and of the expenditure on the project, with an adverse effect on jobs and income – both directly and indirectly.

Moreover, if 'disappointing new geological information' enables a company to withdraw from an obligation, then "encouraging additional geological information" in respect of another block, should give the government the right to stipulate immediately that wells should be drilled additional to those to which a company committed itself under its work programme obligations. This should be irrespective of the company's attitude, which may not necessarily be positive, in the light of commitments and opportunities in other blocks in which it has an interest. If the company declines, then the concession for the block should be withdrawn, on payment of compensation to the company equal only to the cost of the work it has done on the block.

ii) *The question of the initiation of developments in respect of successful discovery wells*
The government, in essence, still appears simply to sit and wait for a company to take the initiative on field development. The company's decision depends not only on the inherent value of the discovery itself, but also on what else the company has under way or under consideration on the UK Continental Shelf and, indeed, what else it has under way in the rest of its international corporate world. But concession agreements made under the 1975 Petroleum Act include the right of the government to order a development, if it judges this to be desirable from the national point of view. A failure by the company to agree to develop, following such a government decision, means that the company's interest in all or part of the licence can be revoked.

A large number of discoveries are not being developed, possibly because of the uncertain economic viability of new field developments from the companies' points of view, or because the companies concerned prefer to concentrate their activities elsewhere – either within, or worse still, outside the British sector. In view of this one wonders why the government has not yet chosen to exercise its rights in respect of ordering development. This would be a more appropriate means of eliminating the risk of a possible decline in production after the mid-1980s than the means proposed by the government – that of depletion controls on fields already in production.

iii) *The nature of the development programme for a field*
Government involvement here to date seems to have been mainly restricted to a go/no-go response: to questions of timing, and to issues of an infrastructural nature (eg. decisions on how best to land the oil from a particular field). This is plainly negative, rather than positive, government.

It is a problem which will, though only in part, be resolved by much increased state participation (through BGC and BNOC). However, even in respect of largely autonomous state corporations, 'prodding' by government may be necessary to secure a field development which is considered desirable from the national and/or regional point of view. This may be seen, for example, in respect of the long delayed development of the Morecambe Bay gasfield. In this case the BGC's marketing policy has been in conflict with the economic needs of the depressed Liverpool and north-west England regions. Their economies would have benefited, both directly and through local multiplier effects, by a decision to develop the field earlier and, even more, by a decision to use the newly available gas to stimulate investment interest in the region – as in the case of Groningen in the Netherlands.

There are several specific reasons why the Department of Energy should be positively involved in development decisions. These include the need to ensure that the whole, rather than a limited part, of a reservoir is developed; and/or that all adjacent and overlapping, but geologically separate, pools are similarly incorporated into a comprehensive development plan; and to make sure that the development plan chosen (it can be one of many considered for a particular field) is the one likely to make the maximum possible contribution to government revenues and/or foreign exchange earnings (when measured in present value terms). The greater the positive government involvement in the long-term development decision for a field, the greater the government's influence in ensuring that the plan is achieved. It also diminishes the degree of company uncertainty over its future prospects from its exploitation of the field. This seems likely to produce a better overall environment for effective government/company co-operation.

iv) *The possible integration of field transport systems with one another*

Steps towards achieving such integration between transport systems in appropriate locational relationships appear not to have been taken by the Department of Energy to date. Such action is possible under the 1975 Petroleum Act so that there are examples in the North Sea where the failure to integrate the separate interests and operations of two or more groups has led to decisions which appear to be sub-optimal from the national point of view. This arises because field developments have been delayed and/or limited either because 'space' could not be obtained in a separately developed pipeline at a price which made field development possible, or because companies, with contrasting interests in exploiting near adjacent newly discovered reserves, were unable to agree on the joint transport system developments which were a prerequisite for the achievement of the necessary economics of scale.

In order to avoid these situations the Department of Energy needs to exercise its legal right to investigate such situations as a matter of course and, where appropriate, to order joint enterprises. This would make a positive contribution to the evolution of the UK's oil production and transport systems.

(v) *The disposal of British oil to third parties (as purchasers of the crude), to refineries and to other countries*
Existing legislation requires British oil to be landed in the UK (except by special permission of the Secretary of State). In addition, there is an 'expectation' by the government (which is not enforceable in a legal sense under present legislation) that up to two-thirds of UK-produced oil be refined in Britain. The 'control' exercised by this expectation is essentially voluntary and so provides a very modest, if indeed any, constraint on companies' decisions on where to refine their British sector North Sea oil production. Moreover, as far as the UK landing requirement is concerned, it neither controls *where* it is landed (except by means of physical planning controls), nor what happens to it once it has been landed. In other words, there is no rationally evolved and comprehensive government-regulated integrated 'downstream' system for the disposal of oil from the UK Continental Shelf to ensure that the national interest is taken into account. Simply leaving BNOC to take care of this is inadequate. It handles only part of the oil involved and it needs the Secretary of State's permission to undertake any oil refining. It is, therefore, limited to trading crude oil, rather than having the greater flexibility of an integrated oil company.

(c) Even after taking account of the UK Continental Shelf holdings of BNOC and the BGC and of the impact of the negotiated participation agreements, the concession system as developed to date means that the ownership of a large part of the oil being produced – and to be produced in the future – lies, and will continue to lie, with the companies concerned. And this, of course, has consequences in respect of both oil pricing and tax calculations – as follows:

(i) Oil pricing is a key element in determining tax liability. This is because the price declared (received) for the oil produced determines the value on which PRT calculations are based. It also helps to determine corporation tax liability. However, the nature of the world oil market is such that unless oil is sold at 'arms's length', its value is not an objectively determinable element.[5] The market for most internationally traded oil is essentially one of negotiated transfer prices (between one company and another of the same group and between companies and government in respect of most OPEC, and some non-OPEC, oil-producing countries), in which the value of a barrel of oil to one company is different from its value to a second or a third company. Similarly, the value attached to a barrel of oil is different as between companies and governments – depending on factors of overall supply/demand situations within corporate and other entities and in the light of the infrastructural contrasts and the different market openings to the various parties. In this sort of situation, almost all governments depending on, or expecting, revenues from oil production have found it necessary to abandon regulations which required, in effect, that each barrel of oil be separately valued, by agreement between the company concerned and the government or its fiscal agency. In order to safeguard revenues such governments found that it was essential to define, unilaterally, a set of tax reference prices as the sole means of valuing the oil produced. Such tax reference pricing is required in respect of the UK's oil production in

5 For a detailed discussion of the point, see 'The Oil Companies and the New World Oil Market' in the 1978 *Yearbook of World Affairs*, University of London, Institute of World Affairs, London, 1978.

order to replace the valuation procedure as set out in the Oil Taxation Act of 1975. Although this Act allows the Inland Revenue to substitute its assessment of market value for that declared by the licensee in respect of sales between affiliates, this can be challenged on appeal by the company concerned. And this creates a situation which must serve to inhibit the valuation assessment by the Inland Revenue.[6]

(ii) The contrasting value of certain quantities and qualities of oil to different parties at specific times and specific places creates a possible 'conflict' between the best interests of the government and the commercial interests of BNOC. The latter, in order to enhance its potential profitability, must seek as low a price as possible from the companies for its purchases of participation oil, as well as for oil it purchases from elsewhere, in a situation in which it is acting as a crude oil trader with oil bought from other parts of the world. Such 'low' price levels could then be used by the oil companies in support of claims for low valuations of their North Sea production. This could be to the detriment of the government's interests in terms of maximising its tax take. The institution of a regulated tax reference price system for offshore oil production would, of course, also serve to eliminate this potential conflict between the government and the state oil corporation. It would, in addition, eliminate the latter's potential embarrassment in possibly having to modify its efforts to secure a 'bargain' because of the effect this could have on government interests.

6 Note that a difference of $1 per barrel between a higher tax reference price and a lower so-called market (transfer) price will, at end-1979 rates of North Sea production, make up to £375 million per year difference to the UK's balance of trade figures and up to £300 million difference to government revenues (given that the high cash flow from higher revenues will comprise mainly additionally collectable taxes). A difference of $1 per barrel is by no means high: in a dispute between the US government and some major oil companies over the costs of moving oil via the trans-Alaska pipeline, the difference between the parties was over $4 per barrel. I have heard the argument, but have not been persuaded by it, that a tax reference price for British oil would upset our tax agreement with the US. The US Treasury has accepted tax reference prices for oil in other countries for many years and I see no inherent reason why the UK should be exceptional. However, given the large and increasing amounts of export earnings/revenues involved, even if there is an exceptional situation, the present method whereby the value of oil is determined for tax purposes seems worthy of examination for changes which would be advantageous to the UK.

(iii) The evaluation and the incidence of costs is, of course, also critical for the tax calculations – as recognised in the Oil Taxation Act of 1975 and its Schedules. Yet, we do not, and indeed cannot, know if it is being monitored/controlled effectively. The question of allowable costs, in respect of specific operations for a specific oil company, is a confidential one for determination by the Inland Revenue alone. This is done on the basis of the general rubric whereby secrecy for individuals' tax returns is maintained. Is it reasonable, in view of the very large revenues expected by the government from a limited number of companies specifically allocated rights to exploit part of the nation's oil resources, and in the context of a highly complex technico-economic situation with many international aspects, that the tax assessment should be secret? Given the nature of the operations and the very few companies involved, ought not the process of tax assessment be open?

In his 1977 Energy Plan, President Carter insisted that there must be openness in the evaluation of the oil companies' responsibility for paying taxes. He added that this must be done in respect of each individual operation by a company and that every such operation (ie. each producing field and each refinery, etc.) should be treated separately for tax purposes. The United States has a relatively much lower degree of financial dependence on the oil tax-take and a much higher degree of domicility of the oil companies concerned within the country. It also has a much longer experience of trying to tax oil companies effectively. In view of this background, the fact that it considers that oil companies must be subject to special rules over tax assessments emphasises the need for such a requirement in the case of the UK.[7] This general proposition may be supplemented by two specific needs in respect of

7 The importance of the requirement is emphasised by the current high degree of uncertainty concerning future oil revenues. Although the widely varying views on future revenues (for example, estimates for the year 1983–84 range from £4,750 million to £8,250 million) relate, in part, to uncertainty over levels of both prices and production, they are also a function of contrasting expectations on the degree to which the companies will actually have to pay PRT and Corporation Tax. The higher estimates fail to recognise the opportunities which the present system gives to tax minimisation procedures by the companies.

monitoring/control over the development of continental shelf oil under the existing pattern of concession/ development legislation, *viz*:

First, for each oilfield development plan the companies should be required to indicate capital and operational costs and their expectations of inflation in respect of both. How, and to what degree, the companies' expectations in these respects are related to the expenses they ultimately claim year by year against their petroleum Revenue Tax (PRT) obligations on individual fields should also be known.

Second, PRT was made chargeable on a field by field basis to prevent the obligation 'disappearing' into operations elsewhere in the offshore area. This has not been done in respect of Corporation Tax which may be payable by a company in respect of its earnings on a particular field. Thus, a company with a near-future obligation to pay Corporation Tax can avoid doing so by making an investment elsewhere in the UK offshore region at the appropriate time. Whilst this is in keeping with the normal rules for calculating any company's liability for Corporation Tax and whilst, in respect of offshore oil operations, it may be a 'useful' device in order to get new fields developed so as to provide continuity in oil flow later in the century, is it reasonable that this should be achieved at the expense of taxes from offshore operations that would otherwise be payable in the short term? This not only discriminates against other existing taxpayers in general, it also discriminates, in particular, against those oil companies which did not make an early strike of developable proportions. And this in itself was not a matter of luck or competence – it emerged out of the preferences shown to certain companies in the early discretionary awards of blocks in the North Sea!

(d) The inherent inflexibility in the application of the concession system, as it has been developed, and the accompanying highly complex associated system of taxation, seem to have produced an unwillingness and/or inability on the part of the companies which have made 'small' finds in the North Sea to develop them. But the designation of a small field as one with less than 100 million barrels or even one with less than 200 million barrels is absurd. Elsewhere in the world of oil, such fields are considered as 'moderate' or even of 'large' size and are eagerly developed. Compare the reaction of the companies to fields of this size in the UK offshore to the recent enthusiastic reaction by Exxon to its discovery of a very modest field (by North Sea standards), estimated to contain some 50 million barrels, in the US Gulf of Mexico. The field, moreover, has been discovered in no less than 1,200 feet of water on the continental slope so it can hardly be said to be in a friendly environment. Indeed, this depth of water compares with 600 feet or less in the North Sea. This offsets, in part at least, the latter region's harsher meteorological and water conditions. Yet, in spite of this and in spite of the fact that the cost to Exxon of developing the field concerned is front-end loaded by the $71 million which it had to pay for the small concession (small, that is, by North Sea block standards), the company moved with speed towards its development. Dozens – perhaps even hundreds – of discoveries of this size have been, or can be expected to be found on the UK Continental Shelf. If they are not developed, then the country's oil potential will be seriously diminished and there will be consequential severe economic disadvantages.

If the nature of the concession and its associated taxation system, as it has evolved to date for Britain's offshore resources, eliminates the possibilities of the development of this oil, as seems to be the case (in spite of the tax concessions which have been or which could be extended to small fields under existing legislation), then the validity of the system, as an appropriate tool for the continental shelf's exploitation, must be seriously in doubt.

(e) In other words, there remain major uncertainties over the degree and the speed of the future exploration and exploitation effort on the UK's Continental Shelf. The way in which the system works also makes it impossible to forecast the overall impact of offshore oil and gas developments on the British economy and society in general, and on specific geographical areas in particular. This is largely because of the incalculable returns from the development in respect, specifically, of government revenues from the taxation system. Continuing modifications to the system, emerging out of the realisation that it is not producing the benefits which were expected, seem likely to fail to make an essential difference to the situation and the outlook.[8]

My recommendation, therefore, is that this conventional, concession-style approach should be abandoned as the basis for organising the exploration and exploitation of the rest of the UK Continental Shelf's oil and gas resources: in the same way as it has been abandoned in every other major oil-producing country outside North America (where special conditions, especially the prior payment of bonuses for concessions, apply). It should be substituted by the establishment of a new type of politico-economic environment for the exploration and exploitation of the country's remaining offshore oil resources. In case this should not be thought to be worthwhile, because so much has already been done in the North Sea, it should be noted that most of the UK's Continental Shelf and adjacent areas

8 In addition to the points made above, there is an additional one which can now be made in light of the decision in 1979 to increase the rate of PRT from 45 to 60 per cent and to reduce from 175 per cent to 135 per cent the 'value' of capital investment in a field's development to set against PRT calculations. We have re-run those parts of our computer programmes relating to the development of the Forties, Piper and Montrose fields (see Odell and Rosing, *The Optimal Developments of North Sea Oilfields*, Kogan Page, London 1976) in which we measure the NPV of government tax-take in the systems as developed, and find that the improvements in government revenues arising from these apparently very significant changes in the tax regime is less than 12 per cent over the life of the fields. And this 'increase' is dependent on the assumption that all Corporation Tax, due from each field's development, will be paid and not postponed, perhaps indefinitely, as a result of the investment of all or part of the profits from a specific field's development elsewhere in the North Sea. Thus, it seems that even apparently formidable changes in the tax rules under the existing system do not make striking difference to the expectations. This is true even in respect of fields already under development and of the investment to which, therefore, the companies concerned have committed themselves as a result of a decision based on previous, lower rates of tax. With the higher tax rates it is likely that a less intensive development plan for a particular field would have been chosen in order that the company could maintain its required rate of return on investment. This would have served to reduce the level of government revenues from the fields concerned. This is also a danger in respect of new developments, given the recent 1980 budgetary decision to increase PRT yet again, this time to 70 per cent.

of the continental slope have yet to be explored. (See Figure IV-4.1). Thus, even if the remaining exploration is only modestly successful, the future development of British oil will take place over a period several times longer than the decade and a half of development we have had to date. It is therefore not too late to consider alternative systems which seem inherently more likely to ensure both the full exploration of the resources and appropriate benefits to the country from the continued development of offshore oil and gas.

D. An Alternative Environment for the Exploitation of the rest of the UK's Offshore Oil and Gas Resources

1. The Auction System

a) Though this is also a system based on concessions it is, nevertheless, one which is essentially different from the discretionary system. The market mechanism (expressed through the opportunity which potential exploration companies are given to evaluate the worth of a particular area) is the regulator of developments and the means whereby the state attempts to collect – in advance – the economic rent which the companies bidding for a concession expect to emerge from the exploitation of the reserves.

The arguments in favour of the auction system have been presented generally in a recent publication by Professor K W Dam[9] and its application to the UK Continental Shelf has recently been recommended by Professor C Robinson and Dr J Morgan[10] and in a paper published by the Bow Group of the Conservative Party.[11] It is thus conceivable that the auction system could be chosen as an alternative basis for future UK policy towards oil exploration and development.

9 See K W Dam, *Oil Resources: Who gets what how?*, University of Chicago Press, Chicago, 1976.

10 In C Robinson and J Morgan, *North Sea Oil in the Future*, MacMillan for the Trade Policy Research Centre, London, 1978.

11 P Lilley, *North Sea Giveaway: The Case for Auctioning North Sea Oil Licences, A Bow Paper*, Bow Publications, London, 1980

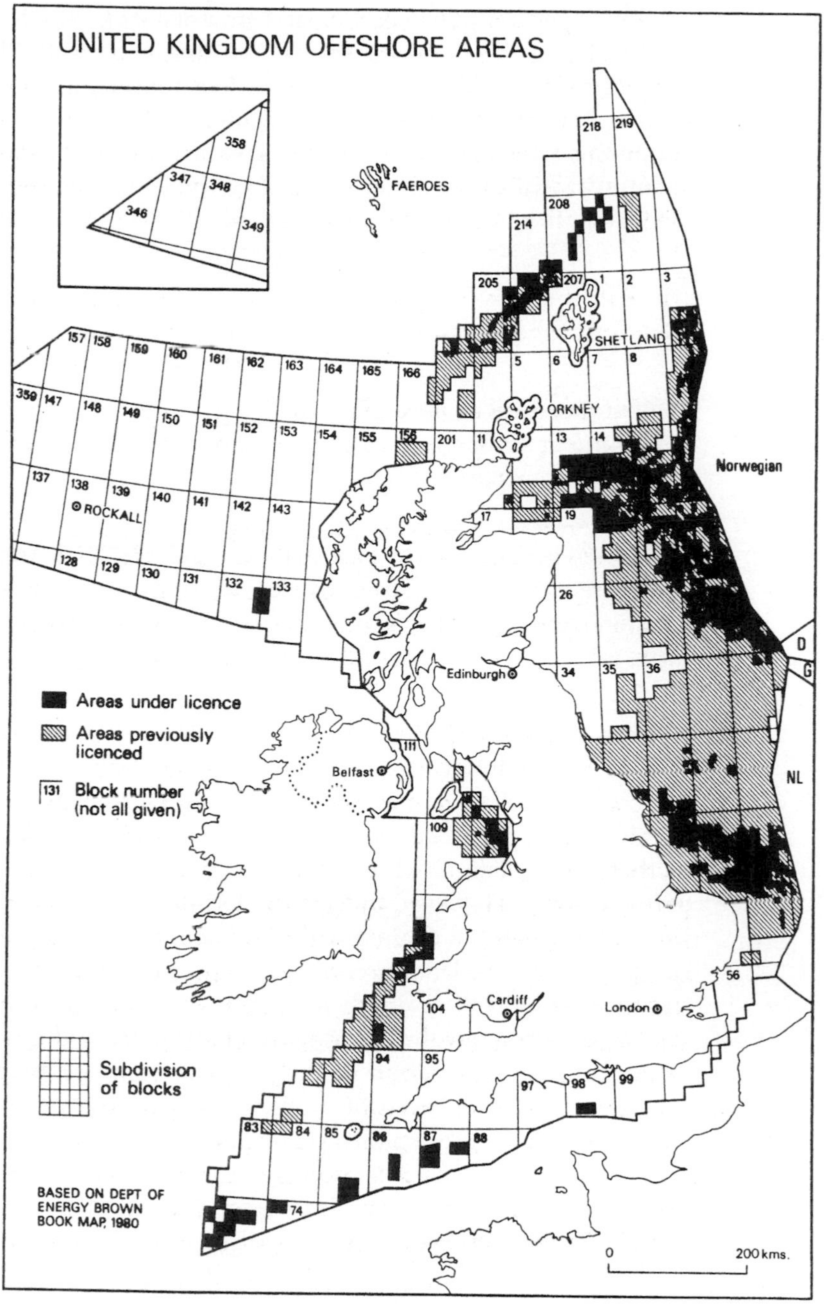

Figure IV-4.1: United Kingdom Offshore Areas, 1980

b) I am not persuaded of its economic and politico-economic validity in general and, in particular, I am not convinced that it is appropriate in the case of the UK Continental Shelf, if the government's wish to maximize its return from offshore oil and gas developments is to be realised. There are two conditions which are necessary to make it possible for the auction system to secure all or most of the economic rent obtainable from oil and gas exploitation, *viz*:

(i) that there is effective competition between many applicants for concessions in the auctioning process;

(ii) that there will be no significant changes in the economic environment subsequent to the allocatory decisions made through the auction.

c) I have presented my conclusions on these – and related – issues elsewhere.[12] In brief, however, my view is that these necessary conditions are not, and indeed cannot, be met in respect of the further development of the UK Continental Shelf.

(i) Even in the United States it is only with some difficulty that the oil industry has been kept on the straight and narrow path of competition for exploration rights. This has involved complex Federal and State legislation and a plethora of regulations which have been evolved over a long period of time. However, outside the US there is a general lack of experience in dealing with the mainly American oil companies and the latter are, as a consequence, in a much stronger position to assert their control or influence over any objective, essentially laissez-faire, allocatory procedures. Given that the completion of these allocatory procedures then leaves the state bereft of influence over what happens next, in respect of the exploitation of any oil discovered, the auction system seems inappropriate for the UK, where efforts over the last ten years have been devoted mainly to evolving a post-discovery set of regulatory conditions. All

12 These can be found in a review article I wrote on Professor Dam's book (op. cit.) in *Energy Policy*, Vol 5, No 3, September 1977, pp 256–7

this experience would be wasted and a new start in regulating the allocatory process would have to be made if an auction system were not to be adopted.

(ii) In the US, the oil and gas industry has been generally isolated from the conditions which apply in the rest of the world – or, rather, in the world of the international oil companies. In particular, companies bidding for concessions in the US can be very confident, not only that they will have a market for any oil found, but also a market at what is virtually a guaranteed price. Thus companies with their prior knowledge of these important economic variables can more easily determine what it is reasonable to bid for a concession. Elsewhere, however, including the UK, oil markets have been and remain open to competition from international oil, so that companies are only able to bid a low figure for a concession – in order to minimise their risk[13]. Indeed, outside the US, the auction system seems unlikely to be able to secure for any one country a set of auction bids which is even as high as the world-wide average level of economic rent expected from all new developments by the companies.

In the case of the UK, the only reasonable auction bids which could have been made by the companies before 1974 would not have secured for the government any part of the much higher economic rent which has emerged out of the increases in the price of oil since then. In such circumstances an auction system would soon have produced bad company/government relationship.[14] This remains the case, given that the world-wide uncertainty over the supply and price of oil continues. Thus, the apparently simple and self-regulating 'auction system' cannot be expected to work effectively for an equitable and effective development of the rest of the oil resources on the UK Continental Shelf.

13 Note that the bids received for a few British blocks which were auctioned in one of the earlier rounds were derisory compared with bids made for acreage on the US Continental Shelf by the same companies.

14 This has even been the case recently in the US where 'old oil' (ie oil from old concessions) has become so profitable for the companies, in spite of lower oil prices in the US, that the government has had to introduce a special petroleum windfall profits tax – and, in so doing, has produced the worst oil company/government relations in the industrialised world!

2. A discretionary system of concession allocations – with high royalties

(a) The lower than expected revenues to the government from the development of the North Sea oilfields which are already in production, appear to arise from the ability of the companies to adjust their policies over prices and over the allocation of costs, so as to minimise their liabilities to PRT and Corporation Tax. In view of this, and if it were thought desirable to maintain the essential framework of the concessionary system as it has been developed, then there is much to be said for calculating the approximate royalty rate which is equivalent to the 70 per cent rate of PRT, and for charging that rate of royalty on production (additional, of course, to the 12½ per cent royalty already levied). This would provide the government with a much more easily calculable – and an inherently less avoidable – level of tax income. The ease of the calculations and the non-avoidability of the amount calculated by the government as due from the companies, would be enhanced if this approach were also combined with a tax reference price valuation for all oil produced.

(b) A royalty rate of, say, 50 per cent (inclusive of the existing 12½ per cent royalty) on the value of the crude produced is one which would give the government 75 per cent of the total profits earned from a field, if it is assumed that costs account for about 33 per cent of the cash flow. What the application of a single rate of royalty would not do, of course, is to distinguish between fields and parts of fields in terms of the variable costs of their development. The result would be that some companies would restrict their developments to the lowest costs fields and to the lowest cost development systems on larger fields, and so still make 'too much' profit. Meanwhile, other companies would risk earning too few profits if they tried to maximize the production of oil by making investments in production facilities with a low productivity. The system would also fail to allow for the way in which costs, relative to prices, change over time and so penalise those companies with a high cost-to-price relationship, whilst favouring those in the reverse position.

(c) The province of Alberta in Canada is one major producing area in which a high royalty rate has been introduced as the means whereby the government collects the bulk of its income from oil exploration and development. The new royalties are basically much higher than they were prior to 1974 (up to 44.33 per cent compared with a maximum of 25 per cent previously). A system has now been introduced whereby the actual rate varies from month to month and from well to well and also according to whether the oil is 'old' or 'new'. The new complex scale starts as low as 8.866 per cent and can, as indicated above, be up to five times as great.[15]

(d) The Alberta formula, involving a well-by-well calculation, could not be directly applied to an offshore area like the UK Continental Shelf where the producing platform and its associated wells, rather than individual wells, is the basic operational unit. It does, however, show what is achievable through a royalty system. Its relevance to a UK situation is certainly worthy of more detailed examination, particularly in respect of the areas which have already been allocated and where it would be impossible to change fundamentally other conditions of development.

(e) It must be noted, however, that there are several reservations arising from this approach to state involvement in oil exploitation. The state certainly achieves efficiency in collecting its revenues, but the system (except in respect of the initial decision to offer concessions) does not imply much else by way of government involvement in the development. In particular, in the case of the UK Continental Shelf developments, it would pose the question of the status of BNOC, in that a high royalty system is not really compatible with direct state involvement in the exploration and development process. If the state entity had to pay the much higher royalties required then it could be inhibited from undertaking nationally desirable activities

15 See Michael Crommelin, 'Government Management of Oil and Gas in Alberts', *Alberta Law Review*, Vol XIII, No 1, 1975, pp 146–211

such as the 'non-commercial' search for additional reserves. If it didn't pay them, then the charge of unfair competition from the private companies would clearly be justified. It also leaves the question of a private company's liability to corporation tax unresolved[16] and one must recognise that, except when the royalty is taken in kind, the system still requires attention in respect of the oil valuation question.[17]

3. The inadequacies of the concession system – even with a state oil company

(a) The nub of the criticism of the concession system and its derivatives lies in the style of relationship between companies and government which it necessarily implies. It is basically a system in which all the positive initiative – except the granting of the concessions in the first place – lies with the companies, to whose actions the government can respond only in a neutral or a negative way. It is, moreover, a system in which the state's collection of revenues from the exploiters of the resources is seen as a 'burden' on the company. It is, in other words, a system which, in politico-economic terms, hardly seems compatible with the concept of the national ownership of a country's oil and gas resources.

(b) The relationship does, not, moreover, change over time though a growing number of regulations certainly serve partially to block loopholes through which potential revenues leak away. However, it also creates an environment in which the concessionary companies continually react against the more stringent conditions by finding new ways of avoiding the worst of their consequences. Thus, in the final analysis, the government does not make the progress it expected to make in controlling the system. Nor does it benefit from sharing in the profits of oil production. It is, in essence, a system of thwarted expectations for the

16 In the case of Alberta there is no corporation tax as this is not a Canadian provincial tax right. There is, however, a Federal Tax on oil companies' profits (as on company profits in general). There seems to be no inherent reason why the royalty and the corporation tax should not be levied in a complementary way by the same authority.

17 The price of oil month by month is one of the variables written into the formula for calculating the monthly basic royalty rates on Albertan oil.

government – and of equally thwarted efforts for the companies which often cannot make the progress they expected, because of the suspicions to which they become subjected by the very workings of the system.

(c) This necessarily unsatisfactory outcome of the concession approach to the development of oil and gas resources has already been partially recognised in the UK. The solution sought has been that of the creation of the British National Oil Corporation which, under legislation introduced the recent Labour government, had to be involved in all new concessions, with a minimum equity interest of 51 per cent. This, of course, gave it a controlling influence over exploration and development, with the intention of providing a counter-balance to the problem of the lack of positive direct government involvement in the concessions. It provides a solution to the problem, however, only insofar as one assumes that the interests of the state company and those of the state remain identical. Given that BNOC is required to act 'commercially' when in partnership with private oil companies, this assumption does not seem to be justified. If not, then the mere creation of a state company to work alongside the private companies does not in itself provide the whole answer to the problem posed for the state by the concession system.[18]

(d) Moreover, in one very important way the existence of a single state entity like BNOC, with an interest in all the concessions, acts as a barrier to the most effective development possible in a large, varied and geologically complex set of opportunities for oil and gas exploration and exploitation as offered by the UK's large offshore areas of continental shelf and slope. This does not have anything to do with the level of technical competence of the state entity or with the arguments on state v. private enterprise approaches to oil development. It arises in a much more simple way, *viz* from the fact that a single interpretation must prevail in any

18 This does not imply, of course, that there are no ways in which a state oil company modifies the states' relationship with the oil companies in a concessionary system. Of particular importance is the 'information' aspect, whereby more effective regulations can be achieved.

organisation about the significance of geological, geophysical and/or drilling information for a region's oil and gas development potential. However, given the existence of many schools of thought on the chances for oil and gas occurrence in respect of all the many different kinds of habitats for hydrocarbons, there is a danger that many good opportunities will be missed because of the limited range of views which any single organisation (no matter how big) can embrace. Indeed, given the necessary hierarchical structure of any exploration and production division in an oil company (state or private) there is likely to be a well-defined, accepted 'company view' of what is, or is not, worth pursuing. As a result, many quite reasonable opportunities will be ignored or dismissed by any single entity and the development of the total oil resource base will be less than complete.[19]

Thus an organisational structure for oil and gas exploration whereby this danger can be avoided, through a 'plurality' of exploration and development efforts, is a 'must' for a system within which there is direct state involvement in the exploration and development system. This suggests that the stage achieved to date in the institutional approach to the development of oil in the UK's offshore areas ought not to be viewed as other than an interim one. The full and profitable development of oil and gas resources depends not only on the replacement of the concession system, but also on a new form of state/oil companies' co-operation which is better able to ensure that the geological complexities of the vast offshore areas in the North Sea and elsewhere on the UK Continental Shelf and slope which remain to be explored, will indeed be tackled.

19 There is evidence that this has already happened with BNOC. Following the views of its chief geological adviser, BNOC appears to 'believe' that all the large North Sea fields have been found and that two-thirds or more of the region's oil resources have been discovered. This attitude must, of course, be a dominant influence in BNOC's exploration/development policy and, as BNOC has been required to be the majority shareholder in all recent concessions, it must also be a dominant influence in the future exploration history of the province. Yet its view of the potential for further North Sea development is by no means generally accepted. Other companies think there are many large fields and extensive resources still to be discovered (eg Phillips). With a single state company view as the only one possible, alternative ideas on the size and nature of the resource base risk never being put to the test – to the country's disadvantage, not simply in respect of its oil sector, but also in more general politico-economic terms.

4. An alternative institutional arrangement: a production sharing system involving joint venture operations

(a) Any part of the continental shelf and slope (involving either one block or contiguous blocks) should be open for a bid for an exploration/drilling licence at any time by any party which sees it as offering an opportunity for potential development.[20] This in no way diminishes the responsibility of the Department of Energy as the final arbiter of what is appropriate in respect of the speed and intensity of development, but it does provide an environment in which the many possibilities for successful exploration are not foregone. Whether or not the party making the bid seems to be 'capable' of undertaking the task[21] ought not to be a prime consideration in the government's response to the bid. An incapable entity will not be able to pursue its intentions very far – and we ought to be prepared to allow such a company to 'lose' its own shareholders' money in respect of the work it asks to be allowed to do in the country's offshore areas. Even a company which proves to be incapable of sustaining an exploration and/or a development programme will generate some employment and income in the economy, produce some geological knowledge and, possibly, generate an element of technological advance in exploration methodologies. There will thus still be a benefit to the community from such private enterprise failures.

(b) The bids, however, should *not* be for a concession to explore for and, if successful, to produce oil, on the value of which the company making the bid will eventually pay taxes (as well as relatively minor sums in respect of lease rentals etc). Instead, a company which is successful will then bid for the right to retain a share of the production of the oil which it has discovered and hopes to produce. The balance of the oil

20 The number of bids accepted for consideration and determination at any point could, however, still reflect both policy and practical considerations. The latter involves the ability of the Department of Energy to handle and process applications; the former implies concern for the overall rate of exploration/exploitation in the light of national needs, in relation to exploration successes to date and the likelihood of 'required' rates of production being achieved.

21 Except, of course, in terms of a technical ability to undertake drilling in a safe way.

to be produced will remain the property of the state (or a designated state company) for it to do with as it pleases in a basically 'no pay' situation. The company, in other words, bids for the privilege of spending its own money on exploring an offshore prospect (with one or more specified exploration wells). If it is not successful with the wells as designated, then the agreement either lapses immediately, or possibly extended, if both company and the Department of Energy agree to more drilling. If the drilling is successful, then development of the field – as initiated and/or approved by the Department of Energy – will take place[22] within the framework of a joint venture between the company concerned and a state entity. The latter will be responsible only for a previously agreed part of the costs. This is a matter for *ad hoc* negotiations by the three parties concerned (company, state entity and Department of Energy) and it *may* be a nil percentage. The state entity will later be responsible for handling and selling that part of the production which is retained by the state – from a minimum of, say, ± 50 per cent (in respect of difficult-to-develop and/or small fields) to 80+ per cent for the lowest-cost-to-develop and/or better located fields.

(c) This partnership system thus entails a series of government/oil companies' joint ventures based on the concept of production-sharing. Each agreement involves the negotiation of the specific joint venture arrangements and implies agreement on the division of the production of oil, as well as on the nature of the developments to be undertaken, and on the division of responsibility for the costs. The company bidding for the venture will, of course, pay all the exploration costs and it may offer to meet most, or even all, of the development expenditure. Thereafter, government regulation of the venture at the politico-economic level is minimised – it will be related simply to questions involving the development plan for the field or

22 Should the drilling be successful, but not 'successful enough' in the company's view to justify development, then all interests in the block will automatically revert immediately to the Department of Energy which may, of course, then receive other bids for the acreage or it may decide to require development by a state entity.

fields[23], but there will be no problems of valuing the oil for royalty calculations and/or for special petroleum taxes. Similarly, as the latter will not exist, there need be relatively little government concern for evaluating the costs which are incurred in developing the field. All or most of the costs will be purely for the account of the company, as agreed in the negotiations.

(d) The state's essential economic interest lies in the share of the oil and gas it retains from the development, and the disposal of which becomes the responsibility of the appropriate state entity – say BNOC, in respect of oil, and BGC, in respect of natural gas. The extent (if any) to which a state entity participates directly in a specific development can also be a matter of negotiation and there is, of course, no reason why the existing public hydrocarbon corporations cannot take the initiative in deciding to seek the right from the Department of Energy to explore/develop a particular prospect. In such flexible circumstances there would seem to be no reason whatsoever for any fixed upper or lower limits to the degree of state involvement – though general government policy requirements will influence the overall degree of state participation which is sought from time to time. Nevertheless, even with a minimum element of direct state involvement in exploration and production there could still be a significant presence by the national oil and gas corporations in this sector of the oil industry. The system would certainly provide the opportunity for the state corporations to develop their expertise and knowledge of the industry. If it is thought desirable, they could also broaden their activities, both functionally and geographically, as indicated below in Section E.

(e) This production-sharing, joint venture approach to the development of Britain's offshore oil wealth does not lie in the state having a specific share – say, 51 per cent – in each

23 Government regulations of a technical character in respect of oilfield practices and environmental considerations etc will, of course, still be required.

development in order to institute an element of control into an otherwise eminently uncontrollable concession style system of the traditional patterns. Under the joint venture, production-sharing agreement the state automatically and directly retains ownership and control over most of the oil produced. This directly provides the equivalent of the royalties and the special taxes of the concession system without the state having to be concerned for the intermediary values of production costs and the 'value' of the oil.

(f) How the government – as opposed to the state entities – will assume its interest in financial terms is a matter for study. The state entities might be permitted to retain enough of the value of the oil[24] to cover their costs (plus a margin as a 'handling commission'), with the balance going to a National Oil Account or its equivalent. The role of the state entities would now be much more orientated to the functions of handling the oil – with up to perhaps 70 per cent of the total amount of oil produced at any time as the initial responsibility of these organisations.[25] In such circumstances the state entities will be major sellers of crude oil on the international market, or they could re-sell to the oil producing companies, or they could become significantly involved in downstream activities – with consequences and opportunities which are discussed in Section E of this study.

(g) Under the joint venture, production-sharing system as described, the private companies secure ownership to some of the crude oil as a return on the privileges extended to them for finding and producing oil from Britain's offshore areas. Most of the companies involved are multi-national oil companies, with international supply, transport and refining operations, the overall optimisation of which the companies

24 The valuation for this purpose would be by the state. There would be no interests of the private companies to take into account.

25 The use of the plural here – and elsewhere – in the discussion of state entities reflects the immediate possibility of both BNOC and the BGC being involved in the new system. It also indicates the future possibility of new state oil entities being created as a mechanism for ensuring a preferred multi-faceted public interpretation of the geological opportunities on the continental shelf (see footnote 19).

seek to achieve through complex computer-based methods, without any particular concern for the specific national interests of the countries in which their operations are located. It therefore may be considered appropriate for this joint venture type of approach to be extended to the transport, refining and associated activities of the oil industry within the UK.

(h) In the first place the joint venture concept could be extended to the transportation of the crude oil produced through common-user pipelines. Each of these lines could thus be jointly developed by several companies, including one or more of the state companies, which would be responsible for handling most of the oil from any group of fields as a result of the production-sharing agreements on the fields concerned. Such common-user joint venture lines will eventually form part of an offshore pipeline system, the need for which reflects the increasing number of fields under production. The development of these ought to be related primarily to the system's overall economics, rather than to the needs of a particular field. The need for a unified approach to the offshore transportation system, in order to avoid the unnecessary duplication of facilities and/or the isolation of some fields which are capable of being developed from an economic means of transport, has previously been analysed (see Chapter II-V in this volume). This is now mentioned in order to show how a common-user pipeline system fits more easily into a joint venture, production-sharing system. As the government, through one or other of its entities, automatically has the responsibility for at least 50 per cent of all crude oil produced – then a common-user, offshore pipeline system is an almost inherent part of the production-sharing system recommended in this report.

(i) Under existing legislation for the exploitation of Britain's offshore oil resources, the state has assumed few powers over the oil once it has been landed in the UK. There is, indeed, nothing more than the 'expectation' that up to two-thirds of all offshore oil production shall be refined in the UK. The Pipelines and Submarine Pipelines Act of 1975 was concerned in part with refineries (Part IV), but only in the context of controlling and authorising their construction and extension by private companies. Authorisation has to be 'consistent with the national policy relating to petroleum' – but 'national policy' was not defined in the Act. The same Act (Part I) also extended to BNOC the powers 'to provide and operate... refineries in connection with petroleum' but the right to exercise such power depends on the specific consent of the Secretary of State (see page 792). The Act, moreover, gives no indication that BNOC's power to provide refineries was to be sought as a means of changing the pattern of oil refining and distribution which emerged in the UK between 1945 and 1974 under the stimulus of a rapidly increasing demand for oil products in the energy and chemical sectors of the economy. A development which was, of course, related to the use of imported oil.

(j) However, the optimal development of the country's off-shore oil resources appears to necessitate a fundamentally restructured oil refining and distribution system. The essential elements in the re-structuring are presented in the next part of this report. Meanwhile, one can summarise as follows the advantages for the UK from the introduction of a production-sharing system in place of the concession system for the areas of the continental shelf and slope which remain to be explored and exploited for their oil and gas resources:

(i) It ensures the necessary and continuing availability to the UK of a broad range of oil companies' exploratory and development expertise whereby the probability of the successful discovery and production of the remaining offshore oil and gas resources can be maximised.

(ii) It provides the opportunity for definitive, firm and continuing government control over the speed and location of developments – within the context of the strategy determined to be most appropriate for any given period.

(iii) It ensures a 'clean' and a guaranteed flow of benefits to the government from the development of oil- and gasfields, thus greatly reducing the currently prevailing uncertainty over the size of the flow of benefits. This uncertainty is a result of the present complex system of concessions, regulations and taxation in which the important cost and value variables, so influential in determining the government tax-take, cannot be readily or easily forecast. The companies, under the production-sharing agreements, also know exactly where they stand in respect of each agreement which they sign.

(iv) It gives the government, through its state entities, access to most of the crude oil produced.

(v) In that most of the oil produced is not transferred to private ownership as in the exploration/exploitation system developed to date, there is a firmer basis for a rational pattern of offshore transportation and for the further evolution of a rationalised refining/distribution system.

(vi) With ownership of most of the produced oil remaining directly in the hands of the state, there is also a firmer base for Britain's use of oil as a 'weapon' in negotiations over energy and over other policies within the EEC; and with other nations or groups of nations.

(vii) The inherent flexibility of the production-sharing system (each joint venture agreement is negotiated individually in the light of the specific circumstances of the discovery and the specific interests of the parties concerned at a particular moment) means that an effective interest in small fields and in the economically less attractive parts of larger fields, can be generated through the negotiations between the state and the company or companies concerned. The system implies a joint state/company evaluation of the prospects for each field discovered and hence a compromise acceptable to both parties is more likely to be reached than in the present inflexible system. At present, the commercial evaluation of the field is completely divorced from the government's evaluation of its interest.

(viii) It establishes a framework for a significant new style of oil company/government relationship which is also applicable in many other parts of the world – notably in the still largely unexplored potentially petroliferous regions of the Third World. The development of these areas is of great importance in future global oil supply/demand relationships. Thus, the experience gained, with the framework as defined, could provide the UK with the opportunity to play a leading role in encouraging and in making possible the development of those resources.

E. The Integration of the Down-stream Activities of the UK Oil Industry into the Production Sharing System

1. The oil industry in the UK – before North Sea developments

a) Until a few years ago, the British oil industry consisted of refineries, product import terminals, inland pipelines and road/rail/canal terminals, distribution facilities, and, of course, gasoline stations and other wholesale and retail outlets. It would be inappropriate to describe in detail the system in this report as I have, in an earlier publication, already attempted to describe, analyse and define possible policy options in respect of the government's role in that system.[26]

b) Before North Sea oil was discovered and even with the low oil prices of the pre-1973 situation, there were elements in the UK's oil industry which seemed to require state intervention. It was needed to ensure the rational use of resources, fair prices and the elimination of unnecessary developments in the industry's infrastructure which is, of course, notoriously disadvantageous for the environment. At that time I suggested that there was a need for a 'National Oil Agency' with particular responsibilities. These would have included, 'the control and direction of British oil

26 See Peter Odell, *Oil: The New Commanding Height*, the Fabian Society, Research Series Pamphlet No 251, December 1965. Edited sections of this are reproduced as Chapter I-1 in this volume.

import policy; the control and direction of refining and distribution within the country; ordering the systematic exploitation of whatever oil wealth may be found in UK offshore areas; and representing Britain's interest in negotiations with other nations of the world with an interest in oil'.[27]

c) The role I suggested for that Agency in respect of the country's offshore oil wealth has, in the meantime, been vested in the Department of Energy and/or BNOC. This report has, however, demonstrated the inadequacy of the system established and has indicated ways and means whereby the nation's systematic exploitation of this wealth can be more effectively achieved. The likelihood of continuing off-shore successes, now makes it even more important than previously to pay attention to 'oil import policy' and to 'refinery distribution'. These aspects of the oil industry's activities are not unrelated to Britain's role as a major new producer of the commodity. Indeed, I would hypothesise the need for a much more positive view of import and refining policies, in order to recognise that part of the benefits to the country from its economy's oil sector are dependent upon changed attitudes and policies towards oil imports, and to refining and distribution. The major downstream issues which it is essential to define and appraise, in order to complete an analysis of the opportunities and challenges to government arising from the development of oil production in Britain's offshore waters, are thus set out below.

2. Oil imports, refining and distribution – with Britain as a major oil producer

(a) The downstream oil sector of the UK economy has traditionally been free from government intervention. This was increasingly inappropriate as oil inexorably became the commanding height of the energy economy in the years between 1951 and 1971. Today, the absence of effective

27 Ibid, p.24

government interest in, and concern for, the industry's downstream activities could lead to restraints on the oil production sector and so create the possibility of wasted investment.

(b) There are several structural reasons for this – organisational and geographical, as well as political and economic – which are derived from the fundamental changes in the country's oil supply position. These changes create new and unconventional problems, arising from the ownership and location patterns of the oil industry's importing, refining, transportation and distribution facilities prior to North Sea oil's availability in sufficient volumes not only to meet domestic demand, but also to provide some oil for export. In essence, the shares of individual companies in the UK oil market – and in its refineries and other infrastructural element – do not match (and are not likely to match for the foreseeable future) the company by company availability of UK-produced oil. For example, refineries which could run in large part on North Sea oil cannot do so because they are owned by companies with inadequate North Sea production. On the other hand, some companies with a large supply of crude oil from North Sea fields which they have developed do not have UK refineries and domestic markets. Even where the problems are not as clear-cut as this, there are, nevertheless, difficulties which arise from the contrast between the historic locations of companies' refining and other facilities, and the requirements for handling and processing crude oil supplies originating in the extreme north-east of the UK.

(c) These infrastructural problems, in respect of both the organisational and geographical aspects, could lead simultaneously, in the short to medium term, to high imports of expensive oil by companies lacking North Sea supplies on the one hand and, on the other hand, to restraints on production from the North Sea fields under exploitation by companies lacking sufficient crude oil outlets in the UK. By the mid-1980s the former companies may be forced by the OPEC nations, from which they draw the bulk of their supplies, to take specified packages of crude oil and oil products, whilst companies in the latter group are forced to shut in potential North Sea oil production. These potential illogical developments in the UK oil system could, however, be sorted out within the context of a joint venture, approach by private companies and state entities to the organisation of the downstream operations of the oil industry in the United Kingdom.

Given such an approach, the use of UK refineries for UK-produced oil could be divorced from the question of the particular company-by-company ownership of the facilities. In addition, an overall rationalisation of the supply, refining and distribution system could be sought. Within such a framework, the relationship between oil exports and oil imports – both crude oil and products – could be established so as to maximize the return to the nation from, for example, the appropriate blending of North Sea crude oil with inherently lower-value imported crude. It would also provide a realistic way of ensuring that the 'expected' up to two-thirds of UK-produced oil is, indeed, refined in the UK. Assuming, of course, that this 'expectation' emerges as the most beneficial one in the context of an industry-wide evaluation of the opportunities available to the recommended joint venture approach to downstream operations. In this way the refineries of the UK could be utilised (with appropriate investment in both expanded and refurbished facilities) to their optimum level. This would produce favourable results from the point of view of the unit cost of refining, whilst the most economic transport facilities for crude oil-to-refineries and for oil products-to-distribution centres could also be developed to the advantage of oil consumers. Such lower refining and transportation costs would also benefit the oil companies involved.

(d) Given the way oil companies have to be vertically integrated, it is clear that BNOC cannot, for very much longer, avoid becoming involved in some way in the oil industry's downstream operations. But, given the proposed integrated state enterprise/private enterprise joint venture system, BNOC ownership of refinery capacity need not be *necessary*. Instead, a re-structured BNOC could simply become a co-ordinating authority for the most appropriate pattern of refining operations. Such co-ordination, however, seems likely to be more effective if the state entity were to have a direct involvement in refining and other downstream activities. This could be achieved in one of several ways, *viz*:

(i) by BNOC being taken into equity and operational partnership by one or more of the existing companies with refining interests.

(ii) by BNOC buying an existing refinery from a company willing to sell.

(iii) by BNOC taking over one or more refineries from a company or companies that did not wish to be involved in a joint venture approach to this sector of the industry.

Its direct ownership of refinery facilities need not, moreover, prevent BNOC from becoming a participant in all the refineries, in a like manner to its participation agreements with the successful concessionary companies in their North Sea oilfields.

(e) The justification for this sort of downstream ownership/organisation structure arises from the fact that the UK, like all other oil producing and exporting countries, will inevitably have to make its refining (and even its petrochemical) industries an integral part of its oil-producing sector. In the not too distant future, for the longer-established oil producing and exporting countries, this seems certain to mean that crude oil exports will only be allowable within the context of an overall oil export policy which requires a greater or lesser emphasis (as determined from time to time) on the exports of oil products. The UK, as an oil exporter, will have to follow suit in order to protect its interests from being undermined by other oil exporting countries. Thus, any country seeking to import oil from the UK would also be expected to take a specified (though variable) percentage of its overall import needs from the UK as oil products.

(f) The location of the oil refining industry was historically and traditionally linked with the production of oil. However, because of factors specific to the international oil situation in the 1950s and the 1960s this locational link between oil production and refining was broken, so that many refineries were established (as were the UK's refineries) in centres of oil demand. The reversion of the general refinery location pattern to the *status quo ante* is now likely to be only a matter of time. It this is so, then there is no inherent reason why the excess refinery capacity of Western Europe should be adjusted downwards to overall current demand levels on an equal basis for all the countries in the region. Within the European Economic Community, given that the supply of a not inconsiderable part of its oil needs is now likely to be met by oil produced from the continent's own offshore resources, then those member countries such as the UK with an availability of indigenous crude oil, must be expected to insist on increasing their share of the total refinery throughput that is required by the overall level of oil demand in the EEC. In this context, the United Kingdom has a justifiable case for an EEC refinery policy which recognises the UK's new locational advantages for refinery developments.

(g) Thus, as far as the UK is concerned, the expansion of its downstream oil activities is part of its use of its oil wealth as an element in its economic policy-making at the EEC level. There are, nevertheless, limitations on the use of oil in this way. If other EEC members are pushed too far, then they could look elsewhere for their oil import needs; and so limit the UK's opportunities in respect of its oil development potential. It is this negotiable situation which provides the basis for a reasonable and realistic oil policy by the UK *vis a vis* the other member countries of the EEC. The development of Britain's considerable offshore oil resources provides the basis for economically rational linkages into expanding downstream activities. The use of these will also benefit the Community in terms of a guaranteed supply of an essential commodity at prices which could also be negotiated to ensure advantages both for the UK and the rest of the EEC. In this respect, too, a positive policy by the UK government as far as oil refining, transport and distribution are concerned would serve to enable the UK to take full advantage of its offshore oil. It would also serve as a means whereby the problem of surplus downstream capacity in the EEC as a whole is divorced from the complications of company supply patterns and put into an economically rational framework for solution by the governments concerned.

An Afterthought on this Study, April 2002

The Report from which this chapter is derived was accepted by the U.K. Department of Energy, headed by the Rt. Hon. Tony Benn, M.P., just a few hours before the Labour government of the day was defeated in the House of Commons and was thus obliged to resign. In the subsequent election the Conservative party, led by Mrs. Margaret Thatcher, was returned to power. Traditionally, reports made by advisers to defeated administrations then have no standing and are summarily relegated to the library of the House of Commons, rather than being considered by the successor Minister.

Given that the recommendations of the Report involved the broadening and strengthening of State intervention in the UK's oil sector, it was, in any case, unlikely that a Conservative Secretary of State for Energy would have wanted to implement much of the advice. Nevertheless, the recommendations for the liberalisation of the process of attracting oil companies to invest in U.K. offshore oil and gas exploitation – and for removing the inherent conflicts between government and companies in decisions on fields' development and on the division of the economic rent between the parties – could well have been attractive to the new government, had tradition allowed consideration of former advisers' analyses.

Instead, the new Conservative government decided to take more or less immediate action significantly to increase the petroleum revenue tax. Furthermore, it also threatened to introduce depletion controls on the rate of exploitation of oil reserves – under the illusion that oil, in general, and UK oil, in particular, was such a scarce commodity that even known reserves should be kept in the ground. Together, these policies succeeded in undermining the companies' willingness to invest in exploration and new fields' development, so that U.K. oil production fell by 25% over the next few years.

In these circumstances the House of Commons Select Committee on Energy determined in the Parliamentary session 1981–2 to investigate the deteriorating situation for the U.K.'s offshore oil industry. The following chapter reproduces the author's invited contribution to that investigation. It complements, albeit at the cost of a little repetition of ideas already set out in this chapter, the author's views on the need for government policies which strive for partnership between oil companies and the state in the exploitation of a country's oil wealth.

Chapter IV – 5

North Sea Oil Depletion Policy*

1. The processes of offshore oil exploration and exploitation on the UK continental shelf and slope are only two decades old and most of the offshore potential remains unexamined. The processes will continue for at least another half century on the assumption that oil and gas remain in demand and that the UK wishes to continue to develop its resources in order to help meet the demand for energy. There is thus justification for an evaluation of the appropriateness of the policies that have been developed in the initial period of exploration and development and for a reappraisal of the validity of the criteria on which the policies are based. Such steps are vital in order to ensure that developments proceed to their full extent and bring the greatest possible benefits to the country.

2. The continuation of the supply of offshore oil and gas, the costs at which they can be produced, and the prices at which they can be marketed clearly constitute the main elements which will shape the energy sector of the British economy for the rest of this century – and into the 21st century. Oil and gas in general, and British produced oil and gas in particular, seem likely to remain the most preferred energy sources. Alternative sources of energy, such as coal and nuclear power, are really only relevant for the UK is as far as

*** A Memorandum submitted to the House of Commons' Select Committee on Energy in the Session 1981–2, H.M.S.O. London, 1982**

they are needed to supplement oil and gas supplies. Polities which deliberately seek to limit, or which have the effect of limiting, the contributions of British oil and gas to the energy economy should thus be treated with great scepticism. This is of immediate importance in light of the very much higher real resource costs which are directly involved in developing the supply of energy alternatives to indigenous oil and gas. It is also important from the point of view of the considerable additional costs, connected with infrastructural and other changes in society, which would be incurred in order to enable the economy to absorb relatively more energy in forms other than oil and gas on which all main users, with the single exception of the electricity generating industry, have become largely dependent.

3. Nevertheless, policies towards oil do seem to be directly concerned with production limitations and/or to have the indirect effect of restricting production potential. The main underlying justification for such action appears to be the unquestioning acceptance of a rather simplistic, but no means self-evident, view of the future of oil at the international level, viz. that the commodity is scarce and getting scarcer, and hence will become increasingly expensive (in real terms) over the time period at which it is necessary to look for policy planning purposes. The Energy Policy Green Paper (HMSO, 1979) stated this belief quite explicitly. 'World oil supplies', one reads in the Paper 'will begin to level off in the late 1980s, reach a peak in the early 1990s and decline thereafter. Oil prices are forecast gradually to rise in real terms at an annual rate of just of 3% per annum'. This view of the future of oil is then used in the Green Paper to justify a restrictive approach to the depletion of the UK's oil resources, even though the forecast doubling of the real price in 20 years is by no means enough to compensate for the losses in wealth which will be suffered by keeping reserves of oil in the ground, rather than producing them. Such production restraint is certainly not an interesting proposition for the companies concerned as they expect a real rate of return on their operations which greatly exceeds the 3% per annum increase in the 'real value' of the unproduced oil. Thus, from the companies' point of view, pressures to restrict output makes their involvement in the UK's oil provinces relatively less attractive than opportunities open to them in many other regions of the world. In particular, controls which restrict the production of oil from individual fields are specially unfavourable. They not only

adversely affect the rate of return (because the discounted value of the reduced output of oil is greater than the reduction in the costs of production), but they also increase the investment risk in that delayed production may not turn out to be possible in the harsh environmental conditions of the North Sea, given the deterioration over time in the safety and effectiveness of the production and/or transport facilities.

4. Production controls are thus uneconomic even in the context of the forecasts made on the future of oil prices by the Department of Energy. It is highly likely, however, (viz. more than 50% probable) that the official view of the development of the international oil supply and price situation is incorrect. Steadily and inevitably increasing real oil prices as the result of an inherent scarcity of oil is one of the least likely futures for oil. 'Scarcity' is unnecessary from the standpoint of the interrelationships of resources and use for a period of *at least* 30 years; there is a 50% chance that scarcity will not become a problem for over 50 years; and there is even a 10% chance that it will not evolve for another 100 years. These data cast doubts on the validity of several aspects of the restricted production policy option, viz.

    (a) One must question the hypothesized inevitability of continuing real price increases. Except in the context of an institutionalised control over oil supply, present oil prices are well above the long-term supply price for oil at the highest future levels of demand that can now be realistically foreseen. Oil use is currently declining and, even with the most optimistic view it is possible to take of the world economy, any increase in oil used over the rest of the century is likely to be extremely modest, at an average of 1–2% per annum.

    (b) The expectation that the value of government revenues per barrel of oil produced will increase is also questionable. With rising production costs (as a result of inflation) and stable, or even declining, prices, unit revenues from oil production will fall, unless rates of tax can be increased. This is almost impossible in the UK, where marginal tax rates are already ± 90%.

(c) The validity of plans which simply assume that all the oil a country chooses to produce will automatically find a market is also open to serious doubt. Given the developing weaknesses in the world oil market, major oil-exporting countries are beginning to take action to try to ensure that their exports have preference in specific markets. The Western European market is the one in which such actions are most likely to be important and could thus undermine the prospects for British oil – in the continued absence of any government efforts to secure guaranteed long-term outlets for it in the EEC.

5. The most likely outlook for the evolution of world oil supply and price, viz. the continuation of a relative easy supply situation and a steadily declining real price under the influence of market pressures, thus indicates that any near-future controls on the level of production of UK oil will have the effect of diminishing the benefits to the country from its exploitation. *The imposition of a depletion policy is thus a high-risk option.* The risk is enhanced, moreover, by the possibility (with a 10–15% degree of probability) that the international oil market will collapse sometime in the later-1980s. The following factors are involved in this possibility;

(a) The continued stagnation, or even decline, in oil demand as a result of the success of deliberate conservation and substitution measures and/or from the failure of the industrialised countries to re-establish condition for sustained economic growth.

(b) The continued development of new and expanded sources of oil supply. These include additional oil (and gas) production in North America (stimulated by de-regulated prices) which is large enough to decimate the US need for imports; enhanced exports of oil (and gas) by the USSR in order to maintain a required level of export earnings, given lower than expected unit revenues; and new oil export potential from a wide range of countries in which oil resources have been found and the development of which is now under way or imminent.

(c) As a result of the consequential continued fall in exports from the OPEC countries, one or more of the members of the organization could become unwilling or unable to stick with the agreed minimum pricing formula. It would thus seek to increase its sales by discounting prices in order to be able to maintain the anticipated and required level of revenues. Other member countries of OPEC would then have to react to protect their own position and matters would be in danger of getting out of hand. Note in this respect the already limited current downside opportunities for OPEC production. The output of the many members of the Organisation which are able to produce more oil has already been squeezed to unacceptably low levels and they are seeking to increase their exports. Meanwhile, Saudi Arabia would not wish to cut-back its production by more than another 1.5 to 2 million b/d; and even this could be more than offset as a result of action by Iraq and Iran both of which have not only the potential, but also the need to expand exports once the fighting between them comes to an end.

I wish to stress that this possibility of a collapse in the world oil market is *not* my 'best guess' of the future of oil (that is presented in para. 4 above). It does, however, have a probability of 10–15%, and thus seems to have a similar chance of happening as the equally high-risk guess of the Department of Energy, viz. the inevitability of scarcity and increasing real prices. They both represent quite extreme views of the geopolitical and geoeconomic realities of the world oil situation.

6. Even though a strong economic case can thus be made for a faster, rather than a slower, rate of production of the UK's oil (and gas) resources, it can still be argued that controlling production in the immediate future will, if nothing else, ensure that the country can be certain of its essential energy supplies in the 1990s and later (note, however, that this argument also depends partly on a belief in an inevitable scarcity of oil within the relatively short-term future). This line of argument, however, presupposes two condition;

   (a) An increasing demand for oil in the UK from the base of the more than 100 million tons which the country used in the mid-1970s. The annual rate of use has, however, now fallen to less than 80 million tons and it is still falling. Even if this decline in use comes to an end (and this is by no means certain even under conditions of economic recovery), the country's cumulative need for oil over the rest of the century will be now more than two-thirds of hitherto expected levels. A realistic upper figure to the rate of cumulative oil use by the year 2000 is about 1500 million tons. This represents only little more than 40% of a conservative estimate of the amount of oil which could have been discovered in UK offshore areas by the end of the century. It is, indeed, less than two-thirds of the reserves that have already been declared proven and probable. Figure IV-5.1 illustrates these points.

(b) A limitation on the likely development of the country's oil and gas resource base. But this is not justified by any reasonable expectations on its continued development, except under conditions of a less-than-complete search for, and/or exploitation of, the reserves that remain to be discovered. In both the United States and Canada, where experience of the oil industry is very much longer than in the UK, the concept of 'undiscovered but discoverable reserves' is accepted as an essential part of the process of evaluating production potential Major studies, sponsored both by Federal and State (Provincial) governments and by the oil industry itself, have been undertaken and are regularly updated, in the context of expanding knowledge and new technology. These show, in detail, what new resources can be expected from varying levels of further exploration and development efforts. This sort of work has simply not been done in respect of the UK's ultimate potential for recoverable oil. The figures for proven, probable and possible reserves from proven fields, other discoveries, and expected discoveries in areas already licensed have no defined economic parameters attached to them, while for the rest of the so far unexplored continental shelf, nothing more than a single figure of potentially recoverable reserves is given; and even that has neither probabilities of the likelihood of its occurrence, nor any technico/economic background information on the methodology of its derivation. This is a woefully inadequate approach to the question of the country's resources. Figure IV-5.1 indicates the magnitude of the possible under-evaluation by the year 2000 of the development of reserves.

It is clear that an element of immense importance to policy decisions on depletion rates is not being effectively presented and considered.

7. There is, however, yet another element in the consideration of the question of the inter-relationship of reserves and rates of production. The latter is usually presented as a function of the former, viz. 'appropriate' rates of production depend upon the size of the reserves declared. *There is, however, another way of looking at the*

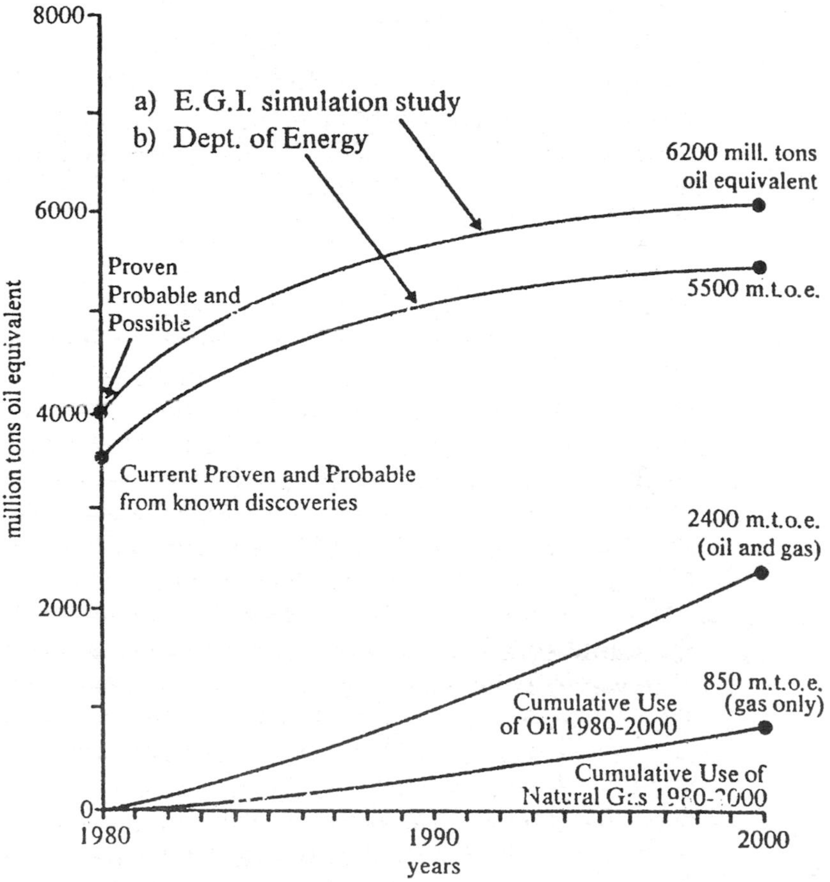

Figure IV-5.1 UK oil and gas use 1980–2000 and forecasts of the development of reserves' potential over this period

inter-relationships, especially in the early days of the development of a new oil province or region; that is the concept that reserves' development is a function of production (use) or, to put it more colloquially, the idea the 'the more oil you use, the more there is to use'. There are two aspects of this concept; first, in relation to the reserves of individual fields and their depletion; and second, in respect of the reserves/use relationship in newly developing oil regions as a whole.

(a) The reserves of individual fields are always very conservatively declared in the first instance, because of the high levels of uncertainty associated both with the complex calculations involved and with the economies of producing the fields. Though mistakes are occasionally made over this initial declaration, so that a field's reserves are later downgraded, world-wide experience shows that "appreciation" of the reserves initially declared normally occurs over time; basically as production experience is achieved and as the complexities of the reservoir are better understood. This process often persists throughout the life of a field for periods of 20 years or more.

The 'oldest' field in UK offshore waters has been producing for less than six years and that field (Argyll) is still in the process of development, with upward re-evaluation of the reserves and production potential still very much under way. The same is true of the five-year old Forties field for which major extensions of the production system are planned or under consideration. Likewise, with most of the more recently developed fields which have already gone far enough into the production phase to make such re-evaluation possible. Appreciation of the declared reserves of the discovered fields in the North Sea is thus already a fact. The unanswered question which remains is the degree of appreciation which will ultimately be achieved and this, paradoxically, is, in part at least, a function of the extent and timing of the process of the production of oil from each field. *Thus, production controls, whether introduced by government as part of its energy policy or by the producing company as a result of commercial considerations, inhibit the appreciation of the reserves of a field simply because they constrain the production process. Such controls on production, when introduced by governments anxious to*

conserve supplies for future use, are thus self-defeating in their objective: they prevent the proving of reserves which could have met future demands and, in the meantime, keep production at levels lower than the already proven reserves could sustain.
The country thus loses out in both the short and the longer term.

(b) The reserves of a newly developing oil province (such as the North Sea still is) are likewise always conservatively estimated, for basically the same reason as with an individual field, viz. the high degree of uncertainty in a situation of little knowledge and limited experience of the province's productivity. This is particularly true in respect of areas, such as the North Sea, which are highly complex geologically, with a large number of so-called 'geological plays', and for basins of a type for which there has been little production experience. The existence of a complex pattern of concession arrangements, with an incomplete allocation of the potentially petroliferous areas, enhances the uncertainties in two ways; first, because knowledge cannot flow freely across the many legal boundaries so created; and, second, because the region cannot be systematically and rationally explored from the most likely, through the less likely and ultimately to the least likely, prospects.

The most rational approach to the exploration opportunities can thus be inhibited by the institutional structure of the system which is created. *All* the factors involved apply in the case of the British sector of the North Sea so that, even after almost 20 years of exploration (in an area which is large by the standards of other major world oil provinces), much uncertainty remains over the possible size of the ultimate resources. The process of reducing uncertainty is not only a long one, but also one which necessitates the continued investment of resources – both human and financial – in the exploration and appraisal phases. Such investment will keep the discovery of new fields and extensions to fields moving along and so ensure the continuation of additions to reserves. The investment will, however, only be made if there are perceived to be prospects for achieving an adequate rate of return on it. *Depletion policies – whether imposed through the medium of too high marginal rates of tax or by government imposed controls on the rate of production – serve to diminish or, in extreme circumstances, to eliminate the prospects for achieving such a rate of return. As a consequence the investment flow will, in turn, be diminished or eliminated.* The end result is an inadequately and incompletely explored region, part of the ultimate resources of which will not be turned into effective oil reserves. *A depletion control policy, at the level of the oil region or province, thus also has the effect of limiting the development of a country's oil reserves. In essence, such a policy serves to create, rather than to solve, the problem of the medium to longer-term scarcity that is feared!*

8. Apart from the generally applicable considerations, discussed in paras. 6 and 7 above, in respect of depletion policies and reserves availability/ production potential, there is also a specific way in which the system which has evolved for taxing the production of oil from North Sea fields, works to limit the size of the recoverable reserves and of the rate of production from an individual field. It thus acts as a depletion control policy instrument, whether intentional or not. This relates to the marginal rate of tax which tends to be progressive with increasing levels of production from a field, in a situation in which the marginal costs per unit of production also tend to rise – given that the easiest-to-recover oil from a field is recovered first. Thus, operating companies have no motivation to invest in production systems which are capable of maximizing the production of oil from a field (by, for example, taking an interest in those parts of a field which lie beyond the so-called drainage area of a set of wells from a platform located above the reservoir). Thus, part of the oil-in-place which is technically recoverable not only remains unproduced, but it also remains outside the calculation of the total volume of declared recoverable reserves of a field. The exploitation of this oil so discounted would involve too high costs and too low profits and it thus remains unproduced and unproducible through the production system which is chosen for a field. Whether or not it will be produced in the future is a matter of uncertainty, depending on both technical and economic factors and so a high risk situation is created. Meanwhile, the company involved is more likely to prefer to invest the profits which it earns from the limited development of the field either elsewhere in the North Sea, in order to benefit from the potential offered by the lowest cost/ highest profit resources of another field: or it will invest them elsewhere in the world, given an assumption that the company does have alternative, foreign options open to it (as is, indeed, the case in

respect of most companies involved in the North Sea). *In either event the British sector of the North Sea is inevitably under-evaluated in terms of its oil resources so that, in effect, a depletion control policy is introduced, but with no guarantee that the resources left behind will be producible later.*

9. This was a view which we advanced some years ago (see P.R. Odell and K.E. Rosing, *Optimal Development of North Sea Oil Fields*, Kogan Page, London, 1976) and which at that time was severely criticized – though in terms of the methodology, rather than of the principles involved. Subsequently, during a period as a special adviser to the Secretary of State for Energy (in 1977–9), I had the opportunity to study, in a limited way, a number of the development plans for individual fields. What I then saw convinced me of the essential validity of our arguments and I so reported to the Secretary of State in August 1978. As I had to sign a set of confidentiality undertakings in respect of that work, (one for each of the companies involved as a partner in one or more of the fields for which a development plan had been submitted and which I was allowed to see), I am unable in this Memorandum to give any information on my evaluation of individual fields' development plans. In general terms, however, I think I may say that the development plans agreed were often based on the development of limited, or even very limited, parts of a field and/or excluded the exploitation of separate reservoirs in close proximity to the reservoir which was being developed. One can, in this context, state the contrasting interests of company and country in the following way;

(a) The company is interested in limited development because this means a limited risk and a limited exposure to capital expenditure, on the one hand, while it ensures an early start to the flow of revenues, on the other.

(b) The country is, or should be, interested in the comprehensive appraisal and exploitation of the total structure (or structures) in order to ensure that all the technically recoverable reserves, the production of which would produce a net return to the country, are in fact recoverable by the system which is installed on the field.

I am not aware of any subsequent announcement of any change in Department of Energy policy and practice on this issue which would mean that this conflict of interest is being resolved in favour of field development plans which aim to maximize the recoverable reserves of a field in the national interest. On the contrary, subsequent changes in the tax structure seem likely to have served to enhance the conflict by diminishing even further the ability of an operating company to go beyond a development plan related to the exploitation of the lowest-cost reserves of a field. There have, indeed, over the last two years, been a number of announcements of companies' withdrawing proposals for the extensions of field development plans. It is thus now all the more important that the ways and means which we have shown to be available (in Odell and Rosing, *ibid.*) whereby the 'conflict' can be resolved in the ultimate best interests of both parties, viz. government and companies, should be implemented in order to eliminate an unintentional depletion control policy and to ensure a more systematic and rational exploitation of the North Sea's resources.

10. This leads on to the final point; viz. that the traditional concession system under which North Sea oil and gas resources have largely been developed to date is as inappropriate for the UK as it has proved to be for almost every other major producing country in the world. Though new and expanded legislation, an increasing number of regulations, and successively more rigorous taxation have served to 'tighten up' the freedoms originally enjoyed by the concessionary companies, I am not convinced that the system as it has now evolved either serves the best interests of the country or meets the basic needs of the companies. On the one hand, the government still lacks the essential means to initiate and guide developments once the concessions have been allocated: the government can – and indeed sometimes does – stop things happening, but it is still largely unable to cause things to happen. There is thus an essentially negative government approach to the development of offshore oil. On the other hand, given that the companies are required to take all the positive initiatives (except the granting of concessions in the first place), they see the involvement of government as a 'burden' on their efforts. They are often thwarted in that they cannot make the progress they expected – and for which they planned – because of the suspicions and reactions to which they are subject, as a consequence of the very nature of the system itself.

This suggests, as I have shown elsewhere (see Chapter IV-4), that the present institutional approach to the development of the UK's offshore oil and gas resources ought not to be viewed as other than an interim one. The full and profitable development of the country's oil and gas resources, a process which is still only in its earliest stages, depends upon the replacement of the outdated and convoluted concession system by a system involving an alternative set of institutional arrangements, viz. a production sharing system involving joint venture operations between the private companies and the state. This would involve the following elements;

(a) Any part of the continental shelf and slope should be open at any time for a bid for an exploration/drilling licence by any party which sees an area as offering an opportunity for potential development.

(b) The company concerned will bid for the privilege of spending its own money on exploring an offshore prospect by offering, if the exploration is successful, a designated share of the oil or gas produced to the state.

(c) The partnership system thus involves a series of government/oil companies joint ventures based on the concept of production sharing.

(d) Once agreement has been reached, government regulation at the politico-economic level is minimized. There will be no problems of valuing the oil for royalty calculations or for special petroleum taxes. The state's essential economic interest will lie in the share of the oil and/or gas it retains from the development. The state's disposal of the oil and gas it receives could be via a state corporation or on a sale-back basis to the companies involved.

(e) The system allows a very wide degree of flexibility in the degree of interest negotiated by the two parties, so that suitable provision can be made for the exploitation of fields of various sizes and other characteristics and in respect of contrasting locations relative to distance from shore, water depth etc.

(f) The joint venture system could, if it were thought appropriate, be extended to cover both transport and refining activities so that these activities could be more effectively integrated with production/

(g) The system will ensure the necessary and continuing availability to the UK of a broad range of oil companies' exploration and development expertise whereby the discovery and production of the country's remaining offshore oil and gas resources is most likely to be achieved.

(h) It provides the opportunity for definitive, firm and effective government control over developments – within the strategy determined to be most appropriate for any given period – without the state being involved in the need for detailed and bureaucratic control procedures.

(i) It can ensure a 'clean' and a guaranteed flow of benefits to the government from the development of new oil and gas fields, so greatly reducing the present uncertainty over the size of the flow of benefits arising from the now very complex system of concessions, regulations and taxation in which the most important variables influencing the level of tax-take cannot be readily or easily forecast.

11. Most of the history of the UK offshore oil and gas industry is yet to come. There is thus more than enough cause and certainly enough time, to change the fundamentals of the system so that the partnership between oil companies and the state, essential for the comprehensive and economic exploitation of the resources, can be put on a footing in which neither party continues to suffer from the frustrations and uncertainties to which the system as developed to date has given rise.

Chapter IV – 6

Natural Gas and Dutch Society*

Introduction

The natural environment of the Netherlands set the country apart from its neighbours and rivals in the modern western world. Alone among the world's modernising countries of the 19th and the early 20th century, it had to fight the elements continuously by means of hard work and effective organisation. Such efforts were required not only to achieve modestly rising levels of welfare in the face of competition from better endowed nations in respect of climatic conditions, mineral wealth, energy resources and richness of soils, but also to keep the very land and its inhabitants safe from the ever-present potential ravishing by the sea.

The success of man against, or in spite of, his environment can seldom have been better portrayed than in the case of the Netherlands where hard work, thrift and respect for the forces of nature were of the essence for the country's progress and the basis for the people's relative prosperity. Even the success of economic activities dependent of natural conditions and phenomena (such as the port of Rotterdam, the country's

*** The Dutch language version of this paper was published in *De Gids*, Amsterdam, January 1987. This particular issue of this Dutch literary review comprised articles commissioned from foreigners living and working in The Netherlands on aspects of Dutch society with which they were professionally involved. Given the nature of the commission, I have previously declined to permit the publication of the article in languages other than Dutch. Now, 15 years on, and given the way in which Dutch attitudes and policies on natural gas can clearly be seen to have significantly influenced policies elsewhere in Europe, its publication in English no longer seems to be inappropriate.**

dairy production and the intensive horticulture of the Westland) necessitated large-scale investments to make the potential realisable. Constant efforts and vigilance was necessary, moreover, not only to contain the forces of nature, but also to find and exploit foreign marketing opportunities.

Thus man-made, rather than God-given, wealth epitomised the evolution of Dutch society in the increasingly competitive world of the last 150 years. In this context there was no scope for relaxation in the continuing battle against the natural odds. Indeed, only recently, as if to warn the Dutch nation not to dare to think of emulating the ways of other nations which elevated individualistic consumerism to the principle on which economic progress should be sought, there came the great storm and disaster of 1953 with its dramatic immediate impact on the country's coastal areas and their populations.

This occurrence, at the very moment in time when some initial relaxation of the country's post-1945 collective rehabilitation in favour of enhancing the wealth and welfare of individuals was becoming possible, could well have been read as a sign from heaven of a requirement to maintain the nation's collective effort and investments needed to keep the country habitable. The response, in terms of the much increased share of the country's financial and human resources which had to be devoted to protecting the land against the water from the sea and rivers, the tides and the wind, was no more and no less than should have been expected from the Dutch nation: this, after all, was what life in the Netherlands had always been about – and so, it seemed, it would, and must, remain.

Groningen Gas

Against this backdrop of a Dutch natural environment which had always generated costs rather than benefits, and in the particularly costly circumstances of the post-1953 natural disaster, the discovery of the Groningen gasfield in 1959 was an unbelievable event. Vast – indeed, initially immeasurable – reserves of a natural resource producible with little effort at a near-zero cost, and marketable at a price which was more than ten times its cost was simply not a phenomenon with which Dutch society was familiar. It was, indeed, a surprising and even a somewhat embarrassing new element for the Dutch economy. It was, moreover, an element with significant and broad political and social implications, both nationally and at the Western European level, given the rapidly rising demand for cheap energy in the country's and continent's expanding and diversifying economies.

How, therefore, could and should the phenomenon be handled? Initially, given the Netherlands's lack of familiarity with natural resource exploitation, and the need for careful and astute consideration of the many implications involved, it was necessary to "buy time" to ensure that no undue commercial or political pressures – national or international – could be brought to bear. Thus, it was necessary to deny the existence of the resource except in a very modest way. This was an easy proposition to implement as the whole of the Groningen oil and gas concession was jointly owned by Shell and Esso – the two largest of the world's seven major international oil companies. Such entities were already highly familiar, from many of their operations elsewhere in the world, with the desirability of keeping secret the real size of their discoveries of oil and gas.

The implications of the exploitation of the Groningen gasfield as the western world's largest gasfield discovered to date, were thus studied jointly by civil servants and their advisers and by technocrats and managers of the two oil companies. This essentially non- or even counter-democratic process led to a decision to implement a partly state owned, partly stated supported monopolistic approach to the exploitation of the resource. The decision was, however, suitably "dressed up" to make it appear compatible with the Netherlands's firm and enthusiastic support for a "free" energy market in Western Europe – in opposition to the statist and/or protectionist policies which were favoured by other member countries of the Common Market. This somewhat hypocritical approach to the exploitation of Dutch gas resources for both domestic and international sales, through the joint state/private monopoly established by legislation, ensured high profits for Shell and Esso and a large flow of revenues to the Dutch exchequer from the late 1960s and through the 1970s.

In essence, Dutch consumers were allowed access to the new source of energy at prices just slightly below those of the alternatives which they had used to date, while foreigners paid a price for the gas at their frontiers with the Netherlands which largely restricted their use of Dutch gas to high-value energy markets. This strategy was calculated by the bureaucrats of the Ministry of Economic Affairs and the managers of Shell and Esso to be a more profitable procedure than the alternative of flooding lower-value energy markets with the cheap gas which was potentially technically producible from the Groningen field. As a result, the rate of depletion of the field was kept significantly lower than would have been the case under conditions of competition.[1] Meanwhile, the state retained ultimate control over the gas resource in a way which reflected the earlier decisions of Dutch society collectively to retain control over the water in, around and under the country.

The idea of "conserving" Groningen gas, later to become an important element in Dutch decision-taking on its production and use, was initially an incidental "spin off" from the application of the more important principle involved, viz. that of maximising the return to the state from the exploitation of the Groningen field, with the proviso that commercial interests of Shell and Esso were also met, These companies were, in effect, given a proverbial licence to print money. The fact that conservation of the gas was not viewed in the early days of its development as a principle is demonstrated in two early responses by the Netherlands to "threats" from Soviet gas. The first of these was the political decision, inspired by NATO pressure channelled through the then Foreign Minister Luns, to sell gas to Italy at what amounted to a discounted price compared with other export contracts, in an unsuccessful attempt to inhibit the Italians from buying Soviet gas. (In the event, Italy, having secured cheap Dutch gas, then went ahead and bought cheap Soviet gas as well.) The second decision was the competitive response by Gasunie and NAM (Shell/Esso) to the offers made by the Soviet Union in the early 1970s to provide cheaper supplies to existing customers for Dutch gas. This response resulted in an increased amount of Dutch gas being committed to foreign markets.

These additional exports, together with a rapidly growing market at home under the stimulus of economic growth (which was itself partly the result of the development of the Groningen gasfield), put the Netherlands on the pathway to highly profitable sales of almost $100x10^9m^3$ or gas per year by the mid-1970s. This was to make the Netherlands the world's largest gas producer (outside the US and the USSR), based on the exploitation of a massive gasfield which offered one of the world's most profitable natural resource ventures ever. This prospect and outcome was a far cry for the Netherlands from its earlier ongoing struggles to contain the costs to its economy and society of utilising its natural environment.

"Limits to Growth" and the Oil Price "Shocks" of the 1970s

The dramatic and formidable sequence of oil price increases between 1973 and 1981 led to a greater than order of magnitude real increase in the price of crude oil. This development in oil prices enhanced the value of the Netherlands's large gas resources by a roughly similar amount, given the way in which the price of oil was used – as a matter of policy – to define the prices of natural gas. The political and economic opportunities which this growth in the value of its natural gas resources engendered for the Netherlands were, however, tempered by the impact of the Club of Rome's contemporary report on *The Limits to Economic Growth*.[2] The report was a "best-seller" in the Netherlands – uniquely so amongst the world's nations – while the Dutch political establishment took the report particularly to heart. Politicians from all parties were indoctrinated to accept what the report purported to show, viz. that the world was in imminent danger of running out of raw materials in general, and out of accessible sources of fossil fuels, in particular. Thus, the politicians argued, those resources which had been discovered should be cosseted and conserved.

In the Netherlands, across the broad spectrum of political opinion and in most of the media and in pressure groups of diverse persuasions, this conclusion was seen to be applicable specifically and emphatically to Dutch gas. Given contracted and expected sales, and on the assumption (which was unjustifiably taken to be an axiom) that no more gas would be added to reserves, it could be shown that the Groningen gas field – as well as the country's other smaller gas fields – would be well-nigh empty by the turn of the century. Thus, the temporary dalliance with resource-based wealth generation was brought to a sudden end. The approach to the exploitation of the country's natural gas could now be put back in the much more familiar context of the proverbial and ongoing battle by Dutch society against the limitations and shortcomings of nature! Back, indeed, to more familiar "first principles" in the "man versus nature" struggle of Dutch economic and social history.

Overnight, the Netherlands became an unwilling seller of gas to its neighbouring countries. It thus became a bad neighbour, given that Western Europe's economic future was, in the mid-1970s, very much at stake because of the increased price of oil and what were then seen to be gross uncertainties as to the likely availability of sufficient oil and gas from elsewhere in the world to keep the energy-intensive European system going. Even at home in the Netherlands itself, the gas-fuelled power stations which had been built over the previous decade and the large-scale gas using industry which had been developed over the same period were warned that their gas-supply contracts would not be extended, and that they must be prepared to change, at six months' to a year's notice, to using other sources of energy.

The idea of conserving the country's proven natural gas reserves was, in effect, elevated to the central principle of Dutch energy policy. Thus, gas which could have been produced remained unused in the ground, whilst Europe's future was in jeopardy because of energy sector uncertainties. This helped to necessitate the introduction of policies which emphasised the development of inherently higher cost energy alternatives, such as nuclear power. The fact that more gas was being added to Dutch reserves year by year was studiously ignored. Indeed, the way in which the gas reserves statistics were presented, both by government and by Gasunie, seemed designed to hide the fact of reserves' growth. Such knowledge would, of course, have prejudiced the maintenance of the principle of conservation: a principle which was argued in ethical terms. Indeed, it was presented as a fundamental matter of (Christian) responsibility for ensuring the well-being of generations yet unborn.

Such a moral approach was not – indeed, it could not be – justified by reference to, and examination of, the gas supply/demand position and outlook. There never was any real doubt (as a few of us argued at the time – to no avail), but that sufficient reserves could, and would, be found to meet the long-term requirements of the markets. The "ethical" stance did, however, by leading to a policy which restricted supply, ensure that the price obtainable for Dutch gas in the energy-short markets of Western Europe in the 1970s and the early 1980s was one which created profits of a level which would justifiably have been declared "obscene" had goods or services provided by a private monopolist been involved.

Indeed, this situation was recognised in the introduction first, of a windfall profits tax on Shell and Esso, in respect of their net share of the very large profits from Groningen gas; and second, in the "gentlemen's agreement" the two companies were obliged to sign guaranteeing the use of some of their remaining profits for investment in the Dutch economy on projects which had to have the approval by the Ministry of Economic Affairs. It is also debatable as to whether it was simply avarice – or an attempt to apply the "conservation principle" through charging much higher prices – that led to the so-called Spierenburg round of negotiations with foreign importers of Dutch gas. Through these negotiations the importers were "persuaded" to have their 25 to 30 year supply contracts rewritten such that the gas still to be supplied could be charged for at a much higher price than would have been allowed under the terms of the original agreements. The higher prices demanded were presented as the only means whereby the continuity of supply could be ensured: an argument which, in the context of the then widely perceived impending "scarcity" of energy, could well be seen as one which was tantamount to a form of economic blackmail.

In any event, whether through purely ethical considerations (related to the need to conserve), or through the application of the more mundane – but highly effective – principle of monopoly pricing, Dutch gas production and sales, in the halcyon days of high energy prices engendered by the OPEC cartel, became a simple money-making machine. The Netherlands was, in effect, an honorary (non-fee paying) beneficiary from that cartel.

Like all profit-maximisation policies with short-term time horizons (policies, that is, with no really justifiable concern for the longer term), the policy of gas supply restraint and high prices sowed the seeds of future difficulties for the Dutch gas economy. The difficulties were not long in emerging. First, the highly profitable opportunities opened to alternative suppliers of gas to the markets in which high-price Dutch gas was sold, attracted competition; notably from the Soviet Union, but also from Algeria, Norway and even Denmark. And second, consumers rebelled against the high prices and reduced their use of gas – in energy markets which became oversupplied with many alternatives.

Within five years Dutch exports fell by over one-third, instead of increasing as had been expected and planned. Even at home, where competition from alternative suppliers was not possible, given Gasunie's monopoly over the gas transmission and distribution systems within the Netherlands, use fell sharply. Overall, instead of the more than $100x10^9m^3$ of annual gas sales planned and expected for the early 1980s, total demand for Dutch gas fell to little more than two-thirds of this amount.

Initially, however, the difficulties and dangers implicit to the overpricing of plentiful Dutch gas were masked, for both companies and government, by the buoyancy of the revenues measured in Dutch guilders (given the high value of the US dollar, on the basis of which gas prices were calculated) between 1982 and early 1985. But this was only the calm before the proverbial storm for, with the fall in the value of the dollar (from mid-1985), followed by the collapse of the oil price (from January 1986), the companies concerned have had their cash-flows seriously reduced, while the government which, as shown above, took most of the "super-normal" profits of the industry, has had its unit revenues from gas reduced by two-thirds. And there was, as also shown above, no compensatory increase in sales, given that the competition both for energy markets generally, and for gas markets in particular, was intensified rather than reduced by the changed situation.

The prospects for Dutch gas re-establishing its earlier large contribution to the national economy are now remote. This is largely because Dutch gas policy makers failed at the time of the perceived energy scarcity, to tie up potential markets with the much greater amounts of gas that customers were then prepared to take. The unjustified "conservation-of-resources" principle has been speedily undermined as a valid approach to resource exploitation in both economic and political terms. It has been equally undermined in ethical terms, if one judges it against the wisdom expressed in "the parable of the talents".[3] The lesson from this parable of the servant who buried the resource given to him by the master went unheeded by Dutch gas policy makers. For a country as steeped in New Testament knowledge as the Netherlands this is surprising. But, given this knowledge, the country itself ought not now to be too surprised by the economic misfortunes that are already flowing from the failure of Dutch society wisely to exploit the country's buried "talent" of natural gas. The "wicked and slothful servant" who buried his talent[4] was, we read, "cast into outer darkness... where there shall be weeping and gnashing of teeth".[5]

The "Dutch Disease"

The tradition, dating back to the Anglo-Dutch wars, of the use in English of the adjective "Dutch" to describe an unfavourable situation or phenomenon has continued into the recent past. One such example is "Dutch Elm Disease". In the United Kingdom, where the population of elms throughout almost all the land has disappeared as a consequence of the rapid spread of the disease, the Dutch part of the name of the disease is thought to indicate the source of the infection, rather than the country where the scientific discovery of the condition was made. Even more emphatically, the term, the "Dutch Disease" has been coined to describe countries – and their governments – in which spendthrift policies, with consequential adverse internal and external effects for the economies concerned, have been implemented as a result of the "easy money" to be made from gas – and oil – revenues. The large Dutch gas revenues – from the years of high gas prices – are thus alleged to have been frittered away in high living and ill-advised social welfare hand-outs to the work-shy and to immigrants from overseas territories. Either through default – or possibly because of political considerations – Dutch economists and others have failed to scotch the essential mythology of the charge. Indeed, there are even cases of its use by Dutch scholars – a case, perhaps, of self-flagellation as the punishment required to compensate for the "luck of the draw" which made the gas-rich Netherlands the one industrial country which, in the days of high oil prices in the 1970s and the early 1980s, had what it took to pay the rising costs of welfare services and of unemployment.

Whether or not one thinks that the Netherlands has spent too much on social welfare is essentially a matter of political philosophy. It is certainly not a self-evident proposition as it depends on what one believes the role of government to be in respect of the weakest – and neediest – members of the national family, and of those invited into a country to share the family's work-load – and its meals. And there is nothing peculiar to the Netherlands concerning these questions. They are issues which have been – and still are – debated throughout Western Europe and in other parts of the industrialised world. Much money has been spent in all countries on welfare payments. The Netherlands hardly stands apart in this respect.

It is, on the other hand, certainly true that Dutch gas revenues enabled the Netherlands to finance such expenditures more easily than other nations in the 1974–85 period. The Netherlands *could* afford to do what it chose to do in this respect. Most other countries of Western Europe and elsewhere could not. Thus, in spite of the expenditures involved, Dutch inflation has been generally and usually lower than that of other small open economies, while the guilder has remained one of the world's strongest currencies. It is thus a perfectly arguable proposition than the exploitation of its natural gas enabled the Netherlands to avoid many aspects of the economic malaise that adversely affected all western nations, to a greater or lesser extent, over the period since the early 1970s. In other words, the "disease" of the western economies is no more a specifically – or even a mainly – Dutch phenomenon, than is the widespread incidence of disease amongst Europe's elm trees over the same period of time.

There is, moreover, another aspect to the Netherlands' use of its gas revenues. This is an aspect which has not, but which most certainly should have, been presented in a highly positive and favourable way: namely, the degree to which gas revenues have been used to up-date and expand the country's physical, social and cultural infrastructure. At one extreme, the country's commitment to the vast public expenditure required for making its dykes and rivers safe against a repetition of the 1953 disaster of tide, storm and water would have been much more difficult to implement in the time available without the flow of gas-derived revenues to the exchequer.

In addition, however, the transportation systems of road, rail and water have been made second-to-none in the Western world; urban renewal has been generously financed with results that are clearly more extensive and qualitatively better than in most other countries; and in cities and towns – and even the villages – of the realm the educational, sporting, social and cultural facilities have been comprehensively attended to. Much of this has been made possible because of the flow of gas revenues, and the whole provides an object-lesson of social democracy at work, viz. the provision of benefits for future, as well as the present, generations. If any of these sorts of developments are included in the definition of what constitutes the 'Dutch Disease' then the phrase has as little validity as others of the same kind that emerged in an earlier era from mistaken impressions of Dutch society arising out of the rivalries and jealousies of the Anglo-Dutch wars.

Conclusions

The exploitation of natural gas in the Netherlands has been marked by technological excellence – in respect of the production, the transmission and the distribution systems which were built to make the country the most intensive gas using region of the world. This has been matched by a generally appropriate deployment of the financial resources created by the sale of large volumes of Dutch gas at a time when it was able to command high prices in both national and international markets – to the undoubted long-term economic and social well-being of the country's population.

On the other hand, lack of experience *vis a vis* natural resources' exploitation, inappropriate advice on depletion policies, bad intelligence on the prospects for the long-term availability of plentiful gas resources and on the most likely evolution of energy prices have cost the country dearly – in terms of the wealth that might otherwise have been created over the 10-year period from 1973 when most countries in Western Europe were willing – indeed, even anxious – potential consumers of large additional volumes of Dutch natural gas. Politicians and bureaucrats in the Netherlands firmly refused to listen to – let alone to act on – the contrary advice (now proven correct) of those of us who recommended an accelerated development of the natural gas industry in such propitious circumstances. The minds of the policy decision makers remained firmly closed to the views of non-conventional wisdom – until it was too late.

For a country renowned for its political disputes over most issues, the left-to-right consensus over gas (a consensus, that is, for conservation and for conservatism), has been extraordinary. Right, centre and left were smitten with the gas-conservationist/gas-scarcity bug. "Saving natural gas" became the watchword of the full spectrum of the political establishment, even though saving discovered gas reserves for future use was neither economically attractive, nor relevant to the likely future availability of energy for succeeding generations. Irrespective of their contrasting political beliefs and persuasions, Dutch politicians and policy makers seemed content to be counted amongst those who preferred to keep the country's God-given resources buried in the ground. Possibly, because, as with the "slothful and wicked" servant in the Gospel, they though that this was the "safest" line to take in an uncertain world full of tough taskmasters. Few expected the market to call them to account so soon for their failure to use the country's resources as effectively as they should have done. But that time has now come and the "weeping" and the "gnashing of teeth", consequent upon the inadequate exploitation of the country's gas resources while the opportunities existed, has begun – in the guise of the search for alternative government revenues and export earnings to substitute for those which could already have been secured had there been a more liberal policy on the exploitation of the Netherlands's gas resources.

References

1 See P.R. Odell's inaugural lecture on his appointment to the Netherlands School of Economics reprinted above as Chapter II-1 in this volume.

2 Meadows, D.H., et al., *The Limits to Growth*, Universal Books, New York, 1972

3 The Holy Bible, *The Gospel according to St. Matthew*, Chapter 25, verses 14–30.

4 *ibid*, verse 26.

5 *ibid*, verse 30.

Chapter IV – 7

Gorbachev's new economic strategy: the role of gas exports to western Europe*

Introduction

Earlier published western assessments of the prospects for Soviet gas exports to western Europe have been relatively pessimistic.[1] Such assessments have, however, now been overtaken by the recent momentous changes in Soviet domestic and external policies – both towards the west, in general, and western Europe, in particular. These changes carry important implications for Soviet-west European trade. Previous studies of Soviet gas exports to western Europe were all made on the assumption that there would be no fundamental politico-economic changes in the Soviet Union itself, and that the Soviet Union would be unwilling to accept – let alone sponsor – the 'zero-zero' option for nuclear arms in Europe. Thus, it was argued, the continuing Soviet inability to increase the level of gas exploitation at home, combined with the west's willingness to accept only limited supplies of Soviet gas in western Europe, because of political and strategic factors, together implied only a slow expansion of Soviet gas exports.

The assumptions on which this view of Soviet gas export prospects to western Europe was based are no longer the most likely. Instead, it is possible to assume a supply/demand relationship that will move to a much higher level than hitherto expected. This development is based, first, on the Soviet Union's intention to create conditions in which its

* **First published in *The World Today*, Royal Institute of International Affairs, London, Vol.43, No.7, July 1987, pp123–125.**

output of natural gas grows rapidly; and, second, on the emergence of a communality of interest between the Soviet Union and powerful political and economic forces in western Europe in favour of a greatly expanded flow of Soviet gas to all parts of western Europe. Clearly, these two expectations – that there will be more Soviet gas and that more of it will be exported to western Europe – are closely related; so much so, indeed, that it is possible to predict that natural gas prospects, potential and policies will provide the lubricant whereby the hitherto slowly turning wheels of change in Soviet international and domestic policies can be accelerated to an unexpectedly rapid pace.

Why gas?

How do all the elements involved come together to produce such a result? *First*, fast Soviet economic progress in the medium and longer term depends on the country's ability in the short term to increase the flows of western capital to its hard-pressed economy. *Second*, the only possible near- and mid-term way of earning most of the hard currency necessary for enhancing the flow of imports is by exporting much more natural gas and the only conceivable markets for such additional gas for the foreseeable future are all in western Europe. *Third*, these markets cannot, however, be exploited without fundamental shifts in Soviet policy on nuclear and other weapons in Europe because of the suspicion of, and the hostility to, the Soviet Union's policies in western Europe, not only among political leaders, but also amongst the overwhelming majority of the population.

Thus, Mikhail Gorbachev's offer of the 'zero-zero' option for nuclear weapons and his strong hints and nods that this will be followed by other military and political action to minimise tension in Europe, can be interpreted as the initial step in the process that would enable the Soviet Union to embark on the restructuring of its own economic system and, thus, of improving its prospects. Much increased sales of Soviet gas in western Europe in a new and more favourable east-west climate thus would enable the long-run process of change in the Soviet Union to begin.

The change in the west European attitude towards the intentions of the Soviet Union is already clearly seen in the opinion polls. These show that west Europeans are increasingly prepared to trust Mr Gorbachev. Most west European politicians are harder to convince, but this is less true in West Germany which is, of course, of critical importance for potential higher Soviet gas sales. With enough goodwill among west Europeans and the prospect for some steps at least towards multilateral nuclear disarmament, a situation will be created in which the economic component in the Soviet strategy can be initiated in the form of an offer of Soviet gas of such dimensions and on such attractive terms that, in the context of the much improved strategic and political atmosphere, a refusal would hardly be feasible – despite the embarrassment and consternation of western Europe's near-cartel of gas buyers.

With the Troll/Sleipner deal under their belts,[2] these buyers have, over the last half year, been congratulating themselves on their success in 'fixing' the gas sector supply/demand relationships in western Europe for the rest of the century in a kind of 'all-problems-solved-until-our-retirement-attitude' towards the western European market. Apart from those entities involved in the activities on the Troll field itself and on its associated transport facilities, the inertia represented by this attitude implies a degree of feather-bedding of the institutions involved which will cost the energy consumer in western Europe very dearly indeed. The consternation within the ranks of the senior management of the continent's gas industry institutions of the implications for their comfort of the gas challenge from the Soviet Union should not, therefore, cause anyone else any loss of sleep.

The nature of the Soviet offer

The essence of the economic attractiveness of the Soviet gas offer lies, however, not so much in its impact on gas and energy prices in western Europe (to be discussed below), but in the prospect of guaranteed markets in the Soviet Union for a much larger volume of goods and services produced in western Europe. West European purchases of Soviet oil and gas have, of course, usually been tied to such guaranteed sales of European products to the Soviet bloc. These have become increasingly important to relevant west European exporters, as market elsewhere have gradually become more saturated as a result of the activities of Japanese and, more recently, other Far Eastern suppliers; leading to endemic unemployment and other underemployed resources in most of western Europe. Competition has been particularly intense in Europe's other main energy supplying area, the Middle East, so presenting the Soviet Union with something of a trump card in its energy deals with west European countries.

There are, moreover, no real indications that international competition for the goods and services which western Europe produces will get less intense over the rest of the century. On the contrary, as more and more industrialising nations enter the markets, such competition seems likely to intensify, to the potential detriment of European industry and commerce. Thus a Soviet offer to triple – or even quadruple – its gas exports over the course of the next 12 to 15 years and, in return, to offer guaranteed markets for European goods and services to roughly the same value would be a development of importance to a broad range of interest groups, from political parties through organised labour, and to the supplying industries themselves. Their combined political clout will easily exceed that of those in the gas and other energy industries who might criticise the Soviet offer as a market-destabilising force – especially since additional Soviet gas would also help lower energy prices consumers have to pay.

Can it be done?

As far as the gas involved itself is concerned, there are two questions to be answered. First, can the supply be made available? And, second, can the markets be found? Doubts over the Soviet Union's supply and deliverability potential – such as found in the recent Joint Energy Programme Study[3] – appear to owe too much to the largely American-inspired views on the technological ineffectiveness of the Soviet oil and gas industry. These, however, fail to recognise that there are ways other than those of the United States in which to organise an oil and gas industry.

On Soviet gas prospects, the newly published report from the International Institute of Applied Systems Analysis (IIASA) in Austria[4] – to which the Soviet Union made a direct contribution, given the joint east-west nature of the IIASA – confirms that the Soviet Union will be able to continue to expand its industry at such a rate that an export potential to western Europe rising from its current level of under 35 billion cubic metres (Bcm) per year to three or four times this level in the fullness of time is well within the bounds of the realistic. The 'fullness of time' is itself variable – with a high possibility of a significant fore-shortening of the period, given, first, the degree of east-west detente envisaged in the strategy described above being implemented; and, second, the consequential promised enhanced supply from western Europe of gas-producing and transporting equipment, as well as technology and know-how, to the Soviet Union.

On the gas-supply side, the development of the enhanced level of production required is guaranteed by the massive proven and prospective gas reserves of the Soviet Union – and by their producibility. What has been missing, hitherto, has been the will and the means to implement expanded production. These, however, may now be assumed to be an integral part of the macro-economic and political strategy described above.

The west European prospect

The more important gas issue at stake in the strategy relates to the prospects for markets in western Europe. To date, the development of the markets has been restrained by a combination of institutional inertia and regulation. These are, however, surmountable limitations, providing that the forces working for gas expansion prove powerful enough. Indeed, western Europe has an energy market structure which is inherently much more favourable to high gas use than the energy markets of either the United states or the Soviet Union. In both of these countries the generally low spatial intensity of energy demand limits gas use to only limited geographical parts of the markets. Nevertheless, in spite of this inherent limitation on markets' developments, gas has achieved a 30 per cent share of the total energy used in the two countries. By contrast, the more densely populated and much more geographically energy-intensive market of western Europe remains gas-use restricted to such a degree that the current approximate 18 per cent share of gas in the west European energy supply is very low compared with the potential.[5]

Inertia and regulation have generally dominated the marketing of gas in western Europe. Gas pricing, both at the point of import and at the burner tip, has reflected these basic conditions. What is needed to overcome these constraints is that gas be made available at an import price which could lead to gas being seen as a potential competitor in almost every sector of the energy economy. In today's and tomorrow's energy market conditions in western Europe this means that imported gas should be available at prices related to those of the lowest-cost alternative, viz. imported coal, rather than to oil prices. Internally, the pricing of gas at the burner-tip could still reflect the specific alternatives' costs, but if gas were available at a basic import price enabling it to break into the bulk steam-raising and power generation markets, as well as into markets currently dominated by other energy sources,[6] then gas' current 18 per cent share of the total west European energy market would become but an interim staging post on the way towards a much higher achievable level – of up to 35 per cent.

Soviet gas export sales will have to be guided, in the context of the strategic considerations described previously, to such a required pricing approach. Thereafter, European energy customers will be able to use much increased volumes of lower-cost gas in the context of the progress towards east-west détente (eventually to overwhelm the propensity of institutions and states to secure the economic rent from marketing cheap gas expensively) and so enable gas in western Europe to achieve its 'rightful' market share. The benefits to be achieved from such an expansion of gas use will be much too large to ignore. Their denial to west Europeans by the restrictive practices of existing energy market institutions will become politically impossible.

The benefits to the Soviet Union of the use of its natural gas as the means of promoting east-west détente within Europe and of enabling the Soviet economy to expand in order to meet the desires of most of its inhabitants are self-evident. But would exporting 100–140 Bcm of gas to western Europe be profitable for the Soviet Union? The Joint Energy Programme Study[7] has expressed doubts about the 'profitability' of such a development, but the study has failed to take into account the possibility of economies of scale and, even more important, of reduced technical costs through application of the best western technology to Soviet production and transmission systems. Such considerations are, however, at the heart of the Gorbachev strategy, and they must be applied to gas at the very beginning of the new political and economic processes now under way in the Soviet Union. This is, indeed, the only way in which the country's only significant potential export in the short-term can be realised and so provide the long-term means for more general progress elsewhere in the Soviet system.

Conclusion

In other words, the foreign-exchange earning power obtainable by the Soviet Union from a tripled or quadrupled export of gas (even when this has to be sold, on average, at a price lower by up to one-third than present Soviet gas export prices in western Europe) is the essential seed-corn on which future Soviet economic 'restructuring' and expansion under the Gorbachev regime depends. Without this, the expansion required seems unlikely to be possible in the time available. For western Europe, the prospect of securing a large part of its total energy supply at a price much lower than hitherto foreseen, in the context of lessened political tension, presents advantages that justify efforts to ensure that its restrictive gas-using policies generally, and its restrictions on Soviet gas imports in particular, are removed.

References

1 For example, the study published under the auspices of the Joint Energy Programme of the Royal Institute of International Affairs and the Policy Studies Institute, by Jonathan P. Stern, *Soviet Oil and Gas Exports to the West: Commercial Transactions or Security Threat?* (Aldershot, Hants. Gower, 1987).

2 For background, see Jonathan P. Stern, 'Norwegian Troll gas: the

consequences for Britain, continental Europe and energy security', *The World Today*, January 1987.

3 *Op. cit.*, Joint Energy Programme Study.

4 H-H Rogner, *The IIASA International Gas Study* (Vienna: International Institute for Applied Systems Analysis, 1986).

5 P.R. Odell, *Natural Gas in Western Europe: a Case Study in the Economic Geography of Resources* (Haarlem: De Erven F. Bohn NV, 1969).

6 American Gas Association, The Outlook for Gas Demand in New Markets 1986–2000 (Arlington, Va.: AGA, 1986).

7 *Op, cit.*, Joint Energy Programme Study.

Chapter IV – 8

The Completion of the Internal Energy Market: On the Need to Distinguish the Hype from the Reality*

1. Introduction

There is a presumption that the establishment of a single European market in energy will significantly reduce energy prices to the direct benefit of individual consumers, in general, and of energy-intensive industries, in particular. The Community's energy industries are also presumed to benefit from consequential (and inevitable) complementarities and from the reduced costs of production, transmission and distribution through integration and rationalisation. Finally, it is argued, the combined effects of the changes will be advantageous to Europe's economic performance and to its global competitiveness.

Studies undertaken to date for the Energy Directorate of the Commission of the EEC, are by no means unequivocal in their conclusions of large net benefits. The results of the quantification attempted have, moreover, been modified by caveats. This should not be considered surprising, given the complexity of the analysis which is required. An effective evaluation requires consideration of energy costs and prices throughout the economic system in which there is no part which is excluded from the effects of energy sector changes. As a result, the studies have stopped short of examining the overall impact of the totality of the process of Europe's conversion to a completed internal energy

*** A Paper presented to the Forum Europe Seminar, *A Single European Community Energy Market: Lower Costs or Uncertainty of Supply*, Brussels, January 1990.**

market and have concentrated instead on evaluating the relatively limited and partial proposals for reducing barriers to energy market competition and trade: as set out in the preliminary reports of the Energy Directorate on the issues. In any case, the possible macro-economic effects of an energy policy relating to the completion of the internal market is only part of the input to policy determination in the energy sector. In addition, questions relating to the Community's external energy sector relationships, to security of supply, to institutional arrangements, to social welfare, employment and regional issues and, last but by no means least, to environmental questions all have to be taken into account to ensure that appropriate energy policies are pursued. Indeed, each of these issues justifies and requires intervention and regulation, so that a fully-liberalised energy sector cannot reasonably be envisaged even in a completed internal energy market at the European level.

In as far as the costs, *first*, of energy security (in terms of supply diversification requirements and of minimum levels of stocks of imported energy); *second*, of protection for communities exposed to a declining demand for traditional deep-mined coal; and *third*, of taxes imposed on energy use to curb demand (in order to achieve reductions in environmentally dangerous emissions), will collectively lead in the 1990s to significantly higher real energy prices to consumers than those which have recently become the norm, there seems unlikely to be much justification for too great a concern for the relatively small overall savings in the economy which the Community might secure from a single market in energy. The achievement in practice of that which is apparent in the realm of the theoretical could well prove to be a "will of the wisp". This conclusion does not mean, however, that major moves towards the establishment of a more integrated energy market are not justified; nor that the energy sector in Europe cannot be made much more efficient and much less costly. This paper is devoted to showing that institutional and other barriers to the efficient production, movement and use of energy have produced higher costs than necessary and that they need to be overcome as a matter of urgency, for economic as well as for other reasons.

2. The Case for the Single Energy Market

In reality, the case for the completion of the internal energy market is related to issues other than the creation of overall net benefits to the macro-economy arising from cost – and price – reductions in the energy sector. It should rather be seen, *first*, as an objective which symbolises Europe's integration, given the high profile which marks energy sector issues, including the existing large policy and price differentials currently existing between the member states of the EEC; and *second*, it is an important, if not, indeed, an essential, objective in the Community's attempt to create "level playing fields" for competing enterprises in the member states. Success in achieving these objectives will be an enhanced degree of European-ness and the equitable (but not necessarily equal) treatment of energy suppliers and users across the Community. These are not trivial objectives in the context of the objective of a single market by 1992 and the subsequent political moves towards the creation of a more-fully integrated Europe over the final decade of the century. Neither are they objectives which stand much chance of being realised simply by accepting the energy sector as it happens to have evolved to date, and/or as it might develop under the influence of "market forces". There are far too many powerful institutional forces at work on both the supply and the demand sides in the energy sector to justify having much faith in the ability of the proverbial "invisible hand" to secure the objectives which are sought. Community-level intervention is a requirement for energy-sector changes whereby the objectives, as defined above, may be achieved.

3. What needs to be done?

a) In respect of Coal and Oil?

For coal, on the basis of which the wheels of Europe mainly turned and its people were kept warm for the first 200 years of the continent's development, there is little to be done which is worth doing (though much more that could be done without it having any real significance for the objectives of Single Energy market). Indigenous coal continues its inevitable decline. This is the result of a combination of increasing geological difficulties in producing the already much depleted reserves and of high and still rising real wage rates in a labour-intensive industry. They jointly make the supply price of the commodity out of line with the cost of internationally traded coal from other parts of the world. This differential seems likely to widen through the 1990s so that levels of coal production in the Community will continue to fall. Thus, the hitherto important issues of subsidies to indigenous coal producers – payable either by governments or by the consumers of the coal in order to protect the communities and regions dependent on this traditional energy resource – will become of decreasing relative importance. In as far as imported coal is continuing to substitute production in Europe and will secure additional markets in the expanding electricity generating sector (though note that both of these hitherto expected growing markets for imported coal are now at risk from the perceived dangers of the greenhouse effect on the world's climate), there is a relatively level playing field for coal in the Community, except for a number of diminishing fiscal constraints and the declining impact of volumetric controls. As has always been the case for coal, it is distance from supply points and the volumes required in specific localities – combining to establish highly variable unit transport costs – which will determine the competitiveness of low-cost imported coal in the market places of the Economic Community: with or without the implementation of the completion of the Internal Market.

The same is largely true for oil, though for very different reasons. Over the whole of its recent and relatively short period of importance in Europe's energy supply (dating back only to the late 1950s and reaching its zenith a mere 20 years later), the industry has, in large part, been organisationally structured to a European-wide market through the dominance of the international oil companies with their predilection for large-scale operations and for supply, refining and distribution systems orientated to their own multi-national organisation. Moreover, Europe's hunger for oil imports in the days, *first*, of energy scarcity after the end of the World War II and, *second*, in the period of very rapidly expanding energy demand from 1955 to 1973, produced a near-zero restraint on oil trade, in respect of both trade to, and within, Europe. Finally, the price of oil to – and in – Europe has been determined exogenously, within the context of the international market, so that prices – excluding local taxes – have not varied greatly from country to country, especially as all national markets expanded and allowed the economies of scale achieved to be passed on to consumers everywhere. As a result of this background there are few elements which restrict the already near-completed internal European oil market.

Even the growth of the Community's own oil production since the mid-70s has not adversely affected this situation. The dominant supplier, the United Kingdom, and one of the smaller suppliers, the Netherlands, have both willingly respected and fully accepted the appropriateness of policies which leave oil supply, pricing and allocation decisions to the producing companies: influenced by their role as countries with important interests in the maintenance of the well-being and the pattern of operations of two of the international oil corporations, Royal Dutch Shell and British Petroleum.

Meanwhile, at the other extreme, the three most recent members of the Community, viz. Greece, Portugal and Spain, each have oil sectors which are state-owned or state-directed and from which competition has hitherto been generally excluded. The elimination of monopoly from the oil sector in these countries, through organisational restructuring and the acceptance of competition from companies established elsewhere in the Community, was, however, negotiated as one of the conditions of entry for the countries concerned. Thus, changes in the oil industries of Greece, Portugal and Spain are already under way; albeit, not at the speed required to achieve the defined objectives of market liberalisation within the time-frame allowed. As previously, however, in the cases of the diminution of state controlled industries in France and Italy, it is not unreasonable to anticipate that the Common Market's new members will eventually have oil sectors not unlike those of the other member countries. The completion of a European international oil market is, in other words, only a matter of time.

The single important unresolved issue is that of continued major variations between countries in the levels of taxes imposed on oil products. Many of the variations arise from purely fiscal considerations and are thus part of the wider problem relating to tax harmonisation. In this context, questions relating specifically to the "levelness" of the oil sector "playing field" are of secondary importance. Other issues of taxes on oil products (such as the tax on fuel oil in the UK to protect indigenous coal, or the taxes on burning oils in Denmark and the Netherlands related to environmental questions), whilst remaining to be solved as part of the completion of the internal market, hardly constitute a serious threat to the essential European-ness of oil or to the creation of the level playing field in the sector.

The same is true of the actual discrimination exercised by the oil-rich countries of the European Community in their allocation of concessions for exploration and production and/or in their requirements for some initial national control of the oil produced. Proximity and familiarity usually provide good enough reasons for national preferences extended by governments in respect of such activities. This, together with a high degree of mutual recognition by all member countries of the special attributes which attach to the exploration for, and the exploitation of, a country's "natural" resources and the absence of any seriously expressed concern by oil companies of other member countries for national preferences to be extended, indicate that the Commission need not worry itself greatly about limitations on the completion of the internal market arising out of the Community's increasing ability to produce its own oil.

To date, and for as long as OPEC pursues production limitation policies which keep the price of oil high enough to ensure a continuing commercial interest by the oil companies in investing in indigenous

European oil exploration and production, the Community does not have to worry about a requirement to protect the relatively high cost production of oil from its member countries (compared, that is, with a long-run supply price for international oil in a fully competitive market estimated at between $5 and $8 per barrel); or of the degree to which such protection might or might not be acceptable to have-not member countries on, say, security of supply grounds. It would thus be somewhat ironic if the question of protecting indigenous oil production were to become a major internal energy market issue in the early to mid-1990s. Indeed, the ideal has not even been discussed, given the Commission's conventional-wisdom view of a tightening international oil market with rising real oil prices over the last decade of the century. Yet the probability of a low international oil price eventually emerging from the continued – and even an expanding – availability of production capacity in the main exporting countries (in relation to the rest of the world's demand for their oil) is by no mean a small one. Indeed, I would currently attach a better then one in four chance of such a development and thus advise early consideration by the Community of the implications of such a development of the single market for energy, in general, and for policy towards indigenous oil, in particular.

b) In Respect of Natural Gas and Electricity?

Natural gas and electricity are neither "Europeanised" already, nor dependent on international markets for the determination of price and other conditions of supply. Thus, there are significant gas and electricity price differentials between member countries of the Community (as shown in Table IV-8.1) as well as major contrasts in market penetration (as seen in Table IV-8.2) which are unrelated to resource supply and demand conditions as such. They are rather a function of institutional forces which are currently permitted, or even required, to influence supply/demand relationships and which often lead to monopolistic conditions – either *de jure* or *de facto*. The resulting tightly controlled and/or regulated regimes, working and organised essentially at the national level (except for negotiations for foreign supplies and/or foreign markets) ensure non-equitable conditions for users of natural gas and electricity across the Community: and so constitute major barriers to the completion of the internal energy market, with its objectives of European-ness and the creation of level playing fields for both producers and consumers.

The current proposals by the Commission for enhancing the technical and economic efficiency of the Community's gas and electricity systems require that their integration, increased inter-country trade and price transparency are achieved in the context of the present organisational structure, with the organisations themselves expected to generate the relevant proposals. As such, they do not get to the heart of the problem. They seem likely, moreover, to be doomed to failure (or, at very least, to long delays in implementation), as a result of powerful institutional opposition to any radical change. If necessary, the institutions seem prepared to use the courts as a means of securing decisions which will sustain their pleas for the maintenance of the *status quo*. In legal terms, the 1987 Campus Oil case decision by the European Court should not be forgotten. In this case the Court ruled that Ireland's security strategy justified the restriction that the Irish government had imposed on oil products imports from refineries elsewhere in the Community as the means whereby the Whitegate refinery in Cork (a state-owned facility) could be made viable. Similar issues of dangers to national security arising from the possible undermining of the rights of national institutions to serve national markets with gas and electricity seem likely to be raised in the event of any attempt by the Commission to declare such national approaches to the supply of these commodities incompatible with the requirement for the completion of the internal market. There are also other arguments – against the imposition of changes in the gas and electricity sectors by Brussels – that may be used to inhibit change, viz. that the distribution of electricity and gas can be considered as constituting natural monopolies: that statutory gas and electricity corporations (whether state or private) have a quasi-public service nature shown by the legal duties on firms to ensure continuity and reliability of supply; and that monopolistic gas and electricity utilities have wider economic and social objectives, the fulfilment of which would be obstructed if markets or if transmission and distribution systems had to open up to competitors.

There are thus clearly issues of substance in political and economic terms – as well as legal considerations that are derived from the Treaty of Rome – which indicate that the barriers to competition and to the free movement of gas and electricity are not going to fall at the first blast of the European 1992 trumpet. Indeed, it seems likely that 1992 will be the year in which the battle over gas and electricity is joined (on the assumption that the Commission does not back down in the meantime), rather than the date when a fully operational free internal market with open competition and rights of Third Party access to transmission systems is established.

Integration of systems, enhanced exchanges and equitable prices, both between countries and between types of consumers, seem more

TABLE IV-8.1

Gas and electricity prices to large industrial users in six common market countries in 1989

a. Gas

i. Prices on 1 June 1989 (in ECUs/1000m^3)

Type of User	Belgium	France	Italy	Neths.	W. Germany	U.K.	Range of Prices
50 million m^3/yr	108.14	122.50	99.45	86.41	99.45–124.99	133.15	86.41 to 133.15
110 million m^3/yr	105.97	111.40	94.55	83.70	94.57–130.96	124.99	83.70 to 130.96
Interruptible (lowest price)	89.13	101.07	89.65	–	99.45	86.96	86.96 to 101.07

ii. Prices on 1 December 1989 (in ECUs/1000m^3)

Type of User	Belgium	France	Italy	Neths.	W. Germany	U.K.	Range of Prices
50 million m^3/yr	115.28	113.98	107.54	102.08	114.47–137.79	111.50	102.08 to 137.79
110 million m^3/yr	112.99	112.99	103.34	98.62	109.53–132.82	107.04	98.62 to 132.82
Interruptible (lowest price)	95.65	112.99	91.69	–	109.53	83.75	83.75 to 112.99

b. Electricity

i. Prices on 1 June 1989 (in ECUs/1000 kWh)

Type of User	Belgium	France	Italy	Neths.	W. Germany	U.K.	Range of Prices
25MW at 60% Load Factor	43.64	44.09	43.94	39.26	70.06	46.81	39.26 to 70.06
80MW at 86% Load Factor	34.73	34.28	30.50	34.28	57.98	43.94	30.50 to 57.98

ii. Prices on 1 December 1989 (in ECUs/1000 kWh)

Type of User	Belgium	France	Italy	Neths.	W. Germany	U.K.	Range of Prices
25MW at 60% Load Factor	47.40	44.39	53.84	44.11	71.10	42.61	42.61 to 71.10
80MW at 86% Load Factor	38.08	34.52	39.05	38.91	58.77	40.00	34.52 to 58.57

Source: From Energy Price Comparisons prepared by Energy Advice Ltd. of London on a quarterly basis for private clients.

likely to emerge (though not by the end of 1992) from a more diversified pattern of supply and from the establishment of consumers' rights to choose their suppliers from amongst the broader set of alternatives. Some member countries – notably the UK, in respect of gas, and the Netherlands, in respect of electricity – are already moving in these directions, though even in these countries progress has so far been slow.

TABLE IV-8.2

Market penetration by natural gas and electricity in the member countries of the E.E.C., 1978–1988

Country	% of Natural Gas in the Primary Energy Supply		% of Electricity* in the Total Final Consumption of Energy**	
	1978	1988	1978	1988
Belgium	20.6	18.0	18.9	29.2
Denmark	0	7.9	30.5	33.1
France	16.1	11.2	28.6	41.8
Greece	0	0.4	24.0	32.5
Republic of Ireland	0	13.4	17.2	33.7
Italy	15.5	22.0	24.9	34.2
Luxembourg	11.9	10.2	19.6	26.7
Netherlands	43.4	40.4	21.6	23.4
Portugal	0	0	31.5	43.7
Spain	2.1	4.0	38.8	36.6
United Kingdom	17.9	23.0	21.6	35.8
West Germany	15.5	16.1	20.4	36.0
E.E.C. Average	15.6	17.4	c.23	c.36
Range of Values	0 to 43.4	0 to 40.4	17.2 to 38.8	23.4 to 43.7

Notes: * *Electricity use value calculated on the basis of the amount of fossil fuel used (or, in the case of primary electricity the amount which would have been required) to produce the electricity. This inflates the apparent share of electricity in final energy use generally, and, in particular in countries (eg. France) where nuclear power and hydro-electricity is important.*

** *Final Energy Consumption excludes wood and other biomass etc.*

Sources: Based on IEA Annual Country Reviews, B.P.'s Statistical Review of World Energy and UNIPEDE data.

This arises from the inbuilt inertia which exists in the systems as a result of the continuity of the power of the pre-existing institutions. An EEC requirement for gas and electricity liberalisation, in the context of the Commission's objectives of progress towards the completion of the internal market, would provide encouragement for change where change is already under way, and bring pressure to bear in countries where all competition within the gas and electricity sectors has, to date, been excluded by national arrangements. "Open access" to power transmission and gas pipeline transport systems could be an integral part of the changes leading to competition, but merely legislating for this does not necessarily lead to any results. This is true even where the legislation is accepted without challenge – as in the case of agreed open access to gas transmission facilities in the UK since 1984. As shown above, such unchallenged proposals in respect of directives emanating from Brussels are unlikely.

It thus seem more likely to be the right of customers to make supply contracts directly with potential new (or old) suppliers which will be the more important early-day tool to break the log-jams of the monopolistic gas and electricity systems with which Europe has been endowed as a consequence of past policies. If necessary (because of the lack of cooperation by transmission companies), or if appropriate (in an economic sense), such supplier/customer agreements require the commitment – as well as the right – of the parties to create new gas and electricity transmission infrastructure independent of pre-existing facilities and their monopolistic operators. The justification for a suggestion which initially might appear to be a higher cost solution to the liberalisation of the market, compared with pressing on with the attempt to make the present operators do what they clearly do not want to do, is set out in the final section of this paper – in respect of natural gas.

Those who are more closely concerned with the electricity industry will be able to judge if my arguments also apply to that industry: two observations would, however, appear to be in order at this stage. First, many of the economic principles and the organisational framework which apply to natural gas in Western Europe appear also to apply in the case of electricity. Second, liberalising supplier/consumer relationships for gas in itself opens up the prospects for the liberalisation of the provision of electricity. This is because natural gas expansion (as implied in the proposal for liberalisation) creates, through its use for power generation in combined heat and power and in combined cycle systems, the opportunities for a dispersed and low-cost pattern of electricity production. Those opportunities provide a means whereby the rigidities of the existing conventional coal, oil and nuclear power based electricity system and, in particular, its requirement for large-scale production and organisation, will be undermined. A multiplication of the number of potential electricity suppliers seems likely to work wonders for the prospects for a liberalised electricity supply and distribution industry.

4. Towards a Liberalised, an Expanded and an Economically Efficient Community Gas Market

Liberalisation, expansion and economic efficiency (as contrasted with the technical efficiency already achieved) in the European gas market constitute a set of desirable changes which, I would argue, can only be achieved collectively: and none of which can be achieved in isolation. The packet is inevitably complex and its implementation clearly fraught with difficulties – both political and legal, as already shown in the examination of the liberalisation elements. These considerations imply that there is no way in which the process can be more than merely started by 1992. Its completion before 2000 seems unlikely. "Sooner begun, sooner completed", however, implies the need for urgency in overcoming the initial difficulty, viz. the existence of an institutionally engendered bottleneck in the expansion of the community's natural gas market.

The facts of this restraint on expansion are set out in Table IV-8.3; viz. a mere 1.7% per annum increase in gas use since the mid-70s and only a 1.1% increase per year since 1980; over a time period, that is, when gas reserves were expanding rapidly (as shown in Table IV-8.4), when oil was not only expensive, but also considered to be an unreliable source of energy and when the production and use of inherently higher cost energy (viz. coal and nuclear power) was being deliberately encouraged. The institutional restraints on increased gas use have been partly related to opposition from competing energy suppliers (especially the coal and nuclear power lobbies) and partly to a widely-held perception of gas scarcity. There were, however, also elements of, at worst, a wanton refusal to expand by the gas industry (or by governments which refused to allow it to expand) and, at best, a lack of entrepreneurial skills and motivation by the gas industry's managers in a competitive energy market. The restraints also included the deliberate limitations of gas use in power generation (even in high efficient C.H.P. and combined cycle plants) and a high price marketing approach to gas used by large and intensive energy users, to the detriment of their international competitive position.

The effective exploitation of these markets by an expanded use of gas would be politically, economically and environmentally advantageous for the European Community. Politically, it would help in respect of security of supply, as it would produce a greater diversification of energy supply by source: the 1989 contribution of gas to the Community's total energy supply is, at under 18%, little changed from a decade ago. It could be increased to almost 27% by 2000 with expansionist policies (see Table IV-8.5). Economically, gas prices to large energy users (including gas for power generation) could be reduced to the advantage not only of the industries concerned, but also to the Community's overall advantage as a consequence of the improved internationally competitive position created for the sectors involved and by the reduction of industrial electricity prices that are also made possible by lower gas prices. Environmentally, the substitution of gas for other fossil fuels will, moreover, reduce the levels of emissions of all atmospheric pollutants as a result of the inherently cleaner burning characteristics of natural gas

TABLE IV-8.3
Natural gas use in (present) member countries of the E.E.C.
1976–1988 (in $m^3 \times 10^9$)

Country	1976	1980	1984	1988
Belgium/Luxembourg	11.3	11.4	9.4	8.8
Denmark	–	–	0.1	1.5
France	20.9	26.0	25.9	26.0
Greece	–	–	0.1	0.1
Republic of Ireland	–	0.6	1.9	1.3
Italy	24.2	25.2	29.2	36.6
Netherlands	36.3	33.5	34.3	32.7
Portugal	–	–	–	–
Spain	1.6	2.0	2.2	3.7
United Kingdom	38.1	45.5	49.6	52.6
West Germany	39.9	48.8	45.2	47.4
Total	*172.3*	*193.0*	*197.9*	*210.7*
% Increase in successive Four Year Periods	12.2%	2.5%	6.5%	

Average Annual Rate of Growth, 1976–1988 = 1.7%
Average Annual Rate of Growth, 1980–1988 = 1.1%

Source: B.P.'s Statistical Review of World Energy *1989 and earlier years' issues of B.P.'s* Statistical Review of the World Oil Industry.

TABLE IV-8.4
Evolution of W. Europe's natural gas production and reserves
1956–1995 (in $m^3 \times 10^9$)

Year	Cumulative Production to date	Cumulative Production over the decade	Decade-End Declaration of Proven/Probable Reserves	R/P Ratio (years)	Total Original Recoverable Reserves	Gross Additions to Reserves in Decade	Net Additions to Reserves in Decade
Pre –1956	50	35	c.500	c.40	550	n.k.	n.k.
1956–1965	225	175	c.1,900	c.76	2,125	1,575	1,400
1966–1975	1,150	925	c.4,350	c.27	5,500	3,375	2,450
1976–1985	2,700	1,550	c.6,900	c.38	9,600	4,100	2,550
Forecast							
1986–1995	c.4,450*	c.1,750*	c.9,250**	c.46	c.13,700	c.3,700	c.2,350

* *Assuming the continuation of present policies.*

** *On the basis of fields discovered by end-1986. Since then more fields have already been found and there is, of course, a near-zero probability that no more fields will be found, given the continuation of an extensive exploration effort in many W. European countries and their off-shore areas.*

Source: Author's research and estimates.

and the fact that it can almost always be burned with a higher overall thermal efficiency than the alternatives.

The speed and intensity with which these significant advantages from an expanded gas market can be achieved depend, however, on the liberalisation of the presently restrained and regulated gas supply and transmission system. Potential suppliers (in the short term comprising those companies which have already discovered gas, but are unable to market it because of demand constraints; and in the longer term by those companies which would be encouraged to accelerate and intensify their gas exploration activities), must have the right to seek customers directly, rather than being forced to depend on the restrained demands of a monopsonistic purchaser. Customers, in turn, must have the right to seek supplies from the most advantageous source, rather than being restricted to the one option of the national or regional monopolistic supplier. Between them, the potential seller and the potential buyer or set of buyers have to be prepared – as well as legally able – to build the necessary transmission pipeline whereby the sales/purchase agreement can be implemented: unless, of course, the existing statutory transmission authorities eventually determine that they can offer the transport facilities at an acceptable price. However, in as far as most of the latter emphatically insist that their systems are already full and in that the demand expansion envisaged is mainly for new markets for gas, then recognition of the immediate appropriateness of new transport capacity for a rapidly growing gas sector in the European community is of the essence. It is also essential that the opportunity to build such new capacity is extended to other than the statutory and other conventionally acceptable entities.

The fact that this proposal is neither economically unattractive, nor a physical impossibility in a densely populated and developed Western Europe is shown by the recent proposals in both the United Kingdom and West Germany for such initiatives. The freedom to undertake such developments – and the Commission's encouragement of them – seems a much more likely bet for effectively liberalising the Community's gas market in the 1990s than through the political and legal battle which will undoubtedly be required to secure "freedom of access" to pre-existing systems. Moreover, the speed with which the owners of the latter systems might well react to the perceived 'threat' to their anticipated revenues from the construction of alternative suppliers/customers dedicated pipelines may produce one of the biggest surprises in Europe's energy sector development in the 1990s. They could well find, after all, that they do indeed have capacity enough to handle the

TABLE IV-8.5

Alternative potential Western European energy supply patterns in 2000 under:
a. continued constraints on natural gas markets
b. the development of a competitive gas market
(Quantities in millions of tons of oil equivalent)

	1988 Energy Supply Pattern			Alternative Patterns in 2000 A.			B.		
	Total	Indigenous	Imports	Total	Indigenous	Imports	Total	Indigenous	Imports
Natural Gas	199	150	49	220	130	90	335	205	130
%	17.4			17.9			26.7		
Oil	594	199	395	595	125	470	580	200	380
%	52.0			48.3			46.2		
Coal etc.	264	183	81	275	150	125	220	120	100
%	23.1			22.4			17.5		
Primary Elect.*	86	86		140	140	–	120	120	–
%	7.5			11.4			9.6		
Total Energy Supply	1143	618	525	1230	545	685	1255	645	610
%	100	54.0	46.0	100	44.3	55.7	100	51.4	48.6

Source: 1988 data from B.P.'s Statistical Review of World Energy, *1989. Alternative patterns in 2000 based on the author's research.*

extra gas on behalf of the sellers/buyers involved – at prices which are attractive enough to justify a deal being made. *De facto* 'Open Access' will, in effect, be achieved.

The final component in the liberalised/expanded/economically efficient market hypothesised for the later 1990s relates to the price of gas. In the existing non-competitive gas market the basis for pricing gas is generally derived from relevant oil prices: in which the central element is the frontier price of Dutch export gas based on a melange of reported Rotterdam values of low sulphur fuel oil and gas oil. Costs to final users are then built up on the basis of charges made by the statutory transmission and distribution monopolies. Even in the Netherlands itself, where transport and other costs are at a minimum by West European standards (though still related to an uncontrollable cost-plus calculation by Gasunie, leading to charges which are high by North American standards), this approach to price formulation makes gas inherently expensive compared with alternative energy sources for most large users. Indeed, for power generation and for much energy intensive industry, it is imported coal (with its internationally determined price) and nuclear power from already completed power stations (from which additional electricity can be offered, in a period of surplus producing capacity, at prices which relate to short-run marginal costs), with which natural gas has to compete. Contracts between non-statutory gas suppliers and potential large-scale users of the gas will necessarily have to take these effective alternatives to gas into account in their gas price discussions and agreements.

Meanwhile, in order to ensure that Europe's energy intensive industries are no longer penalised in respect of their energy costs – in relation to their competitors elsewhere – and in the interests of "levelling the playing field" for such large-scale energy users within the Community, Commission pressure is required to ensure that gas is generally made available by the statutory suppliers to the consumers involved at prices which are the equivalent, at the burner tip, to those at which imported coal or nuclear electricity is already available.

5. Conclusions

The completion of the internal energy market, in the legalistic framework in which it is usually presented, requires attention to a wide range of issues; some are important and some are not; and some of them can be tackled, while other verge on the impossible. This paper has attempted to distinguish the important from the unimportant and the practical from the impracticable and it has suggested that even the important issues that are capable of resolution will take years. It will be well beyond 1992 before the "playing fields" of the sector begin to approach the theoretical "levelness" required by the Single European Act. More positively, however, it has shown that the areas of importance with respect to coal and oil are limited and hardly worth a major effort – unless there should be a near to mid-term future collapse in the oil price. For gas and electricity, on the other hand, the hitherto intensely nationalistic organisation and development of the industries creates massive scope for Europeanisation: but it also shows that any head-on attack on the privileges of, and the discrimination exercised by, the largely state-owned or state-regulated authorities appears likely to lead to prolonged political and legal tussles.

Such an unrewarding and frustrating approach may not, however, even be necessary in order to secure change which goes a long way towards levelling the European energy playing fields and for the establishment of gas and electricity industries which are more European and less nationalistic in their attitudes and policies. Change in the gas industry first is critical in this respect as not only economics, but also politics and the environment, require a fundamentally different approach from the one followed to date for incorporating natural gas into the European economy. In this process of change, liberalisation, expansion of the industry and emphasis on economic efficiency go hand in hand. Once Europe's gas industry has been changed, there would then be the possibility of an alternative, a more dispersed and a potentially more competitive supply system for electricity, the subsequent creation of which would provide a basis on which a more liberal European Community electricity market could also be progressively implemented to the further benefit of individual energy consumers in general, and of energy intensive user industries in particular: and, in so doing, thus fulfil the central aim of the Single European Act.

Chapter IV – 9

Europe's energy: panic over, opportunity knocks*

'Geopolitics' and 'problems' have, in general, a 'strawberries-and-cream' association: in particular, the geopolitics of energy are almost invariably concerned with perceived threats to access to energy resource, disruptions to energy transportation systems and fears that energy production might have a bad impact on the environment. 'Doom-laden' is usually the most apposite description to attach to such analyses.

The European energy situation has had its share of such geopolitical foreboding over the past 25 years: ranging from fears for the economic survival of a region largely dependent on energy imported from one of the world's most unstable regions (namely, oil from the Middle East); through concern from the security of pipelines bringing oil and gas from the Soviet Union during the period of the Cold War; to environmental worries arising from forecasts of high exponential growth rates in energy use.

The problems as defined all turned out to be grossly over-stated. They were, instead, effectively handled by the operation of market forces and/or by appropriate energy sector policies. Even more vital in producing a situation today in which there is an essential absence of European geopolitical problems over energy has been the emergence of important – but generally unforeseen – components in the region's energy supply/demand relationships.

* Reprinted from *The World Today*, (Royal Institute of International Affairs, London), Vol.51, No.10, October 1995, pp.191–192

Western Europe's total energy use has grown by a mere 12 per cent in the 21 years since 1973. Even in the period since 1986, after the collapse of the international oil price, the average annual increase in energy use has been under 0.5 per cent. Moreover, these total energy use figures obscure an even smaller 7 per cent increase in the consumption of fossil fuels since 1973. This has minimised potential geopolitical problems arising from uncertainties over their availability. And it is also constraining the impact of political concern for atmospheric emissions from the combustion of fossil fuels. Europe's success in severely limiting its additional requirements for energy since 1973 has improved its bargaining position in international negotiations over energy-related environmental issues.

Europe's oil supplies

To date, however, the *realpolitik* of European issues has remained supply-side orientated. Thus, the rapid decline in indigenous coal production from the early 1950s – and the coincident rapid growth of energy demand – had produced a serious situation by the time of the oil supply crisis of the early 1970s. An almost 45 per cent energy requirement exposure to Middle East oil, in particular – and a 65 per cent overall exposure to imported oil, in general – constituted a geopolitical problem of significant dimensions.

Twenty-five years later, this has effectively become a non-issue; partly because oil is now less important in the total energy supply, but mainly because dependence on imports has been more than halved by the 15-fold expansion since 1970 of Western Europe's own oil production, from under 20 to almost 300 million tons per annum. Net oil imports are now only 25 per cent of total energy requirements, while imported quantities required have fallen by 55 per cent from 730 to 340 million tons per year. Thus, any short-term supply difficulties, arising from interruptions to supplies for military or political reasons, would not necessitate any demand limitations (other than those engendered by higher prices). A combination of the use of surge capacity in indigenous production facilities, plus a modest run-down of stockpiled oil (now the equivalent of about 245 days of import requirements), would ensure availabilities.

Over any longer period of disruption, Europe's need for imported oil could be reduced through the application of more stringent demand management and of alternative supply mechanisms, based on the intensification of the exploitation of indigenous producing fields and the acceleration of the development of new fields. But, under almost all foreseeable circumstances in the prospective international oil market, the need for such policy action in Europe is highly unlikely. Other existing oil exporters have a strong motivation – and the productive capacity – to ensure that market shortages created by a supply interruption from one or more exporters are expeditiously filled. In effect, the geopolitical dependence phenomenon has now been reversed.

There is, moreover, no time horizon in sight for a reversal to the *status quo ante*. The prospects for continuing high flows of oil from Western Europe's productive areas are good. Investment in the more intensive exploitation of known basins – notably in the North Sea province – continues, while in new basins (such as the Norwegian Sea and West of Shetland) there have already been major discoveries. These have led to estimates of multi-billion barrels of remaining recoverable reserves, even at present price levels, given the significant cost reductions engendered by the application of recently developed technologies. In essence, Western Europe is going to remain roughly as self-sufficient in oil as it has recently become until well into the first decade of the twenty-first century – at the earliest. Except in the context of a collapse in the price of internationally traded oil (in which case investment in Western Europe's upstream oil industry would be curtailed), Europe's oil markets will remain highly competitive – but, nevertheless, very attractive – for an increasing number of external suppliers.

Western Europe's gas market

Unlike oil, Europe's gas demand is expanding. It is expected to be 50 per cent higher than the 1995 level by 2010 and 80 per cent higher by 2020, under conditions of technological advances in gas use for power generation and of environmental advantages from the use of gas. In the anticipated near-stagnant energy market in Western Europe, the contribution of gas will rise from today's 20 per cent to some 27 per cent of energy supply by 2010 and to upwards of 33 per cent by 2020. The volumes of gas required will increase from the present 300 Bcm (billion m^3) per year to over 450 and 550 Bcm, respectively, by 2010 and 2020.

Many gas market analysts and most forecasts have hitherto underestimated the importance and potential of indigenous gas supplies. Indeed, not only were present levels of production not expected to materialise, but the maximum possible levels of production in Western Europe were officially forecast (by the European Union and the International Energy Agency) to occur by 1990. In spite of these past underestimates of the scope for indigenous gas production, a limited view of its post-2005 up-side potential is still widely expressed. Thus, it is argued, the additional expected demand for gas, as indicated above, is capable of being met only through massive increases in imports.

This immediately raises the geopolitical issue of dependence on imported supplies, the availability of which is viewed not in terms of the capability of the foreign reserves to sustain required levels of production, but in terms of the political risks which appear to arise from a widely perceived $200x10^9$ cubic metres (Bcm) need for foreign gas by 2010 (constituting about 50 per cent of supply) and upwards of 400 Bcm (about 70 per cent of supply) by 2020. Geopolitically, future dangers are also seen to be presented by the vulnerability of the pipelines through which such a large percentage of Western Europe's gas requirements will have to pass en route to market.

What such analyses generally fail to treat and evaluate is the capability of indigenous gas to meet the rising demand. Existing fields, initially declared as relatively modest occurrences, have now been proved to be much larger. The Groningen field, for example, has been upgraded many times, so that estimates of its originally declared recoverable reserves have doubled from 1500 Bcm in 1965 to today's 3000 Bcm. This appreciation of reserves, together with continuing successes in the discovery of new fields, have generated rising levels of availability of indigenous gas.

The 40-year process of West European gas development is shown in Table IV-8.4 on p.605. In all respects – in terms of discoveries, reserves and production – each succeeding decade has produced more successful results than the preceding one. By the end of the fourth decade – at the end of this year – remaining recoverable reserves from fields already discovered will be at an all-time high, in spite of a cumulative production since 1956 of over 4400 Bcm. Today's proven and probable reserves offer 47 years of future production at the 1995 production level. And, as indicated in the second footnote to the Table, there is a virtual certainty that more fields will be found. Such high levels of investment in West European gas exploration and exploitation continue to be made that the processes of both reserves' appreciation and discoveries can be practically guaranteed to continue through the fifth and even the sixth decades of the industry's history. As seen from Table IV-9.1, by the year 2000 indigenous production will have increased by 40 Bcm above the 1995 level, and in 2020 by another 75 Bcm per year. The additional gas will be mainly from Britain and Norway, in response to gas prices to producers which will move modestly up from present levels.

As a result, 67 per cent of Western Europe's gas demand could still be derived from indigenous supplies in 2010 and 61 per cent in 2020, thus severely limiting the market potential for external suppliers. Net imports will eventually more than double, but the relatively limited volumes involved will, in the context of the external suppliers' total potential, curb Western Europe's geopolitical exposure to modest proportions. Indeed, the external suppliers (Russia, North Africa, Caspian basin and Middle East) will be permanently in a situation of competing with each other for Western European markets: in a way not dissimilar from that, as described above, of the external suppliers of oil to Western Europe.

TABLE IV-9.1
Gas use, production and trade in the West European gas market, 1995–2020 ($m^3 x 10^9$ = Bcm)

	Estimates for:	Forecasts for:		
	1995	2000	2010	2020
a. Gas use	310	365	460	550
b. Gas production	220	260	310	335
of which				
– used in the countries of production	155	175	205	210
– exported to other West European countries	65	80	95	110
– exported to other countries	0	5	10	15
c. Gas imports				
– Imports from external supply sources	90	110	160	230
– Net imports from external suppliers	90	105	150	215
– Net imports as % of use	(29.0%)	(28.8%)	(32.6%)	(39.1%)

Sources: Author's compilations and forecasts.

Conclusion

Thus, neither oil nor gas requirements in Western Europe post serious potential supply or transportation problems. Taking oil and gas together, then, given their combined domination of the energy market (65 per cent of total supply) and a degree of substitutability between them in terms of their end uses, Europe's energy supply prospects can be viewed with even greater equanimity. Competition between oil suppliers and between gas suppliers – coupled with competition for markets between the two fuels – suggests a high degree of inherent West European power in the evolving situation, making the geopolitics of Europe's energy a matter of opportunities, rather than of problems.

An oil and gas rich Europe, surrounded by a plethora of competing suppliers in the context of a very slowly growing demand, puts Europe firmly in control of its energy destiny and, beyond this, also in a position to determine the prospects – both economic and political – for those energy exporting nations whose well-being will depend increasingly on the markets which Europe offers to them in the context of its limited external energy requirements.

Chapter IV – 10

Political and Structural Change in the former Eastern Europe*

a. Introduction

Eastern Europe's post-1945 adoption of Soviet-style centralised planning, including the maximum possible use of indigenous energy resources and the adaptation of the style of forced industrialisation which had featured in the USSR's Marxist approach to economic development, brought rapid change in the region's energy situation (Dienes and Shabad, 1979; Park, 1979; Voloshin, 1990). On the supply side, this included the expansion of the output of the region's coal, brown coal and lignite resources, and of what indigenous oil there was. The latter was largely restricted to Romania. In marked contrast with the decline in western Europe, eastern Europe's indigenous energy production by 1960 was 194 million tonnes of oil equivalent (mtoe), over 30 per cent up on that of 1937 (Hoffman, 1985). Though imports – mainly of Soviet oil – did increase, these remained largely complementary to national production; and they were not at the expense of indigenous energy resource development, as was the case with western European oil imports (Park, 1979; Hoffman, 1985).

b. Energy Use Developments

In eastern Europe the process of increasingly energy-intensive economic expansion continued uninterruptedly through the 1970s,

* **Extracted from "Energy-Resources and Choices" in D. Pinder (Ed.), *The New Europe: Economy, Society and the Environment*, J. Wiley and Sons, Chichester, 1998, pp.68–72 & 77–79.**

under the pressures of Soviet-style developments and in spite of severe limitations on the growth of private road transport. This region was, moreover, only marginally affected by the 1973/4 international oil price shock and its repercussions, because of the near-complete separation of the Western and Soviet economic systems (Hoffman, 1985). The latter was effectively insulated against the inflation of international oil prices, so that most energy in eastern Europe remained very low cost – and also, as a result, very inefficiently used. This applied as much to indigenous coal and lignite produced in Poland, East Germany and Czechoslovakia, etc. as to the oil and gas imported from the Soviet Union.

Thus between 1973 and 1982 eastern Europe's energy use grew from 330 to 415 mtoe. Its total energy consumption thus rose to only a little under 40 per cent of western Europe's, even though the East's economy was less than 20 per cent of the size of the West's, while car ownership was under 10 per cent of the western level (Hoffman, 1985). Eastern Europe's energy use finally peaked in the mid-1980s, by which time the inefficiencies in use and the underpricing of energy were just being recognised.

Thereafter, the renegotiation of Soviet oil and gas prices led to significant price rises, and thus to the beginning of economic problems relating to energy costs and to an enhanced interest in energy conservation. The question of the degree and speed with which the eastern European centrally planned economies would or could have reacted to this challenge seemed likely to be important because of the ingrained nature of the characteristics of energy supply and use patterns under Soviet-style central planning. The issues were, however, quickly overtaken by much more dramatic events, viz. those leading to the collapse of the politico-economic systems of all the eastern European countries and their substitution by market-orientated regimes. In this process of radical adjustment the economies suffered severe recession in which a significant range of activities was sharply reduced in size or in some instances even closed down – while standards of living were adversely affected. As a consequence energy use fell sharply. From the peak of 420 mtoe in 1986, it declined to 340 in 1990, to 270 by 1994 and to 250 mtoe in 1996.

Paradoxically, the falling use of energy has in many of the countries been accompanied by energy scarcity – particularly of electricity – arising, in part, from the disruption to organisational and infrastructure systems, but also from the difficulties in maintaining supplies of costly foreign exchange imports. These included oil and gas from Russia which, following political change there in 1991, cancelled all the favourable energy supply contracts it had made with eastern Europe. Instead, it began to price its energy exports at internationally orientated prices.

Meanwhile, within eastern Europe itself, the energy supply industries also suffered, first, from the problems of sustaining production in the context of disorder and a lack of investment; and, second, from problems of marketing their output in weak economies which lacked purchasing power. Thus, since 1988 total energy output has declined by almost 40 per cent, from 286 mtoe to under 180 mtoe. Coal overall, but most especially in Poland, has declined most emphatically in the context of an industry which was inherently uneconomic, as effective Western accounting methods have revealed (Coopers and Lybrand, 1991).

Needless to say, now – in the mid-1990s – less than 10 years on from the fundamental break with the previous 50 years' history of energy development in eastern Europe – the process of readjustment continues. The outlook remains highly uncertain, depending, as it largely does, on the speed and effectiveness of implementing Western-style approaches to both production and consumption decisions. These prospects, and the likelihood of integrating eastern Europe's energy systems with those of the rest of the continent, are considered below.

c. Over and Under-Exploitation of its Resources

As shown above, the eastern European centrally planned approach to energy use led to large inputs in highly inefficient modes. This energy-intensive process of development reflected the Soviet Union's own very large energy resource base, on which all eastern European countries eventually came heavily to depend – in the context of the integration of their economies with that of the Soviet Union through COMECON (Park, 1979). Nevertheless, the combination of limitations on Soviet/east European exchanges and the strong national policies encouraging indigenous resource use, meant that each east European country also sought the accelerated and maximum possible development of whatever energy resources it had to hand. Energy production potential within eastern Europe was thus exploited without much consideration of either economic or environmental issues (Hoffman, 1985).

Unhappily, eastern Europe's oil and gas prospects were viewed as extremely limited and, in competition with demands from the coal and lignite industry and from nuclear power, the sector's possible expansion was starved of funds (Hoffman, 1985). Moreover, the Soviet Union discouraged hydrocarbon developments among its allies (except

Romania), in order to ensure that it retained its ability to export oil and gas to east Europe – given the absence of alternative acceptable exports in sufficient volume and value. Even the Romanian oil industry – dating back to the nineteenth century – was starved of investment and new technology so that its output peaked as early as 1976 (Park, 1979). Thereafter, the country struggled to slow down the rate of production decline, but with little success. Production of oil is now little more than one-third of its peak level, while its natural gas output halved between 1976 and the fall of Romania's centrally planned system – and has since continued to decline.

In marked contrast, eastern Europe's coal industry has survived to a much greater extent than that of western Europe. By the early 1960s the latter's coal industry was already over the top, whereas that of eastern Europe then started to go through a period of rapid expansion (Hoffman, 1985). This took production above that of western Europe by 1980 and thereafter the industry continued to grow, albeit more slowly, for almost another decade – until 1988. Its evolution was based on the centrally planned systems' "need" for energy, expressed essentially in volumetric, rather than economic, terms. Thus, by 1982, when coal and lignite output was approaching its peak, the six countries had a total production of 358 million tonnes coal equivalent, accounting for 76 per cent of the total energy output of the region. This was achieved in spite of difficult and worsening geological conditions, the exploitation of deeper horizons for deep-mined coal and increasing ratios of overburden to accessible reserves in the lignite/opencast coal industry. In addition, there was also a general fall in quality of product, while access to new supplies required the relocation of rivers, transport facilities and even villages and towns (Hoffman, 1985). Such development requirements not only seriously increased costs, but also led to deteriorating environmental conditions associated with mining and the use of such heavily polluting sources of energy. In brief, the solid fuel industry in eastern Europe continued to provide the lion's share of total energy required to keep the region's economies expanding – and there were physical resources enough to sustain the policy objective. However, in both economic and environmental terms, coal and lignite production and use increasingly imposed costs on the countries concerned, so that even before the change from the centrally planned systems, there were doubts as to the continuing wisdom of persisting with the policy of maximising the use of solid fuels (Hoffman, 1985).

This was one of the two main reasons which led to the decisions by all the east European countries (except Poland) to build nuclear power stations. The second was the special relationship with the USSR, from which the nuclear technology and the know-how originated (Mountfield, 1991). Such nuclear developments (eventually totalling 30 reactors with a capacity of about 25 GW) moderated the previously very high dependence on coal for power generation and so provided some help in the increasingly desperate attempt by the eastern European countries to sustain the high rate of increase in the use of electricity. This was made more difficult by the fact that output was priced not at rates which reflected the scarcity inherent to the situation, but rather in the context of the Leninist philosophy of "socialism plus electricity equals communism!" (Lenin, 1966). Nevertheless, the environmental impact of the Chernobyl accident in 1986 eventually led to a re-evaluation of the safety and other features of the Soviet-designed nuclear stations in eastern Europe and, after the political changes throughout the region, to even stronger reactions against the expansion of nuclear power. As in most of western Europe, so now in the east; nuclear power is on hold (Mountfield, 1991).

d. Prospects for the New Regimes

The new post-Communist regimes in eastern Europe have thus inherited a set of problems relating to the exploitation and use of energy resources that are as serious as those in any other sector of the new democratised countries. In the mid-1990s they are certainly more difficult to resolve than the current energy sector issues in western Europe. The regional economic and social problems arising from the enforced run-down of the coal industries of east Europe have had their precursors in the West and, as their persistence in the West has shown, they will take decades to resolve properly. This is especially the case because they are, to a much greater degree than in western Europe, bound up with problems of environmental degradation and poor living conditions, to an extent which makes the former mining areas very unattractive for alternative activities. Thus, although coal still provides almost 50 per cent of eastern Europe's total energy, its contribution will continue to fall as the region diversifies to other sources of energy in the new environment of increasingly deregulated and privatised markets.

In contrast, the use of oil, which fell in the aftermath of the political changes, seems now to have bottomed out under the influence of its rising use in transportation, as private cars and trucks become more widely owned in the changing societies. In similar fashion, although gas use also fell in the aftermath of change, it is now a preferred fuel for both economic and environmental reasons and its use is again increasing

(Estrada et al, 1995). In this case, the dynamics of the industry are such that it seems likely to expand quickly in ways which closely parallel those of only a few years previously in western Europe.

The future of electricity depends, first, on the speed at which per capita income increases and engenders demand through the more intensive electrification of households; and, second, on the evolution of a supply system which is reliable, economically attractive and capable of expansion in an environmentally acceptable and safe way. This seems more likely to emerge through the use of natural gas as the prime mover, rather than a resurgence of coal and/or the expansion of nuclear power (Stern, 1995; Estrada et al, 1995).

Overall, eastern Europe's energy economy will grow steadily to resemble that of western Europe, so that the issues of resources and choices will become increasingly similar through the rest of the 1990s. By the year 2000 there seem likely to be far less significant East-West disparities than have existed for several decades. On the other hand, eastern Europe, with its newly acquired political independence, seems more likely to choose to diversify away from its previous overwhelming energy supply ties with Russia. Thus, the latter's share of the east European oil and gas markets will fall to significantly lower levels. This will be an adverse development for Russia even though it may, eventually, be offset by eastern Europe's increased demand for oil and gas at the expense of coal, lignite and nuclear power. For natural gas, however, eastern Europe's dependence on Russian supplies will be cut back by newly-negotiated gas flows from western Europe (notably Norway and the Netherlands) in the interests of supply security through diversification and as a result of the closer integration between the energy systems of the hitherto strongly divided parts of Europe (Odell, 1996b).

e. East/West European Convergence

Though there continue to be important contrasts between west and east European energy systems and structures, major differences are unlikely to persist. Choices between energy sources, the moves towards liberalised and competitive markets, and concerns for geopolitics of energy supply, will gradually apply equally to both parts of Europe. This will steadily reduce the differences which emerged out of the 40 years of separation under strongly contrasting political and economic systems. Such convergence, implying connectivity motivated not only by commercial considerations, but also by formal policies and structures of the EU as eastern European countries seek membership, seems likely to be achieved within a generation.

In particular, following the reduction in Germany's high level of support for its coal industry, together with the rationalisation of the industry in Poland and the Czech Republic, the European coal industry seems likely to be down to little more than half its present size by 2010. Coal imports will continue to grow, though by no means as rapidly as has been expected and is still widely assumed. This is because increasing concern for environmental issues across the continent will slow down the expansion of the market for coal.

Conversely, natural gas will be the major growth element across Europe's energy market for the foreseeable future – for a *melange* of reasons. These include this fuel's ease of supply, its environmental advantages compared with other fossil fuels, technological advances and the emergence of a much ore open and competitive situation for incorporating gas in a broader range of end uses (Odell, 1988 and 1996b). In particular, natural gas use in high-efficiency generating plants will further reduce the level of coal and fuel oil consumption in power stations and industry (especially in eastern and southern Europe respectively). It will also substitute what would otherwise have been nuclear- and coal-based capacity in most other countries. Soon after the turn of the century natural gas will most likely be Europe's single most important energy source, excluding transport fuels. Thereafter, on the assumption that compressed natural gas (CNG) becomes an accepted alternative automotive fuel within the next decade, then natural gas use should replace oil as the overall top source of energy by about 2020. By then, only the most remote parts of Europe and low-populated rural areas will not be connected to the continent-wide and highly reticulated gas transmission and distribution system.

References

Coopers and Lybrand, "Energy in Eastern Europe", *Energy Policy*, Vol.19, No.9, 1991, pp.813–840.

Dienes, H. and Shabad, T. *The Soviet Energy System: Resource Uses and Policies*, Halsted Press, New York, 1979.

Estrada, J. et al., *The Development of European Gas Markets: Environmental, Economic and Political Perspectives*, J. Wiley and Sons, Chichester, 1995.

Hoffman, G., *The European Energy Challenge: East and West*, Duke University Press, Durham, N.C., 1988.

Mountfield, P., *World Nuclear Power*, Routledge, London, 1991.

Lenin, V.I., *Collected Works, Vol.31, April–December 1920*, Progress Publishers, Moscow, 1966.

Odell, P.R., "The West European Gas Market: Current Position and Alternative Prospects", *Energy Policy*, Vol.16, No.5., 1988, pp.480–93.

Odell, P.R.,"The Cost of Longer Run Gas Supply to Europe", *Energy Studies Review*, Vol.7, No.2, 1996, pp.1–15.

Park, D., *Oil and Gas in the Comecon Countries*, Kogan Page, London, 1979.

Stern, J.P., *The Russian Gas Bubble: Consequences for European Gas Markets*, Royal Institute for International Affairs, London, 1995.

Voloshin, V.I., "Electric Power in the COMECON European Countries", *Energy Policy*, Vol.18, No.8, 1990, pp.740–746.

Index

A

B

C

D

E

F

G

H

I

J

K

L

M

N

O

P

Q

R

S

T

U

V

W

Y

Z
